# IMPACT OF MODERN DYNAMICS IN ASTRONOMY

*Proceedings of the IAU Colloquium 172
held in Namur (Belgium), 6–11 July 1998*

*Edited by*

## J. Henrard

*University of Namur, Belgium*

and

## S. Ferraz-Mello

*University of São Paulo, Brazil*

Partly reprinted from *Celestial Mechanics & Dynamical Astronomy*
Volume 73, Nos. 1–4 (1999)

SPRINGER-SCIENCE+BUSINESS MEDIA, B.V.

A C.I.P. catalogue record for this book is available from the Library of Congress.

ISBN 978-0-7923-5842-8      ISBN 978-94-011-4527-5 (eBook)
DOI 10.1007/978-94-011-4527-5

*Printed on acid-free paper.*

# IMPACT OF MODERN DYNAMICS IN ASTRONOMY

*Proceedings of the IAU Colloquium 172*
*held in Namur (Belgium), 6–11 July 1998*

## Analytical and Numerical Tools

## Planets and Satellites

## Planetary Systems

## Tools

## Miscellaneous

# PREFACE

Modern dynamics is increasingly participating in the solution of problems raised by astronomical observations. This new relationship is being fostered on one side by the improvements in the observations, which in recent years contributed several discoveries of new systems, such as the objects in the Kuiper belt, the pulsar and star companions, to speak only of the most striking ones, and, on the other hand, by the progresses in modern dynamics.

The progresses in modern dynamics are due to two factors: the dissemination of fast computers, allowing the numerical studies of very complex systems by a large number of scientists, and the improvement in our understanding of the complex behaviour of Hamiltonian systems. KAM and Nekhorochev theories have shed a light on the subtle and surprizing interplays between regular and chaotic motions; numerical experiments and analytical approximations have shown how these peculiarities are indeed present in astronomically important systems and are instrumental in understanding their formation and evolution.

In view of this it seemed timely to bring together scientists who are applying the tools of modern dynamics to the understanding of various astronomical problems, in order to share their experiences and results. Commission 7 (Celestial Mechanics) of the IAU, supported by Commission 4 (Ephemerides), Commission 20 (Positions and motions of minor planets) and Commission 33 (Structure and Dynamics of Galactic Systems) proposed that a Colloquium be held on this topic. This Colloquium (Colloquium 172 of the IAU) was held in Namur (Belgium) from July 6 to July 11 1998.

Invited lectures (C. Simo, J. Laskar) and contributed papers were devoted to the mathematical and numerical tools underlying the various approaches to astronomical problems.

The two main areas of applications which were discussed are "Stellar Systems" and "Small Bodies in the Solar System". In both cases the concepts of chaotic motion were considered as very important and were fully discussed. S. J. Aarseth and S. Tremaine delivered invited papers on "Stellar Systems" and S. Ferraz-Mello, B. G. Marsden, C. Froeschlé and A. Morbidelli delivered invited papers concerning the dynamics of small bodies in the solar system. Sessions were also devoted to cosmology (invited paper by G. Contopoulos) and planetary systems (invited papers by P. Artymowicz and N. Rappaport).

In addition to the International Astronomical Union, the following local sponsors made possible the organization of such a meeting: the Belgian National Science Foundation, the University of Namur, the Winthertur insurance group, the bank 'Générale de Banque', the regional administration of Wallonia and the SABCA company. Their support is gratefully acknowledged.

J. Henrard                                                                    S. Ferraz-Mello

# CHAOS IN RELATIVITY AND COSMOLOGY

G. CONTOPOULOS[1,2], N. VOGLIS[2] and C. EFTHYMIOPOULOS[1,2]

[1]*Research Center for Astronomy, Academy of Athens*
[2]*Department of Astronomy, University of Athens*

**Abstract.** Chaos appears in various problems of Relativity and Cosmology. Here we discuss (a) the Mixmaster Universe model, and (b) the motions around two fixed black holes. (a) The Mixmaster equations have a general solution (i.e. a solution depending on 6 arbitrary constants) of Painlevé type, but there is a second general solution which is not Painlevé. Thus the system does not pass the Painlevé test, and cannot be integrable. The Mixmaster model is not ergodic and does not have any periodic orbits. This is due to the fact that the sum of the three variables of the system $(\alpha + \beta + \gamma)$ has only one maximum for $\tau = \tau_m$ and decreases continuously for larger and for smaller $\tau$. The various Kasner periods increase exponentially for large $\tau$. Thus the Lyapunov Characteristic Number (LCN) is zero. The "finite time LCN" is positive for finite $\tau$ and tends to zero when $\tau \to \infty$. Chaos is introduced mainly near the maximum of $(\alpha + \beta + \gamma)$. No appreciable chaos is introduced at the successive Kasner periods, or eras. We conclude that in the Belinskii-Khalatnikov time, $\tau$, the Mixmaster model has the basic characteristics of a chaotic scattering problem. (b) In the case of two fixed black holes $M_1$ and $M_2$ the orbits of photons are separated into three types: orbits falling into $M_1$ (type I), or $M_2$ (type II), or escaping to infinity (type III). Chaos appears because between any two orbits of different types there are orbits of the third type. This is a typical chaotic scattering problem. The various types of orbits are separated by orbits asymptotic to 3 simple unstable orbits. In the case of particles of nonzero rest mass we have intervals where some periodic orbits are stable. Near such orbits we have order. The transition from order to chaos is made through an infinite sequence of period doubling bifurcations. The bifurcation ratio is the same as in classical conservative systems.

## 1. Introduction

(a) The subject of chaos in Relativity and Cosmology has attracted much interest in recent years. It started with the papers on the Mixmaster Universe model by Belinskii and Khalatnikov (1969) and independently by Misner (1969). But, despite the large number of papers written on this subject (Hobill et al 1994 and references therein), this problem has not been completely solved up to now.

Some recent results will be discussed in the next sections. Perhaps the most impressive result is that this model is not ergodic, nevertheless it is chaotic in the sense of chaotic scattering.

(b) A problem in General Relativity, where chaos is dominant, is the case of two fixed black holes (Contopoulos 1990, 1991). This problem will be discussed in Section 5 below. It is remarkable that the relativistic problem is chaotic, while the corresponding classical problem of two fixed centers is completely integrable.

Among other problems in Relativity and Cosmology, where chaos plays a role, are the following.

(c) Chaos in Special Relativity (Drake et al. 1996). Chaos was found in the motion of charged particles in a static electric field.

(d) A spinning particle in the Schwarzschild spacetime (Suzuki and Maeda 1997). The Schwarzschild spacetime is a 1-dimensional system, hence it is integrable. But adding further degrees of freedom can make the problem chaotic.

*Celestial Mechanics and Dynamical Astronomy* **73**: 1–16, 1999.

Chaos in perturbed Schwarzschild spacetimes has been found by Bombelli & Calzetta (1992) and by Letelier & Vieira (1997).

(e) The Robertson-Walker spacetime is also a 1-dimensional system (spherical), hence integrable. But the addition of another field makes it non- integrable and chaotic (Calzetta et al. 1993, Cornish & Shellard 1998).

(f) Motion in the field generated by gravitational waves (Varvoglis & Papadopoulos 1992, Chicone et al. 1997, Podolsky & Vesely 1998).

(g) Chaos in Yang-Mills fields of curved spacetime (Darian & Künzle 1996, Barrow & Levin 1998).

In the present paper we will discuss only the first two topics, namely the Mixmaster model and the case of two fixed black holes.

## 2. The Mixmaster Universe Model.

This model is a particular solution of Einstein's field equations (e.g. Landau & Lifshitz, 1971) that leads to three second order differential equations for three scale factors $\alpha$, $\beta$ and $\gamma$:

$$2\ddot{\alpha} = (e^{2\beta} - e^{2\gamma})^2 - e^{4\alpha} \tag{1}$$

and cyclic permutations of it, combined with a zero-energy constrain:

$$H \equiv \frac{1}{16}(p_\alpha^2 + p_\beta^2 + p_\gamma^2 - 2p_\alpha p_\beta - 2p_\beta p_\gamma - 2p_\alpha p_\gamma) \tag{2}$$
$$+ e^{4\alpha} + e^{4\beta} + e^{4\gamma} - 2e^{2(\alpha+\beta)} - 2e^{2(\beta+\gamma)} - 2e^{2(\alpha+\gamma)} = 0 \ ,$$

where

$$p_\alpha = -4(\dot{\beta} + \dot{\gamma}), \quad p_\beta = -4(\dot{\gamma} + \dot{\alpha}), \quad p_\gamma = -4(\dot{\alpha} + \dot{\beta}) \ . \tag{3}$$

The dots are derivatives with respect to the Belinskii- Khalatnikov time $\tau = -\ln t$, where $t$ is cosmological time. Thus when the cosmological time t goes to zero at the big bang, the time $\tau$ goes to infinity.

The Mixmaster equations can be derived from the Hamiltonian (2):

$$\dot{\alpha} = (p_\alpha - p_\beta - p_\gamma)/8, \quad \dot{p}_\alpha = -4[e^{4\alpha} - e^{2(\alpha+\beta)} - e^{2(\alpha+\gamma)}] \ , \tag{4}$$

and cyclic permutations of them.

Another simple set of variables is

$$X = e^{2\alpha}, \quad Y = e^{2\beta}, \quad Z = e^{2\gamma}, \quad p_X = p_\alpha/4, \quad p_Y = p_\beta/4, \quad p_Z = p_\gamma/4 \tag{5}$$

In these variables the equations of motion take the simple form

$$\dot{X} = X(p_X - p_Y - p_Z), \quad \dot{p}_X = X(Y + Z - X), \tag{6}$$

and cyclic permutations of them, while the zero energy constrain is

$$E \equiv p_X^2 + p_Y^2 + p_Z^2 - 2p_X p_Y - 2p_Y p_Z - 2p_X p_Z \qquad (7)$$
$$+ X^2 + Y^2 + Z^2 - 2XY - 2YZ - 2ZX = 0$$

Numerical integrations of the Mixmaster equations strongly suggest that the Lyapunov characteristic number is zero (Hobill et al. 1992).

This, and other theoretical indications, lead to the view that the Mixmaster model is not chaotic (Cushman and Sniatycki 1995). However other indications lead to the conclusion that the Mixmaster is chaotic.

In the present paper we will see how these two views can be made compatible.

One method to find evidence for integrability of a dynamical system is the Painlevé test (Ablowitz et al. 1980). If all solutions of the equations of motion have only poles as movable singularities, then probably the system is integrable.

The first general solution of Eqs. (7), i.e. a solution that depends on 6 arbitrary constants, which satisfies the Painlevé condition, was given by Contopoulos, Grammaticos and Ramani (1993). It is given as Laurent series

$$X = \pm \frac{i}{s} + \delta_1 s + ..., \quad Y = x_2 s + ..., \quad Z = x_3 s + ..., \qquad (8)$$
$$p_X = -\frac{1}{s} + ..., \qquad p_Y = p_2 + ..., \quad p_Z = p_3 + ...,$$

where $s = \tau - \tau_0$, and the arbitrary constants are $\delta_1$, $x_2$, $x_3$, $p_2$, $p_3$, $\tau_0$.

However it was found later that this is not the only general solution of the Mixmaster model. Another solution found by Latifi et al. (1994), by perturbing the well known Taub solution (Taub 1951), does not have the Painlevé property. The same solution was derived by Contopoulos, Grammaticos and Ramani (1995) by starting with a different solution of the Mixmaster equations (6):

$$X = A + CA^2 s + ..., \qquad Y = Z = \frac{1}{As^2} + ..., \qquad (9)$$
$$p_X = \frac{2}{s} + AC + ADs + ..., \quad p_Y = p_Z = \frac{1}{s} + ..$$

that depends on only 4 arbitrary constants $(A, C, D, \tau_0)$. This solution passes the Painlevé test to all orders, but it is special because $Y = Z$ and $p_Y = p_Z$. In order to find a more general solution we set

$$\delta = Y - Z, \quad q = p_Y - p_Z \qquad (10)$$

and find expressions for $\delta$ and $q$ that contain exponential singularities (Contopoulos et al. 1995). This general solution (depending on 6 arbitrary constants) is not of Painlevé type. Therefore the Mixmaster model does not pass the Painlevé test, although it has *one* general solution that is of Painlevé type. As a conclusion the Mixmaster model cannot be integrable with two more integrals besides the energy.

On the other hand Cushman and Sniatycki (1995) noted that the Mixmaster model is not mixing, nor even ergodic. This conclusion is based on a simple equation for the quantity

$$\Omega = e^{-2(\alpha+\beta+\gamma)} ,\tag{11}$$

that can be easily derived from Eqs.(7), namely

$$\ddot{\Omega} = \Omega[(p_\alpha^2 + p_\beta^2 + p_\gamma^2)/8 - E] .\tag{12}$$

An equivalent equation was first derived from the Raychaudhuri (1955) equation (Rugh and Jones 1990; see also Contopoulos et al. 1995).

In Eq. (12) for $E \leq 0$ (and in particular for the Mixmaster model where $E = 0$), the second derivative $\ddot{\Omega}$ has always the same sign as $\Omega$. As a consequence $\Omega$ has only one minimum and tends to infinity as $\tau$ goes to $\pm\infty$. This means that $(\alpha + \beta + \gamma)$ has one maximum at a certain time $\tau = \tau_m$ and decreases monotonically, both for larger $\tau$ and for smaller $\tau$ (Fig.1). The quantity $(\dot{\alpha} + \dot{\beta} + \dot{\gamma})$ is negative for $\tau > \tau_m$ and positive for $\tau < \tau_m$. Thus the values of $(\alpha, \beta, \gamma, \dot{\alpha}, \dot{\beta}, \dot{\gamma})$ cannot come back to their initial values, or close to them. Two conclusions are derived from this fact.

(1) The Mixmaster model does not have any recurrence, therefore it is not mixing, nor ergodic, despite its name (Cushman and Sniatycki 1995).

(2) The Mixmaster model (with $E = 0$) does not have any periodic orbits.

However, the usual notions of ergodicity etc. apply to compact systems, while the Mixmaster model is not compact. As we have noticed already (Contopoulos et al. 1995), the Mixmaster model behaves like a chaotic scattering system.

Christiansen et al.(1995) have found unstable periodic orbits in the case $E > 0$, but this is not the Mixmaster case, which has $E = 0$. Cornish and Levin (1997,1998) have found many unstable periodic orbits in a projected subset of the Mixmaster phase space. The study of these periodic orbits strongly suggests a fractal structure on this subset, and this is a signature of chaos.

Our approach is different. We deal with the full phase space of the Mixmaster model and try to find the sources of chaos in it.

## 3. A Numerical Example.

The fact that $\Omega$ goes to infinity both for $\tau \to \infty$ and for $\tau \to -\infty$ implies that the quantity $(\alpha + \beta + \gamma)$ goes to minus infinity for $\tau \to \pm\infty$ (Fig.1).

This implies that at least one of the quantities $(\alpha, \beta, \gamma)$ goes to minus infinity. But this happens in an oscillatory way, i.e. the quantities $(\alpha, \beta, \gamma)$ are unbounded, but there is no finite time $\tau$, beyond which any of these quantities remains smaller than a given negative number.

As a numerical example we take an orbit with initial conditions: $\alpha, \beta, \gamma, p_\alpha, p_\beta, p_\gamma$ = $\ln 2/2, 0, 0, 0, 0, 8$. In this model the function $(\alpha + \beta + \gamma)$ (Fig.1) has a maximum 0.60 for $\tau = \tau_m = -0.48$, i.e. very close to $\tau = 0$.

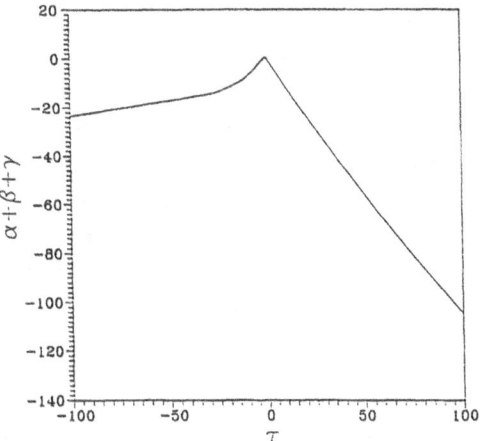

Fig. 1. The variation of $(\alpha + \beta + \gamma)$ as a function of time, for an orbit with initial conditions $\alpha = \ln 2/2, \beta = \gamma = p_\alpha = p_\beta = 0, p_\gamma = 8$.

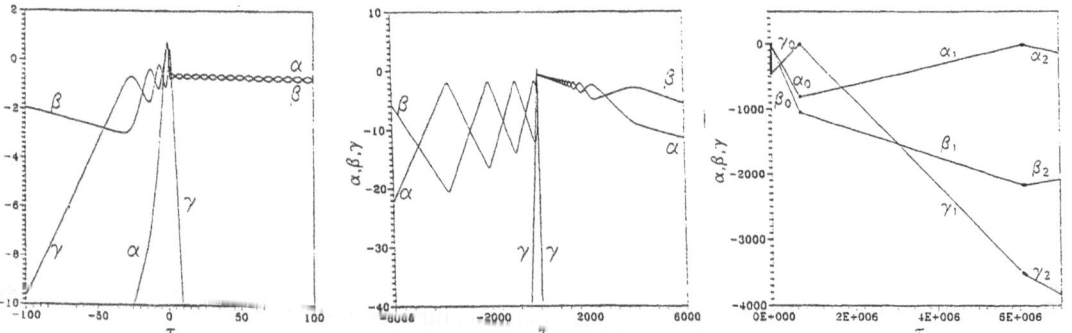

Fig. 2. The variations of $\alpha, \beta, \gamma$ in time $\tau$, (a) from $\tau = -100$ to $\tau = 100$ (b) From $\tau = -6000$ to $\tau = 6000$ (c) From $\tau = 0$ to $\tau = 7x10^8$. Initial conditions as in Fig.1.

In Fig. 2a,b,c we see the variation of the quantities $(\alpha, \beta, \gamma)$ in time, both in the positive and in the negative time direction.

As $\tau$ increases from $\tau_m$ the value of $\gamma$ decreases abruptly. At the same time $\alpha$ and $\beta$ undergo oscillations, first with small period (Fig. 2a) and later with larger and larger period (Figs. 2b, c). The value of $\gamma$ becomes minimum ($\gamma \approx -460$) for $\tau \approx 3500$, and then increases to a maximum close to zero at $\tau \approx 8 \times 10^5$. Later on it decreases and increases an infinite number of times.

If $\tau$ decreases from $\tau_m$, $\gamma$ undergoes some oscillations together with $\beta$ (Fig. 2a), while $\alpha$ goes first to a maximum and then decreases. For more negative $\tau$ the value of $\gamma$ decreases considerably, while $\alpha$ and $\beta$ undergo oscillations (Fig. 2b).

If we exclude a certain interval around $\tau = 0$ (say from $\tau = -100$ to $\tau = +6000$) the quantities $\alpha, \beta, \gamma$ are in general very small, so that the exponential members of Eqs. (5) are negligible. Then $p_\alpha, p_\beta, p_\gamma$ and $\dot\alpha, \dot\beta, \dot\gamma$ are almost constant, i.e. the variations of $\alpha, \beta, \gamma$ are almost linear, except when one of them becomes maximum near zero. In that case a rather abrupt change of the inclination of all three lines $\alpha, \beta, \gamma$, occurs (Fig. 2c).

Every period, in which $\dot\alpha, \dot\beta, \dot\gamma$ are almost constant, is called a Kasner period (Misner et al. 1977, Khalatnikov et al. 1985). We will prove (section 4) that for $\tau$ beyond $\tau_m$ two derivatives among $\dot\alpha, \dot\beta, \dot\gamma$ are negative and one positive. At every transition from one Kasner period to the next two derivatives (e.g. $\dot\alpha$ and $\dot\beta$) change sign, while the third derivative remains negative.

Similar transitions occur for $\tau < \tau_m$. In this case two of the quantities $\dot\alpha, \dot\beta, \dot\gamma$ are positive and one negative.

In order to find the chaotic properties of the orbits we solve the variational equations of Eqs. (5). Namely if we consider a deviation $\xi$ with coordinates $(\Delta\alpha, \Delta\beta, \Delta\gamma, \Delta p_\alpha, \Delta p_\beta, \Delta p_\gamma)$ we have the equations

$$\dot{\Delta\alpha} = \frac{1}{8}(\Delta p_\alpha - \Delta p_\beta - \Delta p_\gamma), \tag{13}$$

$$\dot{\Delta p_\alpha} = -8[(2e^{4\alpha} - e^{2(\alpha+\beta)} - e^{2(\alpha+\gamma)})\Delta\alpha - e^{2(\alpha+\beta)}\Delta\beta - e^{2(\alpha+\gamma)}\Delta\gamma] ,$$

and cyclic permutations of them.

The variational equations (13) are solved, together with the equations of motion, and give the variation

$$\mid \xi \mid = [\Delta\alpha^2 + \Delta\beta^2 + \Delta\gamma^2 + C^2(\Delta p_\alpha^2 + \Delta p_\beta^2 + \Delta p_\gamma^2)]^{1/2} \tag{14}$$

where $C$ is a constant, having the dimension of time. In the present paper we take $C = 1$. Then we find the "finite time Lyapunov characteristic number"

$$L_t = \frac{\ln \mid \xi/\xi_0 \mid}{\tau} \tag{15}$$

whose limit for $\tau \to \infty$ is the Lyapunov characteristic number LCN. We calculate also the "local LCN" (or stretching number)

$$a_i = \frac{\ln \mid \xi_{i+1}/\xi_i \mid}{\Delta\tau} \tag{16}$$

at every step $\Delta\tau$ (Voglis and Contopoulos 1994, Smith and Contopoulos 1996).

We start with two different deviations $\xi_0$ at the initial time $\tau = -15000$. In Fig.3a,b we give the stretching number $a$ for the cases $\xi_0 = (1, 0, 0, 0, 0, 0)$ and $\xi_0 = (0, 1, 0, 0, 0, 0)$ as functions of $\tau$.

Between $\tau = -15000$ and $\tau = -100$ $a$ is very close to zero. Between $\tau = -100$ and $\tau = +100$ $a$ undergoes large variations, especially neat $\tau = 0$, but it is positive most of the time (Fig.3a). Beyond $\tau = 100$ the values of $a$ are again close to zero,

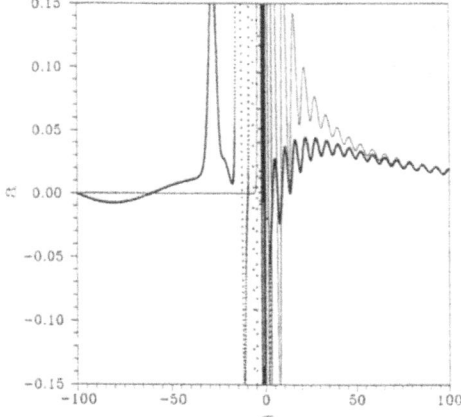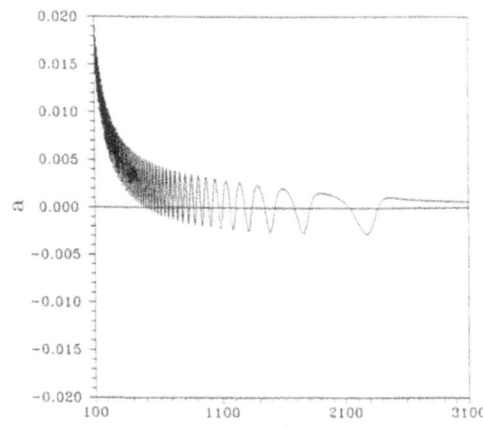

Fig. 3. (a)The variation of the stretching number $a$ as a function of $\tau$ for the orbit of Fig.1 and initial variations at $\tau = -15000$: $\xi = (1,0,0,0,0,0)$ (continuous line) and $\xi = (0,1,0,0,0,0)$ (dots, or heavy line) from $\tau = -100$ to $\tau = +100$. (b) The same functions $a(\tau)$ from $\tau = 100$ to $\tau = 3100$ almost coincide.

but with small scale variations(Fig.3b). The two initial conditions give different curves $a(\tau)$, but for large $\tau$ the two curves come very close to each other.

In Fig.4a,b,c,d we give the spectrum of $a$ for different time intervals. When $\tau$ is between $\tau = -15000$ and $\tau = -1000$ (Fig.4a) the spectrum of $a$ is peaked around $a = 0$, like a delta function. Thus no chaos appears during that period. Between $\tau = -1000$ and $\tau = -100$ (Fig.4b) the spectrum has a small positive part for both initial conditions. Between $\tau = -100$ and $\tau = +100$ (Fig.4c) the spectrum has relatively large positive parts. The two spectra are different, but the average value of $a$ is almost the same: $< a > = 4 \times 10^{-2}$. Therefore the "finite time LCN" during that period is positive. For larger positive times (between $\tau = 100$ and $\tau = 15000$, Fig.4d) the spectrum approaches again a delta function around $a = 0$. The average value of $a$ for large $\tau$ tends to zero. (E.g. in the time interval of Fig.4d it is $< a > = 4 \times 10^{-4}$). Thus the usual LCN is probably zero.

We conclude that the finite time Lyapunov characteristic number is significantly positive only during a period around $\tau = \tau_m$, when $(\alpha + \beta + \gamma)$ is maximum. For large positive and negative times the finite time LCN goes to zero.

Thus chaos appears in the Mixmaster model when $(\alpha + \beta + \gamma)$ is close to its maximum, that is when $\alpha, \beta, \gamma$ are close to zero.

## 4. Theoretical Explanation

When $\alpha, \beta, \gamma$ are negative with absolutely large values, the second members of Eqs. (1) are very close to zero. Thus $\ddot{\alpha} = \ddot{\beta} = \ddot{\gamma} = 0$, i.e. $\dot{\alpha}, \dot{\beta}, \dot{\gamma}$ are almost

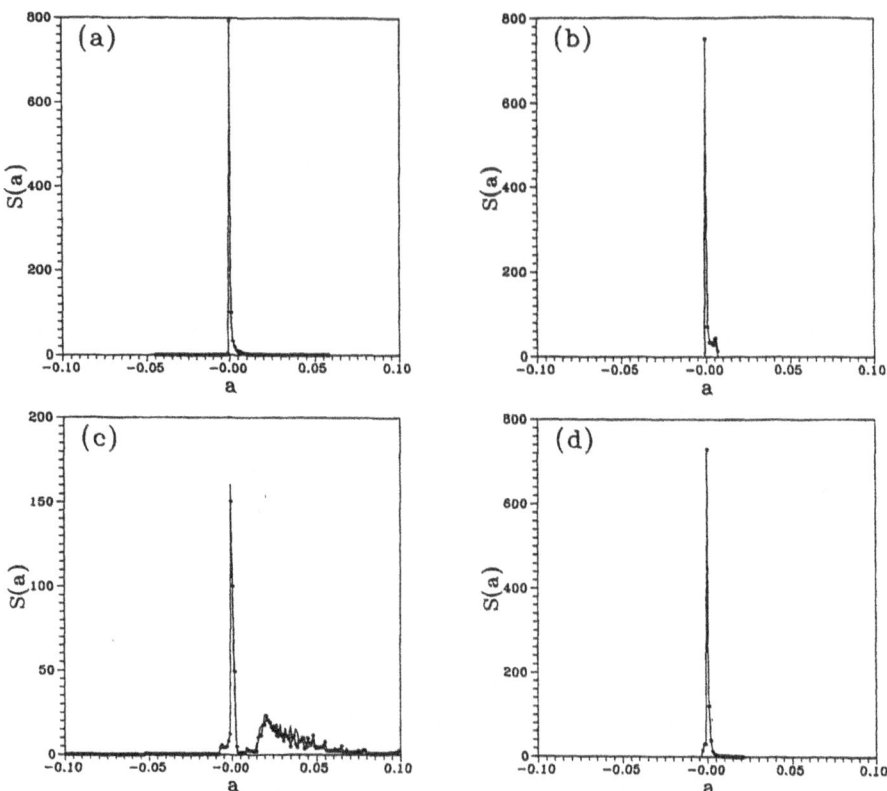

Fig. 4. Spectra of the values of $a$ for the initial conditions of Fig.3, (a)from $\tau = -15000$ to $\tau = -1000$, (b) from $\tau = -1000$ to $\tau = -100$, (c) from $\tau = -100$ to $\tau = +100$, (d) from $\tau = +100$ to $\tau = 15000$.

constant and $\alpha, \beta, \gamma$ vary linearly in time. The zero energy constraint is written

$$-4(\dot{\alpha}\dot{\beta} + \dot{\beta}\dot{\gamma} + \dot{\gamma}\dot{\alpha}) = 0 \ . \tag{17}$$

For $\tau > \tau_m$ we have also

$$\dot{\alpha} + \dot{\beta} + \dot{\gamma} < 0 \ . \tag{18}$$

Equation (17) cannot be satisfied if all three $\alpha, \beta, \gamma$ are positive, or negative. Thus at least one of these quantities is positive, and one negative. Let $\dot{\gamma} > 0$ and $\dot{\alpha} < 0$. Then we will prove that $\dot{\beta}$ is also negative.

In fact from Eq. (17) we derive

$$\dot{\gamma} = -\dot{\alpha}\dot{\beta}/(\dot{\alpha} + \dot{\beta}) \ . \tag{19}$$

Thus, unless $\dot{\beta} < 0$, we must have $\dot{\beta} > 0$ and $\dot{\alpha} + \dot{\beta} > 0$. But then Eq.(18) cannot be satisfied. Therefore $\dot{\beta} < 0$ is the only possible solution. As a conclusion for $\tau > \tau_m$ two quantities among $\dot{\alpha}, \dot{\beta}, \dot{\gamma}$, are negative and one positive.

In a similar way we prove that for $\tau < \tau_m$ two quantities are positive and one negative.

Let us consider now a case with $\dot{\gamma} = \dot{\gamma}_0 > 0$, $\dot{\alpha} = \dot{\alpha}_0 < 0$, $\dot{\beta} = \dot{\beta}_0 < 0$ and $\dot{\alpha}_0 + \dot{\gamma}_0 < 0$. (Fig. 2c). As $\gamma$ increases, it reaches values close to zero, while $\alpha$, $\beta$ are negative with large absolute values. Thus from Eqs.(1) we derive (Landau and Lifshitz 1971,§113):

$$2\ddot{\alpha} = 2\ddot{\beta} = e^{4\gamma}, \qquad 2\ddot{\gamma} + e^{4\gamma} = 0 \ . \tag{20}$$

The last of these equations gives

$$\dot{\gamma}^2 + e^{4\gamma}/4 = \dot{\gamma}_0^2 + e^{4\gamma_0}/4 = R^2/4, \tag{21}$$

where $R$ is a constant, $\gamma_0 < 0$, and $e^{4\gamma_0}$ is very small, hence $\dot{\gamma}_0 \approx R/2$.

The maximum $\gamma$ is slightly above zero

$$\gamma = \ln R \, / \, 2 \tag{22}$$

and then $\dot{\gamma} = 0$. Later $\gamma$ is again negative and the limiting value of $\dot{\gamma}$ is

$$\dot{\gamma}_1 = -R/2 = -\dot{\gamma}_0 \ . \tag{23}$$

Thus the transition through the maximum value of $\gamma$ is approximately a reflection. At the same time $\dot{p}_\alpha$ and $\dot{p}_\beta$ are almost zero, i.e. $p_\alpha$ and $p_\beta$ remain almost constant. Thus we have approximately

$$\dot{\beta}_1 + \dot{\gamma}_1 = \dot{\beta}_0 + \dot{\gamma}_0 \, , \qquad \dot{\alpha}_1 + \dot{\gamma}_1 = \dot{\alpha}_0 + \dot{\gamma}_0 \ , \tag{24}$$

where $\dot{\beta}_0 < \dot{\alpha}_0$ (Fig.2c). Hence

$$\dot{\beta}_1 = \dot{\beta}_0 + 2\dot{\gamma}_0 \, , \qquad \dot{\alpha}_1 = \dot{\alpha}_0 + 2\dot{\gamma}_0 \ , \tag{25}$$

and $\dot{\beta}_1 < \dot{\alpha}_1$. As $\dot{\gamma}_1 < 0$ we must have $\dot{\beta}_1 < 0 < \dot{\alpha}_1$.

The interval during which $\dot{\gamma} = \dot{\gamma}_0 > 0$, is a Kasner period. In the next Kasner period $\dot{\alpha}_1 > 0$ while $\dot{\beta}_1$ and $\dot{\gamma}_1$ are negative, and we observe (Fig. 2c) that $\dot{\gamma}_1 < \dot{\beta}_1 < 0$. Then comes a period where (Fig.2c)

$$\dot{\alpha}_2 = -\dot{\alpha}_1, \qquad \dot{\beta}_2 = \dot{\beta}_1 + 2\dot{\alpha}_1 \, , \qquad \dot{\gamma}_2 = \dot{\gamma}_1 + 2\dot{\alpha}_1 \, , \tag{26}$$

As long as $\dot{\gamma}$ is smaller than $\dot{\alpha}$ and $\dot{\beta}$ we have oscillations between $\alpha$ and $\beta$. But the value of $\dot{\gamma}$ increases at every Kasner period and after several Kasner periods it becomes positive. Then $\gamma$ becomes maximum and we say that a Kasner era is completed. From then on $\gamma$ undergoes oscillations with $\alpha$ or $\beta$.

The successive Kasner periods are longer as $\tau$ increases. In fact from Eqs. (25) we derive $\dot{\beta}_0 < \dot{\beta}_1 < 0$ and $\dot{\alpha}_1 = 2(\dot{\alpha}_0 + \dot{\gamma}_0) - \dot{\alpha}_0$. But $\dot{\alpha}_0 + \dot{\gamma}_0 < 0$, therefore $\dot{\alpha}_1 - \mid \dot{\alpha}_0 \mid = 2(\dot{\alpha}_0 + \dot{\gamma}_0) < 0$, i.e. both $\dot{\beta}_1$ and $\dot{\alpha}_1$ are absolutely smaller than $\dot{\beta}_0$ and $\dot{\alpha}_0$ respectively. Similarly from Eqs. (26) we derive $\dot{\beta}_1 < \dot{\beta}_2 < 0$ and $\dot{\gamma}_1 < \dot{\gamma}_2 < 0$, i.e. $\dot{\beta}_2$ and $\dot{\gamma}_2$ are absolutely smaller than $\dot{\beta}_1$ and $\dot{\gamma}_1$

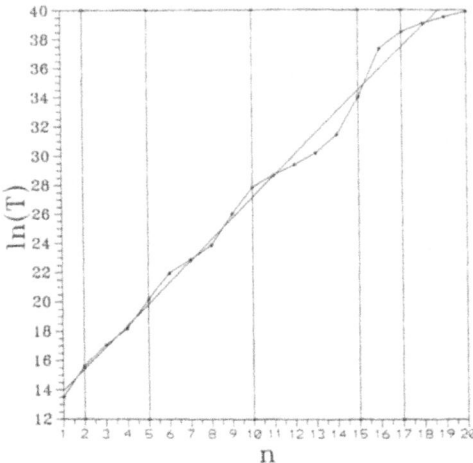

Fig. 5. The length of the Kasner periods $T$ as a function of the order $n$ (dots). The vertical lines separate the various Kasner eras. The straight line gives the best fit (Eq.27).

respectively. In every successive Kasner period similar inequalities show that $\dot{\alpha}, \dot{\beta}$, $\dot{\gamma}$ become always smaller absolutely. Thus the corresponding Kasner periods $T$ become longer. In Fig.5 we give the length of successive Kasner periods. This length increases exponentially according to the formula

$$T = T_n = T_0 e^{\kappa n} \tag{27}$$

where $T_0 = 339558$, $\kappa = 1.43121$ and $n$ is the order of the Kasner period (straight line in Fig.5). This relation implies that $T$ is proportional to $\tau$. (It is approximately $T_n = 3\tau_n$, where $\tau_n$ is the begining of the nth period).

In order to find the source of chaos in the Mixmaster model we solved analytically the variational equations (13) for the various Kasner periods. These give only linear deviations $\xi$ even when one variable among $\alpha, \beta, \gamma$ is close to zero. Only when two, or three variables $\alpha, \beta, \gamma$ are close to zero, we have exponential deviations of $\xi$. The main region where such exponential deviations are observed is near the maximum $(\alpha + \beta + \gamma)$, where all three quantities $\alpha, \beta, \gamma$ are close to zero. In this region the values of the local LCN are different from zero (Fig. 4c), while further away they are very close to zero in general (Figs. 4a,d).

One may ask whether chaos is also produced by the variation of the number of Kasner periods during a Kasner era (Khalatnikov et al. 1985). In fact, a small change of the initial conditions produces a different sequence of Kasner periods, that leads to an exponential deviation of $\xi$ of the form

$$\xi = \xi_0 e^{qi} \tag{28}$$

where $q$ is a constant and $i$ is the order of the Kasner era.

But if $k_i$ is the number of the Kasner periods within a Kasner era, then $n = k_1 + k_2 + ... + k_i$ and the typical length of a Kasner period during the $i$ era is

$$T_i = T_0 e^{\kappa(k_1 + k_2 + ... + k_i)} \approx T_0 e^{i\kappa k} , \tag{29}$$

where $k$ is the average number of Kasner periods within a Kasner era.

The corresponding time $\tau$ is then

$$\tau = kT_0(e^{\kappa k} + e^{2\kappa k} + ... + e^{i\kappa k}) \approx kT_0 e^{i\kappa k} \tag{30}$$

and we derive

$$i \approx \ln\tau \ / \ \kappa k . \tag{31}$$

Thus $\xi$ is given by the *power law in* $\tau$

$$\xi = \xi_0 \tau^s , \tag{32}$$

where $s = q/\kappa k$, i.e. $\xi$ is not exponential in the time $\tau$. Thus this effect is not chaotic in the usual sense.

Another case, where we find exponential deviations $\xi$, is when two quantities among $\alpha$, $\beta$, $\gamma$ happen to be close to zero. Such cases are exceptional (Khalatnikov et al. 1995), and any local increase in $\xi$, that they introduce, is counteracted by the large increase of the length of the Kasner periods.

We conclude that chaos, in the sense of exponential deviation of $\xi$ in time $\tau$, is introduced mainly near the maximum of $(\alpha + \beta + \gamma)$. As a consequence, the finite time LCN is always positive. But, as $\tau$ increases considerably, the finite time LCN tends to zero and the usual LCN is zero. This behaviour is very similar to a chaotic scattering case (Contopoulos et al. 1995).

Of course a different way of measuring time may lead to a positive LCN. In fact it is well known in classical mechanics that a change of the time coordinate leads to a change of the LCN. This was established already by Poincaré (1892) in the case of unstable periodic orbits. A recent discussion of the problem of different times in the Mixmaster model was made by Szydlowski (1997).

In our paper we have used only the Belinskii-Khalatnikov time $\tau$. In this time our conclusions regarding the chaotic behaviour of the Mixmaster model are clear. Further details of this problem will be given in a future paper.

## 5. Chaos in General Relativity.

A particular problem in General Relativity where chaos is predominant, is the motion of photons, or of particles of nonzero rest-mass, around two fixed black holes (Contopoulos 1990, 1991; more recent work has been done by Dettmann et al. 1994, Yurtsever 1995, and Cornish & Gibbons 1997).

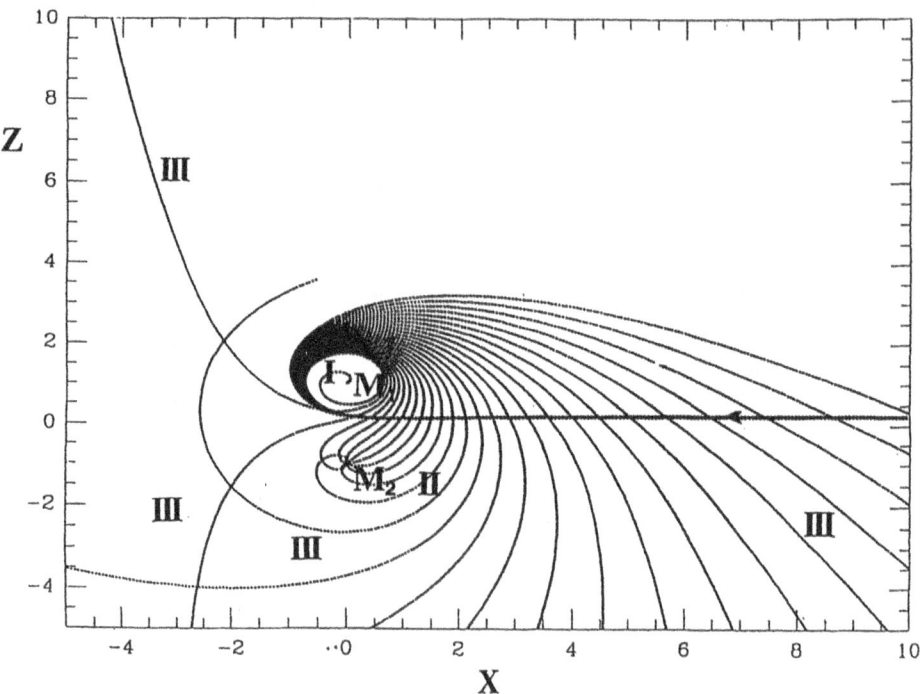

Fig. 6. A thin beam of photons from infinity is split into orbits of type I (falling to $M_1$), II (falling to $M_2$), and III (going to infinity).

This problem is remarkable because the corresponding classical problem of two fixed centers is completely integrable. The metric of the relativistic problem was given by Majumdar (1947) and Papapetrou (1947);

$$ds^2 = U^{-2}dt^2 - U^2(dx^2 + dy^2 + dz^2), \tag{33}$$

where $U$ is the potential due to two black holes of masses $M_1$ and $M_2$, written in the form

$$U = 1 + \frac{M_1}{M_2} + \frac{M_2}{r_2}. \tag{34}$$

We assume $M_1$ and $M_2$ to be on the z-axis at the points $\pm 1$, while the third particle is at distances $r_1$ and $r_2$ from $M_1$ and $M_2$.

A beam of photons coming from infinity (Fig.6) is separated into orbits of 3 types: (I) Orbits escaping to $M_1$; (II) Orbits escaping to $M_2$; and (III) Orbits escaping to infinity.

This is a typical scattering problem. It is chaotic because between an orbit of type I and an orbit of type II there are orbits of type III (Fig.6), between an orbit of type I and an orbit of type III there are orbits of type II and so on. This separation is

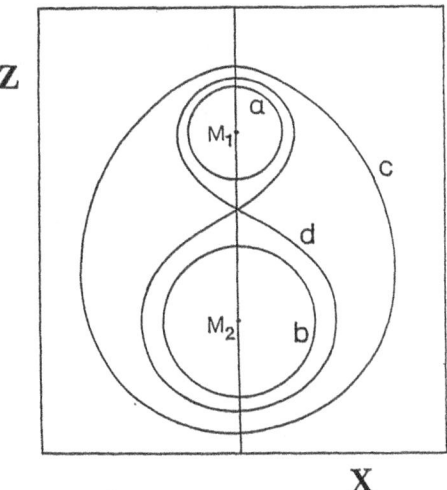

Fig. 7. Four types of unstable periodic orbits: (a) around $M_1$, (b) around $M_2$, (c) around both $M_1$ and $M_2$, and (d) figure eight orbits around both $M_1$ and $M_2$ .

repeated on finer and finer scales. The orbits I,II and III form Cantor sets of finite measure (fractal sets).

Besides these orbits there are inifinite types of unstable periodic orbits that form a set of measure zero. The most important of them are the satellite orbits (a) and (b) around $M_1$ and $M_2$ respectively, and the almost elliptical orbits (c) around both $M_1$ and $M_2$ (Fig. 7).

Around each orbit of a given type (say of type II, Fig.8) there is a continuous set of similar orbits, limited by two asymptotic orbits to the same unstable periodic orbit (orbit b in Fig.8). These asymptotic orbits approach the orbit b from opposite directions. All continuous sets of orbits I,II and III, that are parts of the 3 fractal sets, are limited by asymptotic orbits to the orbits a, b or c respectively.

On one side of such an asymptotic orbit all nearby orbits are of the same type (e.g. orbits of type II in Fig.8), while on the other side there are infinite sets of all 3 types. Orbits close to other periodic orbits (like orbit d in Fig.7, that is of figure eight type) are fractal sets of all 3 types on both their sides.

If we consider orbits starting perpendicularly to the z-axis above $M_1$ , we find that all orbits starting inside the periodic orbit (a) fall into $M_2$ , while all orbits outside the periodic orbit (c) go to infinity. Between the orbits (a) and (c) are the three fractal sets of orbits I,II and III.

In the case of photons there are no stable periodic orbits. However in the case of particles with non-zero rest mass there may be stable periodic orbits of various types. E.g. such orbits appear near the maximum mass $M_1$ of one type of orbits for fixed energy and fixed $M_2$. In Fig.9 we give the characteristic of the orbits of type (a) for $E = \sqrt{0.5}$ and $M_1 = 1$. This characteristic has a maximum $M_1 = M_{1max}$

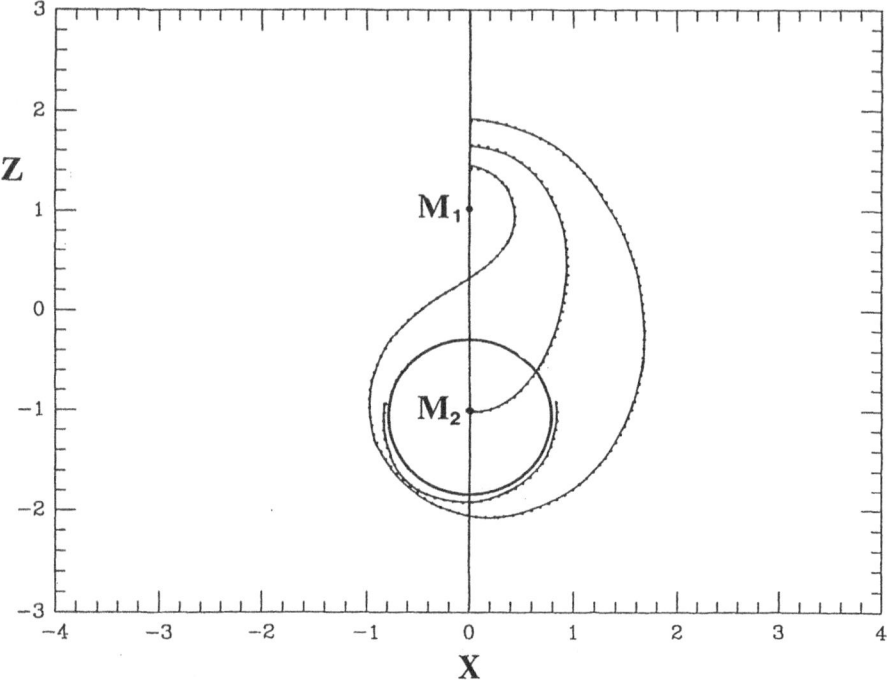

Fig. 8. A set of orbits of type II starting perpendiculary to the z-axis above $M_1$. The set is limited by two asymptotic curves to the unstable periodic orbit b.

and according to a well-known theorem of Poincaré at this maximum start two branches of orbits (a), one stable and one unstable.

In Fig.9 we see some bifurcations of the family (a). The most important bifurcation is the one of double period (2) that appears at $M_1 = M_{1,2}$. The double period family exists for $M_1$ smaller than $M_{1,2}$. It is initially stable and becomes unstable at $M_1 = M_{1,4}$, when a period-4 bifurcation is generated. In the same way there is a period-8 bifurcation at $M_1 = M_{1,8}$ etc.

The most important remark is that the intervals between successive bifurcations decrease almost geometrically with the universal ratio appropriate for conservative systems (Bennetin et al. 1980), namely

$$\frac{M_{1,2} - M_{1,4}}{M_{1,4} - M_{1,8}} \approx \frac{M_{1,4} - M_{1,8}}{M_{1,8} - M_{1,16}} \approx \dots = 8.72 \tag{35}$$

The values of $M_{1,2^n}$ converge to a minimum value $M_{1,\infty}$ beyond which there is an infinity of unstable orbits of multiplicity $2^n$ ($n = 0, 1, 2, \dots$).

Thus in general the orbits of particles of type (a) and their bifurcations are unstable and lead to a large degree of chaos. However there are intervals of stable orbits surrounded by orbits that do not escape to $M_1$, $M_2$ or to infinity. The nonescaping orbits are in general ordered (quasiperiodic), but there are also some chaotic nonescaping orbits (Contopoulos 1991).

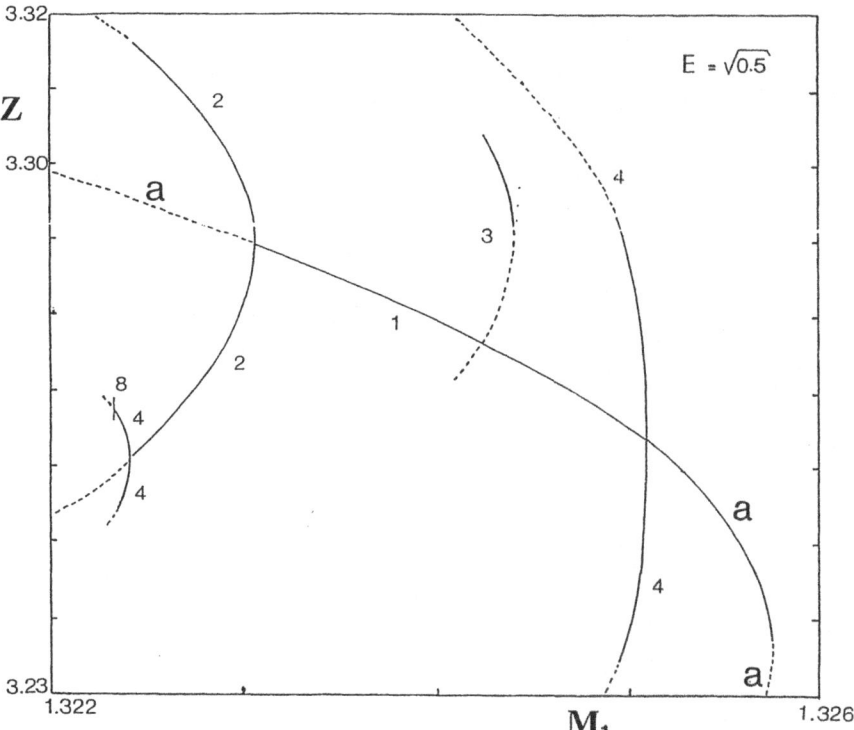

Fig. 9. Characteristics of the two families of orbits of particles of type (a) (stable and unstable) near the maximum $M_1$. From the stable family (a) bifurcate higher order periodic orbits. (—)stable, and (....) unstable orbits.

We conclude that chaos in the above relativistic cases has similar characteristics as chaos and chaotic scattering in classical problems. Even some numerical results are the same, as exemplified by the appearance of the same bifurcation ratio (35) in relativity as in classical dynamics.

## Acknowledgements

This research was supported in part by the Academy of Athens program 200/209. C.E. was supported by the Greek Foundation of State Scholarships (IKY).

## References

Ablowitz, M.J., Ramani, A. and Segur, H.:1980, *J.Math.Phys.*,**21**, 715.
Barrow, J.D., and Levin, J.: 1998, *Phys. Rev. Lett.*, **80**, 656.
Belinskii, V.A. and Khalatnikov, I.M.: 1969, *Sov.Phys.JETP*, **29**, 911.
Benettin, G., Cercignani, C., Galgani, L. and Giorgilli, A.: 1980, *Lett. Nuovo Cim.*, **28**, 1.
Bombelli, L. and Calzetta, E.: 1992, *Class. Quantum Grav.*, **9**, 2573.
Calzetta, E. and El Hasi, C.: 1993, *Class. Quantum Grav.*, **10**, 1825.
Chicone, C., Mashhoon, B. and Retzloff, D.G.: 1997,*Class. Quantum Grav.*, **14**, 699.
Christiansen, F., Rugh, H.H. and Rugh, S.E.: 1995, *J.Phys.A*, **28**, 657.

Contopoulos, G.: 1990, *Proc.R.Soc.Lond. A*, **431**, 183.
Contopoulos, G.: 1991, *Proc.R.Soc.Lond. A*, **435**, 551.
Contopoulos, G., Grammaticos, B. and Ramani, A.: 1993, *J.Phys.A*,**26**, 5795.
Contopoulos, G., Grammaticos, B. and Ramani, A.: 1995, *J.Phys.A*, **28**, 5313.
Cornish, N.J. and Gibbons, G.W.: 1997, *Class. Quantum Grav.*, **14**, 1865.
Cornish, N.J. and Levin, J.J.: 1997, *Phys. Rev. Lett.*,**78**, 998.
Cornish, N.J. and Levin, J.J.: 1998, *Phys. Rev. D*, **55**, 7489.
Cornish, N.J. and Shellard, E.P.S.: 1998,*Phys. Rev. Lett.*, **81**, 3571.
Cushman, R. and Sniatycki, J.: 1995, *Rep.Math.Phys.*, **36**, 75.
Darian, B.K. and Künzle, H.P.: 1996, *Class. Quantum Grav.*, **13**, 2651.
Dettmann, C.P., Frankel, N.E. and Cornish, N.J.: 1994, *Phys. Rev. D*, **50**, R618.
Drake, S.P., Dettmann, C.P., Frankel, N.E. and Cornish, N.J.: 1996, *Phys.Rev. E*,**53**, 1351.
Hobill, D., Bernstein, D., Simpkins, D. and Welge, M.: 1992, *Class. Quantum Grav.*, **8**, 1155.
Hobill, D., Burd, A. and Coley, A. (eds): 1994, *Deterministic Chaos in General Relativity*, Plenum Press, N. York.
Khalatnikov, I.M., Lifshitz, E.M., Khanin, K.M., Shchur, I.N. and Sinai, Ya.G.: 1985, *J.Stat.Phys.*, **38**, 97.
Landau, L.D., and Lifshitz, E.M.: 1971, *The Classical Theory of Fields*, Pergamon Press, Oxford.
Latifi, A., Musette, M. and Conte, R.: 1994, *Phys.Lett. A*,**194**, 83.
Letelier, P.S. and Vieira, W.M.: 1997,*Class. Quantum Grav.*, **14**, 1249.
Majumdar, S.D.: 1947, *Phys. Rev.*,**72**, 390.
Misner, C.M.: 1969,*Phys. Rev. Lett.*, **22**, 1071.
Misner, C.M., Thorne, K. and Wheeler, J.A.: 1977, *Gravitation*, Freeman, San Francisco.
Papapetrou, A.: 1947, *Proc. R. Irish Acad.*,**51**, 191.
Podolsky, J. and Vesely, K. : 1998, *Phys. Rev. D*,**58**, 081501.
Poincaré, H.: 1892, *Les Méthodes Nouvelles de la Mécanique Céleste*, I, Gauthier Villars, Paris.
Raychaudhuri, A.: 1955, *Phys.Rev.*, **98**, 1123.
Rugh, S.E. and Jones, B.J.T.: 1990, *Phys.Lett. A*,**147**, 353.
Smith, H. and Contopoulos, G.: 1996, *Astron. Astrophys.*, **314**, 795.
Suzuki, S. and Maeda, K.: 1997,*Phys. Rev. D*,**55**, 4848.
Szydlowski, M.: 1997, *Gen.Rev.Grav.*,**29**, 185.
Taub, A.H.: 1951,*Ann.Math.*,**53**, 472.
Varvoglis, H. and Papadopoulos, D.: 1992,*Astron. Astrophys.*,**261**, 664.
Voglis, N. and Contopoulos, G.: 1994, *J.Phys.A*,**27**, 4899.
Yurtsever, U.: 1995,*Phys. Rev. D*,**52**, 3176.

# BIANCHI COSMOLOGIES AS DYNAMICAL SYSTEMS

ANDRZEJ J. MACIEJEWSKI

*Toruń Centre for Astronomy, N. Copernicus University,*
*Chopina 12/18, 87-100 Toruń, Poland; E-mail: maciejka@astri.uni.torun.pl*

and

MAREK SZYDŁOWSKI

*Astronomical Observatory, Jagiellonian University,*
*Orla 171, 30-244 Kraków, Poland; E-mail: uoszydlo@cyf-kr.edu.pl*

**Abstract.** We discuss specific properties of dynamical systems originating from cosmology and relativity. In particular, we present results of our study of the Bianchi class A cosmological models. We introduce new variables in which the Hamiltonian constraint for all the class A models is solved algebraically. We present results of dimension reduction of the investigated models.

## 1. Introduction

Models of relativistic cosmology are based on Einstein's theory of gravitation. The Einstein field equations describe the dynamical evolution of spacetime, as well as the motion of matter and physical fields. They provide a system of coupled, non-linear, partial differential equations. Without some simplifying assumptions or idealization they are intractable by analytical tools. The most natural assumption is to postulate a certain symmetry of space-time. Usually, such idealization allows to reduce Einstein's field equations to a system of ordinary differential equations. This reduction gives us a possibility to use the rich theory of dynamical systems. It has to be mentioned, however, that dynamical systems of cosmological (or relativistic) origin have many special features which distinguish them from a 'typical' dynamical system we meet in dynamical astronomy, classical mechanics or physics. Let us mention a few of them. We can assume that systems we meet have the following form

$$\dot{x} = v(x), \qquad x \in \mathbb{R}^n. \tag{1}$$

1. Although the right hand sides of (1) are polynomial, in many cases it is not obvious if the phase flow of this system is complete. It seems that in some cases it is not. It is important to point it out, because for systems with an incomplete flow customary indicators of chaos are not defined, although numerical algorithms do not distinguish between complete and incomplete flows.
2. In many cases the system (1) has a first integral $H$. Usually, only solutions lying on the level
   $$\mathcal{M} = \{x \in \mathbb{R}^n \mid H(x) = 0\},$$
   have a physical interpretation. Thus we have to restrict our system to an invariant set $\mathcal{M}$. However, in many investigations this point is simply ignored. For example, when we ask for the existence of one or more additional first

*Celestial Mechanics and Dynamical Astronomy* **73**: 17–24, 1999.
©1999 *Kluwer Academic Publishers.*

integrals then a negative answer for the non-restricted system is not valid for the system restricted to $\mathcal{M}$.

3. In most cases the equilibria of (1) are degenerated and lie on an one or higher dimensional manifold of equilibria.

4. Frequently the system (1) is Hamiltonian with respect to the canonical symplectic form $\omega$ on $\boldsymbol{R}^n$, $n = 2m$, and the first integral $H$ plays the role of the Hamiltonian function, i.e., $\omega(v, \cdot) = dH$. Usually, the Hamiltonian has the 'natural' form

$$H = \frac{1}{2} g^{ij} p_i p_j + V(q), \qquad x = (q^1, \ldots, q^m, p_1, \ldots, p_m) \in \boldsymbol{R}^{2m},$$

however, the 'kinetic' energy $T = \frac{1}{2} g^{ij} p_i p_j$ is not positive definite. Because of this, there exists no such notion as 'region of possible motion' which plays a fundamental role in studies of natural mechanical systems in classical mechanics. It seems that systems with indefinite kinetic energy possess their own specific properties, however, investigation of these systems is far from being complete.

The last decades gave an immense popularity to the notion of deterministic chaos. Thus, it was natural to look for this phenomenon in the systems mentioned. However, a lot of controversies arose around this subject. On the one hand, these disputes were connected with the numerical character of the obtained results, and, on the other hand, they were caused by some conceptual problems. Here we point out some aspects of this discussion showing our point of view.

The main example of our discussion are class A Bianchi cosmological models. In fact, the controversy about chaotic or non-chaotic behavior concentrates around Bianchi IX model. We describe these models in the next section. In Section 3 we present our original results connected with reduction and simplification of class A Bianchi dynamical systems.

## 2. Bianchi Class A Cosmologies as Dynamical Systems

Let us assume that space-time has a product topology of type $\boldsymbol{R} \times \mathcal{M}^3$, where $\mathcal{M}^3$ is 3-dimensional space-like section admitting the action of simply transitive isometry group, i.e. homogeneity group; then Einstein's equations take the form of a system of ordinary differential equations. The classification of all 3-dimensional homogeneous but anisotropic spaces according to the Lie algebra of the isometry group is called the Bianchi classification (Landau and Lifshitz, 1975).

In this contribution we consider only subclass A of all Bianchi types for which, without any loss of generality, one can assume that the metric of $\mathcal{M}^3$ is diagonal, i.e.

$$ds^2 = (abc)d\tau^2 - \eta_{ab}(t)(e_i^a dx^i)(e_j^b dx^j), \tag{2}$$

where

$$\eta_{ab} = \mathrm{diag}\|a^2(t), b^2(t), c^2(t)\|, \qquad a, b = 1, 2, 3, \tag{3}$$

$\omega^a = e_i^a dx^i$ form the base of invariant 1-forms depending only on spatial variables $(x^i) = (x, y, z)$, while $\eta_{ab}$ is a symmetric metric tensor depending only on the cosmological time $t$ : $dt = abcd\tau$; $\tau$ is called the synchronic time.

The $(a, b)$ components of Einstein's field equations

$$R_k^i = 0, \tag{4}$$

for our case give

$$\begin{aligned}
2(\log a)_{\tau\tau} &= (n_2 b^2 - n_3 c^2)^2 - n_1^4 a^4, \\
2(\log b)_{\tau\tau} &= (n_1^2 a^2 - n_3^2 c^2)^2 - n_2^4 b^4, \\
2(\log c)_{\tau\tau} &= (n_1^2 a^2 - n_2^2 b^2)^2 - n_3^4 c^4,
\end{aligned} \tag{5}$$

where $n_i \in \{0, +1, -1\}$ distinguish the Bianchi type of model. From the $(0, 0)$ component $R_0^0 = 0$ we obtain

$$\frac{1}{2}(\log abc)_{\tau\tau} = (\log a)_\tau(\log b)_\tau + (\log a)_\tau(\log c)_\tau + (\log b)_\tau(\log c)_\tau. \tag{6}$$

After a simple manipulation, we obtain from (5) and (6) the following identity

$$H = H(a, b, c, a_\tau, b_\tau, c_\tau) = 0 ,$$

where $H$ is first integral of (5) of the form

$$H = (\log a)_\tau(\log b)_\tau + (\log a)_\tau(\log c)_\tau + (\log b)_\tau(\log c)_\tau$$
$$+\frac{1}{4}(n_1^4 a^4 + n_2^4 b^4 + n_3^4 c^4 - 2n_1^2 n_2^2 a^2 b^2 - 2n_1^2 n_3^2 a^2 c^2 - 2n_2^2 n_3^2 b^2 c^2). \tag{7}$$

The system (5) forms fundamental dynamical equations for the evolution of three different scale factors $a, b, c$ in three different main directions. In the special case when $a = b = c$ equations (5) represent the Friedmann-Robertson-Walker (FRW) models. Therefore, the Bianchi models are more general than the FRW models (standard models of current cosmology), because we assume only homogeneity of space-like sections. There are different reasons to analyze the evolution of cosmology models which are more general than FRW models (see Wainwright and Hsu, 1989 for details) but the main problem is to find a more realistic description of the very early evolution of universe which leads to the presently observed galactic epoch.

Now, we transform the system (5) to a set of first order homogeneous polynomial equations. To this end let us introduce new variables

$$y_1 = a^2, \quad y_2 = b^2, \quad y_3 = c^2, \quad z_i = \frac{\dot{y}_i}{y_i}. \tag{8}$$

After this change of variables the system (5) reads

$$\dot{y}_i = y_i z_i, \qquad \dot{z}_i = (n_j y_j - n_k y_k)^2 - n_i^2 y_i^2, \tag{9}$$

where a dot denotes differentiation with respect to $\tau$. In the new variables first integral (7) has the form

$$H = \sum_{i=1, i<j}^{3} \left( z_i z_j + 2n_i n_j y_i y_j \right) - \sum_{i=1}^{3} n_i^2 y_i^2. \tag{10}$$

The basic problem is to investigate qualitative properties of the above system.

Dynamical systems methods were first applied in the study of the Bianchi cosmologies by Collins (1971) who considered a number of special cases in which the phase space was two-dimensional. The systematic approach to the study of dynamics of all Bianchi cosmological models was initiated by Bogoyavlenski (1985) who investigated their evolution in 5-dimensional phase space. Another description of the dynamics of Bianchi class A models as a dynamical system was presented by Wainwright and Hsu (1989). They used expansion-normalized variables and an orthonormal frame approach (for details see Wainwright and Ellis, 1997).

There exist several directions in which we can start our study. It seems, however, that the most interesting problems can be extracted from studies of Bianchi IX (Mixmaster) model, for which $n_1 = n_2 = n_3 = 1$. The numerically computed maximal Lyapunov exponent for this system was zero or different from zero depending on the time parametrisation. Because an approximation of the Mixmaster model by a discrete map has strong chaotic properties (Chernoff and Barrow, 1993), it was natural to expect such a behavior in the original system. However, as Cushman and Śniatycki (1995) proved, there exists no recurrence in the system and, thus, no form of standard deterministic chaos is present in it. Interesting numerical works (Cornish, 1997; Cornish Levin, 1997), where 'parametrisation independent' characteristics of chaos were used, do not make big progress in understanding the system. The point is that in these works a certain approximation of the Mixmaster model was investigated, not the Mixmaster model itself. Moreover, one notices several very unprecise notions used in these investigations.

Because a strict proof of chaotic (in a certain sense) behavior of this system seems very difficult, several authors tried to show that the Mixmaster model is not integrable. It must be mentioned that 'integrability' here was understood differently by different authors. Several authors tested if the model passes the standard Painlevé integrability test in the form of the ARS algorithm (Ramani et al., 1989). Contopoulos et al. (1993) indicated that the B(IX) model passes this test. This result was revised (Contopoulos et al., 1994), however, without any strict conclusions concerning integrability. Further studies of Latifi et al. (1995) showed that the B(IX) model does not pass the so called perturbative Painlevé test. The authors of this paper suggest the existence of 'some chaotic regimes' in the system. The strongest result in this direction was obtained in (Contopoulos et al., 1995) where the authors show the existence of movable critical essential singularities in the B(IX) model. This kind of investigations connect a complicated behavior of the system with singularities of their solutions on the complex time plane. It should be mentioned, however, that the relation between Painlevé's test and integrability, e.g. in the Liouville's sense, is not clear. One can notice also that the integrability problem was stated for the global system, not for its restriction to the level $H = 0$.

The strongest and mathematically precise result concerning the Mixmaster model was obtained by J.J. Morales-Ruiz and J.-P. Ramis on the base of their theory connecting Ziglin's method and differential Galois theory (Morales-Ruiz and Ramis, 1997). We state this result shortly. The Mixmaster model can be formulated as a Hamiltonian system. There exists a four dimensional invariant manifold $\mathcal{T}$ on which the system can be integrated explicitly. Solutions of the system restricted to $\mathcal{T}$ are known as Taub solutions. Studying variational equations around Taub solutions J.J. Morales-Ruiz and J.-P. Ramis proved that the Mixmaster model considered as a complex Hamiltonian system is not completely integrable (in the Liouville sense) with rational first integrals.

We make a few remarks about these results. First, it was not excluded that this system possesses one addition rational integral or is integrable in terms of nonrational integrals. Moreover, it can possess an additional integral only on the level $H = 0$ or it can be integrable on it. Thus, by no means, the result of Morales-Ruiz and Ramis closes the subject.

From the above it follows that investigations of system (9) need strong and precise tools, and that the most interesting questions are connected with (partial) integrability of this system.

Although the Mixmaster model seems to be the most attractive, we decided to analyze all class A Bianchi models. Models from this class which have a regular behavior can be used to approximate the complex behavior of Mixmaster models ( Belinskii *et al.*, 1982). It will be shown that the dynamics of the B(I) and B(II) models can be represented on a two dimensional phase space, whereas for the B(VI$_0$) and the B(VII$_0$) dynamical systems are three dimensional. The dynamics of these systems can be described precisely.

## 3. The Bianchi Class A Models in the Reduced Form

The main idea that we propose to apply for the study of the Bianchi class A models is the following. The right hand sides of the system (9) are homogeneous polynomials of degree two. Let us denote the vector field connected with them by $X$. The homogeneity of the system can be considered as its certain generalized symmetry. In fact, if we introduce the Euler vector field

$$E = \sum_{i=1}^{3} y_i \frac{\partial}{\partial y_i} + z_i \frac{\partial}{\partial z_i},$$

then it can be shown that

$$L_E X := [E, X] = X,$$

where $L_E$ denotes the Lie derivative along the vector field $E$, and $[\cdot, \cdot]$ denotes the Lie bracket. This symmetry can be used to lower the dimension of the system. Moreover, we want to restrict explicitly the system to the zero level of first integral $H = 0$, because of the physical interpretation of our system. We will also try to preserve the polynomial form of the right hand sides (Maciejewski and Szydłowski, 1998).

We start from the following change of variables

$$u_1 = z_1/w_3, \quad u_2 = z_2/w_3, \quad u_3 = z_3/w_3, \quad u_4 = w_2/w_3, \quad u_5 = w_3,$$

where

$$w_1 = y_1 + y_2, \quad w_2 = y_1 - y_2, \quad w_3 = y_3. \tag{11}$$

Then the system (9) has the form

$$\frac{du_1}{ds} = p(u_3, u_4, u_5) - 2u_1 u_3, \tag{12}$$

$$\frac{du_2}{ds} = -p(u_3, u_4, u_5) - 2u_2 u_3, \tag{13}$$

$$\frac{du_3}{ds} = \frac{1}{2}\{[(n_2 - n_1)u_5 - (n_1 + n_2)u_4]^2 - 4n_3^2 - 4u_3^2\}, \tag{14}$$

$$\frac{du_4}{ds} = (u_1 + u_2 - 2u_3)u_4 + (u_1 - u_2)u_5, \tag{15}$$

$$\frac{du_5}{ds} = (u_1 + u_2 - 2u_3)u_5 + (u_1 - u_2)u_4, \tag{16}$$

$$\frac{du_6}{ds} = 2u_3 u_6, \tag{17}$$

where

$$p(u_3, u_4, u_5) = \frac{1}{2}[(n_2 - n_1)u_5 + (-n_1 - n_2)u_4 - 2n_3] \times$$
$$[(n_1 + n_2)u_5 + (n_1 - n_2)u_4 - 2n_3],$$

and $ds = \frac{1}{2}u_6 d\tau$.

Let us notice that in equations (12–16) the variable $u_6$ does not appear explicitly, i.e. the dynamical system (12–17) separates into two subsystems, from which system (12–16) is closed. The time dependence of the variable $u_6$ is determined by integration of the equation (17)

$$u_6 = u_{06} \exp \int^\tau u_3(\tau')d\tau'. \tag{18}$$

Therefore instead of studying the 6-dimensional system (12–17) one can study only the system (12–16) because the full information about the dynamics is contained in the reduced system. Thus, our system has five 'true degrees of freedom'.

TABLE I

The collection of constants $n_i$ and the respective forms of the first integral for Bianchi A cosmological models. $k$ denotes the dimension of the reduced system.

| Bianchi type | $n_1$ | $n_2$ | $n_3$ | first integral | $k$ |
|---|---|---|---|---|---|
| B I | 0 | 0 | 0 | $H = \sum_{i<j}^3 u_i u_j$ | 2 |
| B II | 1 | 0 | 0 | $H = 4\sum_{i<j}^3 u_i u_j - u_5^2 - u_4^2 - 2u_4 u_5$ | 2 |
| B VI$_0$ | 1 | -1 | 0 | $H = \sum_{i<j}^3 u_i u_j - u_5^2$ | 3 |
| B VII$_0$ | 1 | 1 | 0 | $H = \sum_{i<j}^3 u_i u_j - u_4^2$ | 3 |
| B VIII | 1 | 1 | -1 | $H = \sum_{i<j}^3 u_i u_j - u_4^2 - 2u_5 - 1$ | 4 |
| B IX | 1 | 1 | 1 | $H = \sum_{i<j}^3 u_i u_j - u_4^2 + 2u_5 - 1$ | 4 |

Moreover, the system (12–17) has the first integral

$$H = \sum_{i<j}^3 u_i u_j - \frac{1}{4}(n_1 - n_2)u_5^2 - \frac{1}{4}(n_1 + n_2)^2 u_4^2 - n_3^2$$
$$+ \frac{1}{2}(n_2^2 - n_1^2)u_4 u_5 + (n_1 n_3 + n_2 n_3)u_5 + (n_1 n_3 - n_2 n_3)u_4, \qquad (19)$$

which in turn can be used to reduce the dimension of dynamical system by one.

Table 1 contains the forms of first integral (19) for all Bianchi class A types. It is easy to see that in all cases one variable can be eliminated, for example in the most general case of the Mixmaster models (B(IX) and B(VIII)) it is possible to eliminate the variable $u_5$ if we take into account that

$$u_5 = \frac{1}{2}n_3(u_4^2 + 1 - \sum_{i<j}^3 u_i u_j), \qquad (20)$$

on the level $H = 0$. For these two systems four equations (12–15) with $u_5$ defined by (20) form a closed system. It defines the reduced dynamics for the B(IX) and B(VIII) models. An important outcome of this reduction is the fact that we are able to prove that neither of these systems (considered on $\mathbb{C}^4$) has an additional analytic first integral. A proof of this fact for the B(IX) model can be found in our paper (Maciejewski and Szydłowski, 1998). For the B(VIII) model the proof is similar.

For the rest of the class A Bianchi models we can perform a next step of reduction. Here we present only some basic facts. Details and complete analysis of the reduced systems will be published elsewhere.

For the Bianchi I model, the first three of equations (12–17) can be integrated explicitly. The last two after redefinition of the time variable $u_1 ds = d\bar{s}$ can be written in the form of two linear differential equations depending on two parameters. Thus, we can perform a complete analysis of this system.

For the Bianchi II model three equations separate from the rest and they have the following form

$$\frac{du_1}{ds} = -2K(u_1, u_2, u_3) - 2u_1 u_3, \tag{21}$$

$$\frac{du_2}{ds} = 2K(u_1, u_2, u_3) - 2u_2 u_3, \tag{22}$$

$$\frac{du_3}{ds} = 2K(u_1, u_2, u_3) - 2u_3^2, \tag{23}$$

where $K(u_1, u_2, u_3) = \sum_{i<j}^{3} u_i u_j$. Because the right-hand sides of the above system are homogeneous functions of order two, we can introduce the projective variables and reduce it to a two dimensional system, which can be integrated explicitly.

For the Bianchi $VI_0$ and $BVII_0$ models the reduction leads to a three dimensional system of the form

$$\frac{dx}{d\bar{s}} = -x^2 + (z+1)y^2, \tag{24}$$

$$\frac{dy}{d\bar{s}} = -4(z+1) + xyz, \tag{25}$$

$$\frac{dz}{d\bar{s}} = -yz(z+2). \tag{26}$$

This system has two two-dimensional invariant manifolds $z = 0$ and $z = -2$. On these manifold we can perform explicit integration. It is unclear if the system (24–26) is integrable or not.

## References

L.D. Landau and E.M. Lifshitz: 1975, *The Classical Theory of Fields*. Pergamon, Oxford.

J. Wainwright and G.F.R. Ellis: 1997, *Dynamical Systems in Cosmology*, Cambridge U.P., Cambridge.

C. B. Collins: 1971 *Commun. Math. Phys.*, **23**, 137 (1971).

O.I. Bogoyavlensky: 1985, *Methods in Qualitative Theory of Dynamical Systems in Astrophysics and Gas Dynamics*. New York, Springer-Verlag.

J. Wainwright and L. Hsu: 1989 *Class. Quantum Grav.*, **6**, 1409.

D. F. Chernoff and J. D. Barrow: 1983, *Phys. Rev. Lett.*, **20**, (1983), 134–137.

R. Cushman and J. Śniatycki: 1995, *Preprint Univ. of Calgary*, .

N.J. Cornish: 1997, in Proceedings of Eight M. Grossmann Meeting, Jerusalem.

N.J. Cornish and J.J. Levin: 1997, in Proceedings of Eight M. Grossmann Meeting, Jerusalem.

A. Ramani, B. Grammaticos and T. Bountis: 1989, *Phys. Reports* **180**, 159–254.

G. Contopoulos, B. Grammaticos and A. Ramani: 1993, *J. Phys.*, **A26**, 5795.

G. Contopoulos, B. Grammaticos and A. Ramani: 1994, *J. Phys. A*, **27**, 5357–5361.

A. Latifi, M. Musette and R. Conte: 1995, *Phys. Lett. A*, **194**, 83–97.

G. Contopoulos, B. Grammaticos and A. Ramani: 1995, *J. Phys. A*, **28**, 5313–5322.

J.J. Morales-Ruiz and J.-P. Ramis: 1997, *Galoisian Obstruction to Integrability of Hamiltonian Systems I, II*, Preprint.

V. A. Belinskii, I. M. Khalatnikov and E. M. Lifshitz: 1982, *Adv. Phys.*, **31**, 639.

A. J. Maciejewski and M. Szydłowski: 1998, *J. Phys. A. Math. & Gen.*, **31**, 2031.

# SLOW AND FAST DIFFUSION IN ASTEROID-BELT
# RESONANCES: A REVIEW

S.FERRAZ-MELLO

*Instituto Astronômico e Geofísico, Universidade de São Paulo,*
*Caixa Postal 3386, São Paulo, SP, Brasil; E-mail sylvio@usp.br*

**Abstract.** This paper reviews recent advances in several topics of resonant asteroidal dynamics as the role of resonances in the transportation of asteroids and asteroidal debris to the inner and outer solar system; the explanation of the contrast of a depleted 2/1 resonance (Hecuba gap) and a high-populated 3/2 resonance (Hilda group); the overall stochasticity created in the asteroid belt by the short-period perturbations of Jupiter's orbit, with emphasis in the formation of significant three-period resonances, the chaotic behaviour of the outer asteroid belt, and the depletion of the Hecuba gap.

## 1. Introduction

This paper reviews the latest advances recorded in the study of the asteroidal mean-motion resonances and the recent discovery of the importance of numerous two and three-period resonances. Following a common usage, we call inner belt, the inner part of the main belt (from 2.0 to 2.9 AU), and outer belt, the domains beyond the 2/1 resonance (i.e. beyond 3.3 AU). The ring between them (from 2.9 to 3.3 AU) will be referred to as outer main-belt. In secs. 2 and 3, we review the role of resonances in the transport of asteroids and asteroidal debris to the inner and outer solar system, and compare results from numerical experiments to the known dynamics of these resonances. In particular, we discuss the contrasts of the depleted 2/1 resonance (Hecuba gap) and the high-populated 3/2 resonance (Hilda group). Sec. 4 gathers some significant results on chaos and resonance in the outer belt. At last (sec.5), we discuss the stochasticity generated by the short-period perturbations of Jupiter's orbit in the asteroid belt, viz. the significant 3-period (or 3-body) resonances and the role of the Great Inequality in the depletion of the central part of the Hecuba gap.

Almost all results in this review came from numerical simulations and analyses. However, it is worth emphasizing that semi-analytical and analytical theories of the main resonances are fundamental in allowing a correct understanding of the results obtained from numerical analyses; a thorough review of these theories was recently published by Michèle Moons (1997) and the reader is referred to it. We also mention that a new paper by Moons et al. (1998), did complement that review in several respects. Moons's review also included a detailed account of the advances recorded before 1996 so that we can restrict ourselves, here, to the discussion of those achieved in the past three years.

## 2. Fast-Diffusion Inner-Belt Resonances

The fast diffusion observed in several inner-belt resonances plays an important role in the transfer of asteroids and asteroid debris to the inner solar system and,

S.FERRAZ-MELLO

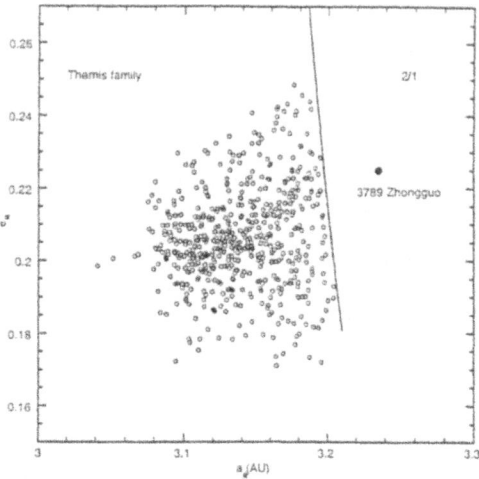

Fig. 1. Themis family together with the leftmost separatrix of the 2/1 resonance. Axes: Resonant proper elements (taken from Morbidelli *et al.*, 1995).

TABLE I
Fast diffusion of asteroids injected in resonances

| Resonance | $< a >$ (AU) | Neighboring Families | Half-Life (Myr) | Impact Sun (%) | expelled (%) | Survivors (%) |
|---|---|---|---|---|---|---|
| 4/1 | 2.06 | | | | | |
| $\nu_6$ (*) | 2.1 | Vesta | 2.3 | 79.1 | 11.8 | 1.8 |
| 3/1 | 2.50 | Vesta, Nysa, Maria | 2.4 | 71.0 | 27.7 | 0.3 |
| 5/2 | 2.82 | Dora, Gefion, Koronis | 0.5 | 7.3 | 91.3 | 0 |

(*) at $i = 7°$

in particular, to the neighborhood of the Earth. This entirely new fact became to appear when the study of the 4/1 and 3/1 resonances, in the frame of the restricted elliptic 3-body problem, showed paths allowing an asteroid to evolve in a short timescale (less than 1 Myr) to very high eccentricities (Wisdom, 1982, 1983; Ferraz-Mello and Klafke, 1991; Klafke *et al.*, 1992; Moons and Morbidelli, 1995). Later, simulations with more complete models, including other outer planets, showed trajectories starting in these regions and falling into the Sun (Farinella *et al.*, 1993, 1994a, 1994b; Morbidelli and Moons, 1995). By the same time, several asteroid families were shown to be intersected by resonances. The best-known example is Themis family (see fig. 1) whose proper elements show a distribution clearly cut by the left border of the 2/1 resonance (Morbidelli *et al.*, 1995). The reconstruction of the boundaries of the set of orbits originated at the fragmentation of the parent body showed that many objects may have been injected into resonances.

What happened to the missing asteroids in each case? An extended investigation

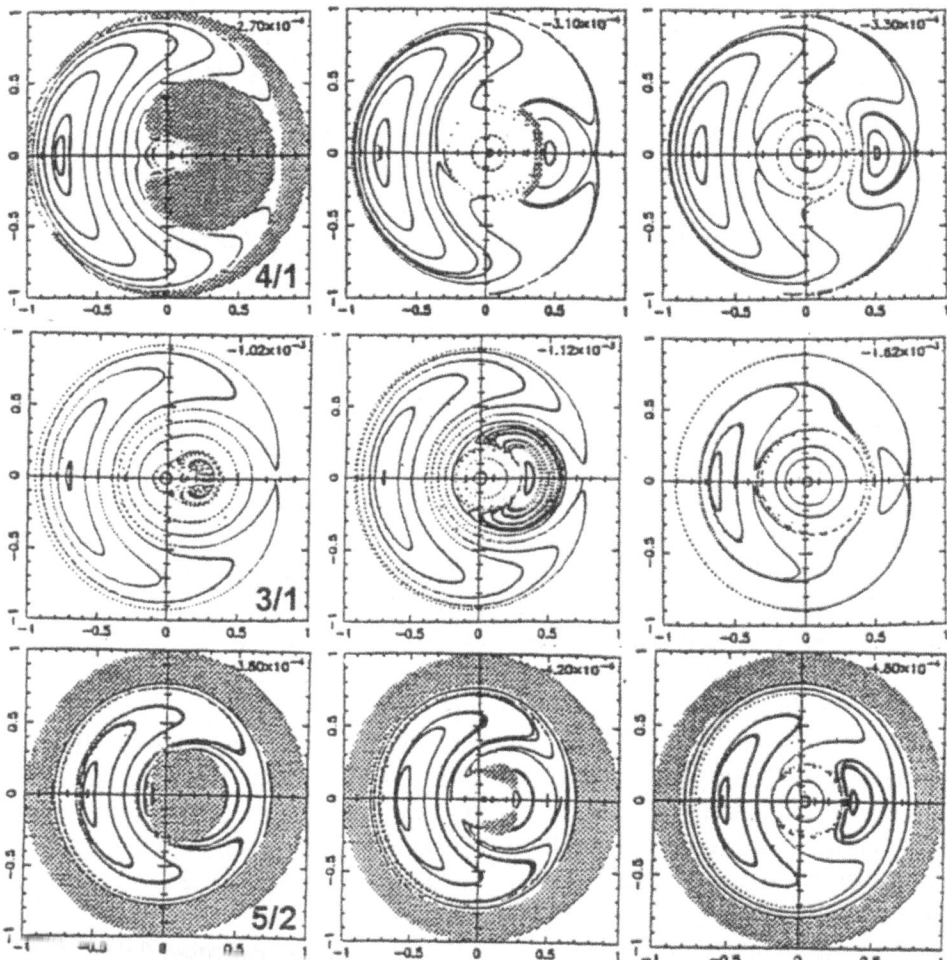

Fig. 2. Surfaces of section of the planar 3-body model of some inner belt resonances. *On top:* 4/1-resonance; *middle:* 3/1 resonance; *bottom:* 5/2 resonance. Coordinates: $x = e . \cos(\varpi - \varpi_{\text{Jup}})$, $y = e . \sin(\varpi - \varpi_{\text{Jup}})$. (adapted from Klafke *et al.*, 1992).

was conducted by Gladman *et al.* (1997) involving 1500 particles with initial conditions chosen in the intersection of the ejection fields of several asteroid families by the main resonances. Three of the families are close to the 3/1 resonance (see Table I). From a total of 393 fictitious asteroids showing an active interaction with this resonance, 279 terminated their evolution falling into the Sun (more than one third of them fell directly in the Sun, without being previously extracted from the resonance by encounters with the inner planets), 108 of them were expelled from the asteroid belt to orbits with aphelions situated beyond Saturn, and a few of them had an impact on one of the planets. The half-life of these populations ($\approx 2.4$ Myr) is in good agreement with the timescale of the large eccentricity jumps seen in simulations using a restricted elliptic model (about 0.5 Myr; see Ferraz-Mello

& Klafke, 1991). It is important noticing that, in general, the median lifetimes found by Gladman *et al.* are one order of magnitude lower than those based on Öpik-type evolution models. The fast diffusion inside the 3/1 resonance shows that the disruption of a parent body close to that resonance shall rapidly increase the number of near-Earth asteroids. But the short half-life of these populations and the high efficiency of the Sun to deplete them are such that these increases do not last for long times; the disruption of a parent-body near the 3/1 resonance will only cause a "shower" lasting some 10 Myr (Zappalà *et al.*, 1998).

The next important resonance in the inner belt is the 5/2 resonance. It lies in a transition zone between fast and slow diffusion resonances. Three of the families studied by Gladman *et al.* are close to the 5/2 resonance. The half-life of the fictitious asteroids injected in this resonance is very short ($\approx 0.5$ Myr), but only 28 evolved to a fall into the Sun. The main fate of the injected particles (349) was to be expelled to orbits whose aphelion is beyond Saturn. This behaviour may be understood from the dynamical features of this resonance. A simple planar 3-body model reveals a dynamics very similar to that of the 3/1 resonance, but, while in the 3/1 resonance the eccentricity jumps go close to 1, in the 5/2 resonance they may only reach $e \approx 0.8$ (fig. 2 *bottom*). As a consequence, almost all 28 fictitious asteroids reaching the Sun have done it only after having been extracted from the resonance by an encounter with one of the inner planets. Important is also that the resonance semi-major axis is larger than half of Jupiter's one and the aphelion of high-eccentricity asteroids goes close to Jupiter's orbit. Thus, once the asteroid is extracted from the resonance in a high-eccentricity orbit, it is bound to encounter that planet in a short time. The most likely result of such encounter is a "swing-by maneuver" in which the asteroid gains energy for a big increase in semi-major axis as did comet P/Oterma during its 1963 aphelic encounter with Jupiter (see the orbit in Carusi *et al.*, 1984). This explains the huge proportion of fictitious asteroids driven to orbits with aphelions beyond Saturn.

To complete the panorama of the fast-diffusion inner-belt resonances, we shall also mention the diffusion in the main branch of the secular resonance $\nu_6$ (resonance 1:1 between the main component of the motion of the asteroid perihelion and $g_6$; $g_6$ is the second most important component of Jupiter's perihelion motion), and the 4/1 mean-motion resonance. Fictitious asteroids from the Vesta family that could have been injected into the $\nu_6$ resonance show fates very similar to those of the 3/1 resonance (see Table I). The diffusion in the 4/1 resonance was not studied. However, as the main branch of the $\nu_6$ resonance almost meets the 4/1 resonance at $i = 0°$, we may expect a diffusion at least as fast as that of the $\nu_6$ resonance. We remind that the planar 3-body model of this resonance (without Saturn and $\nu_6$) shows a dynamics in many points similar to that of the 3/1 and 5/2 resonance, but with many paths leading from lower eccentricities to $e = 1$ (see fig.2 *top*).

TABLE II

Slow diffusion of asteroids injected in resonances

| Resonance | $< a >$ (AU) | Neighbouring Families | Half-Life (Myr) | Impact Sun (%) | expelled (%) | Survivors (%) |
|-----------|--------------|----------------------|-----------------|----------------|--------------|---------------|
| 8/3 | 2.71 | Chloris | 34 | 17.2 | 35.7 | 47.1 |
| 7/3 | 2.96 | Koronis | 19 | 0 | 56.6 | 42.3 |
| 9/4 | 3.03 | Eos | >120 | 2.2 | 25.4 | 72.4 |
| 2/1 | 3.28 | Themis | >100 | 6.5 | 22.9 | 69.9 |
| 3/2 | 3.97 | none | | | | |
| 4/3 | 4.29 | none | | | | |

## 3. Slow-Diffusion Resonances in the Outer Belts

High half-lives were observed by Gladman et al. in several resonances situated in the outermost rings of the main belt (7/3, 9/4, 2/1), as well as in the 8/3 resonance (in the inner belt, between the 3/1 and 5/2 resonances). Moreover, a very stable condition is known to exist in the 3/2 resonance, in the outer belt, where some 30 asteroids with diameters larger than 50 km, presumably primordial, are known. Table II reproduces some of the results of Gladman et al. These results are very different of those summarized in Table I. Now, only a small fraction of the injected asteroids fell into the Sun, and, generally, only after being extracted from the resonance by one planet. On the other hand, the number of survivors is very large (against almost none in the fast-diffusion resonances). Recent analyses of the 9/4 resonance, using well-suited proper elements, showed the existence of several asteroids with increasing eccentricities inside this resonance, probably belonging to the Eos family (bisected by this resonance) (Morbidelli et al., 1995).

In these resonances, in general, 3-body models have very robust structures confining the eccentricity to some tenths (Yoshikawa, 1991; Yokoyama and Balthazar, 1992; Ferraz-Mello, 1994). The minimum models able to show some significant diffusion are 4-body models including both Jupiter and Saturn. Diffusion charts of the 2/1 and 3/2 resonances were computed with a 3-D Sun-asteroid-Jupiter-Saturn model by Nesvorný and Ferraz-Mello (1997), using Laskar's frequency map analysis. The contour lines of these charts (fig. 3) are lines of same $\log_{10} |\rho|$, where

$$\rho = \frac{f_2 - f_1}{f_1}$$

is the relative variation of the perihelion frequency. The two frequencies $f_1$ and $f_2$ come from Fourier spectra obtained in two overlapping time intervals of length $2\Delta t$ and whose medium point separation is $\Delta t$.

The diffusion chart of the 2/1 resonance (fig.3 left) is a composite of 2 parts: one ($e \geq 0.2$) obtained with a $31 \times 51$-points grid and $\Delta t = 133,333$ yr, and the other ($e \leq 0.25$) obtained with a $26 \times 101$-points grid and $\Delta t = 200,000$ yr. In both

S.FERRAZ-MELLO

grids, $\Delta e = 0.01$. (In the domain where the two sets overlap, the one corresponding to the lower part is shown). The initial inclination is $I = 0$, the initial longitudes of the perihelion and ascending node are equal to those of Jupiter and the initial value of the critical angle

$$\sigma = 2\lambda_{\text{Jup}} - \lambda_{\text{ast}} - \varpi_{\text{ast}}$$

is taken equal to zero. ($\lambda$ denote mean longitudes and $\varpi$ the longitude of the perihelion).

The frequency map of the 3/2 resonance (fig. 3 *right*) was obtained with a $51 \times 51$-points grid, $\Delta t = 200,000$ yr, and the same initial value of the angles as above, but, now,

$$\sigma = 3\lambda_{\text{Jup}} - 2\lambda_{\text{ast}} - \varpi_{\text{ast}}.$$

The data used to draw these figures are the same presented as dot maps in Figs. 3 and 10 of Nesvorný and Ferraz-Mello (1997). In this paper, we are only showing the $(a, e)$ maps for $i_0 = 0$; $(a, i)$ maps showing the diffusion of orbits with initial inclinations up to 40° are also given in that paper.

The dispersion (r.m.s. of the time-variations) associated with each $|\rho|$ was determined from the statistical analysis of the corresponding values of $|\rho|$ obtained with two different values of $\Delta t$, in the strip $0.2 \leq e \leq 0.25$ (Ferraz-Mello *et al.* , 1998a). The results were time propagated using a random walk hypothesis. The levels corresponding to the lowest values of $|\rho|$ have a dispersion of the order of $10^{-1}$ in 1 Gyr. The dispersion increases almost linearly with $|\rho|$ (that is, exponentially with $\log_{10}|\rho|$). In the level corresponding to $\log_{10}|\rho| = -2.5$, it may reach the unity in a time not larger than 1 Gyr.

The interpretation of these results is not obvious. In a one-dimensional random walk, a dispersion equal to 1 means a 30 percent probability that the value $|\rho| = 1$ is reached (that is, a 100% change in the frequency value). The level corresponding to a dispersion reaching 1 in about 1 Gyr appears in black in the charts of fig.3. In all domains not encircled by the black strip (painted in green, yellow, orange, red) much higher dispersions are reached and escape from the resonance should be effective to deplete these domains in, at most, a few Gyr. In the domains encircled by the black strip (painted in blue), the fate of a solution is less evident. In the central part of the blue domains, the dispersion is very small: of the order of $10^{-1}$ in 1 Gyr.

In the 3/2 resonance, the low-dispersion area is large and we may expect that the number of solutions starting in its interior and reaching the outside, in the age of the solar system, is very small. Asteroids are not expected to remain there forever, but the escape times are at least one order of magnitude larger than the age of the solar system. In the 2/1 resonance, the low-dispersion area is formed by small disconnected domains; in this case, even a solution with an initial low dispersion can reach the zone of high dispersion and leave the resonance in, at most, some Gyr. These results are consistent with those of numerical simulations done

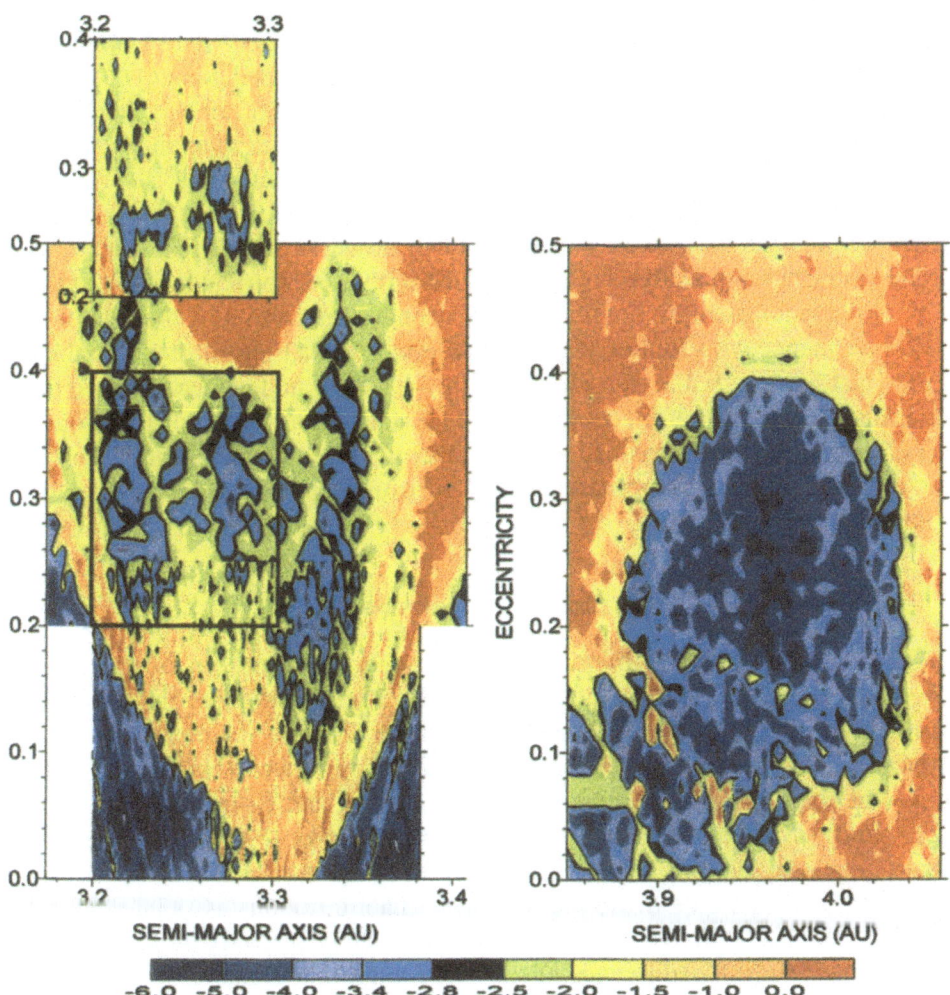

Fig. 3. Diffusion chart of the 2/1 (left) and 3/2 (right) resonances. The plotted parameter is the decimal logarithm of the relative variation of the perihelion motion, $\log_{10}|\rho|$, in 130,000 and 200,000 yr (see text). The inset on top left shows the chaoticity enhancement inside the area marked with a rectangle, when the Great Inequality period is forced to be half of the current one (see sec.5). (Adapted from Nesvorný and Ferraz-Mello, 1997).

by Morbidelli (1996), inside or near these domains, showing a large number of escapes in a 1-Gyr timespan. Some simulated solutions remained in the resonance, but were already showing an increased chaoticity at the end of the integration time.

The results obtained with Laskar's frequency map analysis led to conclusions similar to the ones obtained from the study of the maximum Lyapunov exponents and presented at the ACM-1993 Symposium (Ferraz-Mello, 1994 a, b): There is an extensive stochasticity in the 2/1 and 3/2 resonances, with Lyapunov exponents in the range $10^{-5} - 10^{-3.5}\mathrm{yr}^{-1}$ in the 2/1 resonance, and only $10^{-5.5} - 10^{-7}\mathrm{yr}^{-1}$ in the 3/2 resonance. Those results did unravel that the diffusion rate was the main difference between these two resonances, much slower in the 3/2 resonance than in the 2/1 resonance. An analysis of those results with the empirical formula of Lecar *et al.* (1992) indicated the possibility of a stronger interaction of the asteroids of the 2/1 resonance with Jupiter, and eventual escape, in a time shorter than the age of the Solar System, while those in the 3/2 resonance would, generally, need times larger than $10^{10}$ years before escaping. The great advantage of Laskar's frequency analysis over the calculation of Lyapunov exponents is their easier interpretation, and, moreover, shorter computation times allowing it to be done over large arrays of initial conditions.

## 4. Outer Belt Resonances

To complete the inventory of the acting two-period resonances of the asteroid belt, we have to include those of the outer belt. The domains beyond 3.3 AU are reasonably depleted, except for the Hildas and the Trojans. The chaotic behaviour of the asteroids situated between the 2/1 and 3/2 resonances was extensively studied in the last years (Holman and Murray, 1996; Murray and Holman, 1997; Murray *et al.*, 1998), with many new results. Holman and Murray (1996) have mapped the chaotic region in the $(a, e)$-space. Using appropriated proper elements, they have shown that real asteroids avoid the V-shaped chaotic regions associated with the 11/6, 9/5, 7/4, 12/7, 5/3, 8/5 and 11/7 resonances. However, only those resonances of $2^{nd}$ and $3^{rd}$ orders lead to encounters with Jupiter in less than $3 \times 10^7\mathrm{yr}$. In the higher-order resonances the removal times are much larger than $10^8\mathrm{yr}$ (Murray and Holman, 1997).

Fig.4 shows some calculations of the Lyapunov time (inverse of the maximum Lyapunov exponent) as a function of the asteroid semi-major axis. Results corresponding to three models are shown: (a) Jupiter in a fixed orbit; (b) Jupiter's orbit affected by the main long-period (secular) perturbations; (c) Jupiter's orbit affected also by short-period perturbations Their intercomparison shows that models including only the long-period perturbations of Jupiter's orbit are largely insufficient to give a correct picture of the actual chaoticity of the outer belt. Such chaoticity only appears when short-period perturbations are included. It is worth mentioning that the results obtained with a model fully including the action of the 4 outer planets is very similar to those obtained with model (c) (see Murray *et al.*, 1998).

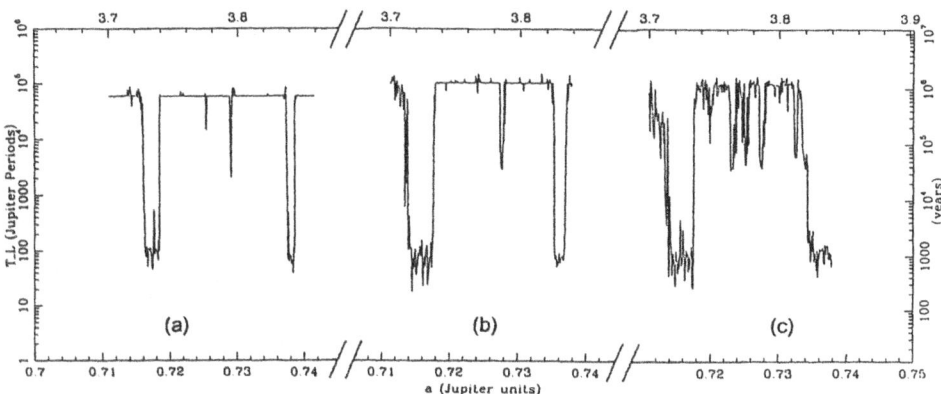

Fig. 4. The Lyapunov time as a function of semi-major axis in the outer asteroid belt in three 3-body models: (a) Jupiter in a fixed orbit; (b) Jupiter's orbit affected by the main long-period (secular) perturbations; (c) Jupiter's orbit affected also by short-period perturbations (adapted from Murray *et al.*, 1998).

In the outer belt, the 3/2 and 4/3 resonances are also found. The 3/2 resonance was already considered in the previous section. The 4/3 resonance was studied with the same tools used to study the 3/2 resonance, and a diffusion chart may be seen in Ferraz-Mello *et al.* (1998a). It is very similar to that of the 3/2 resonance, but with a smaller low diffusion area. However, Nesvorný and Ferraz-Mello (1997) have shown that this resonance is much affected by the outermost planets and a complete model, including also Uranus and Neptune, increases $\rho$ by one order of magnitude.

## 5. The Role of Jupiter's Short-Period Inequalities

A common assumption to many averaged and mapping models is that short-period inequalities (perturbations) of Jupiter's orbit are neglectable. Resonances involving short-period inequalities were, however, since 1996, pointed out as responsible for the depletion of the inner parts of the Hecuba gap (Ferraz-Mello, 1996, 1997; Michtchenko and Ferraz-Mello, 1997; Ferraz-Mello *et al.* 1998 b). In this case, chaos is created by the overlap of the frequency $2n_{Jup} - 5n_{Sat}$ of the Great Inequality of Jupiter's motion, with the frequencies

$$f_\sigma + k_1 f_\varpi + k_2 g_6$$

(complex structure seen in fig.5); $f_\sigma$ is the frequency of libration of the critical angle of the 2/1 resonance, $f_\varpi$ the asteroid's perihelion motion and $k_j$ are integers.

The best way to put this phenomenon into evidence is to play with the critical dependence of the Great Inequality period (currently $\approx 930$ years) on tiny variations of the periods of Jupiter and Saturn. Simulations with the planetary orbits slightly

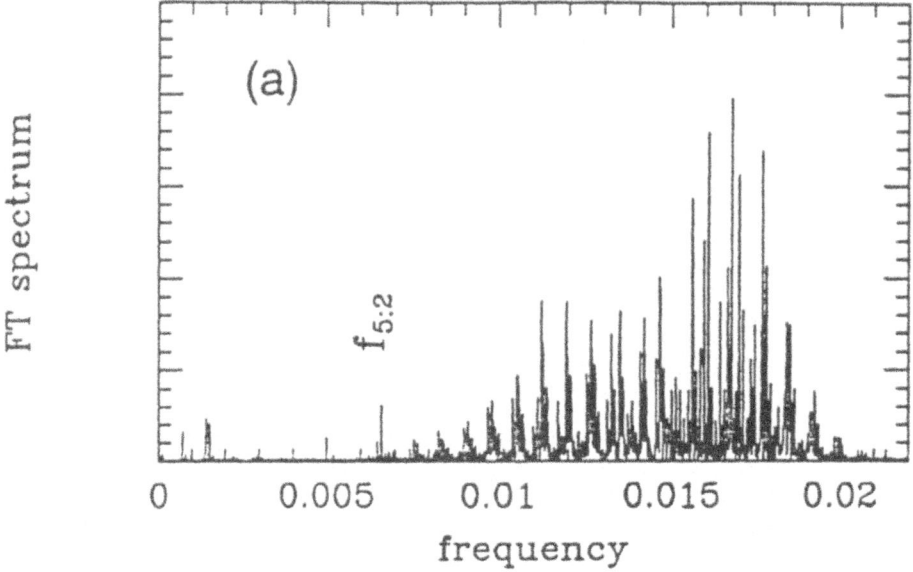

Fig. 5. FFT power spectrum of solutions starting at $e_0 = 0.3$, in the frame of a 4-body model including Saturn. Frequency unit is rad/yr. (Taken from Michtchenko and Ferraz-Mello, 1997).

modified and such that the Great Inequality period is close to the asteroids' libration period (about 440 years), in the middle of the frequency comb of fig.5, have shown a significant enhancement of the diffusion (see the inset in fig.3)

A similar confirmation was obtained by Henrard (1998), but with Saturn's semi-major axis decreased by 2.7 percent. In this case, the period and amplitude of the Great Inequality in longitude are much smaller than the actual ones and, as a consequence, the effects become barely visible. Henrard's work also confirmed the increase of the stochasticity (in a simplified model) at the exact beats of the Great Inequality frequency and the $f_\sigma + k f_\varpi$. A complete study of the dependence of the stochasticity on the period of the Great Inequality was later done by Roig and Ferraz-Mello (1999) with a symplectic mapping allowing fast simulations.

A series of recent papers has shown the role of all Jupiter's short-period inequalities in the formation of three-period (or three-body) resonances involving the periods of Jupiter, Saturn and the asteroid:

$$k_1 n_{\text{Jup}} + k_2 n_{\text{Sat}} + k_3 n_{\text{ast}} \simeq 0.$$

($n$ denote mean-motions and $k_j$ are integers; Murray et al., 1998, Nesvorný and Morbidelli, 1998a,b). Many asteroids are trapped in these resonances and the study of the diffusion in their interior lead to estimate that they can remain trapped there from 100 Myr to some Gyr (Murray et al., 1998). Fig.6 shows the Lyapunov exponents in the outer main belt, just in front of the 2/1 resonance. The peaks

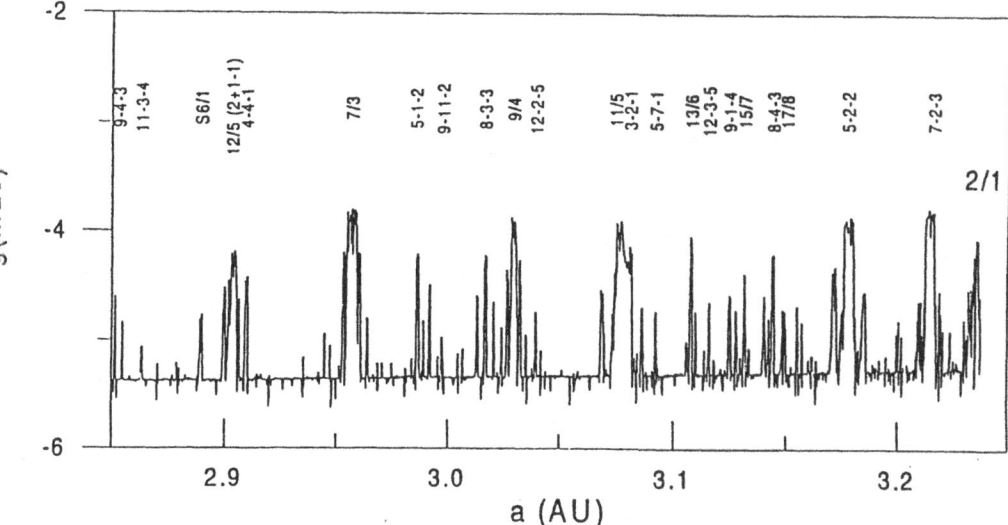

Fig. 6. Maximum Lyapunov exponents (in $yr^{-1}$ units and logarithmic scale) of solutions starting from a regular grid in $a$ (and $e_0 = 0.1$; $i_0 = 0$), in a model including the 4 outer planets. The main resonances are labeled on top of the corresponding peak (taken from Morbidelli and Nesvorný, 1998).

in this plot are identified with several 3-period resonances. These resonances are spanned by the combination of the main short-period perturbations in the motions of the asteroid and Jupiter. For instance, the 5–2–2 resonance is a combination of the perturbations of asteroid and Jupiter whose frequencies are $3n_{Jup} - 2n_{ast}$ and $2n_{Jup} - 2n_{Sat}$, respectively. Among the many asteroids found in these resonances, we mention (490) Veritas in the 5–2–2 resonance, whose chaoticity was first recognized by Milani and Farinella (1994) and is expected to allow an estimate of the age of Veritas family.

## 6. Conclusion

During many years, the success of KAM theory in Mathematics pervaded Astronomy and many results on the dynamics of Hamiltonian systems with 2 degrees of freedom were quickly translated into interpretations of the observed solutions of three-body models. Regularity was expected to be the rule and chaos the marvelous exception. Analytical models were constructed looking for resonances and their overlaps, while numerical simulations were aimed at representing longer and longer times, in the hope of seeing the less visible manifestations of chaos.

Now, the situation is reversed, and we see that chaos is the rule and regularity the exception. Problems as the "unexplained" Hecuba gap, only remained a puzzle for longtime because of insufficient modeling of Jupiter's motion; they started being solved when full 4-body models including Saturn, or 3-body models including the short- period perturbations of Jupiter's orbit, were used. These models allowed the global stochasticity inside the 2/1 resonance to be disclosed (Ferraz-Mello 1994a,b).

On the other hand, it became clear that it is enough to increase the number of planets acting on the asteroids to get an extremely dense network of significant chaos-generating resonances. The role of the short-period inequalities of Jupiter's motion on this phenomenon is well illustrated in fig.4. A similar comparison showing how the depletion rate of the Hecuba gap depends on the adopted model for Jupiter's motion, was given by Ferraz-Mello (1996, 1997).

We not only know the global stochasticity inside the classical resonances, but we find measurable resonances almost all over the asteroid belt. The puzzle is no longer to find a chaotic evolution able to drive an asteroid across the asteroid belt, but to find those asteroids that, like the Hildas, are able to remain in a bounded domain of the belt for times longer than the age of the Solar System.

The lesson to be learned is that in real-world dynamics we have to consider that our models, no matter how sophisticated, never consider all forces acting on the asteroids, and these forces may drive the asteroid to cross the barriers appearing in simplified models (cf. Ferraz-Mello, 1994b).

## Acknowledgements

This investigation was supported by CNPq, CAPES and FAPESP.

## References

Carusi, A., Kresák, L., Perozzi, E. and Valsecchi, G.B.: 1984, Long-term evolution of short-period comets, Inst. Astrof. Spaz. (Rome), Int. Report 12.

Farinella, P., Froeschlé, C. and Gonczi, R.: 1993, Meteorites from the asteroid 6 Hebe, *Celest. Mech. Dyn. Astron.*, **56**, 287-305.

Farinella, P., Froeschlé, C. and Gonczi, R.: 1994a, Meteorites delivery and transport, *IAU Symposium*, **160**, 205-222.

Farinella, P., Froeschlé, C., Froeschlé, C., Gonczi, R., Hahn, G., Morbidelli, A. and Valsecchi, G.B.: 1994b, Asteroids falling into the Sun, *Nature*, **371**, 314-317.

Ferraz-Mello, S.: 1994a, Kirkwood gaps and resonant groups, *IAU Symposium*, **160**, 175-188.

Ferraz-Mello, S.: 1994b, The dynamics of the asteroidal 2:1 resonance, *Astron. J.*, **108**, 2330-2337

Ferraz-Mello, S.: 1996, On the Hecuba gap, *IAU Symposium*, **172**, 177-182.

Ferraz-Mello, S.: 1997, A symplectic mapping approach to the study of the stochasticity in asteroidal resonances, *Celest. Mech. Dyn. Astron.*, **65**, 421-437.

Ferraz-Mello, S. and Klafke, J.C.: 1991, A model for the study of very-high-eccentricity asteroidal motion. The 3:1 resonance, in *Predictability, Stability and Chaos in N-Body Dynamical Systems* (A.E.Roy, ed.), Plenum Press, New York, (*NATO Adv. Stud. Inst. Ser B Phys.*, **272**), 177-184.

Ferraz-Mello, S., Nesvorný, D. and Michtchenko, T.A.: 1998a, Chaos, diffusion, escape and permanence of resonant asteroids in gaps and groups, in *Solar System Formation and Evolution* (D.Lazzaro et al., eds.), ASP, San Francisco, (*ASP Conf. Ser.*, **149**), 65-82.

Ferraz-Mello, S., Michtchenko, T.A. and Roig, F.: 1998b, The determinat role of Jupiter's Great Inequality in the depletion of the Hecuba gap, *Astron. J.*, **116**, 1491-1500.

Gladman, B.J., Migliorini, F., Morbidelli, A., Zappalà, V., Michel, P., Cellino, A., Froeschlé, C., Levison, H.F., Mailey, M. and Duncan, M.: 1997, Dynamical Lifetimes of objects injected into asteroid belt resonances, *Science*, **277**, 197-201.

Henrard, J.: 1998, The effect of the Great Inequality on the Hecuba gap, *Celest. Mech. Dyn. Astron.*, **69**, 187-198.

Holman, M.J. and Murray, N.W.: 1998, Chaos in high-order mean-motion resonances in the outer asteroid belt, *Astron. J.*, **112**, 1278-1293.

Klafke, J.C., Ferraz-Mello, S. and Michtchenko, T.: 1992, Very-high-eccentricity librations at some higher-order resonances, *IAU Symposium* **152**, 153-158.

Lecar, M., Franklin, F. and Murison, M.: 1992, On predicting long-term orbital instability: A relation between the Lyapunov time and sudden orbital transitions, *Astron. J.*, **104**, 1230-1236.

Michtchenko, T.A. and Ferraz-Mello, S.: 1997, Escape of asteroids from the Hecuba gap, *Planet. Sp. Sci.*, **45**, 1587-1593.

Milani, A. and Farinella, P.: 1994, The age of Veritas asteroid family deduced by chaotic chronology, *Nature*, **370**, 40-42.

Moons, M. and Morbidelli, A.: 1995, Secular resonances in mean-motion commensurabilities. The 4/1, 3/1, 5/2 and 7/3 cases, *Icarus*, **114**, 33-50.

Moons, M.: 1997, Review of the dynamics in the Kirkwood gaps, *Celest. Mech. Dyn. Astron.*, **65**, 175-204.

Moons, M., Morbidelli, A. and Migliorini, F.: 1998, Dynamical structure of the 2/1 commensurability with Jupiter and the origin of the resonant asteroids, *Icarus*, **135**, 458-468.

Morbidelli, A.: 1996, On the Kirkwood Gap at the 2/1 commensurability with Jupiter, *Astron. J.*, **111**, 2453-2461.

Morbidelli, A. and Moons, M.: 1995, Numerical evidences of the chaotic nature of the 3/1 mean-motion commensurability, *Icarus*, **115**, 60-65.

Morbidelli, A., Zappalà, V., Moons, M., Celino, A. and Gonczi, R.: 1995, Asteroid families close to mean motion resonances. Dynamical effects and physical implications, *Icarus*, **118**, 132-154.

Morbidelli, A. and Nesvorný, D.: 1998, Numerous weak resonances drive asteroids towards terrestrial planet orbits, *Icarus*, submitted.

Murray, N. and Holman, M.: 1997, Diffusive chaos in the outer asteroid belt, *Astron. J.*, **114**, 1246-1259.

Murray, N., Holman, M. and Potter, M.: 1998, On the origin of chaos in the asteroid belt, *Astron. J.*, **116**, 2583-2589.

Nesvorný, D. and Ferraz-Mello, S.: 1997, On the asteroidal population of the first-order resonances, *Icarus*, **130**, 247-258.

Nesvorný, D. and Morbidelli, A.: 1998a, Three-body mean-motion resonances and the chaotic structure of the asteroid belt, *Astron. J.*, **116**, 3029-3037.

Nesvorný, D. and Morbidelli, A.: 1998b, An analytical model of three-body mean-motion resonances, *Celest. Mech. Dyn. Astron.*, submitted.

Roig, F. and Ferraz-Mello, S.: 1999, A symplectic mapping approach of the dynamics of the Hecuba gap, *Planet. Sp. Sci.*, in press.

Zapallà, V., Cellino, A., Gladman, B.J., Manley, S. and Migliorini, F.: 1998, Asteroid showers on Earth after family break-up events, *Icarus*, **134**, 176-179.

Wisdom, J.: 1982, The origin of Kirkwood gaps: A mapping for asteroidal motion near the 3/1 commensurability, *Astron. J.*, **85**, 1122-1133.

Wisdom, J.: 1983, Chaotic behaviour and the origin of the 3/1 Kirkwood gap, *Icarus*, **56**, 51-74.

Yokoyama, T. and Balthazar, J.M.: 1992, Application of Wisdom's perturbative method to 5/2 and 7/3 resonances, *IAU Symposium*, **152**, 159-166.

Yoshikawa, M.: 1991, Motions of asteroids at the Kirkwood gaps. II, *Icarus*, **92**, 94-117.

# ORIGIN AND EVOLUTION OF NEAR EARTH ASTEROIDS

A. MORBIDELLI

*CNRS, Observatoire de la Côte d'Azur, B.P. 4229, 06304 Nice Cedex 4, France*

**Abstract.** The present paper reviews our current understanding of the origin and evolution of NEAs, at the light of the results of recent quantitative numerical simulations that have revolutioned the previously accepted scenario.

## 1. Introduction

With the discovery of 433 Eros in 1898, the existence of a new population of asteroid–like bodies on orbits intersecting those of the inner planets was established. Explaining the origin of these Near Earth Asteroids (NEAs) was difficult because, at that time, it was not evident which mechanisms could have forced them to evolve from an orbit bounded by those of Mars and of Jupiter –typical of the main asteroid belt– to a planet–crossing orbit. Still in 1976, Wetherill was claiming that most of NEAs must be extinct cometary nuclei.

However, a close look to the distribution of the asteroids in the main belt shows that the belt is structured by several resonances: the existence of the Kirkwood gaps (Kirkwood, 1866) is associated with the main mean–motion resonances with Jupiter (resonances between the orbital periods of the asteroid and of the planet), among which particularly evident are the 3/1, the 5/2 and the 2/1 resonances; the upper bound of the asteroid distribution, when plotted in the semimajor axis $a$ vs. inclination $i$ plane, corresponds with the location of the $\nu_6$ secular resonance, which occurs when the mean precession rates of the longitudes of perihelia of the asteroid and of Saturn are equal to each other (Fig. 1). This implies that somehow the asteroids must be removed from resonant locations, and suggests that this phenomenon could be related to the origin of NEAs.

The first indication that resonances can force bodies to cross the orbits of the planets came from J.G. Williams, in a diagram reported by Wetherill (1979), showing the amplitude of eccentricity oscillations as a function of the distance from the $\nu_6$ resonance: at distances smaller than 0.025 AU, the amplitudes exceed 0.25, forcing the resonant bodies to cross the orbit of Mars at the top of their eccentricity oscillation. Shortly afterwards, Wisdom (1983) showed that the 3/1 mean motion resonance has a similar effect: the eccentricity of resonant bodies can have, at irregular time intervals, rapid and large oscillations whose amplitudes exceed 0.3, the threshold value to become Mars–crosser at the 3/1 location.

Following these pioneering works, several studies confirmed, both analytically and numerically, the role that resonances have in increasing asteroid eccentricities to Mars–crossing or even Earth–crossing values. For a review of the studies on mean motion resonances we recommend the paper by Moons (1997), while a compendium of the investigations on secular resonances can be found in Froeschlé and Morbidelli (1994).

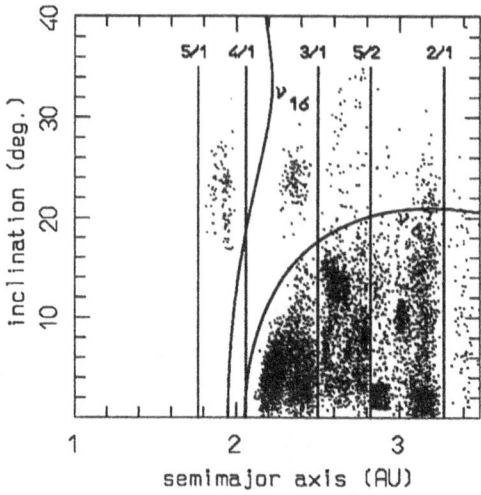

Fig. 1. Orbital distribution of the numbered asteroids that do not cross the orbits of any planet. The two bold curves denote the location of the $\nu_6$ and $\nu_{16}$ secular resonances, while the vertical lines mark the position of the main mean motion resonances with Jupiter associated to evident Kirkwood gaps.

The improved knowledge of resonant dynamics allowed to design a generally accepted scenario on the origin of Near Earth Asteroids, that we will call hereafter "the classical scenario".

## 2. The Classical Scenario on the Origin of NEAs

G.W. Wetherill is the author who most contributed to outlining a coherent scenario for the origin and evolution of NEAs and meteoroids in a long series of papers (Wetherill 1979, 1985, 1987, 1988). His results were reviewed, in an interpreted form, by Greenberg and Nolan in two papers (1989, 1993) that gave a well defined and concised portrait to the classical scenario (Fig. 2).

According to this scenario, collisions in the main belt continuously produce new asteroids by fragmentation of larger bodies. Some of these new asteroids are injected into the $\nu_6$ or the 3/1 resonance by the collisions that liberated them from their parent bodies. Once inside one of these resonances, their average semimajor axes stay constant, while their eccentricities suffer large oscillations, reaching, after a typical time-scale of 1 Myr, values that make the asteroids intersect the orbit of Mars and/or of the Earth.

At this point, close encounters with a planet may occur. Close encounters provide an impulse velocity to the asteroid's trajectory, causing a "jump" of its semimajor

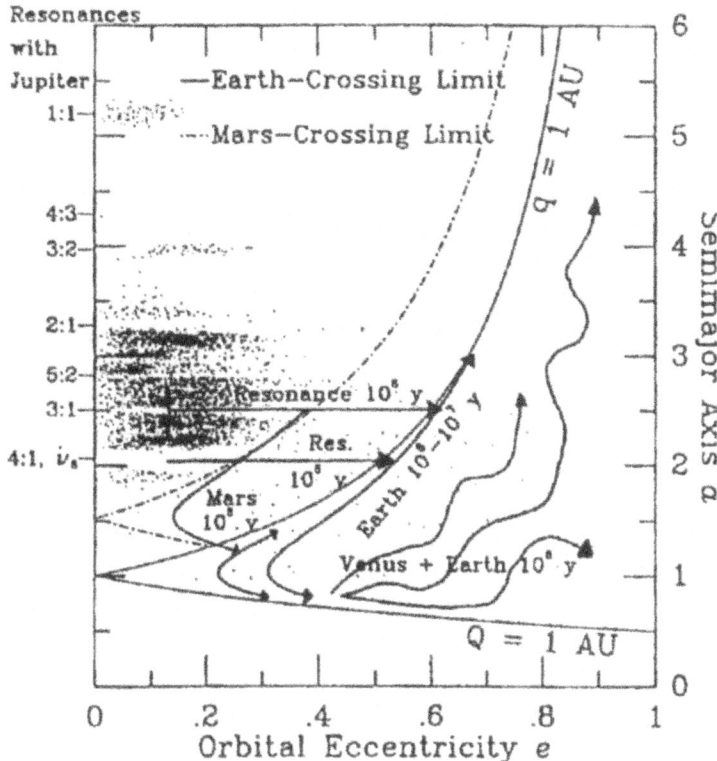

Fig. 2. Schematic representation of the origin and evolution of NEAs according to the classical scenario (from Greenberg and Nolan, 1993). The dash curves denote the set of orbits with perihelion distance $q$ or aphelion distance $Q$ equal to the value of the Martian semimajor axis. Solid curves denote the orbits with $q = 1$ or $Q = 1$ AU. The arrows with labels "Mars $10^8$y" and "Earth $10^6$–$10^7$y" sketch curves of constant Tisserand parameter with Mars and the Earth respectively. See text for further comments.

axis and eccentricity by a quantity depending on the geometry of the encounter and on the mass of the planet. If the jump in semimajor axis is large enough, the asteroid is removed from the resonant location and liberated from the control of the resonance, and then evolves mainly under the solely effects of the subsequent close encounters with the planet.

Repeated encounters with random geometry force the asteroid to random walk on the $(a, e)$–plane. This random walk, however, must follow preferential directions, approximately preserving the so–called Tisserand parameter

$$T = \frac{a_p}{a} + 2\sqrt{\frac{a(1 - e^2)}{a_p}} \cos i$$

relative to the dominating planet with semimajor axis $a_p$. In fact, if the planet is on a circular orbit the Tisserand parameter is equal to $3 - U^2$, where $U$ is the norm of the relative velocity between the asteroid and the planet, which is not changed by the encounter as is well known from the theory of scattering dynamics (Öpik, 1976). Setting the inclination to a constant value, the constancy of the Tisserand parameter defines curves on the $(a, e)$–plane. A few of these curves are sketched in Fig. 2.

If the asteroid has been extracted from the resonance by Mars on a solely Mars–crossing orbit, due to the small mass of the planet its random walk evolution is very slow and eventually may reach, after a typical time of $10^8$ y, an Earth–crossing orbit. If, on the contrary, it has been extracted on an Earth–crossing orbit, its subsequent evolution is faster, requiring $10^6$–$10^7$ years to reach a Jupiter crossing or a Venus–crossing orbit. Combined encounters with the Earth and Venus break the constancy of the Tisserand parameter, since neither planet really may dominate as the "main perturber". This allows the asteroid to go all over the Earth/Venus–crossing space, on a typical time-scale of order $10^8$ y. Encounters with Jupiter, conversely, quickly eject the body from the Solar System.

Undergoing this kind of evolution, a NEA survives as long as it does not collide with a planet or undergo encounters with Jupiter that eject it on an unbound orbit from the Solar System. Monte Carlo codes, which treat in a statistical way the effects of close planetary encounters (Arnold, 1965; Wetherill, 1988), predict that the median lifetime of NEAs, from their original injection into resonance, is of order several tens of Myr (Gladman, personal communication).

## 3. The Solar Sink

The first indication that some qualitatively important dynamical features are missing from the classical scenario came from Farinella *et al.* (1994), who showed that, among the bodies in the $\nu_6$, 3/1 or 5/2 resonances, the collision with the Sun is a fairly common fate. The collision with the Sun happens because the eccentricity increases up to values close to unity, so that the perihelion distance becomes smaller than the solar radius.

The reasons for which the main resonances pump the eccentricity of resonant bodies to unity were quickly understood. The $\nu_6$ resonance location is almost independent of the eccentricity (Williams and Faulkner, 1981), so that the eccentricity may have an infinite regular growth (Morbidelli, 1993). Inside the 3/1 and 5/2 resonances, conversely, secular resonances are present and overlap each other, making most of the resonant phase space chaotic. This allows the eccentricity to evolve in an irregular manner without upper bound (Moons and Morbidelli, 1995).

Farinella *et al.* also showed that NEAs with very large orbital eccentricity –like the Taurids asteroids– may also easily collide with the Sun, despite not being inside any notable resonance. Actually, they may be forced to collide with the Sun by non–resonant secular oscillations of the eccentricity, which are particularly large when all secular arguments are in phase (Levison and Duncan, 1994; Valsecchi *et al.*, 1995).

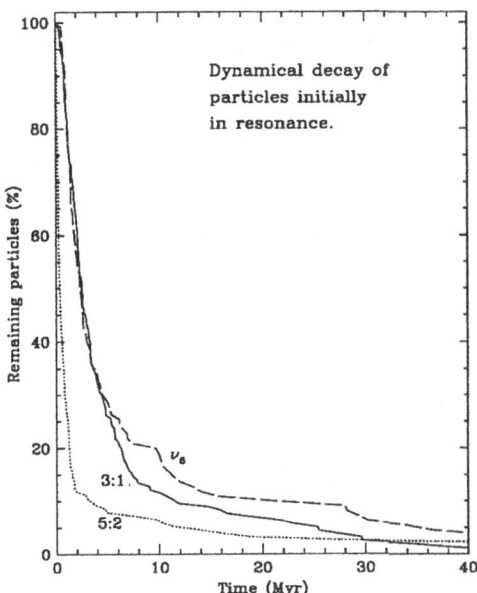

Fig. 3. Decay of populations of particles initially placed in the 3/1, 5/2 and $\nu_6$ resonances (Gladman *et al.*, 1997). The median dynamical lifetime is of order 2 Myr for the 3/1 and $\nu_6$ resonances and less than 1 Myr for the 5/2 resonance.

The Farinella *et al.* simulations, however, were based on too few bodies to conclude on the statistical importance of solar collisions in the overall scenario of the origin and evolution of NEAs. The improvements in computer technology and, in particular, the availability of a new integration algorithm (Wisdom and Holman, 1991; Levison and Duncan 1994) a few years later allowed Gladman *et al.* (1997) to numerically simulate the dynamical evolution of several hundred test particles, initially placed into the $\nu_6$, 3/1 and 5/2 resonances. They found that the median lifetime of the simulated particles is about 2 Myr, while only $\sim 10\%$ of them survive longer than $\simeq 10$ Myr (Fig. 3).

This happens precisely because the resonances tend to pump the eccentricity of almost all the resonant bodies to unity on a Myr time-scale. As a result of the rapid growth of $e$, Mars is able to extract only a few percent of resonant bodies before an Earth-crossing orbit is achieved. Of all the simulated particles extracted by Mars, none was transported to Earth–crossing orbit by successive Martian encounters, conversely to what was expected in the classical scenario. The reason for this is that the 'kicks' provided by Martian encounters are so small that the particles cannot jump over the resonances and (after a typical time ranging from 1 to a few 10 Myr) always find themselves again injected into some resonant mechanism which takes them to Earth-crossing orbits.

In the Gladman *et al.* simulation, Earth and Venus are found to be much more efficient than Mars in extracting bodies from resonances. However, the "mortality" of the extracted particles is still very high. Those which are driven by close encounters to $a > 2.5$ AU are usually ejected by Jupiter on hyperbolic orbits. Many others are injected again into the 5/2, 3/1 or $\nu_6$ resonances and subsequently are forced to collide with the Sun. Only the bodies that reach semimajor axes $< 1.8$ AU may live much longer than the median lifetime because, in this region, although many resonances exist and influence the dynamics (Michel and Froeschlé 1997, Michel 1997), no statistically significant dynamical mechanisms have been found which pump the eccentricities up to Sun–grazing values. Therefore, these bodies die only by colliding with a planet (rare), or after being driven back to $a > 1.8$ AU (after a typical 1-10 Myr journey) and then usually being pushed by a resonance into the Sun.

Finally, the Gladman *et al.* simulations show that particles extracted from the resonances do not evolve by closely following lines of constant Tisserand parameter, as described by the classical scenario. In reality the dynamics is much more complicated: there are many high-order resonances which force the orbits to evolve transversally with respect to the curves of constant Tisserand parameter (see Michel *et al.* 1996), so that the Tisserand parameter is very poorly conserved on a Myr or longer time scale and, in practice, extracted particles are quickly spread all over the Earth- and Venus-crossing region.

Despite the fact that collisions with the Sun quantitatively change the classical scenario regarding dynamical paths and typical lifetime, the basic concept that NEAs are asteroid fragments pushed to Earth–crossing orbits after being injected into the 3/1 or $\nu_6$ resonances could still be considered valid. The resonant bodies' median lifetime of 2 Myr simply implies that, to sustain the NEA population in steady state, the number of asteroids injected into resonance per Myr needs to be roughly equal to 1/4 of the total number of NEAs. This is plausible for small to km–sized bodies. For instance, 2000 NEAs are estimated to exist with diameter of order of 1 km (Morrison, 1992), while the number of 1–km bodies injected into the 3/1 or $\nu_6$ resonances per Myr is estimated to be $\sim$400 by Menichella *et al.* (1996). Moreover, the orbital distribution of NEAs seems to be consistent with the one expected on the basis of the Gladman *et al.* simulations, once observational biases are quantitatively taken into account (Bottke, Jedicke and Morbidelli, work in progress). To have a decent fit between the expected and observed orbital distributions of NEAs, the number of bodies injected into the 3/1 resonance per unit time should be about 5 times larger than the number of bodies injected into the $\nu_6$ resonance (Fig. 4). This ratio is close to the one found by Morbidelli and Gladman (1998) in order to explain the observed orbital distribution of fireballs of chondritic origin, and is reasonable from the point of view of the location of the source asteroids (Farinella *et al.*, 1993).

However, the short dynamical lifetime of 3/1 and $\nu_6$ resonant particles produces inconsistencies for the classical scenario concerning the origin of multi–kilometer NEAs. About 10 bodies larger than 5 km exist on Earth–crossing orbits. To sustain

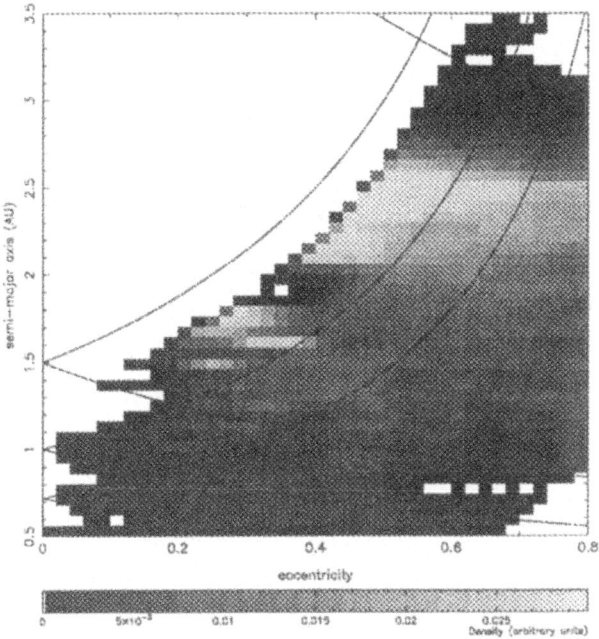

Fig. 4. Expected unbiased orbital distribution of NEAs ($q < 1.3$ AU) coming from the 3/1 and $\nu_6$ resonances, assuming that 5 times more particles are injected into the 3/1 than in the $\nu_6$ resonance per unit time. The grey scale denotes the relative density of NEAs on the $(e, a)$–plane.

this population, 2–3 bodies of this size should be injected into 3/1 or $\nu_6$ resonance per Myr. However, bodies of this size can be injected into resonance only during very energetic and rare break–up events, such as those leading to the formation of asteroid families (Menichella *et al.*, 1996; Zappalá *et al.*, 1998).

## 4. A New Scenario on the Origin of Large NEAs

A hint for a better understanding of the origin of multi–kilometer Earth–crossers comes from the observation that Mars–crossing asteroids of similar size are much more numerous. Taking into account in the Mars–crossing population also the bodies that will intersect the orbit of Mars within the next 300,000 y (namely during their next secular eccentricity cycle), the number of Mars–crossers with diameter larger than 5–km is $\sim$350, i.e. 35 times larger than the number of Earth–crossers of comparable size (Migliorini *et al.*, 1998). Numerical integrations done by Migliorini *et al.* show that Mars–crossers evolve to Earth–crossing orbits on a typical (median) time-scale of 20–25 Myr; this suggests that Mars–crossing asteroids might constitute the intermediate reservoir of Earth–crossers, but moves the fundamental question from the origin of multi–kilometer Earth–crossers to the

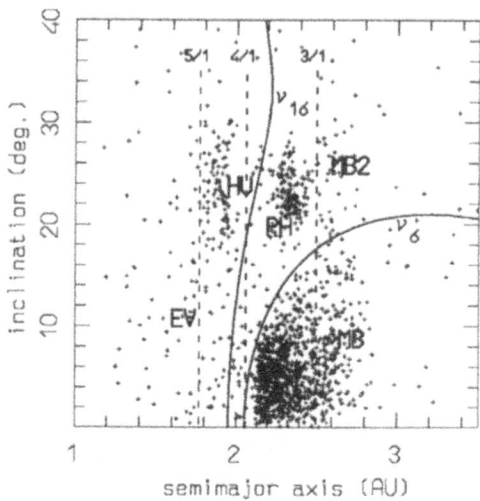

Fig. 5. Orbital distribution of Mars–crossers. Note the similarity with the $(a, i)$ distribution of non planet–crossing main belt asteroids (Fig. 1).

origin of multi–kilometer Mars–crossers.

The orbital distribution of Mars–crossing asteroids shows no concentration around the 3/1 and $\nu_6$ resonances (Fig.5). It reveals the existence of four main groups –denoted by Migliorini *et al.* as MB, MB2, HU and PH– with values of semimajor axis $a$ and inclination $i$ (compare with Fig. 1) similar to those of four populations of non–planet–crosser asteroids: the main belt below the $\nu_6$ resonance, the main belt beyond the 3/1 resonance and above the $\nu_6$, the Hungarias and the Phocaeas. This similarity suggests that these populations continuously lose objects to the Mars–crossing region, sustaining the MB, MB2, HU and PH groups. Only 2% of Mars–crossers larger than 5 km have $a$ and $i$ very different from those of non–planet–crossing asteroids and therefore must have evolved relative to the orbit that they had when they first crossed the orbit of Mars: for this reason they have been denoted as EV.

In order to better understand the mechanism by which the main belt sustains the MB Mars–crossing population, Migliorini *et al.* numerically integrated a sample of 412 asteroids with osculating perihelion distance smaller than 1.8 AU, semimajor axis in the range 2.1–2.5 AU, inclination smaller than 15 degrees and which are not Mars–crosser in the first 300,000 y. Fig. 6 shows the evolution of the proper semimajor axis and eccentricity of the integrated bodies. Very few asteroids exhibit regular dynamics (those which appear as dots in Fig. 6); the vast majority exhibit macroscopic diffusion in eccentricity –that is, a relevant change of proper eccentricity– in agreement with the result of Morbidelli and Nesvorný (1998) that

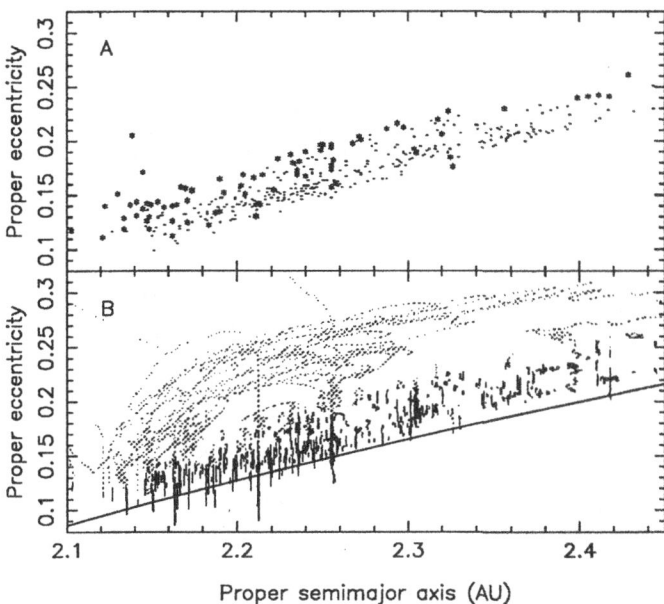

Fig. 6. Origin of MB Mars–crossers from the main belt. (A) the initial proper semimajor axis and eccentricity of the integrated asteroids. Asterisks denote the bodies that will become Mars–crossers within the integration time span. (B) The trace left by their time evolution. Proper elements are constant for regular asteroids, whose orbital elements have quasi–periodic evolution, while change over time for those evolving chaotically. If all asteroids were regular, panel B and panel A would be identical, so that the comparison between the two plots allows to appreciate that most of the asteroids chaotically migrate in the proper element plane. Before becoming Mars–crossers (black), asteroids migrate in proper eccentricity, keeping the proper semimajor axis close to its initial value. During the Mars–crossing regime (grey) asteroids migrate also in proper semimajor axis, as a result of close encounters with Mars which force them to evolve along curves of quasi–constant Tisserand invariant. The solid curve denotes proper perihelion distance equal to 1.92 AU (from Migliorini *et al.*, 1998).

most of the bodies in the inner belt are chaotic, mainly due to the dense location of mean motion resonances with Mars.

In the Migliorini *et al.* simulations, 17% of the integrated bodies become MB Mars–crossers in the first 25 Myr by diffusing to larger eccentricity. Scaling to the number of asteroids larger than 5 km existing in the population sampled by the integrated bodies, this implies that 44 new multi–kilometer asteroids should become MB Mars–crossers in the next 25 Myr. This number could probably be doubled, if one took into account that the chaotic process which leads to the origin of Mars–crossers concerns basically all the asteroids with proper perihelion distance smaller than 1.92 AU (Fig. 6) and that, according to the 1994 update of the proper elements catalog (see Milani and Knežević, 1994), about 1000 asteroids larger than 5 km in the inner belt have proper perihelion distances smaller than this threshold, including parts of the Flora and the Nysa families. On the other hand, in the same time interval (25 Myr), the MB Mars–crossing population should lose $\sim$ 90 bodies (50% of the total population), which evolve to the Earth–crossing region and subsequently collide with the Sun or another planet or are ejected on an hyperbolic orbit. Therefore, the diffusion process should be sufficient to keep the MB Mars–crosser population in a steady state, at least for the next 25 Myr.

If chaotic diffusion is able to sustain the MB and HU populations in steady state, by liberating to the Mars–crossing region a sufficiently large number of asteroids from the main belt and the Hungaria region, then the Migliorini *et al.* simulations show also that the number of asteroids larger than 5 km that should be expected on Earth–crossing orbits and on EV Mars–crossing orbits is very close to that presently observed (10 and 7, respectively). Therefore, the new scenario for the origin of multi–kilometer NEAs appears to be quantitatively able to supply both the Mars–crossing and the Earth–crossing asteroids.

It should be qualified, however, that the Migliorini *et al.* simulations concerned only approximately 1/3 of the known total Mars–crossing population; the integration of the remaining bodies is presently undergoing. It is therefore possible that the numbers provided by Migliorini *et al.* have an uncertainty of a factor 2. Moreover, our knowledge of the Mars–crossing population is probably not complete even for large objects. Most surveys for NEAs, in fact, don't pay care to follow the discovered asteroids that do not appear to be Earth–crosser (Jedicke, private communication), and this could severely reduce the number of Mars–crossers that are eventually listed in the asteroid catalogue. Future work, both from the dynamical and from the observational viewpoint, needs to be done to refine the Migliorini *et al.* scenario.

## 5. Perspectives

While the Migliorini *et al.* scenario brings a credible solution to problem the origin of multi–kilometer NEAs, it also opens new dilemmas. Is the diffusion process from the main belt the dominating mechanism for the origin of NEAs of all sizes? Or do near–Earth bodies smaller than some threshold come mostly from 3/1 and $\nu_6$ resonances, into which they have been collisionally injected? What is this threshold? These questions are intimately related to the size distribution of NEAs (extremely poorly known), because the size distribution of NEAs should be main belt–like in the size–range where the diffusion process is the dominating mechanism, while it might become steeper at sizes such that the contribution of bodies collisionally injected into the main resonances becomes substantial.

A significant advancement in this direction may be achieved by quantitatively fitting the observed orbital distribution of Earth–crossers by the simulated orbital distributions of bodies coming out from $\nu_6$ resonance, 3/1 resonance and Mars–crossing intermediate reservoir. This however requires a reliable knowledge of the observational biases. In principle, this procedure could allow the determination of the relative weight of the $\nu_6$, 3/1 and Mars–crossing sources. On the other hand, an improved knowledge of the collisional mechanisms in the main belt could refine the estimates of the number of bodies injected into the main resonances, as a function of body size.

Another new open question raised by the Migliorini *et al.* results is that the escape rate of asteroids from the main belt seems to decrease, after the first 10 Myr, to about 10% of the population per 100 Myr. This escape rate would be insufficient to keep the present Mars–crossing population in steady state on a $10^8$ y time-scale. In the Migliorini *et al.* simulation this happens because the bodies in the main diffusion tracks escape in majority in the first 10 Myr, so that these tracks result depleted of objects; subsequently, only the bodies in the "diffusion background" contribute to sustain the Mars–crossing population, but can do so only at a much lower rate. However, in reality the main diffusion tracks are not associated with gaps in the distribution of asteroids; this indicates that they must be resupplied with new objects on a $\sim 10$ Myr time–scale by some process(es) that are not taken into account in the Migliorini *et al.* simulation. The important processes that could bring new bodies to the considered large–eccentricity parts of the main diffusion tracks could be *(i)* diffusion from the low eccentricity portion of the belt, *(ii)* injection into resonance by collisions and/or encounters among asteroids, *(iii)* migration in semimajor axis due to some non–conservative force, such as that given by the Yarkovsky thermal re–emission effect (Burns *et al.*, 1979 ; Rubincam, 1995; Farinella *et al.* 1998) which could allow for a 0.01 AU mobility of multi–kilometer asteroids over their collisional lifetime (Farinella and Vokrouhlický, 1998). Until a quantitative analysis of these processes is done, a realistic understanding of the long–term evolution of the asteroid belt will not be achieved, limiting in turn our understanding of the NEA origin process.

The picture of the origin of NEAs is getting more complex, but this is because it is getting more realistic.

## References

Arnold, J.R.: 1965, The origin of meteorites as small bodies II. The model, *Ap. J*, **141**, 1536–1547.

Burns, J.A., Lamy, P.H. and Soter, S. : 1979, Radiation forces on small particles in the solar system, *Icarus*, **40**, 1–48.

Farinella, P., Gonczi, R. , Froeschlé, Ch. and Froeschlé, C.: 1993, The injection of asteroid fragments into resonance, *Icarus*, **101**, 174–187.

Farinella, P., Froeschlé, Ch., Froeschlé, C., Gonczi, R., Hahn, G., Morbidelli, A. and Valsecchi, G.B.: 1994, Asteroids falling onto the Sun, *Nature*, **371**, 315–317.

Farinella, P., Vokrouhlický, D. and W.K. Hartmann, W.K.: 1998, Meteorite delivery via Yarkovsky orbital drift, *Icarus*, **132**, 378–387.

Farinella, P. and Vokrouhlický, D.: 1998, Semimajor axis mobility of asteroidal fragments, *Science*, submitted.

Froeschlé, Ch. and Morbidelli, A.: 1994, The secular resonances in the solar system, I-in *Asteroids, Comets, and Meteors, 1993* (A. Milani, M. Di Martino, A. Cellino, Eds.) Kluwer: Boston, 189–204.

Gladman B., F. Migliorini, F., A. Morbidelli, A., Zappalá, V., P. Michel, P., A. Cellino, A., Ch. Froeschlé, Ch., Levison, H., Bailey, M. and Duncan, M.: 1997, Dynamical lifetimes of objects injected into asteroid main belt resonances, *Science*, **277**, 197–201.

Greenberg R. and Nolan, M.: 1989, Delivery of asteroids and meteorites to the inner solar system, in Asteroids II, eds. R.P. Binzel, T. Gehrels and M.S. Matthews (Tucson: Univ. Arizona Press), 778–804.

Greenberg R. and Nolan, M.: 1993, Dynamical relationships of near–Earth asteroids to main–belt asteroids, in *Resources of Near–Earth space*, J. Lewis *et al.* eds. (Tucson: Univ. Arizona Press), 473–492.

Kirkwood, D.: 1866, in *Proceedings of the American Association for the Advancement of Science for 1866*.

Levison, H.F. and Duncan, M.: 1994, The long term dynamical behaviour of short period comets, *Icarus*, **108**, 18–36.

Menichella M., Paolicchi, P. and Farinella, P.: 1996, The main belt as a source of Near Earth–Asteroids, *Earth, Moon and Planets*, **72**, 133–149.

Michel, P., Froeschlé, Ch. and Farinella, P.: 1996, Dynamical evolution of NEAs: close encounters, secular perturbations and resonances, *Earth Moon and Planets*, **72**, 151–164.

Michel, P., and Froeschlé, Ch.: 1997, The location of secular resonances for semimajor axes smaller than 2 AU, *Icarus*, **128**, 230–240.

Michel, P.: 1997, Effects of linear secular resonances in the region of semimajor axes smaller than 2 AU, *Icarus*, **129**, 348–366.

Migliorini, F., P. Michel, P., Morbidelli, A., Nesvorný, D. and Zappalà, V.: 1998, Origin of Earth–crossing asteroids: the new scenario, *Science*, **281**, 2022–2024.

Milani, A. and Knežević, Z.: 1994, Asteroid proper elements and the chaotic structure of the asteroid main belt, *Icarus*, **107**, 219–254.

Moons M.: 1997, Review of the dynamics in the Kirkwood gaps, *Cel. Mech.*, **65**, 175–204.

Moons M. and Morbidelli, A.: 1995, Secular resonances inside mean-motion commensurabilities: the 4/1, 3/1, 5/2 and 7/3 cases, *Icarus*, **114**, 33–50.

Morbidelli, A.: 1993, Asteroid secular resonant proper elements, *Icarus*, **105**, 48–66.

Morbidelli, A. and Gladman, B.: 1998, Orbital and temporal distribution of meteorites originating in the asteroid belt, *Meteoritics and Planet. Sci.*, **33**, 999–1016.

Morbidelli A. and Nesvorný, D.: 1998, Numerous weak resonances drive asteroids towards terrestrial planets orbits, *Icarus*, submitted.

Morrison, D. Ed.: 1992, *The Spaceguard Survey. Report of the NASA near-Earth object detection workshop*, NASA, Washington, D.C.

Rubincam, D.P.: 1995, Asteroid orbit evolution due to thermal drag, *JGR*, **100**, 1585–1594.

Öpik, E.J.: 1976, *Interplanetary Encounters*, Elsevier Press, New York. 155 pp.

Valsecchi, G.B., Morbidelli, A., Gonczi, R., Farinella, P., Froeschlé, Ch., Froeschlé, C.: 1995, The dynamics of objects in orbits resembling that of P/Encke, *Icarus*, **117**, 45–61.

Wetherill, G.W.: 1976, Where do meteorites come from? A re–evaluation of the Earth–crossing Apollo objects as sources of chondritic meteorites, *Geochimica et Cosmochimica Acta*, **40**, 1297–1317.

Wetherill, G.W.: 1979, Steady–state populations of Apollo–Amor objects, *Icarus*, **37**, 96–112.

Wetherill, G.W.: 1985, Asteroidal sources of ordinary chondrites, *Meteoritics*, **20**, 1–22.

Wetherill, G.W.: 1987, Dynamic relationship between asteroids, meteorites and Apollo–Amor objects, *Phil. Trans. Royal Soc. London*, **323**, 323–337.

Wetherill, G.W.: 1988, Where do the Apollo objects come from? *Icarus*, **76**, 1–18.

Williams, J. G., and Faulkner, J.: 1981, The position of secular resonance surfaces, *Icarus*, **46**, 390–399.

Wisdom, J.: 1983, Chaotic behavior and the origin of the 3/1 Kirkwood gap, *Icarus*, **56**, 51–74.

Wisdom, J. and Holman, M.: 1991, Symplectic maps for the N–body problem, *Astron. J.*, **102**, 1528–1538.

Zappalá, V., Cellino, A., Gladman, B., Manley, S. and Migliorini, F.: 1998, Note: asteroid showers on Earth after family break–up events, *Icarus*, **134**, 176–179.

# SMALL BODIES IN THE OUTER SOLAR SYSTEM

BRIAN G. MARSDEN

*Harvard-Smithsonian Center for Astrophysics*
*Cambridge, Massachusetts, U.S.A.*

This report is a continuation of three earlier reviews (Marsden 1996a, 1996b, 1998) that included a summary of our orbital knowledge of the Kuiper Belt. Presented at conferences held in the middle of 1994, 1995 and 1996, respectively, these reviews showed the steadily developing picture of a system dominated by the *plutinos*, librating in the 2:3 mean-motion resonance with Neptune, and the *cubewanos*, a somewhat more distant population of nonlibrating objects with low orbital eccentricities. The existence of a 3:4 Neptune librator and a 3:5 Neptune librator was also suspected. These librators have now been confirmed, and a possible 4:7 librator and possible second 3:5 librator have also been found. The known and suspected multiple-opposition librators are listed in Table 1. Here it is important to note that the orbital semimajor axes $a$ (in AU), eccentricities $e$ and inclinations $i$ (in degrees with respect to the 2000.0 ecliptic) are mean values that eliminate the large 12-year and 30-year periodicities arising from the indirect perturbations by Jupiter and

**Table 1.** Librating Kuiper Belt objects.

|     |         |              | $a$   | $e$   | $i$  | $H$  | Nep. | Ura. |
|-----|---------|--------------|-------|-------|------|------|------|------|
| 3:4 | (36.41) | 1995 $DA_2$  | 36.39 | 0.073 | 6.5  | 8.0  | 8.0  | 14.5 |
| 2:3 | (39.39) | 1993 RO      | 39.30 | 0.199 | 3.7  | 8.0  | 12.5 | 11.4 |
|     |         | 1995 $YY_3$ ?| 39.36 | 0.220 | 0.4  | 8.5  | 14.7 | 11.1 |
|     |         | 1997 $QJ_4$  | 39.36 | 0.220 | 16.6 | 7.5  | 15.9 | 13.0 |
|     |         | 1993 SB      | 39.38 | 0.320 | 1.9  | 8.0  | 20.1 | 7.5  |
|     |         | 1996 $TP_{66}$ | 39.42 | 0.330 | 5.7 | 6.5  | 22.0 | 6.9  |
|     |         | 1994 $JR_1$  | 39.44 | 0.119 | 3.8  | 7.5  | 11.5 | 15.8 |
|     |         | 1996 $TQ_{66}$ | 39.47 | 0.123 | 14.7 | 6.5 | 14.2 | 14.8 |
|     |         | 1995 $QZ_9$  | 39.52 | 0.148 | 19.6 | 7.5  | 16.5 | 14.3 |
|     |         | 1996 $SZ_4$  | 39.55 | 0.253 | 4.7  | 8.0  | 18.2 | 9.9  |
|     |         | 1994 TB      | 39.55 | 0.318 | 12.1 | 7.0  | 22.8 | 8.8  |
|     |         | Pluto        | 39.55 | 0.250 | 17.1 | −1.0 |      |      |
|     |         | 1993 SC      | 39.59 | 0.186 | 5.2  | 7.0  | 14.9 | 12.4 |
|     |         | 1995 $HM_5$  | 39.60 | 0.253 | 4.8  | 8.0  | 15.2 | 10.6 |
|     |         | 1996 $RR_{20}$ | 39.70 | 0.183 | 5.3 | 7.0  | 11.3 | 13.8 |
|     |         | 1995 $QY_9$  | 39.76 | 0.265 | 4.8  | 7.5  | 11.7 | 9.2  |
| 3:5 | (42.25) | 1994 JS      | 42.29 | 0.217 | 14.0 | 8.0  | 14.4 | 14.4 |
|     |         | 1996 $TR_{66}$ ? | 42.35 | 0.219 | 12.4 | 7.5 | 16.9 | 14.4 |
| 4:7 | (43.65) | 1997 $CV_{29}$? | 43.67 | 0.183 | 8.0 | 7.0 | 11.8 | 16.5 |

*Celestial Mechanics and Dynamical Astronomy* **73**: 51–54, 1999.
© 1999 *Kluwer Academic Publishers*.

BRIAN G. MARSDEN

**Table 2.** Nonlibrating Kuiper Belt objects.

| | | | $a$ | $e$ | $i$ | $H$ | Nep. |
|---|---|---|---|---|---|---|---|
| 5:8 | (41.12) | | | | | | |
| | | 1997 $RX_9$ ? | 41.38 | 0.035 | 30.0 | 8.0 | 10.2 |
| | | 1997 $RT_5$ | 41.71 | 0.082 | 12.7 | 7.0 | 9.2 |
| 3:5 | (42.25) | | | | | | |
| | | 1997 $QH_4$ ? | 42.71 | 0.031 | 13.2 | 7.0 | 11.5 |
| | | 1996 $TK_{66}$ ? | 42.77 | 0.005 | 3.3 | 7.0 | 12.3 |
| | | 1994 $VK_8$ | 42.81 | 0.030 | 1.5 | 6.5 | 11.2 |
| | | 1994 $EV_3$ | 42.94 | 0.041 | 1.7 | 7.0 | 11.3 |
| 7:12 | (43.05) | | | | | | |
| | | 1996 $TO_{66}$ | 43.47 | 0.131 | 27.4 | 4.5 | 10.2 |
| | | 1997 $CU_{29}$ ? | 43.53 | 0.029 | 1.5 | 6.5 | 12.0 |
| 4:7 | (43.65) | | | | | | |
| | | 1993 FW | 43.78 | 0.049 | 7.7 | 7.0 | 12.4 |
| | | 1994 $GV_9$ | 43.79 | 0.061 | 0.6 | 7.0 | 11.1 |
| | | 1997 $CT_{29}$ ? | 43.84 | 0.025 | 1.0 | 5.0 | 12.5 |
| | | 1996 $TS_{66}$ | 43.92 | 0.124 | 7.3 | 6.0 | 9.2 |
| | | 1997 $CS_{29}$ | 43.94 | 0.011 | 2.2 | 5.0 | 13.4 |
| | | 1992 $QB_1$ | 44.00 | 0.071 | 2.2 | 7.0 | 11.1 |
| | | 1996 $RQ_{20}$ | 44.04 | 0.109 | 31.7 | 7.0 | 9.6 |
| | | 1994 $JQ_1$ | 44.14 | 0.049 | 3.7 | 7.0 | 12.0 |
| | | 1995 $DC_2$ | 44.15 | 0.068 | 2.3 | 7.0 | 11.0 |
| 5:9 | (44.48) | | | | | | |
| | | 1997 $CQ_{29}$ ? | 44.71 | 0.081 | 2.9 | 6.5 | 10.9 |
| 6:11 | (45.03) | | | | | | |
| | | 1997 $CR_{29}$ ? | 45.09 | 0.078 | 19.1 | 6.5 | 12.7 |
| | | 1996 $KV_1$ | 45.27 | 0.110 | 8.1 | 7.5 | 10.1 |
| 7:13 | (45.41) | | | | | | |
| | | 1994 $ES_2$ | 45.86 | 0.115 | 1.1 | 7.5 | 10.3 |
| | | 1995 $WY_2$ | 46.53 | 0.129 | 1.7 | 7.0 | 10.4 |
| | | 1995 $DB_2$ | 46.62 | 0.140 | 4.1 | 7.5 | 10.1 |
| 1:2 | (47.71) | | | | | | |

Saturn on sun-centered orbits. The numbers in parentheses are the semimajor axes (in AU) corresponding to the resonances. Following the absolute magnitude $H$, the entries "Nep." and "Ura." show the minimum distances (in AU) from Neptune and Uranus (the latter being of course quite small for the most eccentric 2:3 Neptune librators) within several millennia of the present time.

Table 2 lists the corresponding data (except for the minimum distances from Uranus) for the established cubewanos, which range over $41 < a < 47$ AU, with $e$ rising to 0.14 for the most distant objects. The existence of several high-$i$ objects is an interesting new feature that warrants the need for searches at high ecliptic latitudes. The initial columns in this table show the positions of the resonances with

**Table 3.** Centaurs and Scattered-Disk objects.

| | $a$ | $e$ | $i$ | $H$ | Nep. | Ura. | Sat. | Jup. |
|---|---|---|---|---|---|---|---|---|
| (2060) Chiron | 13.67 | 0.381 | 6.9 | 6.5 | 6.8 | 0.6 | 0.4 | 2.9 |
| (5145) Pholus | 20.30 | 0.572 | 24.7 | 7.0 | 0.3 | 4.9 | 0.6 | 3.5 |
| (7066) Nessus | 24.61 | 0.520 | 15.6 | 9.6 | 4.3 | 4.3 | 1.8 | 6.4 |
| (8405) 1995 GO | 18.05 | 0.620 | 17.6 | 9.0 | 6.4 | 0.8 | 0.2 | 2.1 |
| 1994 TA | 16.79 | 0.303 | 5.4 | 11.5 | 7.9 | 1.4 | 5.0 | 6.4 |
| 1995 DW$_2$ | 25.06 | 0.247 | 4.1 | 8.0 | 1.0 | 0.2 | 8.6 | 12.8 |
| 1997 CU$_{26}$ | 15.78 | 0.170 | 23.4 | 6.5 | 12.7 | 2.8 | 4.4 | 8.1 |
| 1996 TL$_{66}$ | 83.82 | 0.582 | 24.0 | 5.0 | 6.4 | | | |

Neptune (out to sixth order) and possibly give a first indication of the presence of "Kirkwood gaps"—except, of course, for the high-$e$ 3:5 and suspected 4:7 librators shown in Table 1. It is also noteworthy that the population seems to terminate *inside* the 1:2 resonance, at least for objects of the size represented by the observed $H$ values, although this requires further investigation. Sensitivity of the detection probability to distance, and hence to $a$, suggests that the observed plutino-cubewano ratio, 0.65, is about twice what it would be for objects down to a given minimum size.

One of the objects discovered in late 1996 at the general distance of the Kuiper Belt was the perihelic detection (Luu *et al.* 1997) of the single known member, 1996 TL$_{66}$, of what has been termed the "scattered disk", a presumably large population of objects in orbits of rather large eccentricity and moderate inclination that at perihelion can have modest interactions with Neptune. The scattered-disk objects have an obvious dynamical symmetry with the centaurs that exist in rather unstable orbits within the realm of the giant planets and that in some cases interact with Neptune at aphelion. Table 3 gives orbital data for the seven known centaurs (three of which are intrinsically fainter than any of the entries in Tables 1 and 2) and the one known scattered-disk object. For the centaurs, columns 'Sat.' and 'Jup.' also show the minimum distances (again within just millennia of the present) from Saturn and Jupiter. The physical similarity of at least some of the centaurs to the short-period comets has been suspected (Kowal *et al.* 1979) or known (Meech and Belton 1990) for some time, and dynamical analyses (Duncan *et al.* 1995) strongly suggest that the centaurs are objects in transition between the Kuiper Belt (where they presumably originated) and the Jupiter Family of comets with their aphelia generally in the vicinity of Jupiter.

The objects tabulated in this paper have all been observed at more than one opposition. In addition, there are 26 single-opposition Kuiper Belt candidates, at least half of which are lost.

## References

Duncan, M. J., Levison, H. F. and Budd, S. M.:1995, *Astron. J*, **110**, 3073–3081.

Kowal, C. T., Liller, W. and Marsden, B. G: 1979, In *Dynamics of the Solar System*, ed. R. L. Duncombe, *IAU Symp*. No. 81, 245–250.

Luu, J., Marsden, B. G., Jewitt, D., Trujillo, C. A., Hergenrother, C. W., Chen, J. and Offutt, W. B.: 1997, *Nature*, **387**, 573–575.

Marsden, B. G.: 1996a, Searching for comets and planets. In *Completing the Inventory of the Solar System*, ed. T. E. Rettig and J. M. Hahn, *ASP Conf. Ser*, **107**, 193–207.

Marsden, B. G.: 1996b, From telescope to MPC: organizing the minor planets. In *Ephemerides and Astrometry of the Solar System*, ed. S. Ferraz-Mello, B. Morando and J.-E. Arlot, *IAU Symp*. No. 172, 153–164.

Marsden, B. G.:1998, Kuiper Belt: securing adequate orbital data. In *Asteroids, Comets, Meteors 1996*, ed. A. C. Levasseur-Regourd and M. Fulchignoni, COSPAR, in press.

Meech, K. J. and Belton, M. J. S.: 1990, *Astron. J*, **100**, 1323–1338.

# THE USE OF GEOCENTRIC VARIABLES TO SEARCH FOR METEOROID STREAMS AND THEIR PARENTS

CL. FROESCHLÉ

*Observatoire de la Côte d'Azur, B.P. 4229, F-06304 Nice, France*

T.J. JOPEK

*Obserwatorium Astronomiczne Uniwersytetu A. Mickiewicza, Sloneczna 36, PL-60286 Poznań, Poland*

and

G.B. VALSECCHI

*I.A.S. – Planetologia, via Fosso del Cavaliere, I-00133 Roma, Italy*

**Abstract.** A set of geocentric variables suitable for the identification of meteoroid streams has been recently proposed and successfully applied to photographic meteor orbits. We describe these variables and the secular invariance of some of them, and discuss their use to improve the search for meteoroid stream parents.

**Key words:** Meteors, Meteoroid streams

## 1. Introduction

Valsecchi et al. (1999) have recently proposed a new orbital similarity criterion for meteor orbits rather different from the one proposed by Southworth and Hawkins in 1963 and widely used for meteoroid stream identifications ever since.

The criterion by Southworth and Hawkins (1963) is a generalized measure of distance in the 5-dimensional space of the orbital elements $q$, $e$, $i$, $\omega$ and $\Omega$:

$$D_{SH}^2 = [e_2 - e_1]^2 + [q_2 - q_1]^2 + \left[2\sin\frac{I_{21}}{2}\right]^2 +$$
$$\left[\left(\frac{e_2 + e_1}{2}\right)\left(2\sin\frac{\pi_{21}}{2}\right)\right]^2 \tag{1}$$

where

$$\left[2\sin\frac{I_{21}}{2}\right]^2 = \left[2\sin\frac{i_2 - i_1}{2}\right]^2 + \sin i_1 \sin i_2 \left[2\sin\frac{\Omega_2 - \Omega_1}{2}\right]^2 \tag{2}$$

and

$$\pi_{21} = \omega_2 - \omega_1 + 2\arcsin\left[\cos\frac{i_2 + i_1}{2}\sin\frac{\Omega_2 - \Omega_1}{2}\sec\frac{I_{21}}{2}\right]. \tag{3}$$

The formulation of $D_{SH}$, however, does not take into account that meteoroids are observed at heliocentric distance $r \simeq 1$ AU, so that approximately either

$$1 = \frac{a(1 - e^2)}{1 + e\cos(-\omega)} \tag{4}$$

or

$$1 = \frac{a(1 - e^2)}{1 + e\cos(180° - \omega)} \tag{5}$$

depending on whether the meteoroid is at the ascending or at the descending node of its orbit. This implies that although $D_{SH}$ is defined in a 5-dimensional space, meteoroids are actually observed only when they are on a suitable 4-dimensional hypersurface. Moreover the observational errors, in the passage from the observed quantities (time of fall, geocentric velocity and radiant coordinates) to the orbital elements, are redistributed among the latter in a very complicated way. The improved formulations of Southworth and Hawkins' criterion introduced by Drummond (1981) and Jopek (1993) have retained the same basic features, as they also are based on the orbital elements.

Valsecchi et al. (1999) described a set of geocentric quantities that characterize completely a meteor orbit, and introduced an orbital similarity criterion that, when applied to the best photographic meteor orbits, gives significantly different results from $D_{SH}$ in the case of low-inclination streams (Jopek et al. 1999). The geocentric quantities and the criterion based on them are described in Section 2; some of these quantities have near-invariance properties against secular perturbations, as shown in Section 3. The use of these properties to look for parent bodies of meteoroid streams is discussed in Section 4; the conclusions (Section 5) then follow.

## 2. Geocentric Quantities and the Criterion Based on Them

The following treatment is based on the geometric setup of Öpik's theory of close encounters (Öpik 1976; Carusi et al. 1990).

Let us assume that the Earth moves on an unperturbed circular orbit, of radius equal to 1, lying on the ecliptic and that the massless meteoroid is on an orbit, with orbital parameters $a, e, i, \omega$ and $\Omega$, which hits that of the Earth in one of the nodes. Putting the constant of gravity and the mass of the Sun equal to 1, and imposing that the heliocentric velocity of the Earth is $v_\oplus = 1$, we have that the geocentric velocity of the meteoroid when crossing the Earth's orbit is:

$$U = \sqrt{3 - T} \tag{6}$$

where $T$ is the Tisserand parameter:

$$T = \frac{1}{a} + 2\sqrt{a(1 - e^2)}\cos i. \tag{7}$$

In a reference frame centred on the Earth, with the $z$-axis perpendicular to the plane of the ecliptic, the $y$-axis in the direction of the Earth's velocity and the $x$-axis pointing away from the Sun, the unperturbed geocentric encounter velocity

U of the meteoroid has components:

$$U_x = U \sin \theta \sin \phi \tag{8}$$

$$U_y = U \cos \theta \tag{9}$$

$$U_z = U \sin \theta \cos \phi \tag{10}$$

where $\theta$ is the angle between $U$ and the $y$-axis, and $\phi$ is the angle between the $y$-$z$ plane and that containing $U$ and the $y$-axis. If the encounter is at the ascending node $-90° < \phi < 90°$, while at the descending node $90° < \phi < 270°$); moreover, if the meteoroid is receding from perihelion, $0° < \phi < 180°$, and if it is approaching it $180° < \phi < 360°$.

The components of $U$ can be expressed in terms of $a$, $e$ and $i$:

$$U_x = \pm\sqrt{2 - \frac{1}{a} - a(1 - e^2)} \tag{11}$$

$$U_y = \sqrt{a(1 - e^2)} \cos i - 1 \tag{12}$$

$$U_z = \pm\sqrt{a(1 - e^2)} \sin i \tag{13}$$

and $a$, $e$ and $i$ in terms of the components of $U$:

$$a = \frac{1}{1 - U^2 - 2U_y} \tag{14}$$

$$e = \sqrt{U^4 + 4U_y^2 + U_x^2(1 - U^2 - 2U_y) + 4U^2 U_y} \tag{15}$$

$$i = \arctan \frac{U_z}{1 + U_y}; \tag{16}$$

in equation (11) the minus sign must be used for pre-perihelion encounters, and in equation (13) for encounters at the descending node.

A set of variables suitable for our purposes is then composed of the modulus of the unperturbed geocentric velocity $U$, the two angles $\theta$ and $\phi$, given by

$$\theta = \arccos \frac{U_y}{U} \tag{17}$$

$$\phi = \arctan \frac{U_x}{U_z} \tag{18}$$

and of the longitude of the Earth $\lambda_\oplus$ at the time of the meteor observation. These quantities are all directly observed, as $\theta$ and $\phi$ define the direction opposite to the geocentric radiant, in the frame moving with the Earth (and not that of fixed stars).

In practice, one obtains $U, \theta$ and $\phi$, through the components $U_x$, $U_y$, $U_z$, from the geocentric velocity $V_G$ and the equatorial coordinates of the meteor radiant $\alpha_G$ and $\delta_G$, by means of the expression:

$$\begin{pmatrix} U_x \\ U_y \\ U_z \end{pmatrix} = \hat{\mathbf{r}}(\lambda) \cdot \hat{\mathbf{p}}(\varepsilon) \cdot \frac{V_G}{29.7} \begin{pmatrix} -\cos \delta_G \cos \alpha_G \\ -\cos \delta_G \sin \alpha_G \\ -\sin \delta_G \end{pmatrix} \tag{19}$$

where $\hat{p}(\varepsilon)$, $\hat{r}(\lambda)$ are rotational matrices around the $x$- and $z$-axis, respectively, and the angle $\varepsilon$ denotes the inclination of the ecliptic plane to the celestial equator.

The new orbital similarity criterion $D_N$ introduced by Valsecchi et al. (1999) is given by:

$$D_N^2 = [U_2 - U_1]^2 + w_1[\cos\theta_2 - \cos\theta_1]^2 + \Delta\xi^2 \tag{20}$$

where

$$\Delta\xi^2 = \min\left[w_2\Delta\phi_I^2 + w_3\Delta\lambda_I^2, w_2\Delta\phi_{II}^2 + w_3\Delta\lambda_{II}^2\right] \tag{21}$$

$$\Delta\phi_I = \left[2\sin\frac{\phi_2 - \phi_1}{2}\right] \tag{22}$$

$$\Delta\phi_{II} = \left[2\sin\frac{180° + \phi_2 - \phi_1}{2}\right] \tag{23}$$

$$\Delta\lambda_I = \left[2\sin\frac{\lambda_2 - \lambda_1}{2}\right] \tag{24}$$

$$\Delta\lambda_{II} = \left[2\sin\frac{180° + \lambda_2 - \lambda_1}{2}\right] \tag{25}$$

and $w_1$, $w_2$, $w_3$ are suitably defined weighting factors; $\Delta\xi$ is small either if both $\phi_1 - \phi_2$ and $\lambda_1 - \lambda_2$ are small, or if they are both close to 180°. In this case, the two meteors belong to the two showers corresponding to the two node crossings of essentially the same orbit.

Jopek et al. (1999) applied the new criterion, setting the three weighting factors equal to 1, to a set of 865 precisely measured photographic meteor orbits. They grouped the meteors by a single neighbour linking algorithm (Lindblad 1971) not requiring any a priori knowledge of stream orbits, and determined thresholds for $D_N$ by comparison with random samples having the same marginal distributions of the variables (Jopek and Froeschlé 1997), in order to have a reliability level of 99%. For comparison, they made also a classification using $D_{SH}$.

They found that $D_{SH}$ and $D_N$ give similar results for streams of moderate to high inclination, with $D_N$ often adding a few more members to some streams; on the other hand, for near-ecliptical streams the results of the two criteria were markedly different. In particular, 2 streams were identified only with $D_{SH}$ (the $\sigma$ Leonids and the Andromedids), while 5 were identified only with $D_N$ (the $\varepsilon$ Geminids, the Monocerotids, the $\alpha$ Virginids, the Northern $\delta$ Aquarids and the $\varepsilon$ Piscids).

Arithmetic mean orbital and geocentric data concerning the 15 streams identified by Jopek et al. (1999) with $D_N$ are given in Table I.

## 3. Secular Invariance Properties of $U$ and $\cos\theta$

Over not too long timescales, and in absence of planetary close encounters, one can assume that mostly secular perturbations affect meteoroid orbits; the most

TABLE I

Arithmetic mean orbital and geocentric data of the streams found by Jopek et al. (1999) using $D_N$ amongst 865 precise photographic meteor orbits; $i$, $\omega$ and $\Omega$ are given for B1950.0. All angles are given in degrees, $q$ is in AU, and $V_G$ in km/s. For streams identified as single ones, but possessing both a Northern and a Southern branch, the data are given separately for each branch.

| Stream name | $q$ | $e$ | $i$ | $\omega$ | $\Omega$ | $V_G$ | $\theta$ | $\phi$ |
|---|---|---|---|---|---|---|---|---|
| Lyrids | 0.92 | 0.99 | 80 | 214 | 32 | 47 | 119 | 198 |
| $\alpha$ Capricornids (N) | 0.58 | 0.78 | 6 | 268 | 134 | 23 | 89 | 262 |
| $\alpha$ Capricornids (S) | 0.63 | 0.62 | 4 | 89 | 329 | 18 | 89 | 276 |
| Perseids | 0.95 | 0.95 | 113 | 150 | 139 | 59 | 139 | 164 |
| $\kappa$ Cygnids | 0.98 | 0.76 | 39 | 202 | 147 | 25 | 88 | 194 |
| $\varepsilon$ Geminids | 0.81 | 0.96 | 174 | 231 | 203 | 70 | 166 | 258 |
| Taurids (N) | 0.32 | 0.85 | 3 | 299 | 213 | 30 | 104 | 267 |
| Taurids (S) | 0.34 | 0.82 | 6 | 118 | 27 | 28 | 104 | 275 |
| Quadrantids | 0.98 | 0.68 | 72 | 171 | 282 | 41 | 116 | 176 |
| Geminids | 0.14 | 0.90 | 24 | 324 | 261 | 35 | 117 | 258 |
| $\chi$ Orionids (N) | 0.38 | 0.83 | 3 | 291 | 265 | 28 | 101 | 267 |
| $\chi$ Orionids (S) | 0.51 | 0.79 | 5 | 96 | 77 | 25 | 93 | 276 |
| Monocerotids | 0.18 | 1.00 | 37 | 129 | 80 | 43 | 112 | 286 |
| Leonids | 0.98 | 0.92 | 162 | 173 | 235 | 71 | 170 | 168 |
| $\sigma$ Hydrids | 0.24 | 0.98 | 126 | 122 | 78 | 58 | 137 | 295 |
| Orionids | 0.57 | 0.97 | 165 | 83 | 29 | 66 | 155 | 288 |
| $\alpha$ Virginids (N) | 0.41 | 0.83 | 8 | 288 | 34 | 28 | 99 | 263 |
| $\alpha$ Virginids (S) | 0.32 | 0.87 | 7 | 118 | 198 | 31 | 103 | 275 |
| S. $\delta$ Aquarids | 0.08 | 0.97 | 27 | 151 | 308 | 41 | 118 | 278 |
| N. $\delta$ Aquarids | 0.10 | 0.95 | 21 | 328 | 142 | 38 | 117 | 262 |
| $\varepsilon$ Piscids (N) | 0.58 | 0.76 | 5 | 268 | 190 | 22 | 89 | 263 |
| $\varepsilon$ Piscids (S) | 0.61 | 0.73 | 4 | 85 | 5 | 21 | 88 | 276 |
| Draconids | 1.00 | 0.70 | 25 | 177 | 196 | 17 | 71 | 175 |
| Cyclids | 0.83 | 0.12 | 3 | 119 | 26 | 3 | 108 | 296 |
| $\alpha$ Pegasids | 0.97 | 0.68 | 7 | 200 | 230 | 11 | 42 | 226 |

important among them is the one related to the cycle of $\omega$ (Kozai 1962). Assuming that all the planets are on circular coplanar orbits, and that the small body orbit is far from mean motion and secular resonances, it leaves invariant, besides the Kozai integral $K$ (Thomas and Morbidelli 1996), the $z$-component of the orbital angular momentum

$$L_z = \sqrt{a(1 - e^2)} \cos i \tag{26}$$

and the orbital energy

$$E = -\frac{1}{2a}. \tag{27}$$

But then

$$T = \frac{1}{a} + 2\sqrt{a(1 - e^2)} \cos i = 2(L_z - E) \tag{28}$$

is constant, and therefore so is $U$. Finally, if $a$ and $U$ are conserved, so is

$$\cos \theta = \frac{1 - U^2 - 1/a}{2U}. \tag{29}$$

To illustrate the invariance of $U$ and $\cos \theta$ over significant time scales, we have integrated for 100 000 yr a fictitious object on an orbit similar to that of 96P/Machholz 1, in a simplified solar system consisting only of Jupiter and the Earth, both on circular coplanar orbits. The results are shown in Fig. 1, where are given the time behaviours of, from top to bottom, $\Omega$, $\omega$, $i$, $e$, $q$, $U$ and $\cos \theta$. The ordinate range shown for each variable is the total range allowed to an Earth-crossing meteoroid on a non-hyperbolic orbit; the top five panels show that the variables entering $D_{SH}$ ($q$, $e$, $i$, $\omega$ and $\Omega$) undergo large oscillations, spanning in some cases the entire allowed range, whereas the variations of $U$ and $\cos \theta$, in the two lowermost panels are small compared to their allowed ranges. Quantitatively, we have that, in the time span integrated, $0.015 \leq q \leq 1.517$ AU, $0.533 \leq e \leq 0.995$, $10°.86 \leq i \leq 82°.22$, while $1.390 \leq U \leq 1.527$ and $-0.5481 \leq \cos \theta \leq -0.4457$.

Valsecchi et al. (1999) propose to exploit the secular near-invariance of $U$ and $\cos \theta$ by combining them into a reduced criterion:

$$D_R^2 = [U_2 - U_1]^2 + w_1[\cos \theta_2 - \cos \theta_1]^2 \tag{30}$$

where, as before, $w_1$ is a weighting factor. This criterion can be useful to find parent candidates for meteor showers with orbits having different $q$, $e$, $i$, $\omega$ and $\Omega$, due to secular perturbations on one and the same meteoroid stream (Babadzhanov and Obrubov 1993), but one must keep in mind that it gives only a necessary, not a sufficient, condition for such identifications, that must be validated by detailed modelling of the stream evolution.

## 4. Using Geocentric Variables to look for Stream Parents

In their application of $D_N$ to photographic meteor orbits, Jopek et al. (1999) did not relate the streams that they found to the parent bodies. When the orbit of the parent is very similar to that of the stream, this is a straightforward task; things are not so simple when the orbit of the candidate parent body has evolved away from the node-crossing condition.

We have taken the data of Table I and removed the streams composed of only two members (Draconids, Cyclids and $\alpha$ Pegasids), and the single Southern $\alpha$ Capricornid, due to the low significance of average quantities in these cases; we have then applied the reduced criterion of Eq. (30), with weight factor equal to unity. We set as threshold $D_R \leq 0.12$ and, to reduce the number of spurious associations, we stipulated that the difference between the longitudes of perihelion $\varpi = \omega + \Omega$ of the stream and of the candidate parent body should not exceed $17°$. These thresholds were chosen in order to recover, as a possible parent body, comet 96P/Machholz 1 for the Quadrantids (this association is not widely accepted, as

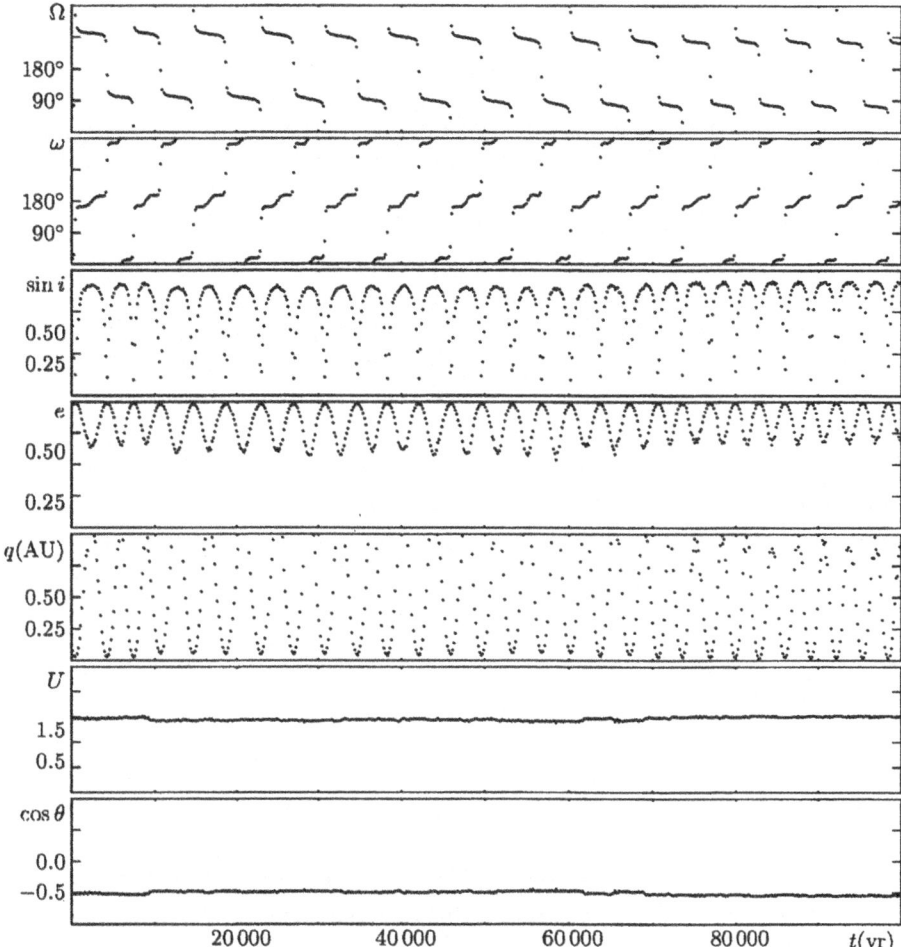

Fig. 1. 100 000 yr of evolution of a fictitious comet with an orbit similar to that of 96P/Machholz 1 in a model Solar System composed of the Sun, the Earth and Jupiter on circular, coplanar, not mutually perturbing orbits. The near constancy of the new variables $U$ and $\cos\theta$ is evident.

discussed by Jenniskens et al. 1997, and by Williams and Collander-Brown 1998). The values of $U$, $\theta$ and $\phi$ for comets and asteroids are computed from the current orbital elements, using Öpik's formulae; note that if the orbit of the comet/asteroid is not currently node-crossing with that of the Earth, the value of $\phi$ computed in this way is different from what it would be at node-crossing, due to the associated changes in $e$ and $i$, and thus is to be considered just as a rough approximation.

Table II gives the streams whose parents have already been suggested in the literature; we have eliminated from the output the many associations that, even to a visual inspection, appear spurious, in order to save space.

TABLE II

Approximate orbital and geocentric data of some of the streams found by Jopek et al. (1999), together with those of the presumed parent body given in literature, if recovered by our procedure. All quantities are as in Table I, except that $V_G$ is substituted by $U = V_G/29.7$, and $\cos \theta$ is given instead of $\theta$. For the parent bodies, the value of $\phi$ in the same quadrant as that of the stream is given; note that this means that two different values of $\phi$ are given for 96P/Machholz 1, for the two streams to which it has been associated.

| Stream name | $q$ | $e$ | $i$ | $\omega$ | $\Omega$ | $U$ | $\cos \theta$ | $\phi$ |
|---|---|---|---|---|---|---|---|---|
| Lyrids | 0.92 | 0.99 | 80 | 214 | 32 | 1.58 | −0.48 | 198 |
| C/1861 G1 Thatcher | 0.92 | 0.98 | 80 | 213 | 32 | 1.58 | −0.48 | 196 |
| Perseids | 0.95 | 0.95 | 113 | 150 | 139 | 1.99 | −0.75 | 164 |
| 109P/Swift-Tuttle | 0.96 | 0.96 | 113 | 153 | 139 | 2.01 | −0.77 | 168 |
| Taurids (N) | 0.32 | 0.85 | 3 | 299 | 213 | 1.01 | −0.24 | 267 |
| Taurids (S) | 0.34 | 0.82 | 6 | 118 | 27 | 0.94 | −0.24 | 275 |
| 2P/Encke | 0.33 | 0.85 | 12 | 186 | 335 | 1.01 | −0.23 | 250 |
| Quadrantids | 0.98 | 0.68 | 72 | 171 | 282 | 1.38 | −0.44 | 176 |
| 96P/Machholz 1 | 0.13 | 0.96 | 60 | 15 | 95 | 1.47 | −0.51 | 111 |
| Geminids | 0.14 | 0.90 | 24 | 324 | 261 | 1.18 | −0.45 | 258 |
| (3200) Phaethon | 0.14 | 0.89 | 22 | 322 | 266 | 1.12 | −0.47 | 252 |
| Monocerotids | 0.18 | 1.00 | 37 | 129 | 80 | 1.45 | −0.37 | 286 |
| D/1917 F1 Mellish | 0.19 | 0.99 | 33 | 121 | 89 | 1.39 | −0.35 | 290 |
| Leonids | 0.98 | 0.92 | 162 | 173 | 235 | 2.39 | −0.98 | 168 |
| 55P/Tempel-Tuttle | 0.98 | 0.90 | 163 | 173 | 235 | 2.35 | −0.98 | 166 |
| Orionids | 0.57 | 0.97 | 165 | 83 | 29 | 2.22 | −0.91 | 288 |
| S. $\delta$ Aquarids | 0.08 | 0.97 | 27 | 151 | 308 | 1.38 | −0.47 | 278 |
| 96P/Machholz 1 | 0.13 | 0.96 | 60 | 15 | 95 | 1.47 | −0.51 | 291 |
| N. $\delta$ Aquarids | 0.10 | 0.95 | 21 | 328 | 142 | 1.28 | −0.45 | 262 |

Our computations did not recover 1P/Halley as possible parent body for the Orionids, due to the large difference in longitude of perihelion (about 60°), and did not give any candidate parent body for the Northern $\delta$ Aquarids.

We could have included, as candidate parent body of both branches of the Taurids, the recently discovered asteroid 1998 QS$_{52}$, since it is associated to them by our procedure. Its parameters are, in fact, $q = 0.31$ AU, $e = 0.86$, $i = 18°$, $\omega = 242°$, $\Omega = 261°$, $U = 1.05$, $\cos \theta = -0.26$ and $\phi = 293°$. It must be noted, however, that this asteroid is currently nearly node-crossing, so that its value of $\phi$ cannot be reconciled easily with those of either branch of the Taurids.

Concerning the remaining streams, we present in Table III the results. In the Table are omitted the candidate parent comets of the near-parabolic streams $\varepsilon$ Geminids and $\sigma$ Hydrids, since all of them have longitudes of the node too far away from the stream for being possible parents; also omitted are the candidate parents of the $\kappa$ Cygnids, since they are long-period comets. Of the candidate parents in the Table, to our knowledge only 45P/Honda-Mrkos-Pajdušáková had been previously suggested. We have started to model the evolution of fictitious streams ejected by all of these candidates to check whether they *can* actually be the parents of the observed streams.

TABLE III

Same as Table II for the streams found by Jopek et al. (1999) with no presumed parent body given in literature. We have omitted the candidate parents of some streams for the reasons given in the text. Note that our procedure associates 1991 GO only to the Northern branch of the $\varepsilon$ Piscids, and 1998 FW$_4$ only to the Southern branch, whereas the three other candidate parents are associated to both branches, and that it associates (8176) 1991 WA to both branches of the $\alpha$ Virginids, but 1998 VP only to the Northern one.

| Stream name | $q$ | $e$ | $i$ | $\omega$ | $\Omega$ | $U$ | $\cos\theta$ | $\phi$ |
|---|---|---|---|---|---|---|---|---|
| $\alpha$ Capricornids (N) | 0.58 | 0.78 | 6 | 268 | 134 | 0.77 | 0.02 | 262 |
| D/1783 W1 Pigott | 1.46 | 0.55 | 45 | 355 | 59 | 0.75 | 0.08 | 211 |
| 45P/Honda-Mrkos-Pajd. | 0.53 | 0.82 | 4 | 326 | 89 | 0.84 | −0.02 | 252 |
| (4596) 1981 QB | 1.08 | 0.52 | 37 | 248 | 155 | 0.72 | 0.03 | 196 |
| 1991 VB | 1.32 | 0.41 | 6 | 135 | 257 | 0.71 | 0.09 | 231 |
| $\kappa$ Cygnids | 0.98 | 0.76 | 39 | 202 | 147 | 0.84 | 0.03 | 194 |
| $\varepsilon$ Geminids | 0.81 | 0.96 | 174 | 231 | 203 | 2.36 | −0.97 | 258 |
| $\chi$ Orionids (N) | 0.38 | 0.83 | 3 | 291 | 265 | 0.94 | −0.19 | 267 |
| 1995 UO$_5$ | 0.56 | 0.64 | 36 | 151 | 39 | 0.90 | −0.25 | 222 |
| 1995 OO | 0.48 | 0.78 | 24 | 211 | 350 | 0.92 | −0.17 | 235 |
| $\chi$ Orionids (S) | 0.51 | 0.79 | 5 | 96 | 77 | 0.84 | −0.05 | 276 |
| 1998 VD$_{31}$ | 0.52 | 0.80 | 10 | 113 | 48 | 0.84 | −0.05 | 296 |
| 1997 VM$_4$ | 0.49 | 0.81 | 14 | 124 | 46 | 0.89 | −0.10 | 299 |
| 1979 XB | 0.65 | 0.71 | 25 | 76 | 86 | 0.80 | −0.05 | 316 |
| $\sigma$ Hydrids | 0.24 | 0.98 | 126 | 122 | 78 | 1.95 | −0.73 | 295 |
| $\alpha$ Virginids (N) | 0.41 | 0.83 | 8 | 288 | 34 | 0.94 | −0.16 | 263 |
| $\alpha$ Virginids (S) | 0.32 | 0.87 | 7 | 118 | 198 | 1.04 | −0.22 | 275 |
| (8176) 1991 WA | 0.56 | 0.64 | 40 | 242 | 67 | 0.94 | −0.28 | 319 |
| 1998 VP | 1.01 | 0.46 | 44 | 101 | 226 | 0.85 | −0.15 | 186 |
| $\varepsilon$ Piscids (N) | 0.58 | 0.76 | 5 | 268 | 190 | 0.74 | 0.02 | 263 |
| $\varepsilon$ Piscids (S) | 0.61 | 0.73 | 4 | 85 | 5 | 0.71 | 0.03 | 276 |
| 1991 GO | 0.66 | 0.66 | 10 | 89 | 25 | 0.65 | 0.05 | 235 |
| 1998 FW$_4$ | 0.68 | 0.73 | 4 | 75 | 3 | 0.66 | 0.12 | 292 |
| 1998 HL$_{49}$ | 0.63 | 0.64 | 11 | 239 | 207 | 0.65 | 0.00 | 306 |
| 1997 YR$_{10}$ | 1.15 | 0.33 | 37 | 191 | 270 | 0.66 | −0.01 | 341 |
| 1991 YA | 1.53 | 0.44 | 44 | 174 | 274 | 0.71 | 0.09 | 329 |

## 5. Conclusions

The geocentric variables, directly linked to observed quantities, proposed by Valsecchi et al. (1999) lead to a stream identification criterion rather different from the classical one by Southworth and Hawkins (1963); the new criterion has been successfully tested on photographic meteor orbits by Jopek et al. (1999), with results disagreeing from those obtained with Southworth and Hawkins' criterion in the case of near-ecliptical streams.

We have exploited in this paper the secular near-invariance of some of the new variables to look for candidate parent bodies for the streams found with the new criterion; we recover all the known parents except 1P/Halley for the Orionids, and find some not previously suggested candidate parents for several of the remaining streams.

# References

Babadzhanov, P.B., Obrubov, Yu. V.: 1993, Dynamics and relationships between interplanetary bodies I. Machholz and its meteor showers, in Meteoroids and their parent bodies, J. Štohl and I.P. Williams, eds., Astronomical Institute of the Slov. Acad. Sci., Bratislava, Slovakia, 49-52

Carusi, A., Valsecchi, G.B., Greenberg, R.: 1990, Planetary close encounters: geometry of approach and post-encounter orbital parameters, *Cel. Mech. & Dyn. Astro.*, **49**, 111

Drummond, J.: 1981, A test of comet and meteor shower associations, *Icarus*, **45**, 545-553

Jenniskens, P., Betlem, H., de Lignie, M., Langbroek, M. and van Vliet, M.: 1997, Meteor stream activity V. The Quadrantids, a very young stream, *A&A*, **327**, 1242-1252

Jopek, T.: 1993, Remarks on the meteor orbital similarity D-criterion, *Icarus*, **106**, 603-607

Jopek, T.J., Froeschlé, Cl.: 1997, *A&A*, **320**, 631-641

Jopek, T.J., Valsecchi, G.B., Froeschlé, Cl.: 1999, Meteor streams identification: a new approach. II. Application to 865 photographic meteor orbits, *MNRAS*, in press

Kozai, Y., 1962, Secular perturbations of asteroids with high inclination and eccentricity, *AJ*, **67**, 591-598

Lindblad, B.A.: 1971, *Smithson. Contr. Astrophys.*, **12**, 1-13

Öpik, E.J.: 1976, *Interplanetary Encounters*, Elsevier, New York

Southworth, R.B., Hawkins, G.S.: 1963, Statistics of meteor streams, *Smithson. Contr. Astrophys.*, **7**, 261-285

Thomas, F., Morbidelli, A., 1996, The Kozai resonance in the outer solar system and the dynamics of long-period comets, *Cel. Mech. & Dyn. Astro.*, **64**, 209-229

Valsecchi, G.B., Jopek, T.J., Froeschlé, Cl.: 1999, Meteoroid streams identification: a new approach. I. Theory, *MNRAS*, in press

Williams, I.P., Collander-Brown, S.J., 1998 The parent of the Quadrantid meteoroid stream, *MNRAS*, **294**, 127-138

# A SYMPLECTIC MAPPING MODEL AS A TOOL TO UNDERSTAND THE DYNAMICS OF 2/1 RESONANT ASTEROID MOTION

JOHN D. HADJIDEMETRIOU

*Department of Physics, University of Thessaloniki, Thessaloniki, Greece,*
*E-mail: hadjidem@physics.auth.gr*

**Abstract.** We present a 3-D symplectic mapping model that is valid at the 2:1 mean motion resonance in the asteroid motion, in the Sun-Jupiter-asteroid model. This model is used to study the dynamics inside this resonance and several features of the system have been made clear. The introduction of the third dimension, through the inclination of the asteroid orbit, plays an important role in the evolution of the asteroid and the appearance of chaotic motion. Also, the existence of the secondary resonances is clearly shown and their role in the appearance of chaotic motion and the slow diffusion of the elements of the orbit is demonstrated.

**Key words**: resonance, chaotic motion, diffusion, secondary resonances.

## 1. Introduction

The explanation of the Kirkwood gaps in the asteroid belt is an old and famous problem in the study of the solar system, which is related to the stability of resonant motion in a nonlinear dynamical system. The interest in this problem was revived after the work of Wisdom (1982,83,85), who showed that the observed gap at the 3:1 mean motion resonance of the asteroid with Jupiter can be explained by purely gravitational forces.

The study of the Kirkwood gaps was extended by Šidlichovský (1988,92) for the 5:2 resonance, by Moons and Morbidelli (1995) for the for the 4:1, 3:1, 5:2 and 7:3 resonances and by Yoshikawa (1991) for the 5:2 and 7:3 resonances, and similar results were obtained for the depletion of asteroids from these regions. Much work has been done recently on the 2:1 gap, and the basic features are now well understood (Henrard et al. 1995, Henrard and Lemaitre 1987, Moons and Morbidelli 1993, Lemaitre and Henrard 1990, Ferraz-Mello 1996, Morbidelli 1996, Morbidelli and Moons 1995, Ferraz-Mello et al. 1998a,b, Moons et al. 1998, Ferraz-Mello and Michtchenko 1997, Michtchenko and Ferraz-Mello 1998, Nesvorny and Ferraz-Mello 1997, Gallardo and Ferraz-Mello 1998, Yoshikawa 1989,1991). A slow diffusion takes place at this resonance, from low to high values of the eccentricity. An extended review on recent work on the Kirkwood gaps was made by Moons (1997).

The problem of the Kirkwood gaps, apart from its particular interest in the study of the solar system, is an interesting problem of nonlinear dynamics, related to the stability of resonant motion and the generation of chaos and its long term effect on the evolution of the system. To understand the dynamics of the system, one can consider a hierarchy of perturbations, starting with the simplest model and going gradually to more complicated models, adding more features and consequently more degrees of freedom. The increase of the degrees of freedom introduces new resonances to the system, the secondary resonances and the secular resonances,

*Celestial Mechanics and Dynamical Astronomy* **73**: 65–76, 1999.
© 1999 *Kluwer Academic Publishers.*

which play an important role in the evolution inside the 2:1 resonance. These resonances may overlap and thus form a bridge to connect low and high eccentricity and inclination regions in phase space (Yoshikawa, 1989, Franklin, 1994, Moons and Morbidelli, 1995 and Henrard et al. 1995).

The purpose of this paper is to contribute to the study of the 2:1 resonant asteroid motion in three dimensions, by making use of a suitable symplectic mapping model. The underlying physical system is the Sun-Jupiter-asteroid system, with Jupiter in a fixed orbit. This mapping model can be used to understand the dynamics at this region and guide us on where to focus our attention in numerical simulations. In particular, we address several aspects of the problem as: (a) The effect of introducing the third dimension in the model. (b) The appearance of the secondary resonances and the generation of chaotic motion through their overlap. (c) The effect of the initial phase on the evolution of the system. This mapping model compares well with other models used in the study of the 2:1 resonance, as we shall see in the following.

## 2. The Mapping Model

The symplectic mapping model that we shall use is based on the averaged Hamiltonian at the 2:1 resonance and corresponds to the elliptic restricted three body problem, in three dimensions, with the Sun and Jupiter as primaries. The planar part of this averaged Hamiltonian has been used in Hadjidemetriou and Lemaitre (1997) and contains all the high eccentricity resonances of the Sun-Jupiter-asteroid model. This was achieved by the introduction of suitable correction terms to the averaged Hamiltonian obtained by the usual perturbation methods. In the present paper we use the averaged Hamiltonian at the 2:1 resonance, in three dimensions, obtained by Šidlichovský (1991). The planar part is the same as that used in the paper by Hadjidemetriou and Lemaitre (1997), and we applied the same correction terms to make the model realistic. We kept first order terms in $e_j$ in the expansion of the averaged Hamiltonian and also first order terms in $\sin^2 \frac{i}{2}$. Since $i'$ is fixed and of the order of $1°$, we ignored all terms of third order in $e$, $\sin \frac{i}{2}$ and $\sin \frac{i'}{2}$.

### 2.1. THE AVERAGED HAMILTONIAN

The averaged Hamiltonian is expressed in the resonant action-angle variables (Moons 1997) $S$, $S_z$, $N$ $\sigma$, $\sigma_z$ and $\nu$, given by

$$S = L - G, \quad S_z = G - H, \quad N = 2L - H,$$
$$\sigma = 2\lambda' - \lambda - \varpi, \quad \sigma_z = 2\lambda' - \lambda - \Omega, \quad \nu = -2\lambda' + \lambda + \varpi'. \quad (1)$$

where

$$L = \sqrt{(1-\mu)a}, \quad G = L\sqrt{1-e^2}, \quad H = G\cos i, \quad (2)$$

and $a$ is the semimajor axis of the asteroid, $e$ the eccentricity, $i$ the inclination, $\lambda$ the mean longitude, $\varpi$ the longitude of perihelion and $\Omega$ the longitude of the node, and the primed quantities refer to Jupiter.

The averaged Hamiltonian is

$$H = H_0(S, S_z, N) + \mu e H_1(\sigma) + \mu e^2 H_2(\sigma) + \mu e^7 H_7(\sigma) + \mu e^7 H_c \tag{3}$$

$$+\mu e_j H_j(S, S_z, N, \sigma, \sigma_z, \nu) + \mu \sin^2 \frac{i}{2} H_i(\sigma, \sigma_z) + \mu e_j \sin^2 \frac{i}{2} H_{ij}(\sigma, \sigma_z, \nu),$$

where

$$H_0 = -\frac{(1-\mu)^2}{2(N - S - S_z)^2} - B(N - S - S_z), \tag{4}$$

$$H_1 = A \cos \sigma, \quad H_2 = C + D' \cos 2\sigma, \quad (D': \text{corrected value}) \tag{5}$$

$$H_7 = P \cos \sigma + Q \cos 3\sigma + R \cos 5\sigma + T \cos 7\sigma, \tag{6}$$

$$H_c = T' \cos 8\sigma, \quad (\text{correction term}) \tag{7}$$

$$H_j = J \cos \nu + \sqrt{\frac{2S}{N}} \left( F \cos(\sigma + \nu) + G \cos(\sigma - \nu) \right) + e_j H \cos 2\nu$$

$$+ e^2 (F_1 \cos(2\sigma + \nu) + G_1 \cos(2\sigma - \nu) + J_1 \cos \nu), \tag{8}$$

$$H_i = C_1 + D_1 \cos 2\sigma_z$$

$$+ e(A_1 \cos \sigma + A_2 \cos(\sigma - 2\sigma_z) + A_3 \cos(\sigma + 2\sigma_z)), \tag{9}$$

$$H_{ij} = J_2 \cos \nu + F_2 \cos(2\sigma_z + \nu) + G_2 \cos(2\sigma_z - \nu)$$

$$e(F_3 \cos(\sigma + \nu) + G_3 \cos(\sigma - \nu)$$

$$+ B_1 \cos(2\sigma_z + \sigma + \nu) + B_2 \cos(2\sigma_z + \sigma - \nu) \tag{10}$$

$$+ B_3 \cos(2\sigma_z - \sigma + \nu) + B_4 \cos(2\sigma_z - \sigma - \nu)).$$

The eccentricity $e$ and the inclination $i$ that appear in the above expressions is a function of $S$, $S_z$ and $N$, obtained from the equations (1) and (2). The coefficients are given by Lemaitre and Henrard (1990) for the planar part and by Šidlichovský (1991) for the three dimensional part as: $A = 1.189, B = 2.00084, C = -0.3866,$ $P = 1.598, Q = 6.964, R = -52.95, T = 51.95, J = -0.4273, F = 0.5739,$ $G = 4.955, H = -3.588, F_1 = 0.7523, G_1 = -13.0959, J_1 = -2.188, C_1 = 1.5483, D_1 = -0.8182, J_2 = 10.0508, F_2 = -0.2138, G_2 = -5.1823, A_1 = -10.7994, A_2 = 3.4926, A_3 = 3.1456, F_3 = -23.7199, G_3 = -70.5988,$ $B_1 = -2.2339, B_2 = 30.8721, B_3 = 12.1473, B_4 = 14.2990.$ The coefficient of the correction term is $T' = -20.0.$ We have also corrected the coefficient $D$ in the second order term $H_2$ of the expansion of the Hamiltonian, taking $D' = -0.30$ instead of $D = -1.691.$ The value of the mass of Jupiter (in normalized units) is taken equal to $\mu = 0.00095387535$ and the value of the eccentricity of Jupiter is taken equal to $e_j = 0.048.$

## 2.2. THE MAPPING EQUATIONS

We consider the mapping which is derived from the generating function

$$
\begin{aligned}
W = \ & \sigma_n S_{n+1} + \sigma_{z,n} S_{z,n+1} + \nu_n N_{n+1} \\
& + T H_0(S, N) && \text{(two-body problem)} \\
& + \mu T W_1(S, N, \sigma) && \text{(2D circ. restr. problem)} \\
& + \mu e_j T W_2(S, N, \sigma, \nu), && \text{(2D ell. restr. problem)} \\
& + \mu \sin^2 \frac{i}{2} T W_3(\sigma, \sigma_z), && \text{(3D circ. restr. problem)} \\
& + \mu e_j \sin^2 \frac{i}{2} T W_4(\sigma, \sigma_z, \nu), && \text{(3D ell. restr. problem)}
\end{aligned} \tag{11}
$$

where $T$ is the period of the 2:1 resonant periodic orbit, which is equal to $T = 2\pi$ in the normalized units we are using and

$$
\begin{aligned}
W_1 &= \mu e H_1(\sigma) + \mu e^2 H_2(\sigma) + f(\mu e^7 H_7(\sigma) + \mu e^7 H_c), \\
W_2 &= H_j(S, N, \sigma, \nu), \quad W_3 = H_i(\sigma, \sigma_z), \quad W_4 = H_{ij}(\sigma, \sigma_z, \nu).
\end{aligned} \tag{12}
$$

through the equations

$$
\begin{aligned}
\sigma_{n+1} &= \partial W / \partial S_{n+1}, \quad \sigma_{z,n+1} = \partial W / \partial S_{z,n+1}, \quad \nu_{n+1} = \partial W / \partial N_{n+1}, \\
S_n &= \partial W / \partial \sigma_n, \quad S_{z,n} = \partial W / \partial \sigma_{z,n}, \quad N_n = \partial W / \partial \nu_n.
\end{aligned} \tag{13}
$$

This mapping is symplectic and it can be proved (Hadjidemetriou 1993) that it has the same fixed points with the same stability index, as the averaged Hamiltonian $H$. Consequently, it has the same fixed points as the Poincaré map of the real system (elliptic restricted problem). This is so because $H$ is corrected in such a way that its fixed points correspond to the families of the periodic orbits of the real system. In addition, we have multiplied the high eccentricity terms $H_7$ and $H_c$ in the generating function $W$ by the coefficient $f = 0.02$, in order to reduce the chaotic regions of the mapping and make them comparable with the chaotic regions of the real system when $e_j = 0$.

The mapping obtained by the generating function (11) has the same topology as the Poincaré map of the real system. Thus, we consider it as a good model to study the long term behaviour of an asteroid close to the 2:1 resonance, because similar dynamical systems have the same generic properties and are expected to behave in the same way.

The equations of the complete mapping are obtained from the generating function $W$, given by (11), by making use of the transformation equations (13)

$$\sigma_{n+1} = \sigma_n + T\frac{\partial W_0}{\partial S_{n+1}} \quad +T\mu\left[\left(\frac{\partial W_1}{\partial S_{n+1}} + e_j\frac{\partial W_2}{\partial S_{n+1}}\right) + (W_3 + e_j W_4)\frac{\partial}{\partial S_{n+1}}\sin^2\frac{i}{2},\right]$$

$$S_n = S_{n+1} \quad +T\mu\left[\left(\frac{\partial W_1}{\partial \sigma_n} + e_j\frac{\partial W_2}{\partial \sigma_n}\right) + \sin^2\frac{i}{2}\frac{\partial(W_3 + e_j W_4)}{\partial \sigma_n}\right]$$

$$\sigma_{z,n+1} = \sigma_{zn} \quad +T\mu(W_3 + e_j W_4)\frac{\partial}{\partial S_{z,n+1}}\sin^2\frac{i}{2},$$

$$S_{z,n} = S_{z,n+1} \quad +T\mu\sin^2\frac{i}{2}\frac{\partial(W_3 + e_j W_4)}{\partial \sigma_{z,n}} \tag{14}$$

$$\nu_{n+1} = \nu_n + T\frac{\partial W_0}{\partial N_{n+1}} \quad +T\mu\left[\left(\frac{\partial W_1}{\partial N_{n+1}} + e_j\frac{\partial W_2}{\partial N_{n+1}}\right) + (W_3 + e_j W_4)\frac{\partial}{\partial N_{n+1}}\sin^2\frac{i}{2}\right],$$

$$N_n = N_{n+1} \quad +T\mu e_j\left[\frac{\partial W_2}{\partial \nu_n} + \sin^2\frac{i}{2}\frac{\partial W_4}{\partial \nu_n}\right]$$

From these mapping equations it is clearly seen how the different degrees of freedom interact. For example, it is readily seen that if $e_j = 0$ (circular orbit of Jupiter), the action $N$ is constant and for $i = 0$ (planar motion), the action $S_z$ is constant. These equations are in implicit form and at each step we have to solve (by the method of Newton-Raphson) the second, fourth and sixth equations (14) to obtain the values of $S_{n+1}$, $S_{z,n+1}$ and $N_{n+1}$ in terms of $\sigma_n$, $\sigma_{z,n}$, $\nu_n$, $S_n$, $S_{z,n}$ $N_n$ and then proceed to the computation of $\sigma_{n+1}$, $\sigma_{z,n+1}$, $\nu_{n+1}$ from the first, third and fifth equations (14). This however does not present any numerical difficulty.

## 3. The Evolution of the Asteroid obtained by the Mapping Model

We shall use now the mapping (14) to study the evolution of a fictitious asteroid which starts inside the 2:1 resonance. The purpose of these computations is to study the different types of evolution and the factors that affect the appearance of chaotic motion. Several interesting aspects of the properties of motion inside this resonance will be made clear and help us to understand the dynamics inside this resonance. The initial conditions of the asteroid are compared with the $a - e$ diagrams, for the inclinations $i_0 = 0$ and $i_0 = 20°$, respectively, given by Moons et al.,1998, (Figures 1a,b), where the regions of secondary resonances and secular resonances are shown. A diagram similar to Figure 2a was also given by Michtchenko and Ferraz-Mello (1997).

### 3.1. EVOLUTION FOR DIFFERENT ECCENTRICITIES

We compute different asteroid orbits, keeping the semimajor axis and the inclination fixed, equal to $a_0 = 3.3\,\text{AU}$ and $i_0 = 1$ degree, and taking three values for the initial eccentricity: $e_0 = 0.02$, $e_0 = 0.12$ and $e_0 = 0.16$. In all cases we take $\sigma_0 = 0$, $\sigma_{z0} = 0$ and $\nu_0 = 0$. These initial conditions are on the line $a_0 = 3.3\,\text{AU}$ in Figure 1a (this Figure is for $i = 0$ but the regions of secondary and secular resonances do

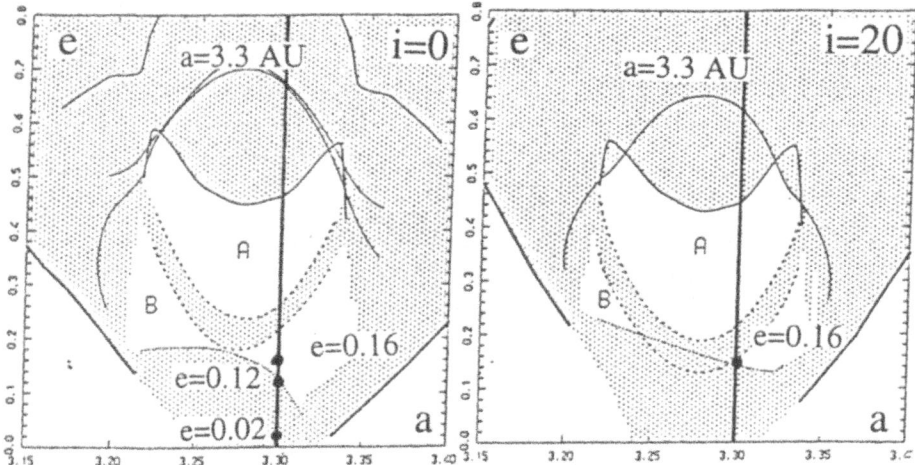

Fig. 1. The regions of secondary and secular resonances: (a) $i = 0$, (b) $i = 20°$ (From Moons et al. 1998). The secondary resonances are situated below the dotted line, in the lower part.

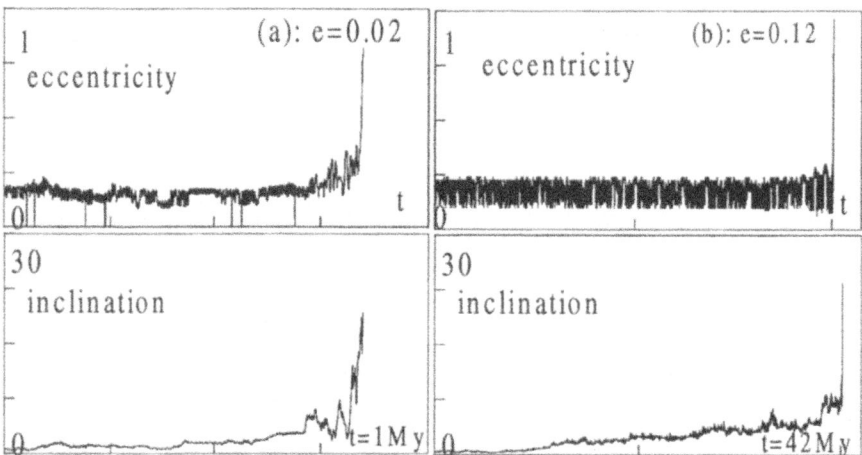

Fig. 2. The evolution of the eccentricity and the inclination, for $a_0 = 3.3$ AU, $i_0 = 1°$ and (a) $e_0 = 0.02$, (b) $e_0 = 0.12$.

not change much for the value $i = 1°$ that we are using). The first two are inside the region of secondary resonances and the third is just outside this region. The evolution of the eccentricity and the inclination is given in Figures 2a,b and Figure 3a. In these computations we give the values of $e$ and $i$ at the points where $\sigma = 0$ and $d\sigma/dt < 0$. We note that chaotic motion appears when we start inside the region of secondary resonances in Figure 1a, (Figures 2a,b) and the eccentricity and inclination jump eventually to high values. On the contrary, the third orbit, which starts outside the secondary resonance region of Figure 1a is ordered and

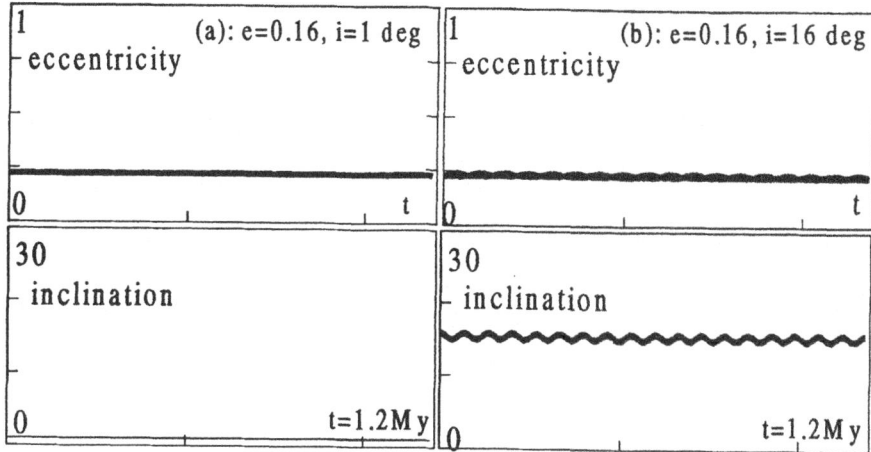

Fig. 3. The evolution of the eccentricity and the inclination, for $a_0 = 3.3$ AU, $e_0 = 0.16$ and (a) $i_0 = 1°$, (b) $i_0 = 16°$.

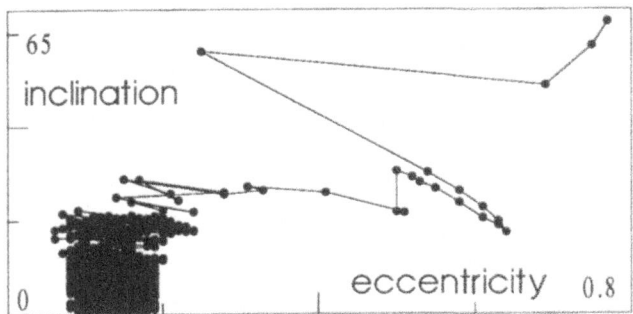

Fig. 4. The evolution shown in Figure 2b for $t = 4.2 \times 10^7$y. The "fountain" pattern is clearly seen.

no significant change takes place (Figure 3a). The chaotic and stable nature of the orbits for $e_0 = 0.12$ and $e_0 = 0.16$, respectively, has been also observed by Michtchenko and Ferraz-Mello (1997), by direct numerical computations.

We note that the third dimension plays an important role in the evolution of the system and cannot be ignored in a realistic model, even if the initial eccentricity is close to zero. From Figures 2a,b we see that it is the inclination that starts first to change in a chaotic way and increase while the eccentricity remains small. It is only after the increase of the inclination to a relatively large value that the eccentricity is excited and starts to increase chaotically, and both the eccentricity and the inclination obtain very high values. This is clearly seen in Figure 4, where we plotted the evolution shown in Figure 2a in the axes $e$-$i$. This type of evolution is typical in many cases and was also found by Michtchenko and Ferraz-Mello 1997 and was called the "fountain pattern".

The case $e_0 = 0.16$, $i_0 = 1°$ of Figure 3a corresponds to ordered motion, as we

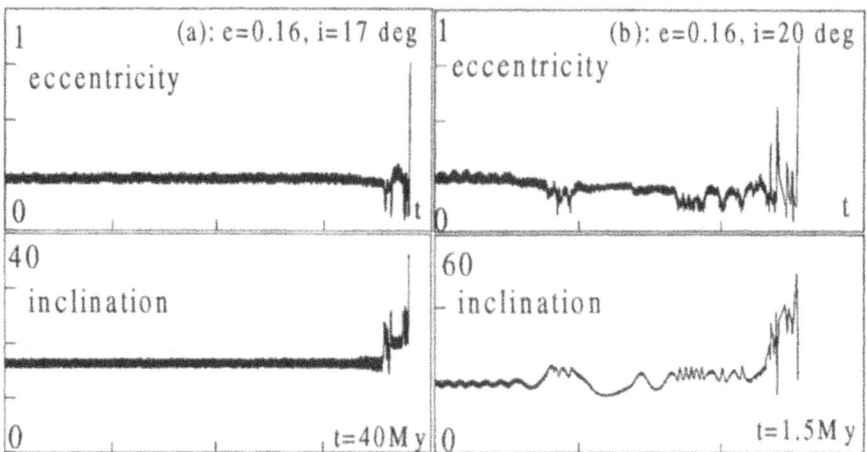

Fig. 5. The evolution of the eccentricity and the inclination, for $a_0 = 3.3$ AU, $e_0 = 0.16$ and (a) $i_0 = 17°$, (b) $i_0 = 20°$.

explained before. We now investigate the effect of increasing the initial inclination. We found that for $a_0 = 3.3$ AU and $e_0 = 0.16$, the motion is the same as that of Figure 3a for inclinations up to $i_0 = 16°$ (Figure 3b). For larger inclinations chaotic motion appears, as we show in Figures 5a,b.

The appearance of chaotic motion as the initial eccentricity increases can be explained by making use of the $a$-$e$ diagram of Figures 1b. For $i_0 < 17°$ we are outside the secondary resonance zone, but as soon as the initial inclination increases further, we enter the secondary resonance region and chaotic motion appears.

### 3.2. THE TRANSITION FROM THE PLANAR TO THE THREE DIMENSIONAL MOTION

As we have mentioned before, the introduction of the third dimension brings new features to the system and may change dramatically the evolution. We present in Figure 6 such a typical case, corresponding to the initial conditions $a_0 = 3.276$AU, $e_0 = 0.32$ and $i_0 = 0°$ (Figure 6a), $i_0 = 1°$ (Figure 6b). In both cases the angles were taken as $\sigma_0 = 0$, $\sigma_{z0} = 0$ and $\nu_0 = 0$. We note that the planar orbit is ordered, but as soon as the initial inclination is nonzero, $i_0 = 1°$, chaotic motion develops. It is worth mentioning that it is the inclination that starts first to change irregularly and increase, while the eccentricity stays at low values, and after a long time interval the motion is driven into a chaotic zone and both $e$ and $i$ increase to high values.

### 3.3. THE EFFECT OF THE INITIAL PHASE

In all previous cases we have started with initial values of the angles $\sigma_0 = 0$, $\sigma_{z0} = 0$ and $\nu_0 = 0$. We will keep now all other initial conditions fixed and change

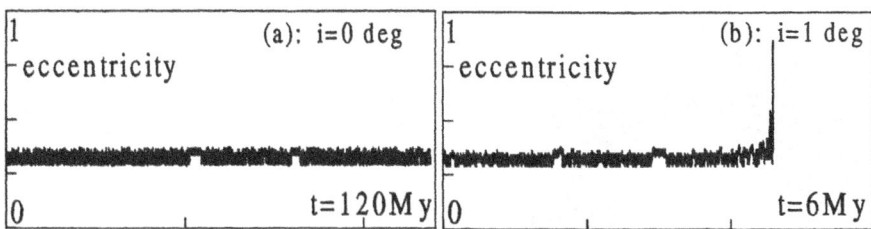

Fig. 6. The transition from planar to 3-D motion. $a_0 = 3.276$, $e_0 = 0.32$ and: (a) $i_0 = 0$, (b) $i_0 = 1°$.

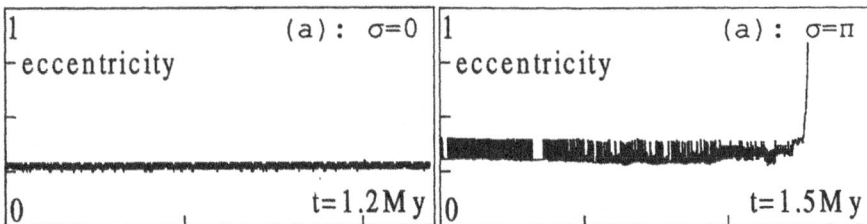

Fig. 7. The effect on the evolution of the eccentricity of changing the initial value of $\sigma$. $a_0 = 3.281$, $e_0 = 0.254$ and: (a) $\sigma_0 = 0$, (b) $\sigma_0 = \pi$.

the initial value of the angle $\sigma$. We consider planar motion and we use the initial conditions $a_0 = 3.281$ AU, $e_0 = 0.254$, $i_0 = 0°$, $\sigma_{z0} = 0$, $\nu_0 = 0$ and $\sigma_0 = 0$ or $\sigma_0 = \pi$. The evolution is shown in Figure 7, where we see that for $\sigma_0 = 0$ we have ordered motion but for $\sigma_0 = \pi$ the motion becomes chaotic. The sensitivity of the evolution to the initial values of the angular elements has been also observed by Ferraz-Mello and Michtchenko (1997).

### 3.4. THE POSITION OF THE SECONDARY RESONANCES

From the results presented in the previous sections, it became clear that the secondary resonances play an important role in the long term evolution of an asteroid inside the 2:1 mean motion resonance.

The secondary resonances correspond to commensurable motion of the averaged Hamiltonian (3) between the libration of the resonance angle $\sigma$ and the rotation of the longitude of perihelion $\varpi - \varpi'$. These resonances correspond in fact to doubly periodic orbits of the original, nonaveraged, system. Inside the 2:1 mean motion resonance it is known that there appear the secondary resonances 2:1, 3:1, 4:1 and 5:1. The existence of the secondary resonances and their role in the evolution of the system have been explored by Wisdom (1985), Henrard and Lemaitre (1987) and Lemaitre and Henrard (1990), using semianalytic methods: In the averaged Hamiltonian, the frequency of the libration of the angle $\sigma$ is much larger, in most cases, than the frequency of rotation of the longitude of perihelion $\varpi$. So, a second averaging can be performed over the "fast" angle $\sigma$. The new averaged Hamiltonian

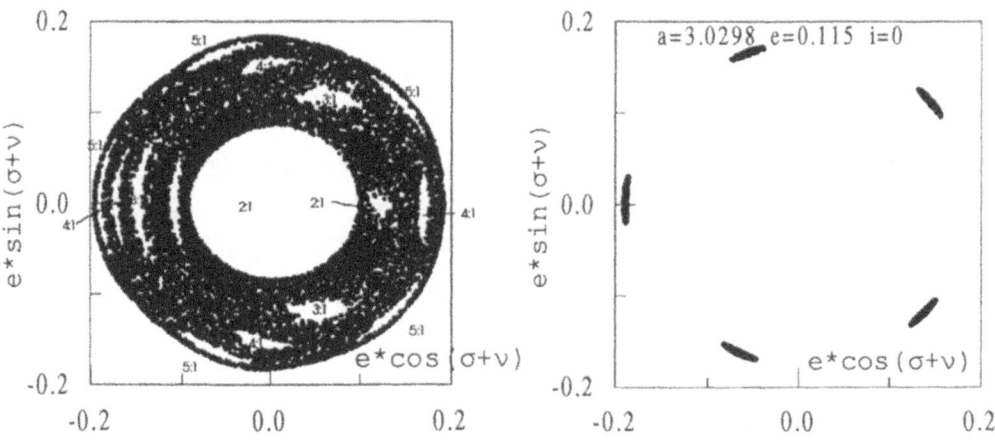

Fig. 8. (a) The secondary resonances, (b) trapping into the 5:1 resonance

contains now all the secondary resonances.

In the present work we use a different, but equivalent, method to find the secondary resonances. We use the mapping (14) and consider a further mapping by taking the points of intersection at $\sigma = 0$ and $\dot{\sigma} < 0$. In this way we obtain a two-dimensional mapping, which we present in the axes $e\cos(\sigma + \nu)$ and $e\sin(\sigma + \nu)$. Note that $\sigma + \nu = -\varpi$. Since the mapping (14) is discreet, the value of $\sigma$ is never equal exactly to 0, and an interpolation would be needed. Since however the frequency of $\sigma$ is much larger than the frequency of $\varpi$, in practice it was enough, for the plot, to take the closest value of $\sigma$ to 0.

The "mapping" of the mapping that we are considering is in fact equivalent to the second averaging of the Hamiltonian (3) over the angle $\sigma$, because by the second mapping we eliminate the libration of $\sigma$.

In Figure 8a we show a typical case of the mapping mentioned above, for the initial conditions $a_0 = 3.298$, $e_0 = 0.122$, $i_0 = 1°$, and $\sigma_0 = \sigma_{z0} = \nu_0 = 0$. The secondary resonances 5:1, 4:1, 3:1 and 2:1 are clearly seen as "islands". These resonances overlap and thus form a chaotic sea, which makes possible the diffusion inside the secondary resonance zone. The same pattern is obtained for all other initial conditions inside the secondary resonance zone (Figure 1a). For special initial conditions, we may have a trapping into a secondary resonance, as is shown in Figure 8b for the case $a_0 = 3.298$, $e_0 = 0.115$, $i_0 = 0°$, and $\sigma_0 = \sigma_{z0} = \nu_0 = 0$.

## 4. Discussion

A realistic symplectic mapping model has been constructed, which is used as a model of the Poincaré map of the three dimensional restricted three body problem Sun-Jupiter-asteroid, valid near the 2:1 mean motion resonance of the asteroid with Jupiter. This is a relatively simple model, which contains all the basic resonances of the physical system. We used this mapping to study the dynamics inside the

2:1 resonance, focusing our attention to the basic features, without being lost in unnecessary details.

Several features of the motion have been made clear. It was demonstrated how the third dimension, through the variation of the inclination, plays an important role in the evolution of the system. It is the inclination that first starts to behave in an irregular way, and this makes, later on, the eccentricity to behave irregularly also. In many cases the orbit is regular in the plane, but chaotic motion appears as soon as the initial inclination takes a small nonzero value.

The role played by the secondary resonances in the dynamics of the system were made clear. The appearance of chaotic motion inside the secondary resonance zone, by the overlap of these resonances, and the slow diffusion of the elements of the orbit, is clearly demonstrated by the mapping model.

The initial phase, as given by the values of the angles $\sigma$, $\sigma_z$ and $\nu$, plays an important role in the evolution of the system. The change from $\sigma = 0$ to $\sigma = \pi$ may change the orbit from ordered to chaotic. This is so, because there exist two 2:1 mean motion resonant periodic orbits, one stable, corresponding to $\sigma = 0$ and one unstable, corresponding to $\sigma = \pi$ (Hadjidemetriou and Lemaitre, 1997). In several cases, not presented here, it was shown that even in the case where the motion is already chaotic for $\sigma = 0$, the change to $\sigma = \pi$ makes the chaotic motion much stronger and the effects appear in a much shorter time interval.

The main aim of the present work is to understand the basic dynamics inside the 2:1 resonance and not to make a complete exploration of the phase space. However, the results obtained throw much light on the evolution inside this resonance and can be used as a guide for simulations of the actual physical model. A complete exploration will be made in a future work, where the gravitational effect of Saturn on the elements of the orbit of Jupiter will be introduced in the model.

## References

Ferraz-Mello, S.: 1997, *A Symplectic Mapping approach to the study of the Stochasticity of asteroidal Resonances*, Cel. Mec. Dyn. Astr. **65**, 421-437.

Ferraz-Mello, S. and Michtchenko, T.A.: 1997, *Orbital evolution of Asteroids in the Hecuba Gap*, in R. Dvorak and J. henrard (eds.) *The Dynamical behaviour of our Planetary System*, Kluwer Acad. Publ., 377-384.

Ferraz-Mello, S., Michtchenko, T.A., Nesvorny, D., Roig, F. and Simula, A.: 1998a, *The depletion of the Hecuba Gap vs. the long-lasting Hilda group*, Planetary and Space Science, in press.

Ferraz-Mello, S., Michtchenko, and Roig, F.: 1998b, *The determinant role of Jupiter's Great Inequality in the depletion of the Hecuba Gap*, Astron. J. **116**, 1491-1500.

Franklin, F.: 1994, *En examination of the relation between Chaotic Orbits and the Kirkwood Gap at the 2:1 Resonance*, Icarus **107**, 1890-1899.

Gallardo, T. and Ferraz-Mello, S.: 1998, *Understanding libration via time-frequency analysis* Astron. J. **113**, 863-870.

Hadjidemetriou, J.D.: 1993, *Asteroid Motion near the 3:1 Resonance*, Cel. Mec. Dyn. Astr. **56**, 563-599.

Hadjidemetriou, J.D. and Lemaitre, A.: 1997, *Asteroid motion near the 2:1 resonance: a symplectic mapping approach*, in R. Dvorak and J. Henrard (eds.) *The Dynamical Behaviour of our Planetary System*, Kluwer Acad. Publ., 277-290.

Henrard, J. and Lemaitre, A.: 1987, *A Perturbative Treatment of the 2/1 Jovian Resonance*, Icarus **69**, 266- 279.

Henrard, J., Watanabe, N. and Moons, M.: 1995, *A bridge between Secondary and Secular Resonances inside the Hecuba Gap*, Icarus **115**, 336-346.

Lemaitre, A. and Henrard, J.: 1990, *On the Origin of Chaotic Behaviour in the 2/1 Kirkwood Gap*, Icarus **83**, 391-409.

Michtchenko, T.A. and Ferraz-Mello, S.: 1997, *Escape of Asteroids from the Hecuba Gap*, Planetary and Space Science **45**, 1587-1593.

Moons, M. and Morbidelli, A.: 1993, *The main Mean Motion Commensurabilities in the Planar Circular and Elliptic Problem*, Cel. Mec. Dyn. Astr. **57**, 99-108.

Moons, M. and Morbidelli, A.: 1995, *Secular Resonances in Mean Motion Commensurabilities: the 4/3, 3/1, 5/2 and 7/3 cases*, Icarus **114**, 33-50.

Moons, M., Morbidelli, A. and Migliorini, F.: 1998, *Dynamical Structure of the 2/1 Commensurability with Jupiter and the origin of the Resonant Asteroids*, Icarus ***

Morbidelli, A.: 1996, *On the Kirkwood Gap at the 2/1 Commensurability with Jupiter: numerical results*, Astron. J. **111**, 2453-2461.

Morbidelli, A. and Moons, M.: 1993, *Secular Resonances in Mean Motion Commensurabilities: the 2/1 and 3/2 cases*, Icarus **102**, 316-332.

Nesvorny, D. and Ferraz-Mello, S.: 1997, *On the asteroidal population of the first-order resonances*, Icarus **130**, 247-258.

Šidlichovský M.: 1988, On the Origin of 5/2 Kirkwood Gap in *Proceedings of the IAU Colloq. 96, The Few Body Problem*, Turku, (ed. Valtonen M.), p. 117.

Šidlichovský M.: 1991, *Tables of the Disturbing Function for Resonant Asteroids*, Bull. Astron. Inst. Czech. **42**, 116-123.

Šidlichovský M.: 1992, *Mapping For the Asteroidal Resonances*, Astron. Astrophys. **259**, 341-348.

Wisdom, J.: 1982, *The origin of the Kirkwood Gaps*, Astron. J. **87**, 577-593.

Wisdom, J.: 1983, *Chaotic Behaviour and the Origin of the 3/1 Kirkwood Gap*, Icarus **56**, 51-74.

Wisdom, J.: 1985, *A Perturbative Treatment of Motion Near the 3/1 Commensurability*, Icarus **63**, 272-289.

Yoshikawa, M.: 1989, *A survey of the Motion of Asteroids in Commensurabilities with Jupiter*, Astron. Astrophys. **213**, 436-458.

Yoshikawa, M.: 1991, *Motions of Asteroids at the Kirkwood Gaps. II. On the 5:2, 7:3 and 2:1 Resonances with Jupiter*, Icarus **92**, 94-117.

# ON STABLE CHAOS IN THE ASTEROID BELT

MILOŠ ŠIDLICHOVSKÝ

*Astronomical Institute Prague, Boční II 1401, 141 31 Praha 4, Czech Republic*

**Abstract.** The twenty most chaotic objects found among first hundred of numbered asteroids are studied. Lyapunov time calculated with and without inner planets indicates that for eleven of those asteroids the strongest chaotic effect results from the resonances with Mars. The filtered semimajor axis displays an abrupt variation only when a close approach to Mars takes place. The study of the behaviour of the critical argument for candidate resonances can reveal which is responsible for the semimajor axis variation. We have determined these resonances for the asteroids in question. For the asteroids chaotic even without the inner planets we have determined the most important resonances with Jupiter, or three-body resonances.

## 1. Introduction

The label "stable chaos" was introduced by Milani and Nobili (1992) for chaos in systems for which no macroscopic changes are observed over a time period longer than 1000 $T_L$ ($T_L$ is the Lyapunov time). Asteroid 522 Helga (with $L_T=6\,900$yr) was the first clear-cut example of stable chaos. Many other asteroids have been found to have $T_L$ much shorter than their orbital stability, and stable chaos now seems to be a common phenomenon. Milani and Nobili (1992) showed that the strongest chaotic effects for Helga are caused by the interaction of the intermediate order resonance 12:7 with secular perturbations acting on the eccentricity and perihelion of the asteroid. Šidlichovský and Nesvorný (1998) applied their modification (Šidlichovský and Nesvorný, 1998) of Laskar method (Laskar, 1990) to the first one hundred of the numbered asteroids and, as a by-product, obtained the Lyapunov times for each studied asteroid. These numerical calculations were performed using Nesvorný's MSI package (Šidlichovský and Nesvorný, 1994) for the Sun, 7 planets (without Mercury and Neptune) and the asteroid.

The integrator was the 12th-order symmetric multistep method by Quinlan and Tremaine (1990). The integration step was one day. Table I shows the twenty most chaotic asteroids of the studied sample. Column 3 shows the Lyapunov time (7 planets used).

The degree of chaos may be studied using many other tools. Froeschlé *et al.* (1998) calculated the so-called Fast Lyapunov Indicators (Lega and Froeschlé, 1997) for 5 400 asteroids without the inner planets. By comparing their list of 848 most chaotic asteroids with ours we found that all their chaotic asteroids (belonging to the first one hundred numbered asteroids) were rather chaotic in our calculation as well, but we also found many others. The most important difference in our calculations was that we took into account Mars, Venus and the Earth. This suggests that the additional chaos was caused by the inner planets, especially Mars.

Holman and Murray (1996) studied the variation of Lyapunov time as a function of the initial semimajor axis throughout the outer asteroid belt in the three-dimensional, elliptic, restricted, three-body problem. They observed $T_L$ in the $a$–$T_L$

*Celestial Mechanics and Dynamical Astronomy* **73**: 77–86, 1999.
© 1999 *Kluwer Academic Publishers.*

## TABLE I

The twenty most chaotic asteroids ($T_L > 32\,000$ yr) of the first one hundred numbered asteroids. Integrations are performed with 7 planets, Column 3 and 4 planets, Column 4. Column 5 shows the resonance with the strongest chaotic effect.

| No. | name | $L_T [10^3$ yr] 7 planets | $L_T [10^3$ yr] 4 planets | resonance |
|---|---|---|---|---|
| 2 | Pallas | 10 | 23 | 18:7 with Jupiter |
| 7 | Iris | 17 | >1000 | 25:49 with Mars |
| 8 | Flora | 30 | >1038 | 19:33 with Mars |
| 10 | Hygiea | 16 | 15 | 8-3-4 Jupiter-asteroid-Saturn |
| 12 | Victoria | 33 | >994 | 29:55 with Mars |
| 15 | Eunomia | 25 | >980 | 7:16 with Mars |
| 23 | Thalia | 25 | >1233 | |
| 33 | Polyhymnia | 10 | 14 | 22:9 with Jupiter |
| 35 | Leukothea | 20 | 17 | |
| 36 | Atlante | 4 | 5 | 4-2-3 Jupiter-asteroid-Saturn |
| 41 | Daphne | 14 | >360 | 9:22 with Mars |
| 46 | Hestia | 30 | >1039 | |
| 50 | Virginia | 10 | 12 | 11:4 with Jupiter |
| 53 | Kalypso | 19 | 14 | 6-2-1 Jupiter-asteroid-Saturn |
| 60 | Echo | 27 | >1159 | |
| 70 | Panopea | 24 | 36 | 2-1-2 Jupiter-asteroid-Saturn |
| 75 | Eurydike | 16 | >932 | |
| 78 | Diana | 13 | 149 | |
| 79 | Eurynome | 32 | 497 | |
| 86 | Semele | 6 | 6 | 13:6 with Jupiter |

graph corresponding to mean motion resonances and verified that the resonant angle makes transitions between circulation in one sense to circulation in the other sense, interspaced with brief periods of librations. Morbidelli and Nesvorný (1998) made similar calculations for the inner belt including the planets (even Mars) and observed peaks in the $a$–maximum LCE graph (which are the same as the dips in the $a$–$T_L$ graph). They could identify many of these peaks with mean motion resonances, but there were many other peaks which were finally identified with three-body mean motion resonances. These resonances correspond to the relation between mean motions of Jupiter, asteroid and Saturn:

$$m_J \dot{\lambda}_J + m \dot{\lambda} + m_S \dot{\lambda}_S \sim 0, \qquad (1)$$

where $m$, $m_J$ and $m_S$ are integers. The analytical model of the three-body resonances was presented by Nesvorný and Morbidelli (1998). Once this relation of resonances to low $T_L$ has been established, we may ask which resonances are mainly responsible for the chaotic behaviour of the most chaotic large asteroids in Tab. 1. This is the question we try to answer and check out by studying the behaviour of the corresponding critical argument with the relation to the variation of the filtered semimajor axis.

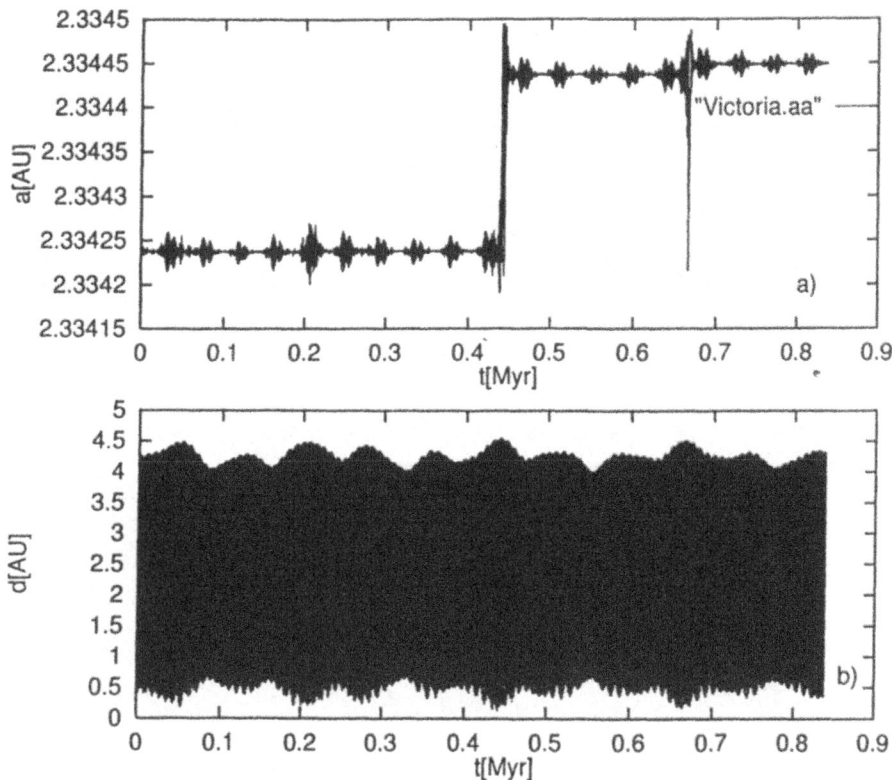

Fig. 1.
a)The filtered semimajor axis of Victoria   b)The distance between Victoria and
Mars

## 2. Effect of Mars

We calculated the Lyapunov time for the sample of twenty asteroids again, but now
without the inner planets. The result is shown in Column 4 of Tab. 1. Our sample
seems to split into two groups:
1) Asteroids with higher degree of chaoticity only when inner planets are considered: Iris, Flora, Victoria, Eunomia, Thalia, Daphne, Hestia, Echo, Eurydike,
Diana, Eurynome
2) Asteroids with higher degree of chaoticity even without considering the inner planets: Pallas, Hygiea, Polyhymnia, Leukothea, Atlante, Virginia, Kalypso,
Panopea, Semele

In studying the first group, we attempted to find some important resonances with
Mars as it is the closest inner planet. We considered the digitally filtered output of
the asteroid semi-major axis $a$ for about 800 thousand yrs. Filter A of Quinn *et al.*

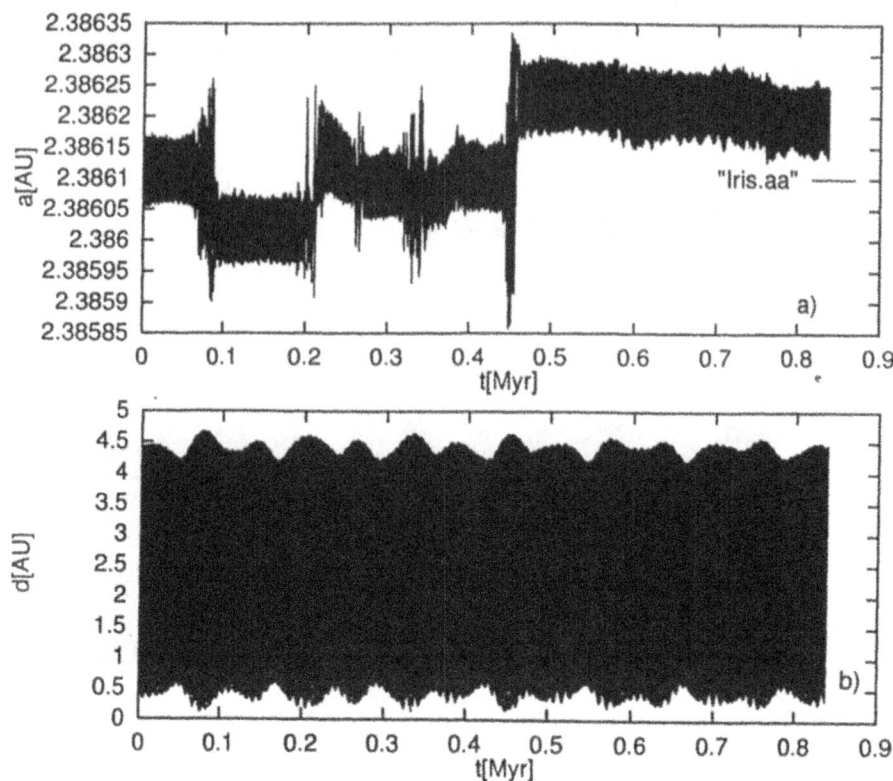

Fig. 2.
a) The filtered semimajor axis of Iris   b) The distance between Iris and Mars

(1991) was used. As our output step in these calculations was 300 days, and filter
A was applied twice, frequencies with periods shorter than 82 days were filtered
out.

Semimajor axis $a$ could vary only very slowly after removing the short period
terms, and, outside the resonance, it should be constant in the theories of second
order in small parameter $\mu$ (Jupiter mass). Figure 1a shows filtered $a$ for asteroid
Victoria. In this period two jumps in $a$ can be seen. The larger is about 0.0002 AU.
The amplitudes of short-period variations in $a$ are of the order 0.005 AU, hence
filtration is required. Figure 1b shows the distance between Victoria and Mars. It
is clear that the abrupt variation in $a$ is correlated to the short approach to Mars. It
is not actually a close encounter as the asteroid stays well outside the Martian Hill
sphere, but a temporary capture into the resonance with Mars is possible.

This is the common behaviour of the asteroids of the first group, except for
Hestia and Eurynome. Figure 2 shows the same graphs for Iris. The jumps in $a$

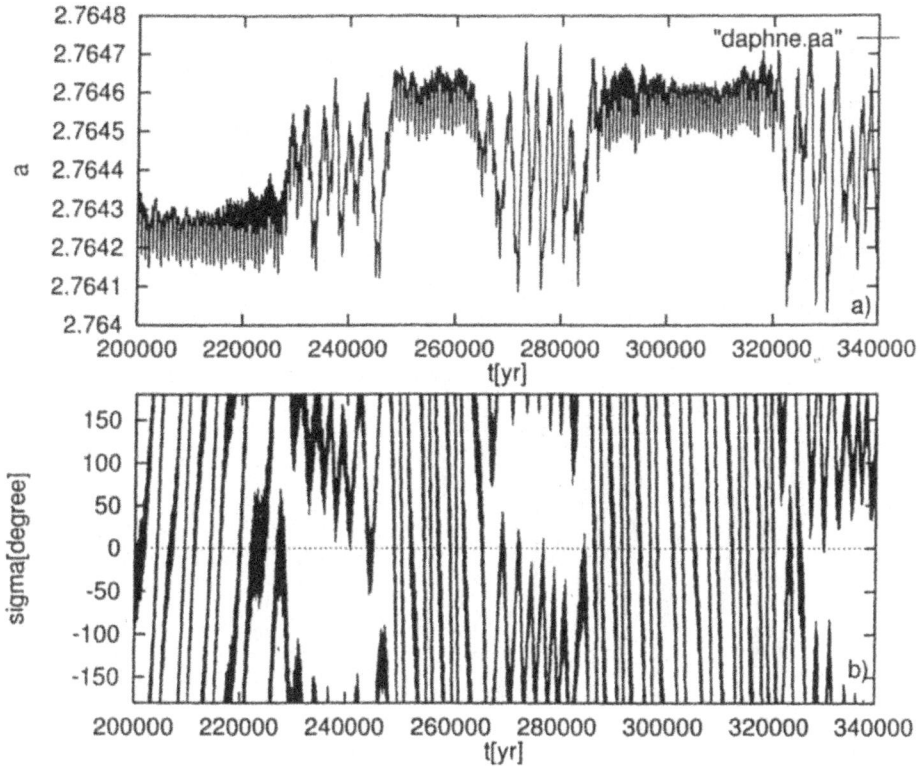

Fig. 3.
a) The filtered semimajor axis of Daphne b) The critical argument of Daphne:
$\sigma = 22\lambda - 9\lambda_M - 13\varpi$

usually start as larger oscillations (presumably resonant) with transition to small amplitude oscillations around the new $a$ value.

As the next step we made extensive tables of position (value $a$) of the resonances with Mars to order 30. For the studied asteroid we can read $a$ from the graph and find the close candidates for the resonance from the tables. We can then check the behaviour of the corresponding critical arguments. Figure 3a shows filtered $a$ for Daphne. Three regions of the resonant $a$ oscillations can be identified. Fig. 3b shows the corresponding resonant argument $\sigma = 22\lambda - 9\lambda_M - 13\varpi$. Here $\lambda$ stands for the mean longitude, $\varpi$ for the longitude of the perihelion, and subscript $M$ for Mars.

We see the transitions between circulation and librations with correlation to the $a$ variations. This is evidence that resonance 9/22 with Mars is the most important for chaotic behaviour of Daphne. Figure 4 shows filtered $a$ and the critical argument

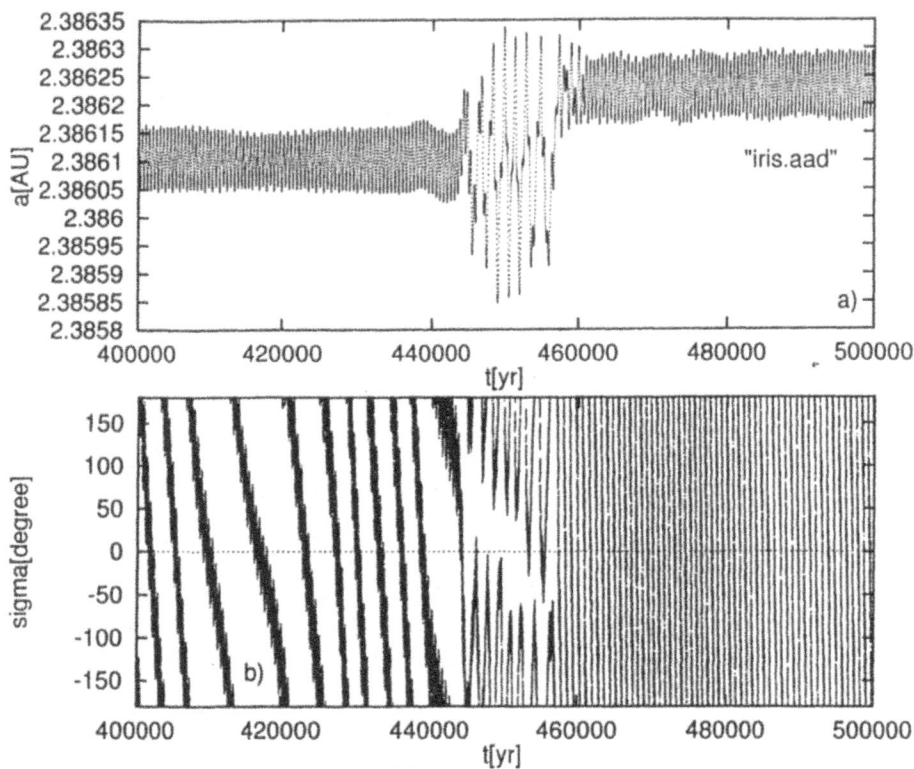

Fig. 4.

a) The filtered semimajor axis of Iris

b) The critical argument of Iris: $\sigma = 49\lambda - 25\lambda_M - 24\varpi$

$\sigma = 49\lambda - 25\lambda_M - 24\varpi$ for Iris. The resonances involved are of very high order and, as the sum of eccentricity powers in the resonant part of the perturbing function is equal to the order of the resonance, resonances of such high order were not suspected to be important. But it can be proved that the coefficients with the resonant terms are very high. For Flora, for instance, Nesvorný (1998) calculated the resonant term using his program in Mathematica:

$$R_{\text{res}} = -1.33\ 10^{10} \frac{M_M}{a_M} e_M^7 e^7 \cos(33\lambda - 19\lambda_M - 7\varpi - 7\varpi_M)$$

The discussion becomes even more complicated as the Laplacian expansion of the disturbing function (Mars-asteroid) may be divergent (Šidlichovský and Nesvorný, 1994).

Figures like Fig. 3 and Fig. 4 cannot be shown here for all the asteroids studied. For the asteroids of the first group (chaos coming from Mars) we found the

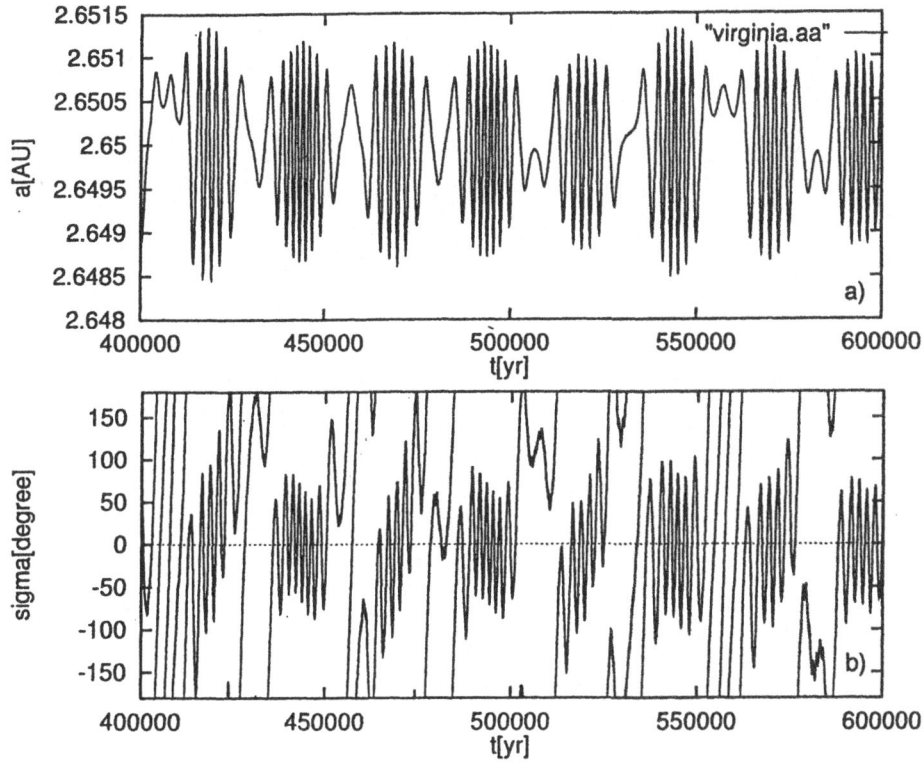

Fig. 5.
a) The filtered semimajor axis of Virginia
b) The critical argument of Virginia: $\sigma = 11\lambda_J - 4\lambda - \varpi_J - \varpi$

following resonances with Mars: Iris 25:49, Flora 19:33, Victoria 29:55, Eunomia 7:16, Daphne 9:22 while for Echo, Eurydike, Diana we did not succeed in finding the responsible resonance. The results are also shown in Column 5 of Tab. 1.

## 3. Resonances with Jupiter

For asteroids of the second group (strongest chaotic effects from the outer planets) we used the same method. Filtered $a$ suggests the candidates for the resonance, and the critical argument is the check. Figure 5 shows filtered $a$ and critical argument $\sigma = 11\lambda_J - 4\lambda - 2\varpi_J - 5\varpi$ for Virginia. We observed the following resonances with Jupiter: Pallas 18:7, Polyhymnia 22:9, Virginia 11:4, Semele 13:6. While for some of these asteroids, e.g., Virginia, the libration of $\sigma$, accompanied by a large variation of filtered $a$, is quite frequent, for some other, e.g., Pallas, only three such

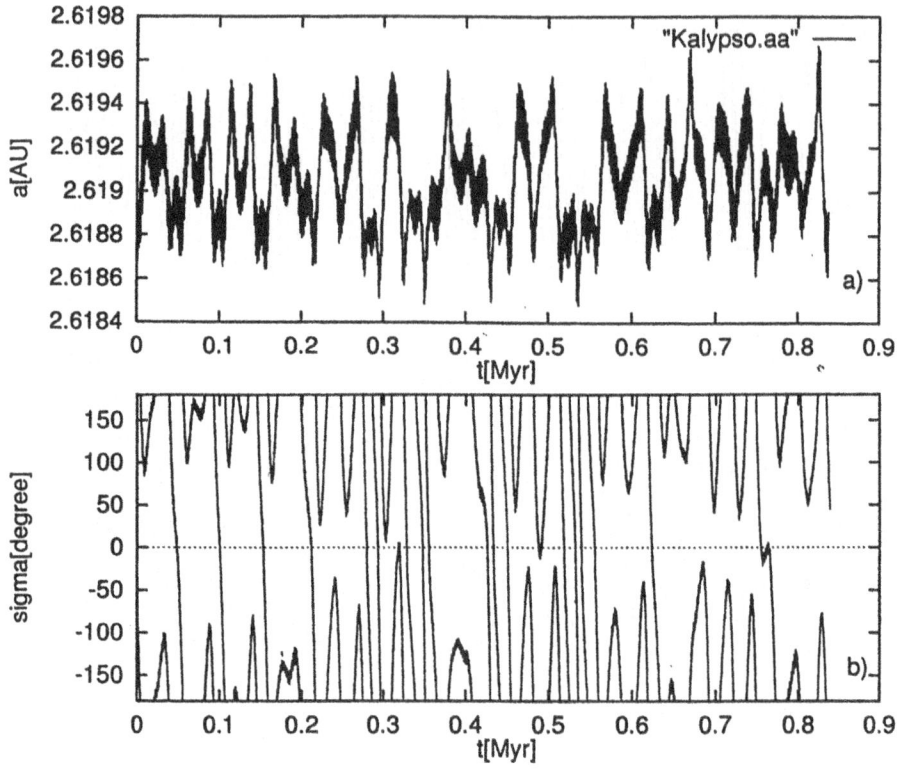

Fig. 6.
a) The filtered semimajor axis of Kalypso
b) The critical argument of Kalypso: $\sigma = 6\lambda_J - 2\lambda - \lambda_S - 3\varpi$

variations (in 800 000 yrs) were found, each accompanied with just two or three libration cycles in $\sigma$.

## 4. Three-Body Jupiter-Asteroid-Saturn Resonances

We will denote the three-body resonance with the critical argument corresponding to Eq.(1) as the $m_J + m + m_S$ resonance. Resonance 6-2-1 has the critical argument $\sigma = 6\lambda_J - 2\lambda - \lambda_S$. Our list of resonances contained three-body resonances so that we could also check the three-body resonant arguments. Figure 6 shows filtered $a$ and critical argument $\sigma = 6\lambda_J - 2\lambda - \lambda_S - 3\varpi$ for Kalypso. We observed the following three-body resonances: Hygiea 8-3-4, Atlante 4-2+3, Kalypso 6-2-1, Panopea 2-1-2.

## 5. Conclusion

For the 20 most chaotic asteroids among the first one hundred numbered asteroids, we studied the resonances with the strongest chaotic effect. The results are shown in the last column of Tab. 1. Many of the resonances are of a very high order, but calculations have shown that the coefficient standing with the resonant argument in the perturbing function can be unexpectedly large.

We checked more neighbouring resonances, but only one was always found important, at least for the period of time studied. We cannot, however, exclude the possibility of the effect of other high-order resonances later. For several sub-resonances of this resonance (the same coefficients with longitudes but different coefficients with perihelion longitudes in critical argument) we usually observed simultaneous libration which indicates overlap of these subresonances.

## Acknowledgements

The support by Grant 205/98/1102 from the Grant Agency of the Czech Republic is gratefully acknowledged.

## References

Froeschlé, Cl., Gonzi, R., Lega, E., Locatelli, U.: 1998, On the stochasticity of asteroid belt, it submitted to Celest. Mech. Dyn. Astron.

Holman ,M. J., Murray, N. W.: 1996, Chaos in high-order mean motion resonances in the outer asteroid belt, *A.J.* **112**, pp. 1278–1293

Laskar, J.: 1990, The Chaotic Motion in the Solar System. A Numerical Estimate of the Size of the Chaotic Zones, *Icarus* **88**, pp. 266–291

Laskar, J., Froeschlé, Cl. and Celetti, A.: 1992, The Measure of Chaos by Numerical Analysis of the Fundamental Frequencies. Application to Standard Mapping, *Physica D* **56**, pp. 253–269

Lega, E. and Froeschlé, Cl.: 1997, Fast Lyapunov Indicators Comparison with Other Chaos Indicators Application to Two and Four Dimensional Maps, in *The Dynamical Behaviour of our Planetary System, Proceedings of the Fourth Alexander von Humboldt Colloquium on Celestial Mechanics*, editors: R. Dvorak, J. Henrard, Kluwer Academic Publishers, Dordrecht, pp. 257–275

Milani, A. and Nobili, A. M.: 1992, An example of stable chaos in the Solar System, *Nature* **357**, pp. 569-571

Milani, A., Nobili, A.M. and Knežević, Z.: 1997, Stable Chaos in Asteroid Belt, *Icarus*, **125**,pp. 13–31

Morbidelli, A. and Froeschlé, Cl.: 1996, On the Relationship between Lyapunov Times and Macroscopic Instability Times, *Celest. Mech. Dynam. Astron.* **63**, pp. 227–239

Morbidelli, A., Nesvorný, D.: 1998, Numerous weak resonances drive asteroids towards terrestrial planets orbits, submitted to *Icarus*

Nesvorný, D.: 1998, private communication

Nesvorný, D., Morbidelli, A.: 1998, An analytic model of three-body mean motion resonances, *Celest. Mech. Dyn. Astron.*, in print

Quinlan, G.D. and Tremaine, S.: 1990, Symmetric Multistep Methods for the Numerical Integration of Planetary Orbits, *A.J.* **100**, pp. 1694–1700

Quinn, T. R., Tremaine, S., Duncan, M.: 1991, A three million year integration of the Earth's orbit, *A.J.* **101**, pp. 2287–2305

Šidlichovský, M., Nesvorný, D.: 1994, Temporary Capture of Grains in Exterior Resonances with the Earth,*Astron. Astrophys.* **289**, pp. 972–982

Šidlichovský, M. and Nesvorný, D.: 1997, Frequency Modified Fourier Transform and its Application to Asteroids, in *The Dynamical Behaviour of our Planetary System, Proceedings of the Fourth Alexander von Humboldt Colloquium on Celestial Mechanics*, editors: R. Dvorak, J. Henrard, Kluwer Academic Publishers, Dordrecht, pp. 137–148

Šidlichovský, M. and Nesvorný, D.: 1998, A study of chaos in the asteroid belt, *The Dynamics of Small Bodies in the Solar System: A Major Key to Solar System Studies, Proceedings of the NATO ASI in Maratea 1997*, editors: A. E. Roy, B. A. Steves, Kluwer Academic Publishers, Dordrecht.

# EXPLICIT SEMIANALYTICAL THEORY OF ASTEROID MOTION

I.V. TUPIKOVA, A.A. VAKHIDOV and M. SOFFEL

*Lohrmann Observatory, Technical University Dresden, Germany*

**Abstract.** A new semianalytical theory of asteroid motion is presented. The theory is developed on the basis of Kaula's expansion of the disturbing function including terms up to the second order with respect to the masses of disturbing bodies. The theory is constructed in explicit form that gives the possibility to study separately the influence of different perturbations in the dynamics of minor planets. The mean–motion resonances with major planets as well as mixed three–body resonances can also be taken into account. For the non-resonant case the formulas obtained can be used for deriving the second transformation to calculate the proper elements of an asteroid orbit in closed form with respect to inclinations and eccentricities.

**Key words:** minor planets, resonances, Hamiltonian systems, Lie series

## 1. Introduction

In order to deal with the long-time behaviour of asteroids, one of the modern fast and accurate integration schemes for the equations of motion in Cartesian coordinates might be employed. In the case when we are interested in some special perturbations, we can calculate the amplitudes at certain frequencies with the aid of fast Fourier transform. But we cannot tell for sure, which effect or interplay of effects has really caused this calculated value of amplitudes (see, for example, the discussion concerning Veritas (Milani *et al.*, 1997; Nesvorny and Morbidelli, 1999). For some problems it might be more efficient or more informative to use a semianalytical approach that gives the averaged equations of motion in explicit form. The averaging procedure is connected with the elimination of short–periodic terms from the initial Hamiltonian describing the motion of asteroids under the influence of selected planets. This procedure can be realized, for example, by means of Lie–series methods ( Hori, 1966; Deprit, 1969). As a result, we obtain the averaged Hamiltonian which contains only secular, long–periodic and resonant terms. The number of these terms (at least in the first two succesive approximations with respect to the small parameter of the problem) is substantially less than the number of short–periodic terms which are eliminated from the Hamiltonian. Finally, we can integrate the system of the averaged differential equations of motion numerically.

The main advantages of semianalytical methods in comparison with the direct numerical integration of equations of motion are connected with the possibilities:

— to eliminate the non–essential short–periodic perturbations from the Hamiltonian (and this is possible up to any desired order in the small parameter of the problem);

— to investigate the "pure" dynamical picture defined only by secular, long–periodic and resonant perturbations;

— to increase substantially the step-size of numerical integration.

 *Celestial Mechanics and Dynamical Astronomy* **73**: 87–96, 1999.
© 1999 *Kluwer Academic Publishers.*

Here, we present a new semianalytical theory which can be used for numerical integration of averaged equations of motion as well as for the construction of the second transformation for the derivation of proper elements of asteroid orbits in the non-resonant case. The theory considers the perturbations of the second order explicitly that allows to select certain resonant harmonics (e.g. certain mixed resonances, see Nesvorny and Morbidelli, 1999) and to reveal a minimal set of perturbations causing certain characteristics in the observed orbital dynamics. Since we have employed Kaula's expansion of the disturbing function, the solution is obtained in closed form w.r.t. inclinations (via inclination functions) and eccentricites (via Hansen coefficients). It is to be understood that the theory can be used only for the case of reasonable convergency with respect to the parameter $\alpha$ (ratio of semi-major axes).

## 2. Averaging procedure

For the construction of perturbation theory of the second order we use the averaging procedure based on the Deprit modification of the Lie–series method (Deprit, 1969); in this procedure the short-periodic perturbations supposedly non-essential are eliminated.

Let us consider the Hamiltonian of the problem in the form

$$F = F_0 + F_1,$$

with the unperturbed part

$$F_0 = \frac{\mu^2}{2L^2} - K,$$

$$\mu = G_0 \mathcal{M}_{\text{Sun}},$$

($G_0$ is the Newton's gravitational constant, $\mathcal{M}$ stands for mass) and with the disturbing function which can be expressed in the classical form (Kaula, 1966; Yuasa, 1973)

$$F_1 = \sum_B \sum_{i=(i_1,\ldots,i_6)} f_i^B(L, G, H, L_B, G_B, H_B) \cos \phi_i^B, \qquad (1)$$

$$\phi_i^B = i_1 l + i_2 l_B + i_3 g + i_4 g_B + i_5 h + i_6 h_B,$$

in the extended phase space of variables $(L, G, H, K, l, g, h, t)$. Here K is conjugate to the time variable $t$ and $L, G, H, l, g, h$ are the usual Delaunay elements. The index $B$ refers to the perturbing body. The motions of the perturbing bodies are supposed to be known functions of $t$; then the disturbing function (1) depends explicitly upon the time only via the elements $L_B, G_B, H_B, l_B, g_B, h_B$.

We assume that $F_1$ is of the order of the small parameter $\mathcal{M}_B/\mathcal{M}_{\text{Sun}}$ which allows to construct a perturbation theory by means of successive approximations.

According to Deprit's method in the first approximation we should solve the differential equation

$$\{F_0, W_1\} + F_1 = F_1^*$$

(where $\{F_0, W_1\}$ are the Poisson brackets in the extended phase space) in order to derive the generating function $W_1$ of the corresponding canonical transformation.

We would like to point out that in the formulae for the determination of the new Hamiltonian and the generating function in all the components the new (mean) elements should be substituted instead of the old (osculating) ones; below we will omit that for simplicity.

We put

$$F_1^* = [F_1]_{sec}$$

as the part of the initial Hamiltonian which contains only secular, long–periodic and resonant terms (in the case of the commensurability between the mean motion of the asteroid $\bar{n}$ and that of the perturbing body $\bar{n}_B$ defined by $\bar{n} : \bar{n}_B \approx \gamma : \gamma_B$, where $\gamma$, $\gamma_B$ are some integers). Then the Hamilton–Jacobi equation for the generating function $W_1$ has the following form:

$$-\bar{n}\frac{\partial W_1}{\partial l} - \frac{\partial W_1}{\partial t} + [F_1]_{short-per} = 0. \tag{2}$$

Here $[F_1]_{short-per}$ is the part of initial Hamiltonian (1) containing only short–periodic terms.

Putting $l_B = \bar{n}_B t + l_B^0$ (here $l_B^0$ is a constant), we find:

$$W_1 = \sum_B \sum_{i=(i_1,\dots,i_6)} \Delta_i^B f_i^B \sin \phi_i^B. \tag{3}$$

To simplify the further computations we introduce here

$$\Delta_i^B = \begin{cases} 0, & \text{for secular, long–periodic} \\ & \text{and resonant (with body "B") terms,} \\ \\ 1/(i_1\bar{n} + i_2\bar{n}_B), & \text{otherwise.} \end{cases}$$

One can show that neglecting the explicit dependence of $g_B$ and $h_B$ upon time when solving equation (2) introduces an error of the third order.

In the second order we have the following Hamilton-Jacobi equation

$$\{F_0, W_2\} + \frac{1}{2}\{F_1 + F_1^*, W_1\} = F_2^* \tag{4}$$

for the generating function $W_2$. In the same way, we define the corresponding part of the second order Hamiltonian which contains only secular, long–periodic and resonant terms as

$$F_2^* = \frac{1}{2}\{F_1 + F_1^*, W_1\}_{sec}$$

to find $W_2$ from

$$-\bar{n}\frac{\partial W_2}{\partial l} - \frac{\partial W_2}{\partial t} + \frac{1}{2}\{F_1 + F_1^*, W_1\}_{\text{short-per}} = 0.$$

Note that $F_2^*$ and $W_2$ contain the mixed terms arising from the interaction of two perturbing bodies with the asteroid.

For two bodies (e.g., Jupiter ($J$) and Saturn ($S$)) we have

$$F_1 = F_1^J + F_1^S, \qquad F_1^* = F_1^{J*} + F_1^{S*},$$

$$W_1 = W_1^J + W_1^S.$$

Then the second Poisson bracket on the l.h.s. of (4) is a sum of four terms:

$$\frac{1}{2}\{F_1^J + F_1^{J*} + F_1^S + F_1^{S*}, W_1^J + W_1^S\} =$$

$$= \frac{1}{2}\{F_1^J + F_1^{J*}, W_1^J\} + \frac{1}{2}\{F_1^S + F_1^{S*}, W_1^S\} +$$

$$+ \frac{1}{2}\{F_1^J + F_1^{J*}, W_1^S\} + \frac{1}{2}\{F_1^S + F_1^{S*}, W_1^J\}. \tag{5}$$

The first two terms describe the direct perturbations of the second order from the perturbing bodies (Jupiter or Saturn, respectively).

With $F_1^B + F_1^{B*}$ written as

$$F_1^B + F_1^{B*} = \sum_{j=(j_1,\ldots,j_6)} \delta_j^B f_j^B \cos\phi_j^B,$$

where

$$\delta_j^B = \begin{cases} 2, & \text{for secular, long-periodic and resonant (with body "B") terms,} \\ 1, & \text{otherwise,} \end{cases}$$

we find

$$\frac{1}{2}\{F_1^B + F_1^{B*}, W_1^B\} =$$

$$= \frac{1}{4}\sum_i\sum_j\sum_{k=-1,+1} f_j^B\Bigg[\left(\frac{\partial f_i^B}{\partial L}j_1 + \frac{\partial f_i^B}{\partial G}j_3 + \frac{\partial f_i^B}{\partial H}j_5\right)\left(\Delta_j^B\delta_i^B + k\Delta_i^B\delta_j^B\right) +$$

$$+ 3f_i^B\frac{\bar{n}}{L}k(\Delta_i^B)^2\delta_j^B i_1 j_1\Bigg]\cos(\phi_i^B - k\phi_j^B). \tag{6}$$

Two last Poisson brackets in formula (5) represent the mixed perturbations:

$$\frac{1}{2}\{F_1^J + F_1^{J*}, W_1^S\} + \frac{1}{2}\{F_1^S + F_1^{S*}, W_1^J\} =$$

$$= \frac{1}{4}\sum_i\sum_j\sum_{k=-1,+1}\Bigg[[f_j^S(\frac{\partial f_i^J}{\partial L}j_1 + \frac{\partial f_i^J}{\partial G}j_3 + \frac{\partial f_i^J}{\partial H}j_5)$$

$$+k\, f_i^J \Big(\frac{\partial f_j^S}{\partial L} i_1 + \frac{\partial f_j^S}{\partial G} i_3 + \frac{\partial f_j^S}{\partial H} i_5\Big)\big](\Delta_j^S \delta_i^J + k\Delta_i^J \delta_j^S) +$$

$$+3 f_i^J f_j^S \frac{\bar{n}}{L} k i_1 j_1 ((\Delta_i^J)^2 \delta_j^S + (\Delta_j^S)^2 \delta_i^J)\Big] \cos(\phi_i^J - k\phi_j^S). \tag{7}$$

Let us write

$$F_2^* = F_2^{J*} + F_2^{S*} + F_2^{JS*},$$

then $F_2^{B*}$ is given by

$$F_2^{B*} = \frac{1}{2}\{F_1^B + F_1^{B*}, W_1^B\}_{\text{sec}}, \tag{8}$$

that is, as the part of (6) which contains secular and long–periodic terms determined by the following conditions:

$$i_1 - kj_1 = i_2 - kj_2 = 0$$

and in the resonant case $\bar{n} : \bar{n}_B \approx \gamma : \gamma_B$ also the resonant terms satisfying the equation:

$$\frac{i_1 - kj_1}{i_2 - kj_2} = -\frac{\gamma_B}{\gamma}.$$

The function $F_2^{JS*}$ is determined by

$$F_2^{JS*} = \frac{1}{2}\{F_1^J + F_1^{J*}, W_1^S\}_{\text{sec}} + \frac{1}{2}\{F_1^S + F_1^{S*}, W_1^J\}_{\text{sec}}, \tag{9}$$

that is, as the part of (7) containing only secular, long-periodic and resonant terms. Notice, that the trigonometric arguments in (7)

$$\phi_i^J - k\phi_j^S = (i_1 - kj_1)l + i_2 l_J - kj_2 l_S + (i_3 - kj_3)g + (i_5 - kj_5)h +$$
$$+ i_4 g_J + i_6 h_J - kj_4 g_S - kj_6 h_S$$

give the resonant components not only in the case of mean motion commensurabilities but also for mixed resonances.

Similarly, let us write $W_2$ as

$$W_2 = W_2^J + W_2^S + W_2^{JS},$$

then $W_2^B$ $(B = J, S)$ is given by

$$W_2^B = \frac{1}{4} \sum_i \sum_j \sum_{k=-1,+1} \Delta_{ij}^B f_j^B \times$$

$$\times \Big[\Big(\frac{\partial f_i^B}{\partial L} j_1 + \frac{\partial f_i^B}{\partial G} j_3 + \frac{\partial f_i^B}{\partial H} j_5\Big)\Big(\Delta_j^B \delta_i^B + k\Delta_i^B \delta_j^B\Big) +$$

$$+3 f_i^B \frac{\bar{n}}{L} k \Delta_i^{B^2} \delta_j^B i_1 j_1\Big] \sin(\phi_i^B - k\phi_j^B) \tag{10}$$

with

$$
\Delta_{ij}^{B} = \begin{cases} 0, & \text{for secular, long-periodic} \\ & \text{and resonant with "B" terms,} \\ \\ 1/((i_1 - kj_1)\bar{n} + (i_2 - kj_2)\bar{n}_B), & \text{otherwise.} \end{cases}
$$

Finally,

$$
W_2^{JS} = \frac{1}{4} \sum_i \sum_j \sum_{k=-1,+1} \Delta_{ij}^{JS} \Big[ [f_j^S (\frac{\partial f_i^J}{\partial L} j_1 + \frac{\partial f_i^J}{\partial G} j_3 + \frac{\partial f_i^J}{\partial H} j_5) +
$$
$$
+ k f_i^J (\frac{\partial f_j^S}{\partial L} i_1 + \frac{\partial f_j^S}{\partial G} i_3 + \frac{\partial f_j^S}{\partial H} i_5)](\Delta_j^S \delta_i^J + k\Delta_j^J \delta_j^S) +
$$
$$
+ 3 f_i^J f_j^S \frac{\bar{n}}{L} k i_1 j_1 ((\Delta_i^J)^2 \delta_j^S + (\Delta_j^S)^2 \delta_i^J) \Big] \sin(\phi_i^J - k\phi_j^S), \tag{11}
$$

where

$$
\Delta_{ij}^{JS} = \begin{cases} 0, & \text{for secular, long-periodic and} \\ & \text{resonant terms,} \\ \\ 1/((i_1 - kj_1)\bar{n} + i_2\bar{n}_J - kj_2\bar{n}_S), & \text{otherwise.} \end{cases}
$$

### 3. Application to Kaula's Expansion of the Disturbing Function

For asteroids moving relatively far from the perturbing bodies we will employ Kaula's expansion of the disturbing function (Kaula, 1966; Fominov, 1980) for each perturbing body "B" ($\alpha_B = a/a_B$):

$$
F_1^B = \frac{\mu_B}{a_B} \sum_{n=2}^{\infty} \sum_{m=0}^{n} \sum_{p=0}^{n} \sum_{h=0}^{n} \sum_{q=-\infty}^{\infty} \sum_{j=-\infty}^{\infty} \frac{\alpha_B^n}{2n+1} \times
$$
$$
\times \bar{F}_{nmp}(I) \bar{F}_{nmh}(I_B) H_{npq}(e) G_{nhj}(e_B) \cos \Phi_B ,
$$

where

$$
\Phi_B = (n - 2p + q) M - (n - 2h + j) M_B +
$$
$$
+ (n - 2p)\omega - (n - 2h)\omega_B + m(\Omega - \Omega_B),
$$

and

$$
H_{npq}(e) = X_{n-2p+q}^{n,n-2p}(e), \quad G_{nhj}(e_B) = X_{n-2h+j}^{-n-1,n-2h}(e_B).
$$

Here, $a, e, I, \Omega, \omega, M$ and $a_B, e_B, I_B, \Omega_B, \omega_B, M_B$ are Keplerian elements of the asteroid and disturbing body, respectively; $\bar{F}_{nmp}(I)$ are normalized Kaula inclination functions, $X_q^{n,p}(e)$ are the classical Hansen coefficients and $\mu_B$ is the gravitational constant of the perturbing body: $\mu_B = G_0 \mathcal{M}_B / \mathcal{M}_{\text{Sun}}$.

With the aid of the averaging procedure described above we can construct the averaged Hamiltonian up to the second order

$$F^* = F_0^* + F_1^* + F_2^*,$$

$$F_1^* = F_1^{J*} + F_1^{S*}, \qquad F_2^* = F_2^{J*} + F_2^{S*} + F_2^{JS*}$$

and the generating function

$$W = W_1 + W_2,$$

$$W_1 = W_1^J + W_1^S, \qquad W_2 = W_2^J + W_2^S + W_2^{JS}.$$

The part of the Hamiltonian of the first order containing only secular and long–periodic terms reads

$$F_1^{B*} = \frac{\mu_B}{a_B} \sum_{n=2}^{\infty} \sum_{m=0}^{n} \sum_{p=0}^{n} \sum_{h=0}^{n} \frac{\alpha_B^{n}}{2n+1} \bar{F}_{nmp}(I)\bar{F}_{nmh}(I_B) X_0^{n,n-2p}(e) \times$$

$$\times X_0^{-n-1,n-2h}(e_B) \cos[(n-2p)\omega - (n-2h)\omega_B + m(\Omega - \Omega_B)]. \quad (12)$$

Resonant terms appear in development (12) in the case $\bar{n} : \bar{n}_B \approx \gamma : \gamma_B$, if

$$\frac{n-2p+q}{n-2h+j} = \frac{\gamma_B}{\gamma}.$$

In this case the corresponding resonant terms have to be included in $F_1^{B*}$. Note, that in (12) there are no summations with respect to indices $q$ and $j$ which reduces the problem of convergency of the series with respect to Hansen coefficients.

According to (3), the generating function $W_1^B$ of the first order is

$$W_1^B = \frac{\mu_B}{a_B} \sum_{n=2}^{\infty} \sum_{m=0}^{n} \sum_{p=0}^{n} \sum_{h=0}^{n} \sum_{q=-\infty}^{\infty} \sum_{j=-\infty}^{\infty} \alpha_B^n \times$$

$$\times \Delta_{npqhj}^B \frac{1}{(2n+1)} \bar{F}_{nmp}(I)\bar{F}_{nmh}(I_B) H_{npq}(e)G_{nhj}(e_B) \sin \Phi^B,$$

where

$$\Delta_{npqhj}^B = \begin{cases} 0, & \text{for secular, long–periodic} \\ & \text{and resonant (with B) terms,} \\ \\ 1/((n-2p+q)\bar{n} - (n-2h+j)\bar{n}_B), & \text{otherwise.} \end{cases}$$

Using the expressions for the partials ($s = \sin I/2$)

$$\frac{\partial}{\partial L} = 2\sqrt{\frac{a}{\mu}} \frac{\partial}{\partial a} + \frac{1-e^2}{eL} \frac{\partial}{\partial e},$$

$$\frac{\partial}{\partial G} = \frac{1}{L\sqrt{1-e^2}}(-\frac{1-e^2}{e}\frac{\partial}{\partial e} + \frac{1-2s^2}{4s}\frac{\partial}{\partial s}),$$

$$\frac{\partial}{\partial H} = -\frac{1}{L\sqrt{1-e^2}}\frac{1}{4s}\frac{\partial}{\partial s},$$

$$\frac{\partial \bar{n}}{\partial L} = -\frac{3\bar{n}}{L},$$

from (6) and (8) we get $F_2^{B*}$ in the form

$$F_2^{B*} = \frac{1}{4L}\frac{\mu_B^2}{a_B^2}\sum_{nmphqj}\sum_{n'm'p'h'q'j'}\sum_{k=-1,+1}^{*}\alpha_B^{n+n'}\frac{1}{2n+1}\frac{1}{2n'+1}\times$$

$$\times \bar{F}_{nmh}(I_B)\bar{F}_{n'm'h'}(I_B)G_{nhj}(e_B)G_{n'h'j'}(e_B)\bar{F}_{n'm'p'}(I)H_{n'p'q'}(e)\times$$

$$\times [\bar{F}_{nmp}(I)H_{npq}(e)\Delta_2^B +$$

$$+(\bar{F}_{nmp}(I)H_{npq}^1(e) + \frac{1}{\sqrt{1-e^2}}H_{npq}(e)\bar{F}_{nmp}^1(I))\Delta_1^B] \times$$

$$\times \cos[(n-2p+q-k(n'-2p'+q'))M + (n-2p-k(n'-2p'))\omega -$$

$$- (n-2h+j-k(n'-2h'+j'))M_B - (n-2h-k(n'-2h'))\omega_B +$$

$$+ (m-km')(\Omega - \Omega_B)].$$

Here,

$$\sum_{nmphqj} = \sum_{n=2}^{\infty}\sum_{m=0}^{n}\sum_{p=0}^{n}\sum_{h=0}^{n}\sum_{q=-\infty}^{\infty}\sum_{j=-\infty}^{\infty},$$

$$\sum_{n'm'p'h'q'j'} = \sum_{n'=2}^{\infty}\sum_{m'=0}^{n'}\sum_{p'=0}^{n'}\sum_{h'=0}^{n'}\sum_{q'=-\infty}^{\infty}\sum_{j'=-\infty}^{\infty},$$

$$H_{npq}^1 = [(n'-2p'+q')(1-e^2) - (n'-2p')\sqrt{1-e^2}]\frac{1}{e}\frac{\partial H_{npq}}{\partial e},$$

$$\bar{F}_{nmp}^1 = \frac{1}{4}[(n'-2p')(1-2s^2) - m']\frac{1}{s}\frac{\partial \bar{F}_{nmp}}{\partial s},$$

$$\Delta_1^B = \Delta_{n'p'q'h'j'}^B\delta_{npqhj}^B + k\Delta_{npqhj}^B\delta_{n'p'q'h'j'}^B,$$

$$\Delta_2^B = 2n(n'-2p'+q')\Delta_1^B +$$
$$+ 3k\bar{n}(n-2p+q)(n'-2p'+q')(\Delta_{npqhj}^B)^2\delta_{n'p'q'h'j'}^B,$$

$$\delta^B_{npqhj} = \begin{cases} 2, & \text{if } n - 2p + q = n - 2h + j = 0 \\ & \text{or } (n - 2p + q) : (n - 2h + j) = \gamma_B : \gamma \\ & \text{in the resonant case,} \\ \\ 1, & \text{otherwise,} \end{cases}$$

$\Delta^B_{n'p'q'h'j'}$ and $\delta^B_{n'p'q'h'j'}$ are defined by the formulae for $\Delta^B_{npqhj}$ and $\delta^B_{npqhj}$ by putting primes at the indices, and $*$ at the summation symbol means that one should take only the terms with the indices satisfying the conditions: $n - 2p + q - k(n' - 2p' + q') = n - 2h + j - k(n' - 2h' + j') = 0$ or

$$\frac{n - 2p + q - k(n' - 2p' + q')}{n - 2h + j - k(n' - 2h' + j')} = \frac{\gamma_B}{\gamma}.$$

Finally, by means of (7) and (9) we get

$$F_2^{JS*} = \frac{1}{4L} \frac{\mu_J}{a_J} \frac{\mu_S}{a_S} \sum_{nmphqj} \sum_{n'm'p'h'q'j'} \sum_{k=-1,+1}^{**} \alpha_J^n \alpha_S^{n'} \times$$

$$\times \frac{1}{(2n+1)} \frac{1}{(2n'+1)} \bar{F}_{nmh}(I_J) \bar{F}_{n'm'h'}(I_S) G_{nhj}(e_J) G_{n'h'j'}(e_S) \times$$

$$\times \Big[ \bar{F}_{n'm'p'}(I) \bar{F}_{nmp}(I) H_{npq}(e) H_{n'p'q'}(e) \Delta_2^{J,S} +$$

$$+ \bar{F}_{nmp}(I) \bar{F}_{n'm'p'}(I) [H^1_{npq}(e) H_{n'p'q'}(e) + k\, H_{npq}(e)^1_{n'p'q'}(e)] \Delta_1^{J,S} +$$

$$+ \frac{1}{\sqrt{1 - e^2}} H_{npq}(e) H_{n'p'q'}(e) \times$$

$$\times [\bar{F}^1_{nmp}(I)) \bar{F}_{n'm'p'}(I) + k\, \bar{F}_{nmp}(I) \bar{F}^1_{n'm'p'}(I)] \Delta_1^{S,J} \Big] \times$$

$$\times \cos[(n - 2p + q - k(n' - 2p' + q'))M + (n - 2p - k(n' - 2p'))\omega -$$

$$- (n - 2h + j)M_J + k(n' - 2h' + j')M_S - (n - 2h)\omega_J + k(n' - 2h')\omega_S +$$

$$+ (m - km')\Omega - m\Omega_J + km'\Omega_S)],$$

where

$$\Delta_1^{J,S} = \Delta^S_{n'p'q'h'j'} \delta^J_{npqhj} + k\Delta^J_{npqhj} \delta^S_{n'p'q'h'j'},$$

$$\Delta_2^{J,S} = (2n(n' - 2p' + q') + 2kn'(n - 2p + q))\Delta_1^{J,S} +$$

$$+ 3k\bar{n}(n - 2p + q)(n' - 2p' + q')[(\Delta^J_{npqhj})^2 \delta^S_{n'p'q'h'j'} + (\Delta^S_{n'p'q'h'j'})^2 \delta^J_{npqhj}],$$

$$H^1_{n'p'q'} = [(n - 2p + q)(1 - e^2) - (n - 2p)\sqrt{1 - e^2}] \frac{1}{e} \frac{\partial H_{n'p'q'}}{\partial e},$$

$$\bar{F}^1_{n'm'p'} = \frac{1}{4}[(n - 2p)(1 - 2s^2) - m] \frac{1}{s} \frac{\partial \bar{F}_{n'm'p'}}{\partial s}.$$

Here $**$ by sums means that one should take only the terms with the indices satisfying the conditions $n-2p+q-k(n'-2p'+q') = n-2h+j = n'-2h'+j' = 0$ or satisfying the conditions of a mean-motion or three-body resonance. The components of the generating functions $W_2^B$ can be easily deduced from formula (10) and the function $W_2^{JS}$ – from formula (11).

These expressions give the explicit second-order solution for the secular, long-periodic and resonant (also mixed–resonant) perturbations.

Presently, we are in the process to implement the perturbations in the motion of disturbing bodies and then to apply this formalism to the problem of mixed–resonances and other perturbations of the second order in the motion of asteroids. We are also planning to develop explicitly the second transformation to exclude the remaining angular variables and to get the analytical solution in the non-resonant case.

## Acknowledgements

The authors are very grateful to Dr. S.Klioner and Dr. M.Šidlichovsky for some useful discussions concerning our work. We would like to express also our gratitude to Prof. R.Dvorak and Dr. E.Pilat–Lohinger for the help in our research and many useful suggestions. We thank also Dr. C.Ron for his attention to our work. A.V. thanks also the German Academician Exchange Service (DAAD) for financial support of this research.

## References

Milani, A., Nobili, A., Knezevic, Z.: 1997, *Icarus*, **125**, 13.
Nesvorny, D., Morbidelli A.: 1987, *Astron. J*, in press.
Hori, G.I.: 1966, *Publ. Astron. Soc. Japan*, **18**, 287.
Deprit, A.: 1969, *Celest. Mech.*, **1**, 12.
Kaula, W.M.: 1966, *Theory of Satellite Geodesy*, Blaisdell Publ. Co., Mass.
Yuasa, M.: 1973, *Publ. Astron. Soc. Japan*, **25**, 399.
Fominov A.M.: 1980, *Bull. ITA*, **10**, 621.

# ON THE MOTION OF TRAPPED PARTICLES
# IN THE VICINITY OF COROTATION CENTERS

C. BEAUGÉ

*Observatorio Astronómico, Universidad Nacional de Córdoba,*
*Laprida 854, (5000) Córdoba, Argentina*

and

A. LEMAÎTRE and S. JANCART

*Département de Mathématique, FUNDP, 8 Rempart de la Vierge,*
*5000 Namur, Belgium*

**Abstract.** In the present paper we analyse the motion of a massless particle during the capture process in an exterior mean-motion resonance under the effects of an external dissipative force. In particular, we study the orbital evolution from its initial approach to the commensurability up to the final nesting place in the periodic orbit around the equilibrium solution.

## 1. Introduction

In a recent work (Beaugé et al., 1998; hereafter BLJ98) we presented a new model for the averaged equations of motion of the capture problem, based on a Lie perturbation method (Kamel, 1969) truncated to second order. The application of this model to the case of Stokes drag allowed us to determine the equilibrium solutions of the system (i.e. corotation centers) with a significant increase in precision with respect to previous works.

However, when these results are compared with numerical simulations of the exact equations, we find that the real motion of the particles is not restricted to the equilibrium value but, due to the effects of the short-period terms, will define a periodic orbit around the fixed point. We call such orbit a limit cycle (Poincaré, 1885), and its period is simply the synodic period of the resonance. Fortunately, these cycles may also be reproduced with the second-order model. By means of the inverse transformation to the non-averaged variables we can find approximate analytical expressions for the periodic orbits which, when compared to the numerical simulations, show a very good agreement. Consequently, we currently have a fairly good idea of the final resting place of the trapped particles. The next step is to study the road that takes them there from the initial entrance to the resonance region. In other words, the motion of the body in the vicinity of the limit cycle itself.

This question has received a boost in interest from a recent work by Gomes and Mothe-Diniz (1998). In this paper, the authors performed several studies, via numerical simulations, of the evolution of two different particles (of radii $r$) trapped in the same corotation point. They found that, under a wide range of the initial conditions, the relative distance between both bodies diminished exponentially to values of the order of machine precision in times of the order of $10^4$ years. Their relative velocity also decreased to extremely low values ($\sim 10^{-10} m/s$ at $\Delta x \sim 2r$), which means that these particles experience more a "smooth approach" than an actual collision. These results are of considerable importance to those cosmogonic theories in which resonance trapping is present. Not only do they ensure the actual attraction of different bodies, but they also indicate that any collision will result in accretion. This could help explain how particles of very small size could accrete, even when the mutual gravity is virtually non-existent.

So stated, we wonder whether it would be possible to use our second-order model, together with the expressions for the limit cycles, to obtain approximate analytical solution of the complete non-averaged system during the capture process itself. This is the question we wish to address in this communication.

The manuscript is organized in the following manner: Section 2 presents the problem and reproduces the main results of the second-order theory, together with the expressions for the periodic orbits. Section 3 discusses the linearization of the system around the fixed point and the general solution of the eigensystem. Results and comparisons are presented in Section 4 and conclusions close the paper in Section 5.

## 2. The Second-Order Model

### 2.1. STATEMENT OF THE PROBLEM

As usual, let us suppose the planar elliptic restricted three-body problem in the vicinity of an exterior $(p + q)/p$ mean-motion resonance. The external dissipative force will be modeled by a Stokes drag with $\alpha = 0.995$ (Adachi et al., 1976). Let $C$ denote the drag coefficient which has units of unity over time. The set of resonant variables is chosen to be $(L, e, \sigma, \sigma_1, \sigma_2)$, where $L = \sqrt{\mu a}$ and the angular variables are defined through

$$
\begin{aligned}
q\sigma &= (p + q)\lambda_1 - p\lambda - q\varpi \\
q\sigma_1 &= -(p + q)\lambda_1/p + \lambda \\
\sigma_2 &= \lambda_1/p.
\end{aligned}
\tag{1}
$$

Here $\mu$ is the gravitational constant, $a$ and $e$ denote the semimajor axis and eccentricity of the body, $\lambda$ stands for the mean longitude and $\varpi$ the longitude of perihelion. Similar quantities, with subindex 1, hold for the perturbing planet of mass $m_1$.

Defining the position vector of this non-canonical set as $x = (L, e, \sigma, \sigma_1, \sigma_2)$, the system of differential equations governing the motion of the particle can be succinctly written as

$$\frac{dx}{dt} = \sum_{i_1,i_2,i_3,i_4,i_5} \left( f_0^{(0)} + m_1 \, f_1^{(0)} \right) L^{i_1} e^{i_2} E^{\sqrt{-1}(i_3\sigma + i_4\sigma_1 + i_5\sigma_2)} \tag{2}$$

where $f_0^{(0)} = f_0^{(0)}(i_1)$ marks the two-body gravitational contribution and $f_1^{(0)} = f_1^{(0)}(i_1, i_2, i_3, i_4, i_5)$ groups the complete perturbing forces (per unit of $m_1$), including the dissipative part. Here we have used the notation $E^\theta = \exp\theta$. We must note that all the $f_0^{(0)}$ and $f_1^{(0)}$ coefficients are constant throughout the phase space (see BLJ98 for further details).

## 2.2. THE AVERAGED SYSTEM

We search for a transformation of variables $W(y; m_1, C) : x \to y$ to new variables $y = (y_1, y_2, y_3, y_4, y_5) = (\bar{L}, \bar{e}, \bar{\sigma}, \bar{\sigma}_1, \bar{\sigma}_2)$ such that the new equations of motion do not depend explicitly on the transformed synodic angle $y_5$:

$$\frac{dy}{dt} = \sum_{i_1,i_2,i_3,i_4} \mathcal{A}_{i_1,i_2,i_3,i_4} L^{i_1} e^{i_2} E^{\sqrt{-1}(i_3\sigma + i_4\sigma_1)}. \tag{3}$$

Expanding the transformation function $W$ in a power series of the perturbing mass $m_1$, we can explicitly write the relationship between both sets of variables as:

$$x = y + m_1 W_1(y; C) + \frac{1}{2} m_1^2 W_2(y; C) + \ldots \tag{4}$$

where the new $W_i(y; C)$ are independent of the planetary mass. Via the Lie transform method (Kamel, 1969; Henrard, 1970) we find that the constant coefficients of (3) can be approximated by

$$\mathcal{A}_{i_1,i_2,i_3,i_4} = f_0^{(0)} + m_1 \, f_0^{(1)} + \frac{1}{2} m_1 C \, f_0^{(2)} + \ldots \tag{5}$$

where the new $f_0^{(k)} = f_0^{(k)}{}_{i_1,i_2,i_3,i_4}$ are obtained through the averaging process itself (Triangle rule). The fixed points of system (3) automatically yield the corotation centers of the averaged system. Denoting these values as $y_c = y_c(m_1, C)$, we have that $\dot{y}_c = 0$.

Next, to obtain the limit cycles around $y_c$, we must perform the inverse transformation. Retaining only the lowest-order terms, the periodic orbit in the old variables can be written as

$$x \simeq y_c + m_1 W_1(y_c; C) = y_c + \sum_j \omega^{(j)} E^{\sqrt{-1}(j\sigma_2)} \tag{6}$$

where the $\omega^{(j)} = \omega^{(j)}(y_c; m_1, C)$ coefficients are also given by the averaging process.

## 3. Linearization around the Fixed Point

### 3.1. THE EIGENSYSTEM

The usual way to study the evolution of the system in the vicinity of the periodic orbit is through Floquet theory. In this, we linearize the equations of motion around the periodic orbit itself, determine the monodromy matrix, and search for the eigenvalues of the resulting system at fixed times. However, the steps taken in the averaging process described in the previous section allow us to take a simpler road.

Since the averaging process relates the limit cycle in $x$ with the fixed point in $y$, we can directly study the linear system around the corotation center. The solution of the eigensystem (in $y$) can then be transformed to the old variables directly by means of equations (6). Thus, we avoid the complications of the linearization around a periodic solution and can work directly on the fixed point (see Hale, 1969). This is the aim of the present section.

Let us define $\xi = y - y_c$ and write the linearized system of (3) as $\dot{\xi} = \mathcal{B}\xi$. From (3) the expression for the Jacobian matrix is trivial, since all the coefficients $\mathcal{A}$ are constant and the derivatives can be performed explicitly. Thus, we can write

$$\mathcal{B} = \sum_{i_1,i_2,i_3,i_4} \mathcal{A}'_{i_1,i_2,i_3,i_4} L_c^{i_1} e_c^{i_2} E^{\sqrt{-1}(i_3\sigma_c+i_4\sigma_{1c})} \tag{7}$$

where $\mathcal{A}'$ denote the new coefficients after the differentiation.

Knowing $\mathcal{B}$, we can construct the characteristic equation and solve for the eigenvalues $\lambda_i$ and the eigenvectors $\psi_i$. Then, the general solution will be given by:

$$y = y_c + \sum_{i=1}^{4} K_i \psi_i E^{\lambda_i t}. \tag{8}$$

The complex constants $K_i$ are determined as solutions of the linear system $\psi K = y_0 - y_c$, where $\psi$ is the matrix of eigenvectors components and $y_0$ mark the initial conditions.

### 3.2. THE COMPLETE SOLUTION

It is now possible to introduce the solution (8) of the averaged system into the inverse transformation (6). In order to obtain an explicit expression, we approximate

$$W_1(y) \simeq W_1(y_c) + \left(\frac{\partial W_1}{\partial y}\right)_{y=y_c} (y - y_c). \tag{9}$$

Writing the derivative of the transformation function (evaluated at the fixed point) as

$$\left(\frac{\partial W_1}{\partial y}\right)_{y=y_c} = \sum_j \Omega_j E^{\sqrt{-1}j\sigma_2} \tag{10}$$

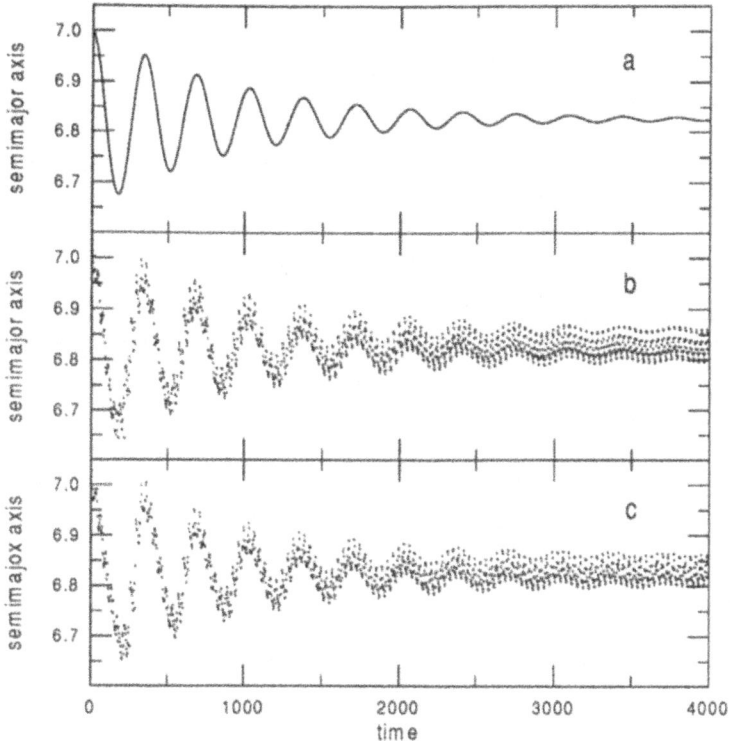

Fig. 1. Temporal evolution of the semimajor axis during the capture of a particle in a corotational solution in the 2/3 resonance. (a) Evolution of the mean semimajor axis (i.e. $y_1$) according to equation (8). (b) Idem, but for the non-averaged semimajor axis (i.e. $x_1$) as obtained by equations (11). (c) Numerical simulation of the exact equations.

and remembering expression (8) for $(y - y_c)$, we can finally obtain the solution of the complete system around the limit cycle as

$$
\begin{aligned}
x(t) = y_c &+ \sum_{i}^{4} K_i \phi_i E^{\lambda_i t} + \sum_{j} \omega^{(j)} E^{\sqrt{-1} j (\nu_2 t + \sigma_{20})} \\
&+ \sum_{i,j} \Omega_j K_i \phi_i E^{(\lambda_i t + j \sqrt{-1}(\nu_2 t + \sigma_{20}))}
\end{aligned}
\tag{11}
$$

where $\nu_2 = \sigma_2(y = y_c)$ and $\sigma_{20}$ is the initial value of the synodic angle.

## 4. Results

Equation (11) finally gives the evolution of the system variables as a function of time. The value of $y_c(m_1, C)$ is determined as zeros of (3) and the initial

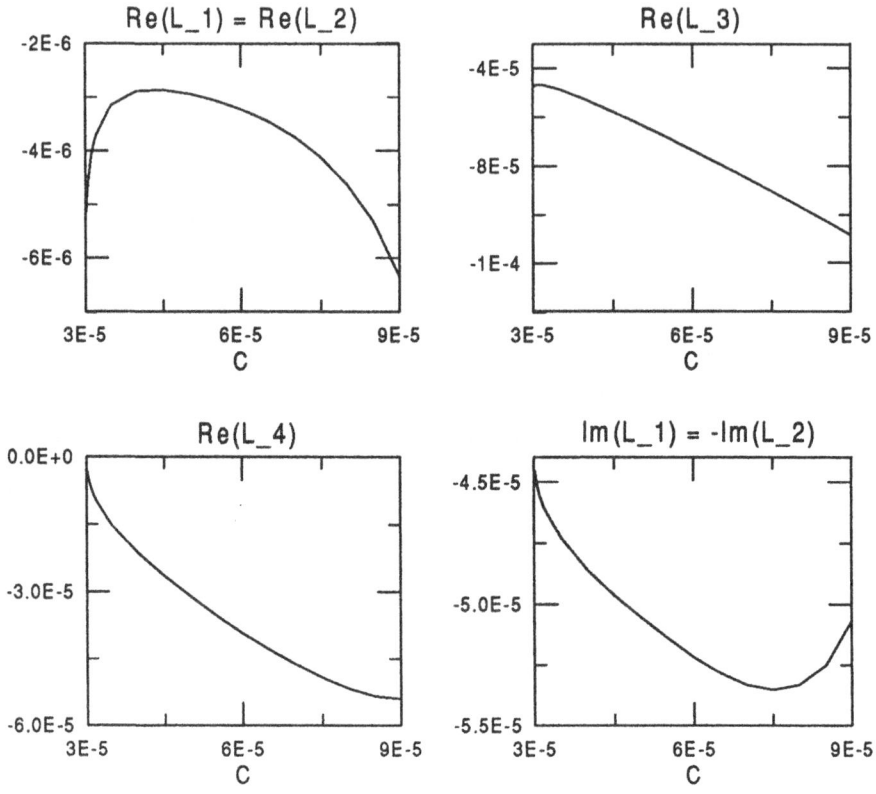

Fig. 2. Real and imaginary parts of the eigenvalues of the variational equations for the 2/3 resonance, as a function of the drag coeffcient $C$.

conditions enter through coefficients $K$. The next step is to compare this model with the numerical results.

We begin with a direct comparison of $x(t)$ with that obtained from numerical simulations. In Figure 1 we present the semimajor axis as a function of time for a body initially located at $a = 7$ U.A. On the top graph we show the behaviour of the mean variable, as determined by the averaged solution (8). We notice the damped oscillation and the final corotational value for large values of $t$. Since the eigenvalues $\lambda_i$ are such that $\Im(\lambda_1) = -\Im(\lambda_2)$, and $\lambda_3$ and $\lambda_4$ contain no imaginary parts (i.e. $\Im(\lambda_3) = \Im(\lambda_4) = 0$), the frequency of oscillation is identical for all the orbital elements. The middle plot in Figure 1 now shows the temporal evolution of the osculating semimajor axis, as determined through (11). This can now be compared with the numerical results, seen in the bottom graph in Figure 1. The agreement between them is very good.

Having tested the analytical solution, we can concentrate our study on the eigen-

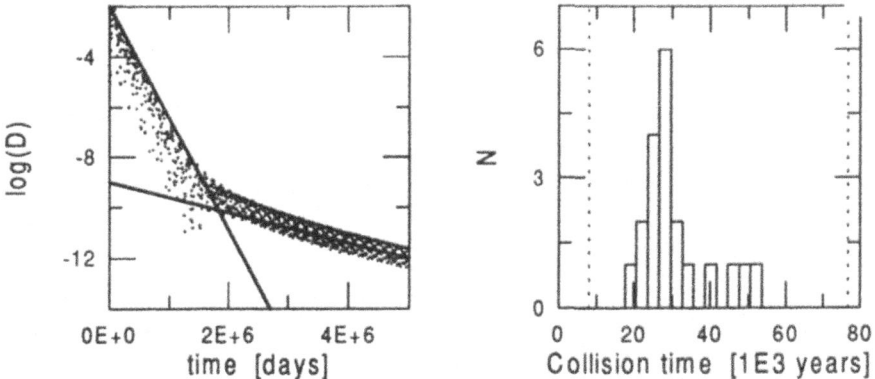

Fig. 3. (a). Logarithm of $\Delta x$ as function of time. Thick lines show linear decrease according to $\lambda_1$ (small $t$) and $\lambda_4$ (large $t$). (b). Distribution of collision times $T_{col}$ for 20 particles with different initial conditions. $C = 3.02 \times 10^{-5}$ in both cases.

values and their relationship with the convergence rate of nearby orbits. Figure 2 shows the variation of $\lambda_i$ as a function of the drag coefficient $C$ for the 2/3 resonance. As was mentioned in the previous paragraph, the first two values are complex conjugates and contain the only non-zero imaginary parts. Thus, any solution of the linearized system will display a damped oscillation with a single fundamental frequency in all coordinates. This frequency will be given by $\Im(\lambda_1)$ which, in principle, has no relationship at all with the inverse of the synodic period. Hence, the final limit cycle will only become evident for $t >> -1/\Re(\lambda_1)$. Second, we can see that $\lambda_4 \to 0^-$ as $C \to C^* \simeq 3.2 \times 10^{-5}$. This critical value of the drag coefficient marks the end of the corotational regime and the beginning of the libration zone. All capture with $C < C^*$ will thus evolve asymtotically to periodic motions in the averaged system, and not to point attractors.

Next, we can focus on the orbital convergence itself, as described in Gomes and Mothe-Diniz (1998). Although exponential convergence of nearby orbits is expected from the stability condition of the equilibrium solutions, the present model can give us additional information, such as the expected accretional timescale $T_{col}$ for a given population of planetesimals and its dependence on the drag coefficient or with the initial conditions. Let us then suppose two particles of equal radius $r$ (and thus equal drag coefficient $C$) and on initial conditions $x(t = 0)$ and $x'(t = 0)$ chosen such that both are captured in the same corotational solution. Then, from equations (8)-(11), and once we have eliminated all periodic modulations, we can estimate the vectorial difference between them as

$$x(t) - x'(t) \equiv \Delta x(t) \propto \sum_i^4 \Lambda_i E^{\lambda_i t} \tag{12}$$

where the coefficients $\Lambda_i$ are function of the initial conditions of both particles.

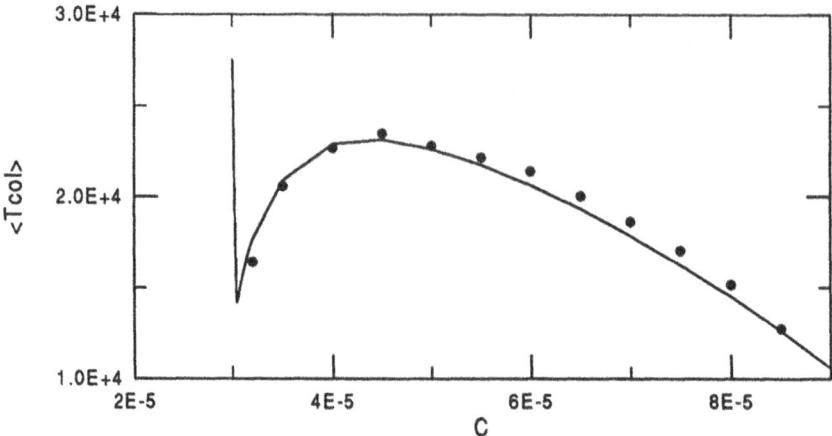

Fig. 4. Real and imaginary parts of the eigenvalues of the variational equations for the 2/3 resonance, as a function of the drag coefficient $C$.

From this expression we can see that the convergence is proportional to four exponentials, each weighted by coefficients that depend on the initial conditions. Now, even though for small timescales the different terms may be of similar magnitude, for extremely large values of $t$ the temporal evolution of $\Delta x(t)$ will tend asymtotically to an exponential proportional to the smallest eigenvalue. In other words, $\Delta x(t) \propto E^{\min(\lambda_i)t}$ (see Fig. 3a).

Thus it seems that, independently of the initial conditions, the convergence rate is given by the lowest eigenvalue of the system. However, this is valid for $t \to \infty$ (which implies $\Delta x \to 0$) but not necessarily true for finite values of $t$ corresponding to a collision between both bodies (i.e. $\Delta x = 2r$). If $T_{col}$ is sufficiently small, the convergence rate will no longer be unique but function of the initial separations. In order to see this in more detail, we once again make use of equation (12). With it, we can simulate a variation of initial conditions via modifications of the numerical values of $\Lambda_i$, and determine collision times in each case.

This calculation was performed for 20 different sets of coefficients and the resulting distribution of $T_{col}$ is shown in Figure 3b. In broken vertical lines we also show the theoretical times if only $\lambda_4$ (left line) and $\lambda_1$ (right line) where present. $\lambda_3$ is too large and can be ignored in the model. We can see that the real values lie between both limits and a certain tendency is noted towards faster accretion. Nevertheless, the dispersion is very significant, covering more than half an order of magnitude. From a cosmogonical point of view this result may be of importance, since it means that even though a swarm of equal-massed particles will in fact converge towards the same corotation point, there is no unique timescale for the accretion process.

Notwithstanding this last fact, it is possible to determine a mean collision time

$\langle T_{col} \rangle$, defined as the average over the different $\lambda_i$ (see (12)). Even though individual times may vary, this quantity can be thought as a characteristic timescale of the system for the given drag coefficient. The result of this calculation, as a function of $C$, is shown as a continuous line in Figure 4. Notice the steep increase in collision times for $C$ close to the librational limit ($T_{col}$ at this point). For comparison with numerical results of the exact equations, for each value of $C$ we took 20 different initial conditions and determined directly $\langle T_{col} \rangle$. These were then plotted in full circles. We can see that the agreement with the analytical results is very good.

## Acknowledgements

This work is part of collaboration initiated during C. Beaugé's one month visit to Namur. The authors are grateful to the FUNDP for the partial financing of the stay. Additional support from CONICET, CONICOR and Secyt/UNC is also greatly appreciated.

## References

Adachi, I., Hayashi, C., and Nakazawa, K.: 1976, *Prog. Theor. Phys.*, **56**, 1756-1771.
Beaugé, C., Lemaître, A. and Jancart, S.: 1998, *Planet. Space Science*, in press.
Gomes, R. and Mothe-Diniz, T.: 1998, in preparation.
Hale, J.K.: 1969), *Ordinary Differential Equations*, Wiley-Interscience, J. Wiley & Sons, New York.
Henrard, J.: 1970, *Celest. Mech.*, **3**, 107-120.
Kamel, A.A:. 1969, *Celest. Mech.*, **1**, 190-199.
Poincaré, H.: 1885, *J. de Math. pures appl.*, ser. 4, 1, 167. Reprinted in *Oeuvres de Henri Poincaré*, I, 90, Gauthier-Villars, Paris, (1928).

# MIGRATION OF TRANS–NEPTUNIAN OBJECTS TO THE EARTH

SERGEI I. IPATOV

*Institute of Applied Mathematics, Miusskaya sq. 4,
Moscow 125047, Russia; E-mail: ipatov@spp.keldysh.ru*

**Abstract.** Migration of trans–Neptunian objects under their mutual gravitation influence and the influence of the giant planets is investigated. These investigations are based on computer simulation results and on some formulas. We estimated that about 20 % of near–Earth objects with diameter $d \geq 1$ km may have come from the Edgeworth–Kuiper belt.

## 1. Introduction

The first object of the Edgeworth–Kuiper belt (EKB) was found in 1992, and now orbits of about 90 trans–Neptunian objects (TNOs) are known. The object 1996 $TL_{66}$ with diameter $d \approx 500$ km moves around the Sun at a larger distance than other TNOs. It's orbital elements are $a = 85$ AU, $e = 0.6$, and $i = 24°$. Luu *et al.* (1997) considered that there are about $10^4$ such objects between 40 and 200 AU with a total mass of $0.5 m_\oplus$, where $m_\oplus$ is the mass of the Earth. Semimajor axes of other observed TNOs lie between 35 and 49 AU. Their diameters are between 100 and 400 km, and their characteristic mass is about $5 \cdot 10^{-12} M_\odot$, where $M_\odot$ is the mass of the Sun. Jewitt *et al.* (1996) and Levison and Duncan (1997) supposed that the inferred number of such objects is 70,000. Jewitt *et al.* (1996) estimated the total mass of the present–day belt in the 30–50 AU region to be about $(0.06 - 0.25) m_\oplus$ at $a \sim 30 - 50$ AU. As Pluto, approximately 35 % of the known TNOs reside in the 2:3 resonance with Neptune, but Jewitt *et al.* (1998) concluded that this fraction is overrepresented as a result of observational selection and only $10 - 20$ % of TNOs in the 30–50 AU region are Plutinos. Jewitt *et al.* (1996) assumed that for the 100–400 km diameter objects the differential size distribution obey $n(r)dr = \Gamma r^{-q} dr$, with $\Gamma$=const and $q = 3$. The number of bodies with $d > 1$ km is predicted to be about $10^{10}$ for $30 \leq a \leq 50$ AU. Weismann (1995) supposed that the dynamically inactive region beyond 45 AU may extend out to 1000 AU or more and may contain up to several times $10^{13}$ objects with the total mass of several hundred Earth masses.

The asteroid and Edgeworth–Kuiper belts are considered to be the main sources of near–Earth objects (NEOs). Collisions of some NEOs (such as the Tunguska object) with the Earth can destroy a large city. The comet origin of some NEOs and meteorites was suggested by Öpik (1963). According to Wetherill (1988, 1989) and Weissman *et al.* (1989), it is difficult to explain the number of objects of the Apollo and Amor groups and features of their orbits (for example, their mean inclinations, which are larger than those in the main asteroid belt), if one considers only asteroidal sources. Wetherill (1991) considered that NEOs should come from the EKB, but not from the Oort cloud in order to supply present inclinations of orbits of NEOs. Fernandez (1980) first supposed that the trans–Neptunian belt is a source of short–period comets.

Many papers devoted to the EKB were published last years (see a review by Weissman (1995) and a later paper by Levison and Duncan (1997)). Stern (1995, 1996a,b), Davis and Farinella (1997) considered the collisional evolution of TNOs. It was obtained that the population of bodies with $d > 100$ km did not change much during the age of the Solar System. Most of bodies with diameter of several kilometers are debris of larger bodies. Davis and Farinella (1997) considered that collisions of TNOs can provide a significant influx of Chiron–sized bodies.

Migration of various bodies under the gravitational influence of planets was considered by Duncan *et al.* (1988), Hahn and Bailey (1990), Levison and Duncan (1994), and many other authors. Initial orbits of most of these bodies crossed the orbits of the giant planets. The median dynamical lifetime of short–period comets was obtained to be equal to 0.45 Myr and that for Chiron–type objects is about 1.1–1.4 Myr.

The main aim of the present paper is to appreciate the number of TNOs migrating to the Earth.

## 2. Interactions of Trans–Neptunian Objects

Ipatov (1980, 1988, 1995a,b, 1997b, 1999a) considered variations in $a$ and $e$ of bodies due to the gravitational influence of the large ($d > 100$ km) TNOs using new formulas for characteristic times elapsed up to close encounters. These formulas are different from those for the Öpik's approach used by other scientists. The case of variable inclinations was also considered. The evolution of three gravitating bodies with masses equal to that of Pluto was investigated by Ipatov (1980) with the use of a spheres' method and by Ipatov (1988) by numerical integration. The sphere of action (i.e., the Tisserand sphere) of radius $r_{sa} \approx R(m/M_\odot)^{2/5}$ was used, where $R$ is a distance from the Sun, $m$ is the mass of a TNO. Ipatov (1995a,b, 1997b, 1999a) made some appraisals of the evolution of a disk of TNOs. Basing on this appraisals, below we make some conclusions about interactions and the origin of TNOs.

Fernandez (1980), Ip and Fernandez (1997) in their investigations of mutual gravitational influence of TNOs used spheres smaller than $r_{sa}$ and so decreased the role of mutual gravitational influence. Fernandez (1980) investigated the variations in perihelion distance and made his calculations when masses of TNOs were unknown. Ip and Fernandez (1997) considered the gravitational influence of TNOs only on the orbital evolution of bodies located in the 2:3 resonance with Neptune.

We obtained that the characteristic time elapsed up to a close encounter (up to $r_{sa}$) for two objects with masses $m \sim 5 \cdot 10^{-12} M_\odot$ is equal to $7 \cdot 10^{10}$ yr. If there are about $7 \cdot 10^4$ objects with diameter $d \geq 100$ km in the belt, then during the age $T_{ss}$ of the Solar System a body with a semimajor axis $a \approx 40$ AU takes part in about 3000 close encounters with TNOs with $d \geq 100$ km. The probability of that during $T_{ss}$ an object with $d \geq 100$ km collided with one of objects of the same size equals to 0.005. If the number of objects with $d \geq 10$ km is greater than that with $d \geq 100$ km by a factor of 100, then a probability of a collision during $T_{ss}$ of an object with $d \geq 100$ km with some object with $d \geq 10$ km equals to 0.15. As the average eccentricity of TNOs is about 0.1, variations in $a$ for smaller colliding bodies can exceed several AU. Probably, less than 1 % of 100–km bodies had such large variations in $a$ due to collisions during $T_{ss}$.

Our numerical estimates showed that the mean variation in $a$ at one close encounter is $\delta a \sim (1 - 3) \cdot 10^{-5} a$ for $m = 5 \cdot 10^{-12} M_\odot$ and mean eccentricities and inclinations of the observed TNOs. The variation in $a$ of an object located in the middle of the EKB probably is proportional to $\delta a \sqrt{N}$ (where $N$ is the number of close encounters of this object with large TNOs) and usually does not exceed 0.1 AU during $T_{ss}$. For many objects at the inner part of the belt (at $a < 39$ AU), $a$ could vary more monotonously and could decrease by more than 1 AU during $T_{ss}$. Such decrease of $a$ may be a reason of that a region ($36 \leq a \leq 39$ AU) with small values of $e$ and $i$ is unpopulated, though, as it was shown by Duncan et al. (1995), it is dynamically stable under the gravitational influence of planets for $T_{ss}$. At some very close encounters, semimajor axes of TNOs could change by several AU. Several percents of TNOs could take part in such very close encounters during $T_{ss}$. The number of such close encounters is, by an order of magnitude, larger than the corresponding number of collisions, and so the role of gravitational interactions in variations of orbital elements is larger than that of collisions. On average, during $T_{ss}$ a TNO had several close (up to $r_s$) encounters with Pluto and changed its $a$ by $\sim 0.1$ AU at these encounters.

Due to the mutual gravitational influence of TNOs, their mean eccentricity and mean inclination could not reach their present values at the present–day mass of the EKB, but could reach them only for a total mass of the belt of about several $m_\oplus$ (Ipatov, 1995a,b). The results of numerical simulation of the process of planetary accretion (Ipatov, 1987, 1993) showed that the total mass of planetesimals that entered the trans–Neptunian region from the feeding zone of the giant planets during their accumulation could exceed tens $m_\oplus$. These planetesimals could increase initially small eccentricities of 'local' TNOs formed in the EKB, and swept most of the local TNOs, which total initial mass $M_\Sigma$ could exceed $10 m_\oplus$. A small part of such planetesimals could left beyond the orbit of Neptune in eccentric orbits (such as that of 1996 TL$_{66}$).

Eneev (1980) supposed that large TNOs were formed from rarefied ice–gas condensations and smaller objects are their debris. Stern (1996a), Stern and Colwell (1997) simulated the formation of 100–1000 km objects from the 1–10 km planetesimals. They obtained that such objects could be formed only for $M_\Sigma > 10 m_\oplus$ and $e < 0.002$. Kenyon and Luu (1998) obtained runaway growth in 100 Myr for $e = 0.001$ and in 700–2000 Myr for $e = 0.01$ at $M_\Sigma \geq 10 m_\oplus$. These times are greater than the time of formation of the massive Jupiter, which does not exceed several tens of Myr. We consider that such small eccentricities could not exist during a large time span, both due to the mutual gravitational influence of TNOs and due to the gravitational influence of the forming giant planets. Therefore, probably, most of large ($d \geq 100$ km) local TNOs were formed mainly by the compression of condensations but not by the accumulation of smaller planetesimals. Some smaller objects could be formed directly from condensations and a considerable part of them are debris of larger objects.

## 3. Migration of Trans–Neptunian Objects under the Gravitational Influence of the Giant Planets

Duncan et al. (1995) found regions of the values of $a$ and $e$, for which TNOs can migrate to the orbit of Neptune under the gravitational influence of planets. Gladman and Duncan (1990), Torbett and Smoluchowski (1990) found that due to gravitational influence of the giant planets the orbits of bodies of the inner part of the EKB could begin to cross the orbit of Neptune. Holman and Wisdom (1993), Levison and Duncan (1993), and Duncan et al. (1995) investigated times survived by test TNOs before they became Neptune–crossers. They finished their calculations of the evolution of test TNOs when their orbits began to cross the orbit of Neptune or the body entered inside the Hill sphere of Neptune. Levison and Duncan (1997) considered the motion of TNOs to the Jupiter's orbit. Morbidelli et al. (1995) investigated evolution of some resonant TNOs that did not get deep inside the Solar System.

Ipatov (1997a) considered migration of objects from the EKB under the gravitational influence of the giant planets not only to the orbits of Neptune and Jupiter, as other authors, but also further to the orbit of the Earth. Below we present more wide results of these runs. The gravitational influence of the giant planets was taken into account with the use of the RMVS2 program of symplectic method from the Swift integration package worked out by Levison and Duncan (1994). The time step was the same as in their test, i.e., equaled to 1 yr. Initial orbits of planets also were taken from the test. This integrator is by an order of magnitude faster than previous methods of integration. We considered various initial eccentricities $e_o$, inclinations $i_o$, and orbital orientations of orbits. Initial values $a_o$ of semimajor axis were varied from 35 to 50 AU. Orbital evolution of one hundred bodies was considered. The considered time span $T$ usually equaled to 20 Myr. In some runs $T$ reached $100 - 150$ Myr.

In the case without close encounters for some typical orbits, we compared results obtained with the use of the RMVS2 integrator with those obtained with the integrator by Bulirsh and Stoer (1966). The plots of variations in orbital elements with time are presented by Ipatov (1999b). It was shown that limits of variations in $a$ during 1 Myr differed by less than 5 %, and differences in $e$ and $i$ were smaller. A more detailed comparison of the results obtained by these integrators is presented below in section 4.

We found that some bodies of the EKB can migrate deep inside the Solar System. The minimal perihelion distance of 2, 8, and 10 considered migrating bodies was in the regions [6, 9.5], (9.5, 20), and [20, 30) AU, respectively. Four

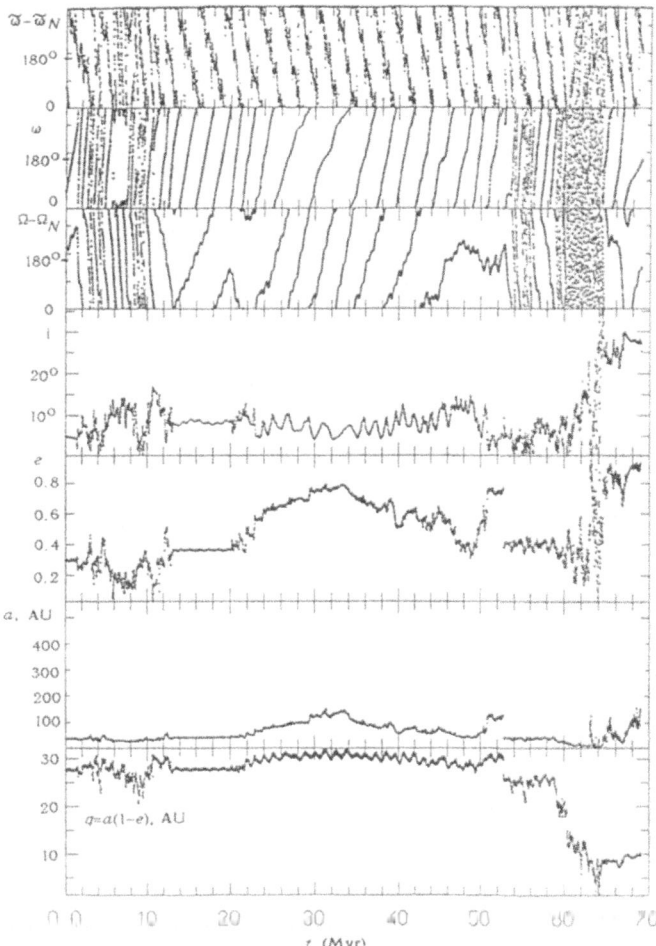

Fig. 1. Time variations in the semimajor axis $a$, eccentricity $e$, perihelion distance $q = a(1 - e)$, inclination $i$, the difference between the longitudes of the ascending node for a body and Neptune, $\Delta\Omega = \Omega - \Omega_N$, the argument of perihelia, $\omega$, and the difference between the longitudes of perihelia for a body and Neptune, $\Delta\tilde{\omega} = \tilde{\omega} - \tilde{\omega}_N$. Initial data: $a_0 = 40$ AU, $e_0 = 0.15, i_0 = 5°, \Omega_0 = \omega_0 = M_0 = 60°$.

bodies reached the Jupiter's orbit during evolution and two of them almost reached the Earth's orbit. At $i_0 = 5°$ and initial values of the longitude of ascending node, the argument of perihelion, and the mean anomaly equal to $\Omega_0 = \omega_0 = M_0 = 60°$, for $a_0 = 40$ AU and $e_0 = 0.15$ and for $a_0 = 39.3$ AU and $e_0 = 0.3$, the perihelion distance $q$ decreased from 34 and 27.5 AU to 1.25 and 1.34 AU in 25 and 64 Myr, and these bodies were ejected into hyperbolic orbits in 30 and 70 Myr, respectively (Figs. 1 and 2). Maximum inclinations $i_{max}$ in these runs were equaled to 57° and

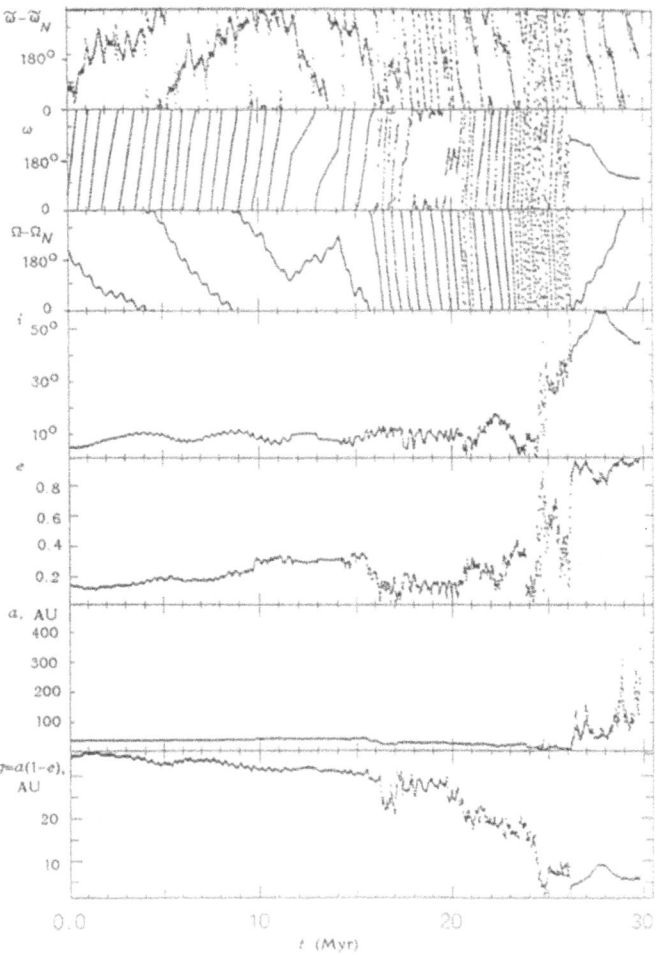

Fig. 2. Same as figure 1; initial data: $a_0 = 39.3$ AU, $e_0 = 0.3$, $i_0 = 5°$, $\Omega_0 = \omega_0 = M_0 = 60°$.

$36°$, respectively, and there were large variations in $e$ and $i$ when $q$ was close to minimum. As orbital elements were sampled with a time step $\Delta t = 20{,}000$ yr, the actual minimal values of $q$ can be smaller than those presented above. The time interval, during which $q$ decreased from 10 to 1.3 AU, equaled to 0.3–0.5 Myr, and that for $q$ decreased from 5 AU to 1.3 AU was considerably smaller. Plots for some other initial data are presented by Ipatov (1999b). For one run at $a_0 = 39.3$ AU, $e_0 = 0.15$, and $i_0 = 5°$, the time interval, during which $q$ decreased from 30 AU to 3 AU, was less than 2 Myr.

For some runs $q$ exceeded 10 AU during all time span before the ejection of a body into a hyperbolic orbit. For example, at $a_o = 39.3$ AU, $e_o = 0.15$, and $i_o = 5°$, a test body was ejected into a hyperbolic orbit after 41.5 Myr, and $q > 15$ AU and $i \leq 33°$ during this time span. For $a_o = 40$ AU, $e_o = 0.05$, and $i_o = 5°$, a test body was in an elliptical orbit with $q > 10$ AU during the considered time span $T = 100$ Myr, though the variations in orbital elements were large and maximum eccentricity $e_{max}$ exceeded 0.8. For $a_o = 38.9$ AU, $e_o = 0.15$, $i_o = 5°$, and $T = 150$ Myr, we have $q \geq 15$ AU, $e_{max} = 0.76$, and $i_{max} = 27°$. During the first 30 Myr, variations in $a$ were small in this run. For larger values of $e_o$, the portion of bodies that cross the orbit of Neptune during evolution is larger. The time $T_h$ elapsed until the ejection of a body into a hyperbolic orbit is smaller, as a rule, for smaller values of $i_o$. For example, at $a_o = 39.3$ AU and $e_o = 0.3$, it was obtained that $T_h \approx 70$ Myr for $i_o = 5°$ and $T_h \approx 7$ Myr for $i_o = 0$.

Limits of variations in $a$, $e$, and $i$ may differ considerably for runs with the same values of $a_o$, $e_o$, and $i_o$, but with different values of $\Omega_o, \omega_o$, and $M_o$. Such influence of initial orbital orientations was obtained both for resonant and nonresonant values of $a_o$. For example, at $a_o = 39.3$ AU, $e_o = 0.15$, and $i_o = 5°$, we investigated evolution of 12 orbits with various values of $\Omega_o, \omega_o$, and $M_o$ during $T = 20$ Myr. All these objects moved in the 2:3 resonance with Neptune. It was obtained that $e_{max} < 0.2$ for 5 runs and $e_{max} > 0.3$ for 3 runs. For nonresonant value $a_o = 40$ AU at $e_o = 0.15$, $i_o = 5°$, $\omega_o = M_o = 60°$ and the same time span, we obtained $e_{max} \approx e_o = 0.15$ at $\Omega_o = 0$ and $e_{max} = 0.36$ at $\Omega_o = 60°$. For $\Omega_o = 60°$, the object was in the $\nu_8$ resonance at $10 \leq t \leq 12$ Myr and in the Kozai resonance at $18 \leq t \leq 20$ Myr. For $a_o = 42$ AU, $e_o \doteq 0.05$, $i_o = 5°$ and $T = 20$ Myr, we have $e_{max} = 0.07$, $i_{max} = 5.7°$, and the minimum value of $q$ is equal to $q_{min} = 39$ AU at $\Omega_o = \omega_o = M_o = 60°$, and $e_{max} = 0.33$, $i_{max} = 9.6°$, and $q_{min} = 31$ AU at $\Omega_o = \omega_o = M_o = 0$. For the last run at $t < 10$ Myr, the object was in the $\nu_{18}$ resonance ($\Omega-\Omega_N$ librated around 180°, where $\Omega_N$ is the value for Neptune), $\tilde{\omega}-\tilde{\omega}_N$ ($\tilde{\omega} = \Omega+\omega$) varied between 120° and 370°, and $e$ grew monotonously from 0.05 to 0.2. The above examples show that limits of variations in orbital elements of TNOs can highly depend on their initial orientations of orbits. Therefore, small variations in orbital elements due to collisions and mutual gravitational influence of TNOs can cause large variations in orbits under the gravitational influence of the giant planets.

For all considered runs and $T = 20$ Myr, the variations in $a$ and maximum values of $e$ and $i$ during evolution exceeded 0.6 AU, 0.05 and 3°, respectively. Therefore, the values of $e$ and $i$ of TNOs can not be small for a long time.

Several types of variations in orbital elements for the 2:3 resonance with Neptune were considered by Ipatov and Henrard (1999). Ipatov (1995b, 1996) investigated migration of small bodies under the gravitational influence of all planets with the use of the method of spheres of action. Hundreds of bodies were considered in these runs, and it was obtained that individual bodies decreased their aphelion distances $Q$ from the trans-Neptunian zone to the values even less than 1 AU. The first object (1998 DK$_{36}$) with $Q \leq 1$ AU was found in 1998. For 1998 XX$_2$ and

1999 $AO_{10}$, $Q \approx 1.013$ AU.

## 4. The Number of Trans–Neptunian Objects Reaching the Orbit of the Earth

The estimates of the number of TNOs migrating to the Earth can be made on the basis of simple formulas and results of numerical integration. The number of TNOs, which reached the Jupiter's orbit during the considered time span $T$, equals to $N_J = N \cdot P_N \cdot p_{JN}$, where $N$ is the number of objects in the belt, $P_N$ is the portion of TNOs migrating to the Neptune's orbit and leaving the belt during this time $T$, and $p_{JN}$ is the portion of Neptune–crossing objects leaving the belt, which reaches the Jupiter's orbit during their lifetimes. The number of TNOs, which are now Jupiter–crossers, equals to $N_{Jn} = N_J \Delta t_J / T$, where $\Delta t_J$ is the average time, during which the object crosses the Jupiter's orbit. Our results of numerical simulations showed that $\Delta t_J \approx 0.2$ Myr.

It is considered that there are about $10^{10}$ objects with $d > 1$ km and $30 \leq a \leq 50$ AU. Duncan et al. (1995) obtained that the portion $P_N$ of TNOs that left the EKB during $4 \cdot 10^9$ yr under the influence of the giant planets is about 0.1–0.2. As mutual gravitational influence of TNOs also takes place, we take this portion equal to 0.2. The portion of Neptune–crossing objects, which reach the Jupiter's orbit during their lifetimes, was obtained by Duncan et al. (1995) to be equal to 0.34. So we have $N_J = 6.8 \cdot 10^8$ for $T = 4 \cdot 10^9$ yr and $N = 10^{10}$, and also $N_{Jn} = 3.4 \cdot 10^4$ at $\Delta t_J = 0.2$ Myr.

The number of TNOs, which reached the orbit of the Earth during the considered time $T$, equals to $N_E = N_J \cdot p_{JE} = N \cdot P_N \cdot p_{JN} \cdot p_{JE}$ , where $p_{JE}$ is the portion of Jupiter–crossing objects that reach the orbit of the Earth during their lifetimes. The ratio of the number $N_{EN}$ of Earth–crossing objects (ECOs), which came from the EKB, to the total number $N_{ECO}$ of ECOs at the considered time moment equals to $P_{NE} = N_{EN}/N_{ECO} = N \cdot P_N \cdot p_{JN} \cdot p_{JE} \cdot \Delta t_E / (T \cdot N_{ECO})$, where $N_{NE} = N_E \cdot \Delta t_E / T$, $\Delta t_E$ is the mean time, during which a Jupiter–crossing object crosses the orbit of the Earth. The number of collisions of TNOs with the Earth during time $T$ equals to $N_{col} = N_E \cdot \Delta t_E / T_E$, where $T_E$ is a characteristic time elapsed up to a collision of an ECO with the Earth.

Ipatov and Hahn (1997, 1999a, 1999b) investigated the evolution of orbits close to the orbit of the object P/1996 R2 (Lagerkvist) under the gravitational influence of all planets. This object has a Jupiter–crossing orbit ($a \approx 3.79$ AU, $e \approx 0.31$, $i \approx 2.6°$). Results of investigations with the use of the RMVS3 integrator (which is the modification of the RMVS2 integrator) from the Swift package by Levison and Duncan (1994) for initial integration step equal to 30 days were close on the whole to those obtained with the BULSTO integrator by Bulirsh and Stoer (1966). Median lifetimes (before ejections into hyperbolic orbits) in series of 12 runs were equal to several hundreds of thousands years. Using the BULSTO integrator, we obtained that the minimal distance from the Sun $q_{min} < 1$ AU for 8 among 24 runs (i.e., for 33 % of runs) for integration into the future and for 2 among 24 runs for integration into the past (in total, this inequality is fulfiled for 10 among 48 runs, i.e., for about 20 %). Using the RMVS3 integrator, we obtained $q_{min} < 1$ AU for 3 among 24 runs and for 1 among 24 runs for integration into the future and into the past, respectively. In our runs Jupiter–crossing bodies reached the orbit of the Earth more rarely for the RMVS3 integrator. It is mainly caused by the main difference between the results obtained by the RMVS3 and BULSTO integrators, which is that test bodies more often moved in resonances with planets, when we used the BULSTO integrator. So actual migration of TNOs to the Earth can be even larger than that obtained with the use of the RMVS2 integrator and presented in Section 3.

Basing on the above runs, we estimated $p_{JE} = 0.2$ and $\Delta t_E \approx 5000$ yr. According to (Ipatov, 1995b), $T_E \approx 75$ Myr, and so the probability of a collision of an ECO with the Earth during 5000 yr was obtained to be about $7 \cdot 10^{-5}$. It is considered that there are about 500–1000 ECOs with $d > 1$ km. For $d \geq 1$ km, $N_{ECO} = 750$, and $T = 4 \cdot 10^9$ yr, we have $P_{NE} \approx 0.2$, $N_E \approx 1.4 \cdot 10^8$, $N_{NE} \approx 170$, and $N_{col} \approx 10^4$. For $d \geq 100$ km and $N = 7 \cdot 10^4$, the probability of a collision with the Earth during $T = 4 \cdot 10^9$ yr is about 0.05. The above estimates are very approximate, but they show that the number of TNOs hitting the Earth is not small.

As it is more easy to destruct icy bodies than stone or metal bodies (for example, bodies can break up at close encounters with planets on their way to the Earth), the portion of TNOs among ECOs for bodies with $d \leq 100$ m (for example, for Tunguska–type bodies) may be greater than that for 1 km bodies and can exceed 20 %, but small icy bodies disappear in the atmosphere and can not reach the surface of the Earth. The above estimates were obtained for bodies, which migrated from distances 30–50 AU from the Sun. Bodies with $a > 50$ AU moving in highly eccentric orbits also can migrate to the Earth.

## 5. Conclusion

Probably, TNOs with $d > 100$ km moving in low eccentrical orbits were formed directly by the compression of large rarefied condensations, but not by the accretion of smaller solid planetesimals. Small variations in orbital elements of TNOs due to their mutual gravitational influence can cause large orbital variations due to gravitational influence of the giant planets. About 20 % of NEOs with $d \geq 1$ km may have come from the EKB.

### Acknowledgements

This work was supported by the Russian Foundation for Basic Research, project no 96-02-17892, and by the Russian federal scientific and technical program Astronomy. Computer simulations of the orbital evolution of trans–Neptunian objects under the gravitational influence of the giant planets were made in 1995 during the visit to FUNDP (Namur, Belgium) supported by the ESO grant no. B-06-018.

# References

Bulirsh, R. and Stoer, J.: 1966, *Numer. Math.*, **8**, 1–13.

Davis, D.R. and Farinella, P.: 1997, *Icarus*, **125**, 50–60.

Duncan, M., Quinn, T., and Tremaine, S.: 1988, **328**, L69–L73.

Duncan, M.J., Levison, H.F., and Budd, S.M.: 1995, *Astron. J.*, **110**, 3073–3081.

Eneev, T.M.: 1980, *Sov. Astron. Letters*, **6**, No 5, p. 295–300 in Russian edition.

Fernandez, J.A.:1980, *Monthly Notices Roy. Astron. Soc.*, **192**, 481–491.

Gladman, B. and Duncan, M.: 1990, *Astron. J.*, **100**, 1680–1696.

Hahn, G. and Bailey, M.E.: 1990, *Nature*, **348**, 132–136.

Holman, M.J. and Wisdom, J.: 1993, *Astron. J.*, **105**, 1987–1999.

Ip, W.–H. and Fernandez, J.A.: 1997, *Astron. Astrophys.*, **324**, 778–784.

Ipatov, S.I.: 1980, *Preprint of Inst. of Applied Mathematics*, N 43, Moscow, 33 P. (in Russian).

Ipatov, S.I.: 1987, *Earth, Moon, and Planets*, **39**, 101–128.

Ipatov, S.I.: 1988, *Kinematics Phys. Celest. Bodies*, **4**, N 6, 76–82.

Ipatov, S.I.: 1993, *Solar System Research*, **27**, 65–79.

Ipatov, S.I.: 1995a, *Solar System Research*, **29**, 9–20.

Ipatov, S.I.: 1995b, *Solar System Research*, **29**, 261–286.

Ipatov, S.I.: 1996, *Earth, Moon, and Planets*, **72**, 211–214.

Ipatov, S.I.: 1997a, *LPSC XXVIII*, 615–616.

Ipatov, S.I.: 1997b, *LPSC XXVIII*, 617–618.

Ipatov, S.I.: 1999a, 'Migration of Kuiper–belt objects inside the Solar System', *"Planetary systems — the long view"*, Proc. 9th Rencontres de Blois (June 22–28, 1997), Editions Frontieres, Gif. sur Yvette Cedex, Celnikier, L.M. and Tran Thanh Van, J., (eds.), 157–160

Ipatov, S.I.: 1999b, *Migration of celestial bodies in the solar system*, URSS Publishing Company, Moscow, in press.

Ipatov, S.I. and Hahn, G.J.: 1997, *LPSC XXVIII*, 619–620.

Ipatov, S.I. and Hahn, G.J.: 1999a, *"Planetary systems — the long view"*, Proc. 9th Rencontres de Blois (June 22–28, 1997), Editions Frontieres, Gif. sur Yvette Cedex, Celnikier, L.M. and Tran Thanh Van, J., (eds.), 179–180

Ipatov, S.I. and Hahn, G.J.: 1999b, *Solar System Research*, **33**, in press.

Ipatov, S.I. and Henrard, J.: 1999, *Solar System Research*, **33**, in press.

Jewitt, D., Luu., J., and Chen J.: 1996, *Astron. J.*, **112**, 1225–1238.

Jewitt, D., Luu., J., and Trujillo, C.: 1998, *Astron. J.*, **115**, 2125–2135.

Kenyon, S.J. and Luu, J.X.: 1998, *Astron. J.*, **115**, 2136–2160.

Levison, H.F. and Duncan, M.J.: 1993, *Astrophys. J.*, **406**, L35–L38.

Levison, H.F. and Duncan, M.J.: 1994, *Icarus*, **108**, 18–36.

Levison, H.F. and Duncan, M.J.: 1997, *Icarus*, **127**, 13–23.

Luu, J., Marsden, B.G., Jewitt, D., Trujillo, C.A., Hergenrother, C.W., Chen, J., and Offutt, W.B.: 1997, *Nature*, **387**, 573–575.

Morbidelli, A., Thomas, F., and Moons, M.: 1995, *Icarus*, **118**, 322–340.

Öpik, E.J.: 1963, *Adv. Astron. Astrophys.*, **2**, 219–262.

Stern, S.A.: 1995, *Astron. J.*, **110**, 856–868.

Stern, S.A.: 1996a, *Astron. J.*, **112**, 1203–1211.

Stern, S.A.: 1996b, *Astron. Astrophys.*, **310**, 999–1010.

Stern, S.A. and Colwell, J.E.: 1997, *Astron. J.*, **114**, 841–849.

Torbett, M. and Smoluchowski, R.: 1990, *Nature*, **44**, 722–729.

Weissman, P.R.: 1995, *Annu. Rev. Astron. Astrophys.*, **33**, 327–357.

Weissman, P.R., A'Hearn, M.F., McFadden, L.A. and Rickman, H.: 1989, *Asteroids II*, Tucson: Univ. Arizona Press, Binzel, R.P., Gehrels, T. and Matthews, M.S. (eds.), 880–919.

Wetherill, G.W.: 1988, *Icarus*, **76**, 1–18.

Wetherill, G.W.: 1989, *Meteoritics*, **24**, 15–22.

Wetherill, G.W.: 1991, *Proc. of the IAU Colloquium 121 "Comets in the post–Halley era"*, Amsterdam: Kluwer Acad. Publ., Newburn, R.L., Rahe, J. and Neugebauer, M. (eds.), 537–556.

# TROJANS IN STABLE CHAOTIC MOTION

E. PILAT-LOHINGER, R. DVORAK AND CH. BURGER

*Institut für Astronomie, Universität Wien,*
*Türkenschanzstraße 17, A-1180 Vienna, Austria*

**Abstract.** The orbits of 13 Trojan asteroids have been calculated numerically in the model of the outer solar system for a time interval of 100 million years. For these asteroids Milani et al. (1997) determined Lyapunov times less than 100 000 years and introduced the notion "asteroids in stable chaotic motion". We studied the dynamical behavior of these Trojan asteroids (except the asteroid Thersites which escaped after 26 million years) within 11 time intervals – i.e. subintervals of the whole time – by means of: (1) a numerical frequency analysis (2) the root mean square (r.m.s.) of the orbital elements and (3) the proper elements. For each time interval we compared the root mean squares of the orbital elements (a, e and i) with the corresponding proper element. It turned out that the variations of the proper elements $e_p$ in the different time intervals are correlated with the corresponding r.m.s.(e); this is not the case for $\sin I_p$ with r.m.s.(i).

**Key words:** Trojan asteroids – long term evolution – proper elements

## 1. Introduction

The Trojan asteroids librating in the vicinity of Jupiter's $L_4$ or $L_5$ point are well known examples for the Lagrangian solutions of the three-body problem. The interest in this kind of motion began only in 1906 when Max Wolf discovered the first asteroid – (588) Achilles – moving near $L_4$ of Jupiter's orbit. More than 400 Trojan asteroids[1] are known up to now, whereof, 246 have been found in the vicinity of $L_4$ and only 167 near $L_5$. This difference in the number of Trojan asteroids around the two equilibrium points is possibly caused by long term perturbations of Saturn as it was shown by a numerical study of hypothetical Trojans (cf. Barber, 1986). Actually it was believed that these asteroids are not very numerous but nowadays several thousands are thought to exist with diameters $\geq 15km$ (cf. Shoemaker et al., 1989).

To study the problem of the Trojans theoretically with the aid of perturbation theory one has to take into account that the inclinations of these asteroids can reach large values (up to about 40°); this makes the problem different to the study of the main belt asteroids which are mostly confined to low inclined orbits. Using simplified models several theoretical studies of the motion of the Trojans have been carried out; the most complete one has been developed by Érdi (1984, 1988, 1996, 1997). Several numerical investigations have been carried out by Schubart & Bien (1984 and 1987) and by Bien & Schubart (1984 and 1987). Milani (1993, 1994) computed the orbits of 174 Trojans in the model of the outer solar system for $10^6$ years and in some cases for $5 \cdot 10^6$ years. In a more recent study by Levison et al. (1997) the orbits of 270 fictitious $L_4$ Trojans and of 36 real Trojans have been computed in the model of the outer solar system up to $10^9$ years and $4 \cdot 10^9$ years, respectively.

---

[1] These asteroids are called Trojans since all of them are named after heros of the Trojan war.

 *Celestial Mechanics and Dynamical Astronomy* **73**: 117–126, 1999.
©1999 *Kluwer Academic Publishers.*

In this paper we present the orbital evolution of a sample of 13 Trojan asteroids listed in table 1, which was taken from a paper by Milani et al. (1997). In their numerical study positive Lyapunov exponents for these asteroids were computed which indicate chaotic behavior; the Lyapunov time is less than $10^5$ years (column 6) in all cases. Since former computations established the dynamical stability of these orbits over a much longer time interval the notion *asteroids in stable chaotic motion* (ASC) was introduced by these authors[2]. In our computations 12 of the 13 Asteroids are stable over the time interval of 100 million years and thus are stable for more than $10^3$ times the Lyapunov time.

Our previous computations of all known Trojan asteroids over 10 million years showed that the Trojans *in stable chaotic motion* have larger variations in the semimajor axis (cf. Dvorak & Pilat-Lohinger, 1998) and one of these asteroids – (5144) Achates – has exceptionally strong variations in the inclination. The continuation of the computations of 13 Trojans (table 1) up to $10^8$ years yielded to one escape orbit – i.e. (1868) Thersites. The goals of this long term study of the dynamics are as follows:

1. to study the long term evolution of the orbital elements of the 13 Trojans with short Lyapunov times
2. to compute the proper elements for selected time intervals within the 100 Million years integration
3. to compute for the same time intervals the root mean square (r.m.s.)[3] of the orbital elements semimajor axis, eccentricity and inclination
4. to compare the qualitative and quantitative behavior of the proper elements with the respective r.m.s.

## 2. The Computations

The asteroids of table 1 were integrated over a time interval of $10^8$ years whereby the equations of motion have been computed by means of the Lie integration method (cf. Lichtenegger, 1984, Hanslmeier & Dvorak, 1984). The outer solar system (i.e., Sun and the planets Jupiter through Neptune) has been used as dynamical model, where the Sun's mass has been increased by the masses of the inner planets in order to approximate the perturbations by the inner planets; relativistic terms were not taken into account.

From the whole time interval of 100 million years we determined 11 time intervals, where each covers 1 million years. These "short-time intervals" are defined as follows:

$$I_k = [k, (k+1)] \text{ Myrs with } k = 0, 10, 20, \ldots, 100$$

[2] Although the notation of ASCs is not accepted by all scientists it seems – at least for the authors – that it hits the important points of indicating chaos via the relatively large positive Lyapunov exponent on one hand, and being stable for more than 1000 times the Lyapunov time on the other hand.

[3] The root mean square is defined as the square root of the mean or average of the square of an argument, i.e. $\sqrt{E[(x-m)^2]}$ with $m = E(x)$.

TABLE I
Trojan asteroids in stable chaotic motion (cf. Milani et al., 1997).

| Asteroid | $a_{ini}$ | $e_{ini}$ | $i_{ini}$ | LCE$\times 10^{-5}$ | LT [ $10^3$ yrs] |
|---|---|---|---|---|---|
| (1868) Thersites | 5.290 | 0.110 | 16.8 | 1.12 | 89 |
| (1869) Philoctetes | 5.303 | 0.065 | 4.0 | 1.49 | 67 |
| 1988 AK | 5.305 | 0.064 | 22.1 | 1.07 | 93 |
| (4543) Phoinix | 5.082 | 0.098 | 14.7 | 1.11 | 90 |
| 4523 P-L | 5.236 | 0.048 | 0.9 | 2.12 | 47 |
| 5187 T-2 | 5.131 | 0.031 | 8.6 | 1.24 | 81 |
| 1991 HN | 5.098 | 0.011 | 8.3 | 1.73 | 58 |
| (1173) Anchises | 5.326 | 0.137 | 6.9 | 2.04 | 49 |
| (2594) Acamas | 5.113 | 0.086 | 5.5 | 2.90 | 34 |
| (3451) Mentor | 5.086 | 0.070 | 24.7 | 1.90 | 53 |
| (5144) Achates | 5.232 | 0.273 | 8.9 | 1.10 | 91 |
| 1988 RN11 | 5.269 | 0.096 | 1.4 | 1.99 | 50 |
| 1989 UX5 | 5.104 | 0.024 | 4.3 | 6.38 | 16 |

The evolution of the orbits within these 11 intervals was studied in a first step by the computations of the r.m.s. of the semimajor axes $(a)$, the eccentricities $(e)$ and the inclinations $(i)$.

In a second step we calculated the proper elements $d$ (the libration amplitude of the semi-major axis, i.e. $a - a_J = d \sin \theta + ...$, where $\theta$ is the argument of libration), $D$ (the libration amplitude of the critical argument, i.e. $\lambda - \lambda_J = \chi + D \cos \theta + ...$, where $\chi$ denotes the position of the libration point), $e_p$ (the proper eccentricity was derived from the nonsingular variables $h = e \sin \varpi$ and $k = e \cos \varpi$) and $\sin(I_p)$ (the proper inclination was derived from the nonsingular variables $p = \sin(I_p/2) \sin \Omega$ and $q = \cos(I_p/2) \sin \Omega$. Since proper elements are known as "quasi invariants", we checked their values within these eleven intervals defined above (each covers 1 million years). For their determination we used a numerical method similar to the one by Milani (1993, p. 61ff):

—   The short periodic terms were eliminated using the method of Labrouste (cf. Bien & Schubart, 1987 and Burger, 1998).

—   The long periodic terms were determined with the aid of a numerical frequency analysis (=NFA) provided by Chapront (1995) and substracted from the signal of $h$ and $k$ for the proper eccentricities respectively of $p$ and $q$ for the proper inclinations.

—   Using again the NFA we derived the proper elements $e_p$ and $\sin I_p$ with the two frequencies in the order of $300''/y$ (for the eccentricities) and $10''/y$ (for the inclinations) depending on the asteroid.

Milani in his articles already determined proper elements for all – in those days available – Trojans with reliable orbits. Fig. 1 shows the agreement between the

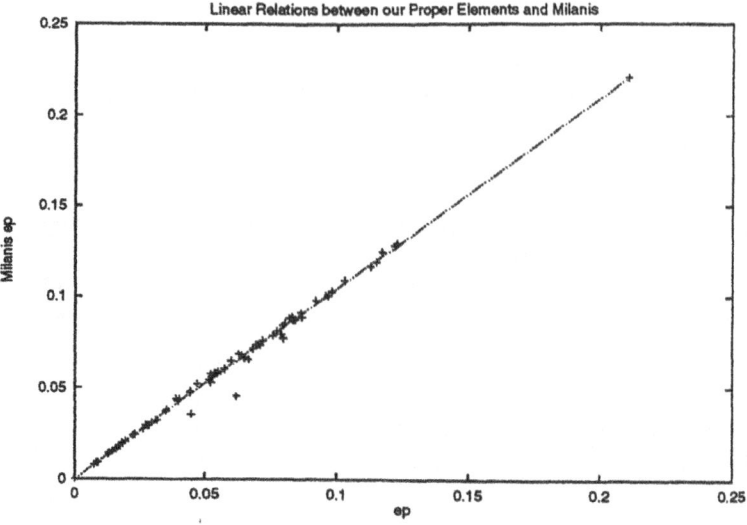

Fig. 1. Correlation between our proper eccentricities (x – axis) and the values determined by Milani (y – axis)

two determinations for the proper elements $e_p$.

Additionally in table 2 we show the proper elements for our sample of 13 Trojans determined for the first million years ($1^{st}$ interval of time). For Achates it was not possible to determine a proper $\sin I_p$ because of the very large variations in inclination (compare figure 4b).

### 3. Long Term Orbital Evolution

In a numerical study of Levison et al. (loc.cit) 21 of the 36 real Trojans turned out to be unstable in less than 4 billion years. In fig. 2 we show a similar plot as it is given by Levison et al. where all known Trojan asteroids are plotted in the plane of the proper elements D (i.e. libration amplitude) and $e_p$ (i.e. proper eccentricity). The full squares represent the asteroids of table 1 and the dotted line is Rabe's stability curve (cf. Rabe 1965) – which was determined in the simple model of the circular restricted three body problem. Shoemaker et al. (1989) concluded from the existence of objects above this stability curve that the true stability curve must be above Rabe's. In the study of Levison et al. the stable region exceeds Rabe's stability curve slightly, nevertheless there are still Trojans above Levison's stability boundary – like the asteroid Achates. Fig. 2 shows that 12 ASCs within Rabe's stability boundary have in general large D values ($D > 20^o$) and small proper eccentricities ($e_p < 0.1$). But no conclusion in what concerns their short Lyapunov time may be drawn from that.

Figs. 3 and 4 show the dynamical behaviour of 4 selected asteroids of the

TABLE II

Proper elements for the $1^{st}$ time interval:

| Asteroid | $d$ | $D$ | $e_p$ | $\sin I_p$ |
|----------|-----|-----|-------|------------|
| (1868) Thersites | 0.1189 | 24.80 | 0.929 | 0.2850 |
| (1869) Philoctetes | 0.1112 | 22.16 | 0.506 | 0.0618 |
| 1988 AK | 0.1085 | 22.79 | 0.0226 | 0.3737 |
| (4543) Phoinix | 0.1375 | 28.26 | 0.0455 | 0. 2594 |
| 4523 P-L | 0.1302 | 25.89 | 0.0637 | 0.0379 |
| 5187 T-2 | 0.1089 | 21.99 | 0.0266 | 0.1758 |
| 1991 HN | 0.1443 | 29.14 | 0.0270 | 0.1076 |
| (1173) Anchises | 0.1281 | 25.39 | 0.0.874 | 0.1374 |
| (2594) Acamas | 0.1673 | 33.543 | 0.0546 | 0.0950 |
| (3451) Mentor | 0.1345 | 28.76 | 0.0239 | 0.3983 |
| (5144) Achates | 0.0677 | 13.23 | 0.2175 | – |
| 1988 RN11 | 0.1254 | 24.90 | 0.21275 | 0.0464 |
| 1989 UX5 | 0.1536 | 31.04 | 0.1053 | 0.0481 |

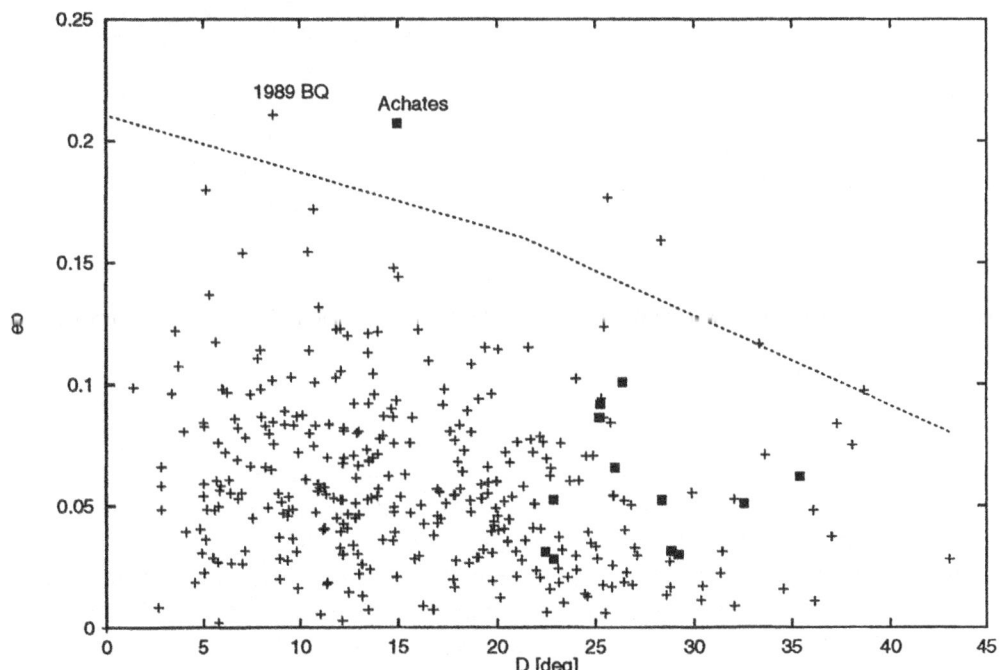

Fig. 2. All known Trojan asteroids in the proper elements plane ($D$, $e_p$). The proper elements were calculated over 10 million years. The full squares represent the asteroids of table 1 and the dotted line is Rabe's stability curve (cf. Rabe 1965).

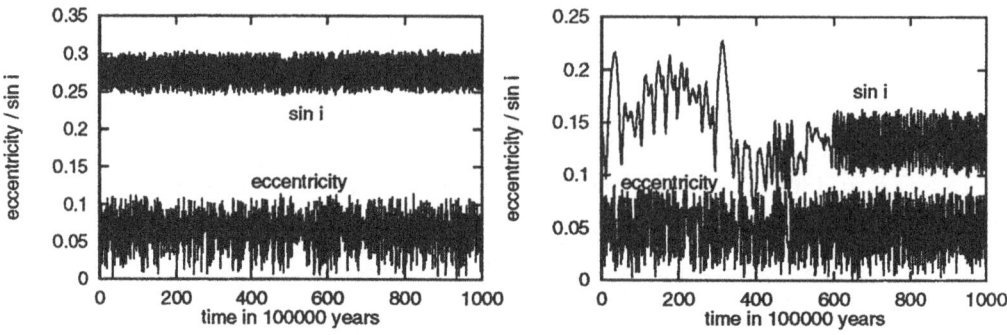

Fig. 3. Orbital evolution for 100 Myrs of two $L_4$ ASCs: (4543) Phoinix with $e_0 = 0.098$ and $i_0 = 14.7$ (left) and 1991 HN with $e_0 = 0.011$ and $i_0 = 8.3$ (right). Note that on the y-axis we plotted the eccentricity and the sine of the inclination.

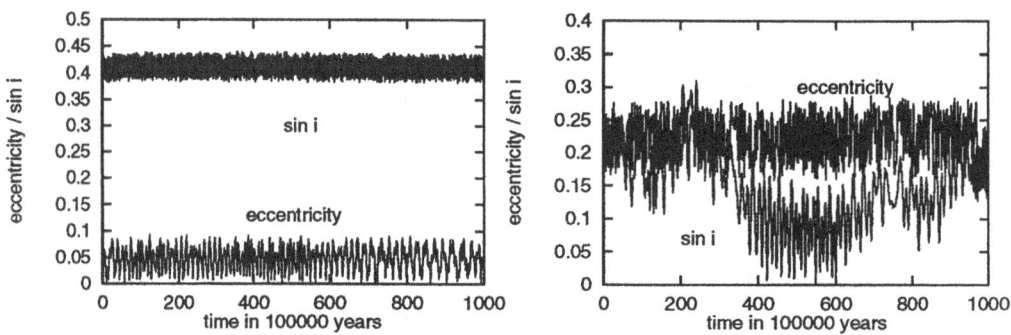

Fig. 4. Orbital evolution for 100 Myrs of two $L_5$ ASCs: (3451) Mentor with $e_0 = 0.07$ and $i_0 = 24.7$ (left) and (5144) Achates with $e_0 = 0.273$ and $i_0 = 8.9$ (right). Note that on the y-axis we plotted the eccentricity and the sine of the inclination.

Trojan ASCs for the whole time interval of 100 Million years. The evolution of eccentricity and inclination does not show significant changes over the whole time interval for the $L_4$ asteroid Phoinix and the $L_5$ asteroid Mentor. The dynamical evolution is quite different for 1991 HN and for Achates: while the first one – an $L_4$ asteroid – is suffering strong and irregular perturbations of the inclination for the first 60 million years and then it is in a very regular looking orbit the second one – an $L_5$ asteroid – shows strong variations in the inclination from 30 million years on. Both effects may be caused by second order resonances inside the 1:1 resonance with Jupiter; this will be studied more detailed in connection with the interesting escape of the asteroid Thersites, which escaped after about 26 million years. Therefore, we excluded this Trojan from our numerical analysis.

## 4. Comparison of the Proper Elements with the r.m.s.

In fig. 5 the proper elements $e_p$ are compared to the r.m.s.(e) for the same 11 intervals; one recognizes how well the two curves agree. The correlation coefficients are 0.993 for the $L_4$ and 0.978 for the $L_5$ asteroids. A comparison of the proper element $\sin I_p$ with the r.m.s. of the inclinations for the time intervals does not show such an agreement. In fig. 6 we show two examples namely the $L_4$ Trojan 88AK and the $L_5$ Trojan 88RN11. The calculated correlation coefficients of 0.883 and 0.693 are small and thus these two quantities are obviously not correlated. The similar behavior of the r.m.s.(a) and the proper element $d$ is obvious through the pragmatic definition of $d$.

In the former mentioned paper by Levison et al. (loc.cit.) fig. 2 shows the variations of D and $e_p$ for a "typical Trojan" in an unstable orbit over a time scale of $1.5 \times 10^9$ years. The changes of the proper eccentricity are in the same order of magnitude in their example ($0.06 < e_p < 0.08$) as in our calculations for the ASCs (e.g. for Achates we found the following variations: $0.217 < e_p < 0.226$).

## 5. Conclusions

In this numerical investigation we examined the long term stability of a sample of 13 especially selected Trojans with relatively small Lyapunov times covering a time interval of 100 million years.

The results can be summarized as follows:
- the proper elements are slowly varying quantities for most of the asteroids examined, with the exception of 1991 HN and Achates (large variations of the inclination)
- the asteroid Thersites escaped after 26 million years
- the proper elements $e_P$ are correlated with the corresponding r.m.s. of the eccentricities and thus the r m s (e) may – as a first rough estimation – serve as proper element indicator for $e_p$; also the r.m.s.(a) may serve as proper element indicator for $d$
- the proper element $\sin I_p$ is seemingly not correlated with the r.m.s. of the inclination.

In summary no special dynamical evolution for these asteroids has been found with the exception of Thersites. So it remains an unsolved problem why these asteroids have such short Lyapunov times; more work has to be done to answer this interesting question.

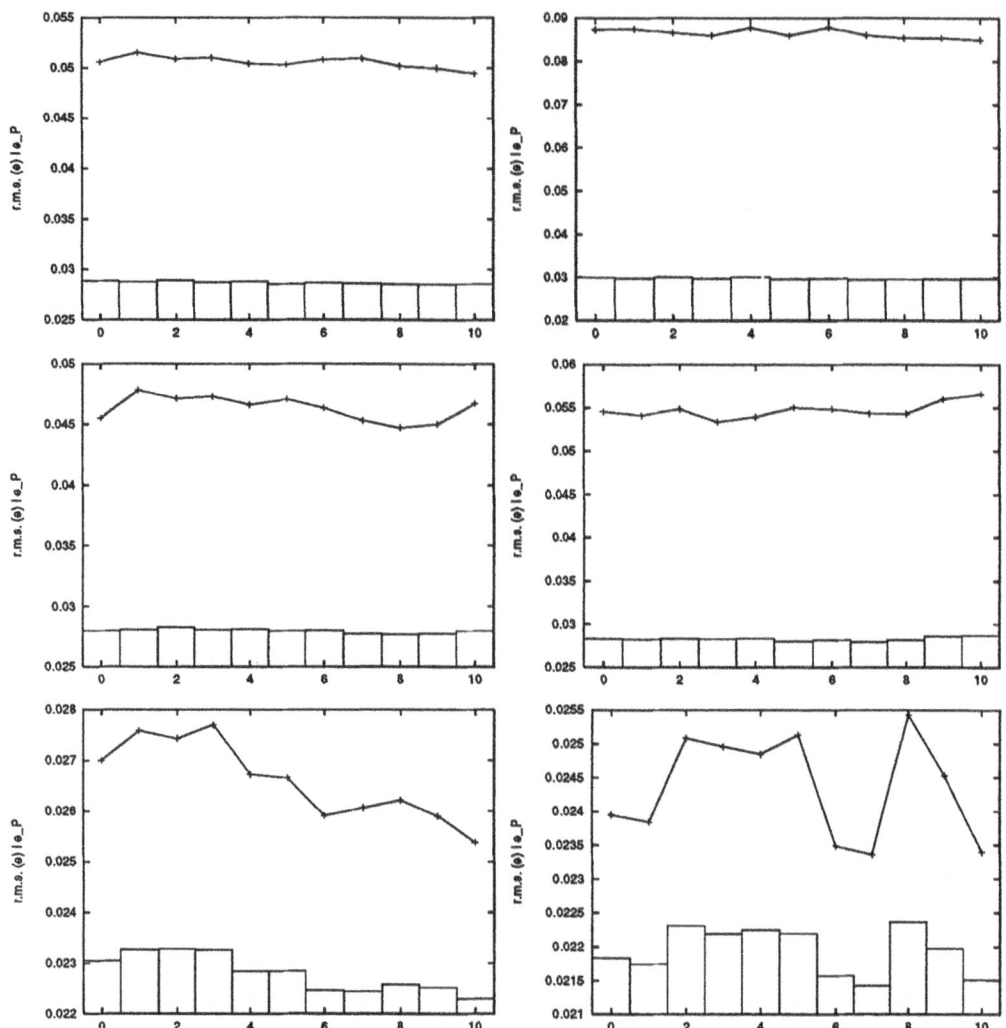

Fig. 5. r.m.s.(e) (boxes) and proper element $e_p$ (lines) versus 11 different time intervals of $\Delta t = 1$ Million years: [0,1], [10,11], ... [100,101] for the $L_4$ Trojans: Philoctetes (upper left), Phoinix (middle left), 1991 HN (lower left) and for the $L_5$ Trojans: Anchises (upper right), Acamas (middle right) and Mentor (lower right).

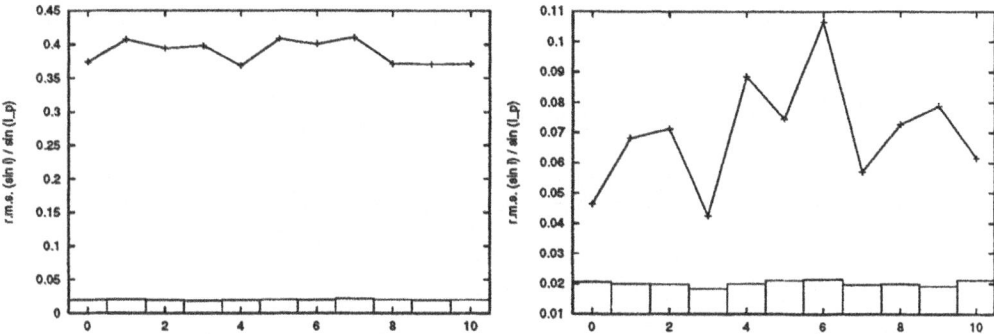

Fig. 6. r.m.s.(i) (boxes) and proper element $sin I_p$ (lines) versus 11 different time intervals of $\Delta t = 1$ Million years: [0,1], [10,11], ... [100,101] for the $L_4$ Trojan 88AK (left) and for the $L_5$ Trojan: 88RN11(right)

## Acknowledgements

We are grateful to the critical comments by Dr. Morbidelli and also by Dr. Ferraz-Mello who helped to improve the paper significantly. Special thanks go to Dr. Wodnar for many discussions on the problem of proper elements.

The work presented here was carried out within the framework of the *Jubiläums-fondsprojekt* Nr. 6446 of the Österreichische Nationalbank. Two of the authors (EP-L and CB) wish to thank Prof. J. Henrard for the financial support. EP-L received a travel grant from the *Österreichische Forschungsgemeinschaft*.

## References

Barber, G.: 1986, "The Orbits of Trojan Asteroids", in C.I. Lagerkvist, B.A. Lindblad, H. Lundstedt and H. Rickman (eds.), *Asteroids, Comets, Meteors II*, University of Uppsala, 161

Bien, R., Schubart, J.: 1984, "Trojan orbits in secular resonances", Cel.Mech & Dyn. Astro., **34**, 425.

Bien, R., Schubart, J.: 1987, "Three characteristic parameters for the Trojan group of asteroids", *Astron.Astrophys.*, **175**, 292.

Burger Ch.: 1998, "The Method of Labrouste with an Application in Celestial Mechanics", Diploma thesis, University of Vienna, 75 p.

Chapront, J.: 1995, "Representation of planetary ephemerides by frequency analysis. Application to the five outer planets", *Astron. Astrophys.*, **109**, 181.

Dvorak, R., Pilat-Lohinger, E.: 1998, "Les astéroides troyens: une nouvelle analyse" in D. Benest & C. Froeschlé (eds.) *Aux Frontières de la Dynamique chaotique des systèmes gravitationelles* O.C.A. Observatoire de Nice, 19.

Erdi, B.: 1984, "Critical inclination of Trojan asteroids ", Cel.Mech & Dyn. Astro., **34**, 435.

Erdi, B.: 1988, " Long periodic perturbations of Trojan asteroids ", Cel.Mech. & Dyn. Astro., **43**, 303.

Erdi, B.: 1996, "On the Dynamics of Trojan Asteroids ", in S. Ferraz-Mello, B. Morando, and J.E. Arlot (eds.), IAU Symposium 172, *Dynamics, Ephemerides and Astrometry in the Solar System*, 171.

Erdi, B.:1997, "The Trojan Problem", Cel.Mech & Dyn. Astro., **65**, 149.

Hanslmeier, A., Dvorak, R.: 1984, "Numerical Integration with Lie Series", *Astron. Astrophys.*, **132**, 203.

Lemaitre, A.: 1993, "Proper Elements: What are they", Cel.Mech. & Dyn. Astro., **56**, 103.

Levison, H. F., Shoemaker, E.M. and Shoemaker C.S.: 1997, "Dynamical Evolution of Jupiter's Trojan asteroids",*Nature*, **385**, 42.

Lichtenegger, H.: 1984, "The Dynamics of Bodies with Variable Masses", Cel.Mech. **34**, p. 357.

Milani, A.: 1993, "The Trojan Asteroid Belt: Proper Elements, Stability, Chaos and Families", Cel.Mech & Dyn. Astro., **57**, 59.

Milani, A.: 1994, "The Dynamics of the Trojan Asteroids", in A. Milani, M. di Martino, A. Cellino (eds.) IAU Symposium **160**, ACM 1993, Kluwer Academic Publishers, The Netherlands. 159.

Milani, A., Nobili, A., Knežević, Z.: 1997, "Stable Chaos in the Asteroid Belt", *Icarus*, **125**, 13.

Pilat-Lohinger E., Dvorak, R., Burger Ch.: 1998, " Asteroids in stable chaotic motion" in R. Dvorak, H.Haupt and K. Wodnar (eds.): *Modern Astrometry and Astrodynamics*, Verlag der Österreich.Akad.d.Wiss., Wien, 29.

Rabe, J.: 1965, "Limiting Eccentricities for Stable Trojan Librations", *Astron. J.*, **70**, 687.

Schubart, J., Bien, R.: 1984, "An application of Labrouste's method to quasi-periodic asteroidal motion", Cel.Mech & Dyn. Astro., **34**, 443.

Schubart, J., Bien, R.: 1987, "Trojan asteroids – Relations between dynamical parameters ", *Astron. Astrophys.*, **175**, 200.

Shoemaker, E.M., Shoemaker, C.S., Wolfe, R.F.: 1989, "Trojan asteroids – Population, dynamical structure and origin of the L4 and L5 swarms." in *Asteroids II*, Tucson, Univ. of Arizona Press, 487

# STAR CLUSTER SIMULATIONS: THE STATE OF THE ART

SVERRE J. AARSETH

*Institute of Astronomy, Madingley Road, Cambridge, England*

**Abstract.** This paper concentrates on four key tools for performing star cluster simulations developed during the last decade which are sufficient to handle all the relevant dynamical aspects. First we discuss briefly the Hermite integration scheme which is simple to use and highly efficient for advancing the single particles. The main numerical challenge is in dealing with weakly and strongly perturbed hard binaries. A new treatment of the classical Kustaanheimo-Stiefel two-body regularization has proved to be more accurate for studying binaries than previous algorithms based on divided differences or Hermite integration. This formulation employs a Taylor series expansion combined with the Stumpff functions, still with one force evaluation per step, which gives exact solutions for unperturbed motion and is at least comparable to the polynomial methods for large perturbations. Strong interactions between hard binaries and single stars or other binaries are studied by chain regularization which ensures a non-biased outcome for chaotic motions. A new semi-analytical stability criterion for hierarchical systems has been adopted and the long-term effects on the inner binary are now treated by averaging techniques for cases of interest. These modifications describe consistent changes of the orbital variables due to large Kozai cycles and tidal dissipation. The range of astrophysical processes which can now be considered by N-body simulations include tidal capture, circularization, mass transfer by Roche-lobe overflow as well as physical collisions, where the masses and radii of individual stars are modelled by synthetic stellar evolution.

**Key words:** Numerical methods, KS-regularization, N-body problem

## 1. Introduction

The study of self-gravitational $N$-body systems by direct integration poses many technical challenges which must be addressed. However, progress during the last decade now enables such problems to be tackled with confidence. In this personal review of recent developments, we concentrate on four main numerical tools which appear to be sufficient for the task in hand. The corresponding algorithms may be summarized under the following headings:

- Hermite integration
- Two-body regularization
- Chain regularization
- Hierarchical systems

These topics are discussed briefly in the subsequent sections, together with an outline of current applications. Given adequate tools, a massive effort is still required in order to develop an efficient star cluster simulation code but these aspects are beyond the scope of the present contribution.

## 2. Hermite Scheme

Although the Hermite integration scheme was developed for the special-purpose HARP computer (Makino 1991), it is also proving highly effective for standard workstations as well as conventional supercomputers. Since coding is now some-what simpler than for the traditional divided difference formulation (Ahmad and Cohen 1973, Aarseth 1985), it should be considered the method of choice for direct

*Celestial Mechanics and Dynamical Astronomy* 73: 127–137, 1999.
© 1999 *Kluwer Academic Publishers*.

N-body simulations. It may also be remarked that Hermite integration is actually more accurate than divided differences for the same order.

The main idea is again to employ a fourth-order force polynomial but now the *two* first terms are evaluated by explicit summation over all $N$ particles, thereby enabling two corrector terms to be formed. At first sight it may seem rather expensive to extend the full summation to the force derivative since this also requires prediction of velocities. However, simplicity as well as increased accuracy combine to outweigh the drawback of extra operations, particularly if block-step predictions are introduced. We expand a Taylor series for the force $\mathbf{F}$ and its first derivative $\mathbf{F}^{(1)}$ for each particle up to the third derivative about the reference time $t$ as

$$\mathbf{F} = \mathbf{F}_0 + \mathbf{F}_0^{(1)}t + \tfrac{1}{2}\mathbf{F}_0^{(2)}t^2 + \tfrac{1}{6}\mathbf{F}_0^{(3)}t^3, \tag{1}$$

$$\mathbf{F}^{(1)} = \mathbf{F}_0^{(1)} + \mathbf{F}_0^{(2)}t + \tfrac{1}{2}\mathbf{F}_0^{(3)}t^2. \tag{2}$$

Substituting $\mathbf{F}_0^{(2)}$ from (2) into (1) and simplifying then yields the third derivative corrector

$$\mathbf{F}_0^{(3)} = (2(\mathbf{F}_0 - \mathbf{F}) + (\mathbf{F}_0^{(1)} + \mathbf{F}^{(1)})t)\frac{6}{t^3}. \tag{3}$$

Similarly, substituting (3) into (1) gives the second derivative corrector

$$\mathbf{F}_0^{(2)} = (-3(\mathbf{F}_0 - \mathbf{F}) - (2\mathbf{F}_0^{(1)} + \mathbf{F}^{(1)})t)\frac{2}{t^2}. \tag{4}$$

Using $\mathbf{F}_0$ and $\mathbf{F}_0^{(1)}$ evaluated at the beginning of a time-step, the coordinates and velocities are first predicted to order $\mathbf{F}^{(1)}$ for all particles. Following determination of the new $\mathbf{F}$ and $\mathbf{F}^{(1)}$ by summation over all the contributions, the two higher derivatives are obtained by (3) and (4). This gives rise to corrector terms for coordinates and velocities given by

$$\Delta\mathbf{r}_i = \tfrac{1}{24}\mathbf{F}_0^{(2)}\Delta t^4 + \tfrac{1}{120}\mathbf{F}_0^{(3)}\Delta t^5,$$

$$\Delta\mathbf{v}_i = \tfrac{1}{6}\mathbf{F}_0^{(2)}\Delta t^3 + \tfrac{1}{24}\mathbf{F}_0^{(3)}\Delta t^4. \tag{5}$$

Given the high-order derivatives, individual time-steps can now be assigned in the usual way from some suitable convergence criterion.

The overheads of predicting $N$ coordinates and velocities at each time-step can be reduced considerably by adopting so-called hierarchical time-steps (McMillan 1986), where the indicated values are truncated to be factor 2 commensurate. The apparent inefficiency of just a few particles sharing the same (small) step and yet requiring one full prediction is compensated by having a distribution of discrete levels (typically 16 for $N \simeq 10^4$) such that the number of predictions is significantly reduced with respect to the continuous case (say by factor of 100). This scheme is particularly suitable for the special-purpose HARP computers but lends itself equally well to other architectures, including parallel supercomputers (Spurzem

1998). Somewhat surprisingly, the workstation code NBODY6 which is based on the Ahmad-Cohen (1973) neighbour scheme (Makino and Aarseth 1992, Aarseth 1994) is in fact slightly faster and more stable than the older NBODY5 code for $N = 1000$ single particles and the same number of steps.

## 3. Two-Body Regularization

The early 1970's saw the introduction of the Kustaanheimo-Stiefel (1965) regularization for treating close encounters and hard binaries in $N$-body simulations (Bettis and Szebehely 1972, Aarseth 1972) and the elegant KS method has proved to be very resilient. However, even a regularized two-body solution is subject to small but systematic errors when studied over long times. In order to avoid this undesirable feature, the concept of energy stabilization has been tried for weak perturbations (Aarseth 1985). Although this procedure ensures that the orbit is constrained to have the correct energy arising from the perturbation, the corresponding angular momentum is no longer conserved so well.

The subsequent exploitation of adiabatic invariance (Mikkola and Aarseth 1996) by the so-called slow-down principle tends to alleviate this imperfection since now one KS orbit may represent a number of physical periods by augmenting the perturbation itself and neglecting short-period effects. As for the earlier claim that a time-symmetric KS method would be superior (Funato et al. 1996), it now appears that the requirement of variable time-steps for perturbed orbits cannot be accommodated (Kokubo et al. 1998). So far there is no evidence that the resulting eccentricities of cluster binaries studied by the stabilization scheme cannot be trusted, especially bearing in mind that the long-term evolution of most binaries is predominantly subject to discrete changes of a random nature. The case of long-lived hierarchical systems deserves special consideration, however, but here additional effects should also be considered, as discussed in a subsequent section.

An alternative KS regularization scheme has been presented recently (Mikkola and Aarseth 1998) which achieves a high accuracy without extra cost. This new approach is based on the idea of a truncated Taylor series, where additional correction terms represent the neglected higher orders and which yields exact solutions in the unperturbed case. The new algorithm is again of Hermite type and will be outlined in the following.

First, coordinates and velocities of the perturbers are predicted in the usual way (i.e. to first order), whereas the regularized coordinates and velocities $(U, U')$ are predicted to highest order. Here $U^{(4)}, U^{(5)}$ include the modified Stumpff (1962) functions

$$\tilde{c}_n(z) = n! \sum_{k=0}^{\infty} \frac{(-z)^k}{(n+2k)!},$$ (6)

where the argument is related to the time-step by $z = -\frac{1}{2}h\Delta\tau^2$ and $h$ is the specific binding energy. These coefficients only deviate slightly from unity and a twelfth-order expansion (re-evaluated every step) appears sufficient. After transforming the physical coordinates and velocities to global values, the predictor cycle is completed by evaluating the perturbing acceleration $\mathbf{F}$ as well as its explicit derivative $\dot{\mathbf{F}}$.

Because of the insufficient accuracy of the predicted deviation from unperturbed motion at the end of a step, the corrector cycle employs an iteration. Setting $\Omega = -\frac{1}{2}h$, the basic equation of motion takes the familiar form

$$\mathbf{U}^{(2)} = -\Omega\mathbf{U} + \tfrac{1}{2}r\mathcal{L}^T\mathbf{F}, \tag{7}$$

where $\mathcal{L}(\mathbf{U})$ is a $4 \times 3$ linear matrix and $r = \mathbf{U} \cdot \mathbf{U}$ is the separation. We express the new KS acceleration and its derivative (where $\mathbf{F}' = r\mathbf{F}$) at the start of a step as

$$\mathbf{U}_0^{(2)} = -\Omega_0\mathbf{U}_0 + \mathbf{f}_0^{(2)}, \tag{8}$$

$$\mathbf{U}_0^{(3)} = -\Omega_0\mathbf{U}_0' + \mathbf{f}_0^{(3)}, \tag{9}$$

where $\mathbf{f}_0^{(2)} = \frac{1}{2}r\mathbf{Q}$, with $\mathbf{Q} = \mathcal{L}^T\mathbf{F}$, is the perturbed force contribution evaluated after the *previous* predictor cycle.

The two next Taylor series terms are constructed from the Hermite scheme. Using the current value of $h$ (and $\Omega$), predicted to fourth order, we form the new perturbative functions at the end of the step

$$\mathbf{f}^{(2)} = (\Omega_0 - \Omega)\mathbf{U} + \tfrac{1}{2}r\mathbf{Q}, \tag{10}$$

$$\mathbf{f}^{(3)} = (\Omega_0 - \Omega)\mathbf{U}' - \Omega'\mathbf{U} + \tfrac{1}{2}r'\mathbf{Q} + \tfrac{1}{2}r\mathbf{Q}', \tag{11}$$

from which the corrector derivatives $\mathbf{f}_0^{(4)}, \mathbf{f}_0^{(5)}$ are recovered by the Hermite rule (Makino 1991).

The expressions for $\mathbf{U}_0^{(4)}$ and $\mathbf{U}_0^{(5)}$ are readily formed in analogy with Eqs. (8) and (9) which yield

$$\mathbf{U}_0^{(4)} = -\Omega_0\mathbf{U}_0^{(2)} + \mathbf{f}_0^{(4)}, \tag{12}$$

$$\mathbf{U}_0^{(5)} = -\Omega_0\mathbf{U}_0^{(3)} + \mathbf{f}_0^{(5)}. \tag{13}$$

From Eqs. (8) - (13), the provisional solution for $\mathbf{U}, \mathbf{U}'$ is then obtained by the general expression (cf. Mikkola and Aarseth 1998), which contains the Stumpff functions. The treatment of the energy remains the same as for standard Hermite based on $\Omega' = -\mathbf{U}' \cdot \mathbf{Q}$ and the physical time is obtained from integrating $t' = \mathbf{U} \cdot \mathbf{U}$ which also involves Stumpff functions. Substituting for $\mathbf{U}^{(2)}$, we write the second derivative as

$$\Omega^{(2)} = \Omega_0\mathbf{U} \cdot \mathbf{Q} - \mathbf{f}^{(2)} \cdot \mathbf{Q} - \mathbf{U}' \cdot \mathbf{Q}'. \tag{14}$$

The two corrector terms constructed from $\Omega'$ and $\Omega^{(2)}$ are added to the predicted value *without* any Stumpff functions to yield an improved solution for $\Omega$ at the start of the next iteration or at the end point.

Subsequent iterations repeat the procedure above, starting from Eq. (10) without re-evaluating the physical perturbation and its derivative. Thus the new values of Eqs. (10) and (11) are based on the improved solution for $\mathbf{U}$, $\mathbf{U}'$ and $r$, $r'$, as well as the new $\Omega$. In the present treatment, one iteration yields a significant improvement for modest perturbations and experience so far indicates that this may also be sufficient for strong interactions because of the shortening of the stepsize $\Delta\tau$ (cf. Aarseth 1994).

The corrector cycle ends by specifying new derivatives for use in the next prediction, as well as saving the perturbative derivatives (10) and (11) required for the Hermite scheme. This is completed by re-initializing Eqs. (8) and (9) at the end point, substituting $\mathbf{f}^{(2)}$, $\mathbf{f}^{(3)}$ as well as the iterated values of $\Omega'$ and $\Omega^{(2)}$. It is advantageous to employ the corrected values of $r$ and $r'$ for this purpose; the re-evaluation of $\mathbf{f}^{(2)}$ and $\mathbf{f}^{(3)}$ is fast and also benefits the final quantities $\mathbf{U}^{(2)}$ and $\mathbf{U}^{(3)}$ to be used in the next prediction. A more accurate expression of the fourth KS derivative at the *end* of the interval is obtained by including the next order by

$$\mathbf{U}^{(4)} = \mathbf{U}_0^{(4)} + \mathbf{U}_0^{(5)}\Delta\tau, \tag{15}$$

and similarly for the *third* derivative of the energy,

$$\Omega^{(3)} = \Omega_0^{(3)} + \Omega^{(4)}\Delta\tau. \tag{16}$$

The above scheme has been implemented in the state of the art codes NBODY4 and NBODY6 and has proved itself in large-scale simulations. Accuracy tests obtained by a toy code shows that high accuracy can be obtained with 30 steps per orbit for relatively weak perturbations, which is about half that required by the old stabilization scheme. A significant part of this gain is due to the modifications by the Stumpff coefficients, although the basic Taylor series (or Encke-type) formulation is also considerably more accurate than the standard method. The number of operations for a typical step is not much larger in the new method, including the overhead for the Stumpff functions and one iteration in the corrector. Hence the computational effort is less for typical calculations, although this depends on the actual number of perturbers. Finally, we remark that the Stumpff method also includes the slow-down scheme in exactly the same way as before.

## 4. Chain Regularization

The concept of chain regularization is simple, yet the mathematical formulation is quite technical and this has acted as an impediment to wider usage. However, it enables new types of problems to be studied and is therefore worth the extra effort. The basic idea is a generalization of three-body regularization (Aarseth and Zare 1974) which treats two perturbed KS solutions with respect to a common reference body, where each two-body solution is described by regular equations. Thus an extension to four participating bodies merely introduces one more perturbed KS solution, although the formalism is somewhat different (Mikkola and Aarseth

1990). Once the step from three to four particles has been mastered, the general case becomes feasible (Mikkola and Aarseth 1993).

The essential feature of chain regularization is that dominant interactions along the chain itself are treated as perturbed KS solutions and all the other attractions are included as perturbations. Hence it becomes imperative to select the chain vectors in such a manner as to minimize the perturbations. Since we are dealing with dynamical interactions, the chain vectors need to be redrawn in response to changing configurations. Fortuitously, all the relevant decision-making constitute a minor overhead here since the integration is carried out by the high-order Bulirsch-Stoer (1966) scheme and a certain elasticity is tolerated as regards switching to more favourable chain vectors.

The equations of motion are derived from a regularized Hamiltonian of the form

$$\Gamma^* = g(H - E), \tag{17}$$

where $H$ is expressed in terms of the coordinates and momenta and $E$ is the internal system energy. Here the function $g$ is given by the corresponding time transformation

$$dt = g d\tau \tag{18}$$

and choosing the inverse Lagrangian energy ($L = T + \Phi$) ensures regular solutions for any chain separation $R_k$.

The treatment begins by selecting a compact subsystem of three or four particles; i.e. so-called $B + S$ or $B + B$ type. External perturbers are chosen in analogy with the KS implementation and the internal integration includes any perturbation effect which also tends to change the total energy according to its separate equation of motion. At the same time, the c.m. motions are advanced by the standard Hermite scheme with due attention to the slightly modified form of the corresponding acceleration which requires a differential correction.

The analogy with KS does not hold in one important respect since the chain membership may change before termination occurs. Thus an initial subsystem of four members may lose one member due to ejection, or an approaching perturber - a single particle or binary - may be added. Alternatively, the membership may also change through physical collision. All the relevant corrections and re-initializations are performed *in situ*. Hence the use of chain variables is also highly beneficial for the evaluation of nearly singular quantities. Chain termination usually occurs when a binary becomes well separated from one or two other members in which case the binary is accepted for KS treatment, whereas the remaining membership is initialized by the Hermite scheme or even as a second KS system. The actual decision-making also takes into account the cluster environment and is therefore quite involved.

Cluster simulations of primordial binaries frequently involve interactions of two binaries where the size of one is much less than the other. In such cases even the powerful chain method becomes prohibitive because the shortest period is a small fraction of the local crossing time. Fortunately the principle of slow-down applied to weakly perturbed KS binaries can also be employed here (Mikkola and Aarseth 1996). This permits a consistent study of binaries with arbitrarily short periods which would otherwise have to be treated as inert systems. The implementation itself differs from the KS case since here we adjust the slow-down factor continuously according to the maximum apocentre perturbation exerted by the other chain members, rather than choosing an appropriate discrete level (factor of 2) at each apocentre passage.

Since the strong interactions studied by the chain method are usually of short duration, the simulation code only allows one such case to be considered at a time for technical reasons. However, there is provision for studying one triple as well as one quadruple system by *unperturbed* three-body (Aarseth and Zare 1974) and chain (Mikkola and Aarseth 1990) regularization. Given a few hundred critical events in a typical cluster simulation, the latter procedures are usually not needed but this may change with the addition of more primordial binaries.

## 5. Hierarchical Systems

The Solar neighbourhood contains many examples of multiple systems where the inner component of a binary is itself a binary, and levels of higher multiplicity also exist. Likewise, hard binaries in star clusters may acquire an outer component with sufficiently small eccentricity to be stable over many orbits. Hierarchical triples may be formed by the classical three-body capture mechanism in which the binary itself acts mainly as a point-mass. However, in clusters with significant binary populations such systems are more likely to form in strong interactions between two binaries since this involves two-body encounters. The second formation process was already identified in scattering experiments with colliding binaries which yielded a high percentage of positive outcomes (Mikkola 1983). Thus one way for such triples to become stable requires the impact parameter to exceed some critical value and yet be sufficiently small for the weakest binary to be disrupted, but other processes are also favoured, including exchange.

Given a newly formed hierarchical triple, the question of long-term stability naturally arises. Depending on the period ratio, the direct calculation of a perturbed inner binary can be quite time-consuming even with KS regularization. However, since the corresponding semi-major axis may not be subject to any secular effects it becomes possible to adopt the centre-of-mass approximation and thereby only neglect cyclical changes of the eccentricity. Various empirical criteria have been obtained by fitting the results of systematic three-body calculations for a restricted set of parameters (Harrington 1977, Eggleton and Kiseleva 1995). Based on these results, the so-called merger procedure has been employed for some time (Aarseth 1985). Thus provided the stability condition is satisfied, the inner binary is replaced by its combined mass to facilitate KS treatment of the *outer* orbit.

A more rigorous approach based on correspondence with the chaos boundary in the binary-tides problem (Mardling 1995) has yielded a semi-analytical stability criterion which holds for quite large mass ratios and arbitrary outer eccentricities (Mardling and Aarseth 1998). Here the critical outer pericentre distance is given in terms of the inner semi-major axis, $a_{in}$, by

$$R_p^{out} = C \left[ (1 + q_{out}) \frac{(1 + e_{out})}{(1 - e_{out})^{1/2}} \right]^{2/5} a_{in}, \tag{19}$$

where $q_{out} = m_3/(m_1 + m_2)$ is the outer mass ratio, $e_{out}$ is the corresponding eccentricity and $C \simeq 2.8$ is determined empirically. This criterion is only valid for coplanar prograde motion and still ignores a weak dependence on the inner eccentricity. However, the general case of inclined orbits exhibit increased stability so that Eq. (19) represents an upper limit. Further tests suggests an inclination correction factor $f = 1 - 0.3i/180$ (with $i$ in degrees) which has been adopted in practical simulations; this is also in qualitative agreement with the original stability condition for retrograde orbits (Harrington 1972). The merger treatment is only allowed while the pericentre condition is satisfied, after which the inner binary is re-initialized.

A further refinement is included when the outer component itself is a binary. In the case of a $B + B$ configuration, the smallest binary plays the role of the outer body in a triple. Since the corresponding chaos boundary is not very sensitive to a second extended object (Mardling 1991), we adopt an additional correction factor $f_1 = f + 0.1min(a_{in}/a_2, a_2/a_{in})$, with $a_2$ representing the second semi-major axis. We also mention here that even double hierarchies may be formed, where a system of type $B + S$ or $B + B$ itself acquires an outer bound component. Such configurations do occur occasionally and procedures have therefore been developed for their special treatment.

The criterion (19) above is concerned with long-term stability and hence the absence of escape. However, it is also of interest to consider the possibility of exchange between the outer component and one member of the inner binary. According to classical developments (Zare 1977, Szebehely and Zare 1977), the critical value for exchange in a coplanar prograde triple is given by

$$(c^2 E)_{crit} = -\frac{G^2 f^2(\rho) g(\rho)}{2(m_1 + m_2 + m_3)}, \tag{20}$$

where $c$ is the angular momentum and the functions $f(\rho), g(\rho)$ are expressed in terms of the masses, with $\rho$ determined by iteration from a fifth-order algebraic equation for the collinear equilibrium points. Numerical tests show that the chaos boundary given by Eq. (19) lies above the exchange boundary when the masses are comparable and the latter only begins to overlap above $q_{out} \simeq 5$. Application of the exchange criterion is therefore less useful in practical calculations. We also note that once an exchange occurs the final evolution will inevitably lead to escape.

The long-term evolution of a hierarchical triple is characterized by cyclic oscillations of the inner eccentricity where the amplitude depends on the inclination.

The so-called Kozai effect (Kozai 1962) has received much attention recently in connection with external planetary systems but there is also an early example from N-body simulations (van Albada 1968) which points to the relevance for star clusters. Various analytical tools have been employed in order to model this process in some detail, including tidal dissipation for high eccentricities (Mardling and Aarseth 1999). Among the useful quantities which can be calculated theoretically (Heggie 1995) are the time-scale for a complete oscillation, $T_K$, as well as its maximum value, $e_{max}$.

Since the time-scale for the Kozai cycle is usually much greater than the Kepler period, the merger scheme for hierarchical triples lends itself particularly well to a semi-analytical treatment. At present only systems with $e_{max} > 0.8$ are considered since smaller amplitudes are less likely to result in tidal activity. We have used a double averaging procedure (Eggleton 1997, Mardling and Eggleton 1998) to calculate the evolution of such systems in terms of the inner Runge-Lenz vector and angular momentum vector. Thus some examples show that inclinations near 90° may induce tidal circularization even if oblateness effects are included. Clearly further developments of this experimental approach is needed in order to improve the modelling of these complicated processes.

## 6. Astrophysical Applications

The realistic simulation of star cluster dynamics requires a variety of astrophysical processes to be considered. In particular, the implementation of consistent stellar evolution enables the study of mass loss and finite-size effects. This is an ongoing project which has been outlined elsewhere (Aarseth 1996) and now employs an improved description of Roche mass transfer and physical collisions (Tout et al. 1997). Particular emphasis has been devoted to the modelling of chaotic motions and tidal circularization which form a link between an initial binary distribution and the Roche stage (Mardling and Aarseth 1999). In particular, it is found that very high eccentricities ($e > 0.999$) are produced in stable hierarchies or by exchange and these in turn lead to orbital shrinkage by tidal dissipation. Primordial binaries also leave an imprint in the form of high-velocity escapers. At the same time, more general cluster simulations have yielded much insight into dynamical evolution (McMillan et al. 1992, Aarseth and Heggie 1998, Portegies Zwart et al. 1998).

The modelling of synthetic stellar evolution is based on fast look-up tables for the radius, luminosity and type as a function of the initial mass and age (cf. Tout et al. 1997). Instantaneous mass loss due to stellar winds or supernovae explosions are adopted for the advanced stages. An energy-conserving integration scheme is preserved by including relevant corrections and re-initializations. The standard open cluster model includes $10^4$ single stars with 5% primordial hard binaries. Once the most massive single stars have evolved, the binaries dominate the mass segregation and increase their central abundance significantly with increased disruption probability. Even so, the original binary population is not depleted preferentially such that there is always an energy source which prevents core

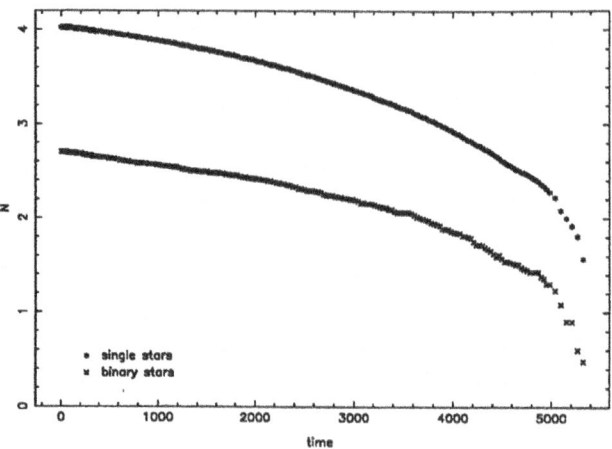

Fig. 1. Logarithmic plot of single stars and binaries as functions of time in $Myr$.

collapse. This behaviour is illustrated well in the figure which displays the bound membership.

In conclusion, the algorithms presented above have proved highly efficient for star cluster simulations. Hopefully these numerical tools will also play a part in future efforts involving more powerful hardware.

## References

Aarseth, S.J.: 1972, *Gravitational N-Body Problem*, ed. Lecar, M. (Reidel), 373.
Aarseth, S.J.: 1985, *Multiple Time Scales*, ed. Brackbill, J.U. and Cohen, B.I. (Academic Press, New York), 377
Aarseth, S.J.: 1994, *Galactic Dynamics and N-Body Simulations*, ed. Contopoulos, G., Spyrou, N.K. and Vlahos, L. (Springer-Verlag), 277.
Aarseth, S.J.: 1996, *Dynamical Evolution of Star Clusters*, ed. Hut, P. and Makino, J. (Kluwer), 161.
Aarseth, S.J. and Heggie, D.C.: 1998, *Mon. Not. R. astr. Soc.*, **297**, 794.
Aarseth, S.J. and Zare, K.: 1974, *Celest. Mech.*, **10**, 185.
Ahmad, A. and Cohen, L.: 1973, *J. Comput. Phys.*, **12**, 389.
van Albada, T.S.: 1968, *Bull. Ast. Inst. Neth.*, **19**, 479.
Bettis, D.G. and Szebehely, V.G.: 1972, *Gravitational N-Body Problem*, ed. Lecar, M. (Reidel), 388.
Bulirsch, R. and Stoer, J.: 1966, *Num. Math.*, **8**, 1.
Eggleton, P.P.: 1997, personal communication.
Eggleton, P.P. and Kiseleva, L.K.: 1995, *Astrophys. J.*, **455**, 640.
Funato, Y., Hut, P., McMillan, S. and Makino, J.: 1996, *A.J.*, **112**, 1697.
Harrington, R.S.: 1972, *Celest. Mech.*, **6**, 322.
Harrington, R.S.: 1977, *A.J.*, **82**, 753.
Heggie, D.C.: 1995, personal communication.
Kokubo, E., Yoshinaga, K. and Makino, J.: 1998, *Mon. Not. R. astr. Soc.*, **297**, 1067.
Kozai, Y. 1962, *A.J.*, **67**, 591.
Kustaanheimo, P. and Stiefel, E.: 1965, *J. Reine Angew. Math.*, **218**, 204.
Makino, J.: 1991, *Astrophys. J.*, **369**, 200.

Makino, J. and Aarseth, S.J.: 1992, *Publ. Astron. Soc. Japan*, **44**, 141.

Mardling, R.A.: 1991, Ph.D. Thesis, Monash University.

Mardling, R.A.: 1995, *Astrophys. J.*, **450**, 722.

Mardling, R. and Aarseth, S.: 1998, Proceedings NATO ASI, Maratea, ed. Roy, A.E., Steves, B. and Barnett, A.D. (Kluwer), in press.

Mardling, R.A. and Aarseth, S.J.: 1999, submitted.

Mardling, R.A. and Eggleton, P.P.: 1998, in preparation.

McMillan, S.L.W.: 1986, *The Use of Supercomputers in Stellar Dynamics* ed. Hut, P. and McMillan, S., (Springer-Verlag), 156.

McMillan, S.L.W., Hut, P. and Makino, J.: 1990, *Astrophys. J.*, **362**, 522.

Mikkola, S. 1983, *Mon. Not. R. astr. Soc.*, **203**, 1107.

Mikkola, S. and Aarseth, S.J.: 1990, *Celest. Mech. Dyn. Astron.*, **47**, 375.

Mikkola, S. and Aarseth, S.J.: 1993, *Celest. Mech. Dyn. Astron.*, **57**, 439.

Mikkola, S. and Aarseth, S.J.: 1996, *Celest. Mech. Dyn. Astron.*, **64**, 197.

Mikkola, S. and Aarseth, S.J.: 1998, *New Astron.*, **3**, 309.

Portegies Zwart, S., Hut, P., Makino, J. and McMillan, S.L.W.: 1998, *Astron. Astrophys.*, **337**, 363.

Spurzem, R.: 1999, *The Journal of Computational and Applied Mathematics*, ed. Riffert, H. and Werner, K. (Elsevier), in press.

Szebehely, V. and Zare, K.: 1977, *Astron. Astrophys.*, **58**, 145.

Stumpff, K.: 1962, Himmelsmechanik, Band I, (VEB, Berlin).

Tout, C.A., Aarseth, S.J., Pols, O.R. and Eggleton, P.P.: 1997, *Mon. Not. R. astr. Soc.*, **291**, 732.

Zare, K.: 1977, *Celest. Mech.*, **16**, 35.

# DIMENSIONALITY OF STABLE AND UNSTABLE DIRECTIONS IN THE GRAVITATIONAL $N$-BODY PROBLEM

R. H. MILLER

*University of Chicago, e-mail: rhm@oddjob.uchicago.edu*

**Abstract.** The gravitational $n$-body problem is chaotic. Phase trajectories that start very near each other separate rapidly. The rate looks exponential over long times. At any instant, trajectories separated in certain directions move apart rapidly (unstable directions), while those separated in other directions stay about the same (stable directions). Unstable directions lie along eigenvectors that correspond to positive eigenvalues of the matrix of force gradients. The number of positive eigenvalues of that matrix gives the dimensionality of stable regions. This number has been studied numerically in a series of 100-body integrations. It continues to change as long as the integration continues because the matrix changes extremely rapidly. On average, there are about $1.2n$ unstable directions out of $3n$. Issues of dimensionality arise when the tools of ergodic studies are brought to bear on the problem of trajectory separation. A method of estimating the rate of trajectory separation based on matrix descriptions is presented in this note. Severe approximations are required.

**Key words:** Chaos, phase space, Lagrangian displacements

Extreme sensitivity to initial conditions, one of the accepted signatures of chaotic systems, was demonstrated for the gravitational $n$-body problem (Miller, 1964) at about the same time as Lorenz's famous paper (Lorenz, 1963) in meteorology. The effect is physical, an example of Krylov's instability (Krylov, 1979). It was demonstrated analytically (Miller, 1966), and the interaction between trajectory separation and integrals of the motion was explored numerically by integrating the matrix equations Eq. (3) explicitly as an $n$-body system developed (Miller, 1971). A plea was also made to the numerical analysis community for help in trying to understand how valid physical conclusions might be inferred from numerical experiments in view of the underlying chaotic properties of phase trajectories (Miller, 1974). Some help may be on the way. Quinlan and Tremaine (1992) constructed shadowing orbits in a restricted form of the gravitational $n$-body problem. Their results suggest that real phase trajectories are so wild that a few can be found which remain near almost any "reasonable" computed trajectory for surprisingly long times.

Several attempts have been made to determine growth rates for trajectory separation numerically ((Kandrup et al., 1994); (Goodman et al., 1993), with references to earlier work in each of these papers) and analytically (Gurzadyan and Savvidy, 1986), (Kandrup, 1990). The numerical results lead to a general consensus that the growth rate is a few per crossing time (e.g., some multiple of $(G\rho)^{1/2}$). This same dependence had been found analytically, based on rather crude arguments in (Miller, 1966).

*Celestial Mechanics and Dynamical Astronomy* **73**: 139–147, 1999.
© 1999 *Kluwer Academic Publishers.*

## 1. Formulation

Let the pair of $3n-$vectors $(\mathbf{p}(1), \mathbf{q}(1))$ be the instantaneous phase coordinates of an $n-$particle system in the $6n-$dimensional phase space ($\Gamma-$space), and let $(\mathbf{p}(2), \mathbf{q}(2))$ represent a second system in the same space. Presume the two phase points to be "near" each other and let $\delta\mathbf{p} = \mathbf{p}(2) - \mathbf{p}(1)$ and $\delta\mathbf{q} = \mathbf{q}(2) - \mathbf{q}(1)$ represent the difference-vectors in the same space. The motion of either system is determined by the usual Hamiltonian equations, governed by a Hamiltonian $\mathcal{H}(\mathbf{p}, \mathbf{q})$.

Equations of motion for $\delta\mathbf{p}$ and $\delta\mathbf{q}$ can be constructed by forming a Taylor series expansion for $\mathcal{H}(\mathbf{p}(2), \mathbf{q}(2))$ in terms of $\mathcal{H}(\mathbf{p}(1), \mathbf{q}(1))$ and powers of $\delta\mathbf{p}$ and $\delta\mathbf{q}$. The linearized form of these equations is

$$\frac{d}{dt}(\delta\mathbf{q}) = \frac{\partial^2\mathcal{H}}{\partial\mathbf{q}\partial\mathbf{p}}\delta\mathbf{q} + \frac{\partial^2\mathcal{H}}{\partial\mathbf{p}^2}\delta\mathbf{p} \qquad \text{and}$$

$$\frac{d}{dt}(\delta\mathbf{p}) = -\frac{\partial^2\mathcal{H}}{\partial\mathbf{q}^2}\delta\mathbf{q} - \frac{\partial^2\mathcal{H}}{\partial\mathbf{p}\partial\mathbf{q}}\delta\mathbf{p},$$

$$(1)$$

where the derivatives are to be evaluated at $(\mathbf{p}(1), \mathbf{q}(1))$. Some care is required to remain in a regime in which the linear terms of the expansion suffice.

Write the $3n$ $\delta\mathbf{p}$'s and the $3n$ $\delta\mathbf{q}$'s as elements of a $6n-$vector:

$$\xi = \begin{pmatrix} \delta\mathbf{q} \\ \delta\mathbf{p} \end{pmatrix}. \tag{2}$$

Then the equations (1) can be written in matrix form,

$$\frac{d\xi}{dt} = \mathcal{M}\xi, \tag{3}$$

where the elements of $\mathcal{M}$ can be read off from Eq. (1):

$$\mathcal{M} = \begin{pmatrix} \dfrac{\partial^2\mathcal{H}}{\partial\mathbf{q}\partial\mathbf{p}} & \dfrac{\partial^2\mathcal{H}}{\partial\mathbf{p}^2} \\ -\dfrac{\partial^2\mathcal{H}}{\partial\mathbf{q}^2} & -\dfrac{\partial^2\mathcal{H}}{\partial\mathbf{p}\partial\mathbf{q}} \end{pmatrix}. \tag{4}$$

This matrix has dimension $6n \times 6n$, and it breaks into four $3n \times 3n$ blocks.

The matrix equation (3) is to be thought of in a Lagrangian sense. The matrix $\mathcal{M}$ is to be evaluated at the phase point currently occupied by the unperturbed system (system 1), while $\xi$ gives the displacement of the second system relative to the first. The matrix $\mathcal{M}$ depends implicitly on the time through the motion of the phase point of the unperturbed system. It changes pretty drastically and pretty rapidly – on the time scale of nearest particles moving past each other.

In Cartesian coordinates, all the elements in the blocks that contain mixed second derivatives vanish, leaving those two blocks completely filled with zeroes.

The off-diagonal elements are another story: the upper right block is diagonal. With equal masses, it is just $1/m$ times the $3n \times 3n$ identity. Unequal masses present no problem of principle, but they mess up the notation, so they'll not be worked out in this note.

The remaining $3n \times 3n$ block, that in the lower left, contains the gradient of forces. (Minus the) first derivative of the Hamiltonian gives the force acting on each particle, so (minus the) second derivative gives the gradient of the forces, $\mathcal{G} =$ (grad **F**). Its structure is worked out in detail in §1.2, but for now we need only note that all its elements are real and that it is symmetrical. A small displacement in which every particle in the system is moved infinitesimally adds an increment to the force acting on each particle by an amount given by (grad **F**).

## 1.1. REDUCE $\mathcal{M}$ TO DIAGONAL FORM

Since the $\mathcal{G}$ is real symmetric, it can be brought to diagonal form by a $3n \times 3n$ orthogonal matrix, $\mathcal{O}$:

$$\mathcal{O}^{\mathrm{T}} \mathcal{G} \mathcal{O} = \mathrm{Diag}$$

(superscript T denotes transpose: since $\mathcal{O}$ is orthogonal, its transpose is its inverse). However, the identity matrix remains an identity matrix under the same transformation, and the scalar multiple, $1/m$, does no harm. Thus we can apply $\mathcal{O}$ to *both* nonzero blocks, bringing both to diagonal form. In the language of $3n \times 3n$ blocks, this reads

$$\begin{pmatrix} \mathcal{O} & 0 \\ 0 & \mathcal{O} \end{pmatrix}^{\mathrm{T}} \begin{pmatrix} 0 & \frac{1}{m}\mathcal{I} \\ \mathcal{G} & 0 \end{pmatrix} \begin{pmatrix} \mathcal{O} & 0 \\ 0 & \mathcal{O} \end{pmatrix} = \begin{pmatrix} 0 & \frac{1}{m}\mathcal{I} \\ \mathrm{Diag} & 0 \end{pmatrix}.$$

The resulting matrix, which is diagonal in the upper right and lower left $3n \times 3n$ blocks, with null blocks on the diagonal, can be reduced to a block diagonal form with $3n$ blocks, each of dimension $2 \times 2$, by means of row and column interchanges. Each of the $2 \times 2$ blocks is of the form:

$$\begin{pmatrix} 0 & 1/m \\ \lambda_K & 0 \end{pmatrix}. \tag{5}$$

Here, $\lambda_K$ is one of the eigenvalues of $\mathcal{G}$. Those $2 \times 2$ matrices can each be diagonalized to give

$$\Lambda_K = \pm\sqrt{\lambda_K/m}, \tag{6}$$

on the diagonal. A pair of eigenvalues of the full $6n \times 6n$ matrix corresponds to each $\lambda_K$. There are $6n$ in all – the required number.

Positive $\lambda$'s give real square roots, while negatives give pure imaginary square roots. The real roots lead to unstable solutions, while the pure imaginary roots give stable solutions. Thus the topic of determining the dimensionality of the stable and unstable domains reduces to counting the number of positive (or negative) eigenvalues of the $3n \times 3n$ matrix, $\mathcal{G} = (\text{grad } \mathbf{F})$. The matrix, $\mathcal{M}$, is not self–adjoint.

The original $6n \times 6n$ matrix has trace 0 (the Liouville theorem!), and so the diagonal form must as well. It's reassuring that it does.

Lagrangian displacements have allowed this problem to be reduced to an exercise in diagonalizing two matrices simultaneously. All the standard texts on classical mechanics show the procedure in their sections on small oscillations.

## 1.2. THE MATRIX OF FORCE GRADIENTS

Eigenvalues of $\mathcal{M}$ can be worked out once the eigenvalues of $\mathcal{G} = (\text{grad } \mathbf{F})$ are in hand. The matrix $\mathcal{G}$ has the explicit form

$$\mathcal{G}_{ij}^{(\alpha\beta)} = \frac{1}{r_{\alpha\beta}^3} \left[ \delta_{ij} - 3 \frac{\left( x_i^{(\beta)} - x_i^{(\alpha)} \right) \left( x_j^{(\beta)} - x_j^{(\alpha)} \right)}{r_{\alpha\beta}^2} \right], \tag{7}$$

where

$$r_{\alpha\beta}^2 = \left( x_k^{(\beta)} - x_k^{(\alpha)} \right) \left( x_k^{(\beta)} - x_k^{(\alpha)} \right), \tag{8}$$

$\alpha, \beta = 1, 2, 3, \cdots, n$ are particle indices, and $i, j = 1, 2, 3$ are coordinate indices. Summation over repeated $i, j, k$ pairs is implied, but not over repeated $(\alpha, \beta)$ pairs. An $i, \alpha$ pair runs $1, 2, 3, \cdots 3n$; the matrix is best thought of as having $i, \alpha$ as its first index and $j, \beta$ as the second index. These expressions apply for $\alpha \neq \beta$; the $\alpha$-$\alpha$ element is simply the sum of the elements with a minus sign

$$\mathcal{G}_{ij}^{(\alpha\alpha)} = -\sum_{\beta \neq \alpha}' \mathcal{G}_{ij}^{(\alpha\beta)}$$

$$= -\sum_{\beta \neq \alpha}' \frac{1}{r_{\alpha\beta}^3} \left[ \delta_{ij} - 3 \frac{\left( x_i^{(\beta)} - x_i^{(\alpha)} \right) \left( x_j^{(\beta)} - x_j^{(\alpha)} \right)}{r_{\alpha\beta}^2} \right], \tag{9}$$

so the row (or column) sums are zero. The matrix is symmetric under coordinate index $(i, j)$ and under particle number $(\alpha, \beta)$ interchanges separately. The quantities $x_i^{(\alpha)}$ are the coordinates of the individual particles. All particles have been taken to have the same mass; generalization to different masses does not change the character of the problem.

The $(i, \alpha)$ notation facilitates decomposition into an $n \times n$ matrix of $3 \times 3$ blocks; each $3 \times 3$ block runs through $i, j = 1, 2, 3$, and the $n \times n$ block matrix has a block entry for each particle pair (with special entries on the block diagonal to make the row– or column-sums be zero). Each of the $3 \times 3$ blocks is real symmetric and has trace zero. The $n \times n$ block matrix is also symmetric, so the full $3n \times 3n$ matrix is real symmetric and has real eigenvalues. The trace of the $3n \times 3n$ is zero, since that of each of the $3 \times 3$ submatrices is zero. Thus the sum of the eigenvalues is zero. There are nonzero eigenvalues, so some must be positive and some negative.

## 1.3. Eigenvalues of $\mathcal{G}$

All eigenvalues of $\mathcal{G}$ are real, since the matrix is real symmetric. It has three zero eigenvalues, corresponding to parallel displacements of all particles:

$$x_i^{(\alpha)} \longrightarrow x_i^{(\alpha)} + X_i.$$

Nothing more is known analytically beyond this and the fact that all eigenvalues must sum to zero.

The eigenvalues have been studied experimentally. Some matrices were generated in which coordinates $x_i^{(\alpha)}$ were chosen randomly within the unit cube, the force gradient matrix computed according to Eq. (7) and Eq. (9), and the eigenvalues found. Cases were tried for various values of $n$. In all cases, the expected three zero eigenvalues were smaller than the nonzero values by a factor near $10^{10}$ (reasonable for the diagonalization routine used), and were easily distinguishable. No other zero eigenvalues were found.

Eigenvalues have also been studied experimentally by forming the matrix from configurations generated as snapshots of a $100$–body systems advanced by direct integration. Properties of these sets of eigenvalues are essentially indistinguishable from those generated by the random number loads except that closest pairs tended to be closer in the actual integrations. Numerical results quoted in the remainder of this note are based on these $100$–body integrations.

The number of positive eigenvalues changes as an $n$–body system develops. There is no clear trend toward more (or toward fewer) positive eigenvalues during an integration.

No repeated eigenvalues were found, apart from the three zeroes always present.

The number of positive eigenvalues averages about $1.2n$ for systems containing 10 particles or more.

The most positive eigenvalue is well approximated by $4/r_{12}^3$ and the most negative by $-2/r_{12}^3$, where $r_{12}$ is the separation of the closest pair of particles.

## 2. Trajectory Separation

Configuration and momentum components remain distinct in the formulation so far presented. Further discussion is facilitated by eliminating the momenta in favor of velocities to get a coupled set of second-order ODE's:

$$\frac{d^2}{dt^2}(\delta \mathbf{q}) = \frac{1}{m}\mathcal{G}(\delta \mathbf{q}). \tag{10}$$

These equations decouple if $G$ is brought to diagonal form. This decoupling holds only instantaneously, since the matrix $G$ changes very rapidly with time. Significant changes occur in the time it takes the closest pair of particles to pass each other. Eq. (10) is also valid only for the tiny displacements allowed in writing the equations in matrix form, Eq. (3). Nonetheless, the picture presents a useful way of looking at the physics of trajectory separation. The rest of this section refers to "snapshots" that last only for these very short times.

Eigenvectors of $G$ represent displacements of all particles in the system such that the change in the force acting on each particle lies along the vector by which that particle was displaced. The particle is accelerated along its vector displacement if the corresponding eigenvalue is positive, decelerated if it is negative.

### 2.1. "NORMAL COORDINATES"

Eigenvectors of $G$ are orthogonal and span the configuration space. They can serve as a basis in which to express $\delta q$ within the little patch on which we are concentrating. They are "normal coordinates," and each corresponds to a "normal mode." There are $3n$ of them. They would be normal modes in the usual sense if the matrix $G$ did not vary with time. Two degrees of freedom are associated with each "mode," accounting for the required $6n$ degrees of freedom. Most are stable, but about 40% of them are unstable, and the three with zero eigenvalues are neutral.

Apart from the three zero eigenvalues that correspond to rigid displacement of all the particles together, the only degeneracies arise when $G$ is evaluated at a point that lies on one of several special hypersurfaces of reduced dimension. These hypersurfaces are sets of measure zero. Degeneracies present no problem in practice. The three zero eigenvalues reflect the six first integrals of the centroid motion, which are conserved in this picture.

### 2.2. THE RATE OF TRAJECTORY SEPARATION

Much of the interest in trajectory separation is to estimate the rate at which trajectories separate. Special interest attaches to its dependence on the number of particles, with the goal of finding how it might act with very large numbers of particles. Any attempt to address rates through the formulation presented here involves lots of approximations. The linerization introduced at Eq. (1) is the only approximation used so far in this note.

Estimates for $e$-folding rates of trajectory separation entail averages of $G$ over realistic trajectories as the unperturbed system explores the phase space. The principal problem is to find some suitable average. The present discussion is included to illustrate the kinds of approximations involved. The description provides useful insights into the physical processes at work in trajectory separation.

Several steps are involved. A new assumption is introduced at each step. One way to go about this is sketched here.

First, the separation vector $\delta\mathbf{q}$ is referred to the basis instantaneously provided by eigenvectors of the current matrix $\mathcal{G}$. While $\delta\mathbf{q}$ itself may be fairly stable in direction, changes in $\mathcal{G}$ make it appear to vary. Eigenvectors of $\mathcal{G}$ swing around without any special relation to the direction of $\delta\mathbf{q}$, so $\delta\mathbf{q}$ appears to take random directions relative to the changing basis. Thus the amplitude of each degree of freedom is approximated as $\pm|\xi|/\sqrt{6n}$. The trajectory separation has been analyzed into "normal modes" with amplitudes split between the two degrees of freedom within each "normal mode."

Next, growths in $\xi$ can be approximated by a random walk with unequal steps. Only unstable "modes," those with positive eigenvalues of $\mathcal{G}$, contribute. Each unstable "mode" grows like $\exp(\Lambda_K t)$ (Eq. (6)). Both momentum and configuration parts in the $K^{\text{th}}$ "normal mode" grow together. The stable and neutral "normal modes" don't grow at all. Since $\Lambda_K$'s are as often negative as positive, the expected growth is zero, a consequence of microscopic reversibility. But the expected square is nonzero. When growths over all unstable "normal modes" are added together, the mean square growth during a short time $\Delta t$ is

$$\left(\frac{|\Delta\xi|^2}{|\xi|^2}\right) = \left[\ln\left(\frac{|\xi_{i+1}|}{|\xi_i|}\right)\right]^2 \approx \frac{1}{6n}\sum_K [\Lambda_K \Delta t]^2 = \frac{2(\Delta t)^2}{6nm}\sum\nolimits^+ \lambda_K. \qquad (11)$$

The second form indicates the change between successive "snapshots." The last form, through Eq. (6), involves the sum over positive eigenvalues of $\mathcal{G}$. The factor 2 appears because each $\lambda_K$ appears twice, once for each sign in Eq. (6). Eq. (11) gives the variance, $\sigma^2$, of $\Delta\ln|\xi|$, since the mean is zero. Each "normal mode" acts like the free motion of an harmonic oscillator. Energy is conserved because there is no forcing term. Estimates according to Eq. (11) areexamples of the "spectrum of stretching numbers" of Contopoulos and Voglis (1996).

Third, the random walk is asymptotically Gaussian if enough steps are involved. Steps take the spatial directions of unstable "normal modes" within one snapshot. Since the logarithm tends to a normal distribution, mean growth over this interval is the mean from a log–normal distribution, which is $e^{\sigma^2/2}$. It is always greater than unity.

Fourth, the quadratic dependence on time interval (the $(\Delta t)^2$ in Eq. (11)) arises because the growth is treated as coherent over the interval $\Delta t$. After a coherence time, this must be terminated, to be replaced by a new randomization of the separation vector. The time for the nearest particles to orbit by a radian might be used for a coherence time. This gives $(\Delta t)^2 \approx d^3/(2Gm)$ where $d$ is the separation of the nearest pair of particles. We then have $s = T_{\text{cr}}/\Delta t$ independent steps, each growing by $e^{\sigma^2/2}$, leading to a total estimated growth per crossing time of $e^{(s\sigma^2)/2}$.

Happily, growth rates estimated in this manner from our 100–body integrations are around 1.5 to 2 per crossing time, which agrees reasonably well with values found by Kandrup et al. (1994) and earlier papers referenced therein. The greatest positive eigenvalue contributed about 40% of the sum of positive eigenvalues, and it is very nearly $4/r_{12}^3$ for the nearest pair. This implies that the contributions of the "mean field" and of the nearest pair to the growth are nearly the same, as observed by Kandrup *et al*. (1992).

## 3. Discussion

1. The number of positive eigenvalues of $\mathcal{G}$ changes as particles move about. Changing numbers of positive and negative eigenvalues have been known for some time (Miller, 1972). They indicate a changing dimensionality of the stable and unstable regions. There are always three zero eigenvalues, which correspond to rigid displacements of the entire system in any of three orthogonal directions.

2. The number of positive eigenvalues is about $1.2n$, where $n$ is the number of particles. This ratio is attained for $n > 10$.

3. There are no duplicate eigenvalues, save for the three zero cases. Occasional duplications require special configurations that reduce them to a set of measure zero.

4. The formulation presented here takes the discrete particle nature of $n-$body systems into account explicitly. Many-body interactions are included up to the full number of particles. Systems with large numbers of particles show exponential trajectory separation, up to and including the infinite numbers implied in some analytic models.

5. The sum of all eigenvalues is zero (trace of $\mathcal{G}$ is zero). This means that there are some positive, as well as some negative, eigenvalues for any $n-$body system.

6. The greatest positive eigenvalue is well approximated by $4/d^3$, where $d$ is the separation of the nearest pair of particles anywhere within the configuration. It contributes about 40% of the sum of all positive eigenvalues. The most negative eigenvalue is near $-2/d^3$.

7. This formulation shows why separations projected onto the configuration space and onto the velocity space each grow at the same rate, as noted by Kandrup and Smith (1991). They're tightly coupled, since the full system is actually described by the $6n \times 6n-$matrix $\mathcal{M}$.

8. There is nothing intrinsic in the arguments presented here to indicate that basic $e-$folding rates should be a few per crossing time. Estimates for $e-$folding rates of trajectory separation entail averages of $\mathcal{G}$ over realistic trajectories as the unperturbed system explores the phase space.

9. An heuristic argument permits an estimate of the $e-$folding rate for trajectory separation. It yields estimates of a few per crossing time, like those of (Kandrup et al., 1994) and earlier papers cited there. Severe approximations are required, but it provides a different way of looking at the problem of trajectory separation.

10. The changing numbers of positive (and of negative) eigenvalues along an actual trajectory demonstrates that the gravitational $n-$body system is *not* an "Anosov system" (such systems are sometimes called "C-systems", (Arnold and Avez, 1968)).

Computations were made possible by means of a grant of computing services from the Numerical Aerodynamic Simulation (NAS) Facility at NASA-Ames, which is gratefully acknowledged. Funds for partial support of this work have been allocated by the NASA-Ames Research Center, Moffett Field, CA, under Cooperative Agreement NCC 2-265 with the University of Chicago.

## References

Arnold, V. I. and A. Avez: 1968, *Ergodic Problems of Classical Mechanics*. New York: W. A. Benjamin.

Contopoulos, G. and N. Voglis: 1996, 'Spectra of Stretching Numbers and Helicity Angles in Dynamical Systems'. *Celest. Mech. & Dyn. Astron* **64**, 1–20.

Goodman, J., D. C. Heggie, and P. Hut: 1993, 'On the exponential instability of $N$-body systems'. *ApJ* **415**, 715–733, Fiche 196–F11.

Gurzadyan, V. G. and G. K. Savvidy: 1986, 'Collective Relaxation of Stellar Systems'. *A&A* **160**, 203–210.

Kandrup, H. E.: 1990, 'Divergence of nearby trajectories for the gravitational $N$-body problem'. *ApJ* **364**, 420–425, Fiche 201–B13.

Kandrup, H. E., M. E. Mahon, and H. Smith, Jr.: 1994, 'On the sensitivity of the $N$-body problem toward small changes in initial conditions. IV'. *ApJ* **428**, 458–465, Fiche 135–G12.

Kandrup, H. E. and H. Smith, Jr.: 1991, 'On the sensitivity of the $N$-body problem to small changes in initial conditions'. *ApJ* **374**, 255–265, Fiche 101–C9.

Kandrup, H. E., H. Smith, Jr., and D. E. Willmes: 1992, 'On the sensitivity of the $N$-body problem to small changes in initial conditions. III'. *ApJ* **399**, 627–633, Fiche 205–F9.

Krylov, N. S.: 1979, *Works on the Foundations of Statistical Physics*. Princeton, N. J.: Princeton University Press.

Lorenz, E. N.: 1963, 'Deterministic Nonperiodic Flow'. *J. Atmos. Sci.* **20**, 130.

Miller, R. H.: 1964, 'Irreversibility in Small Stellar Dynamical Systems'. *ApJ* **140**, 250–256.

Miller, R. H.: 1966, 'Polarization of the Stellar Dynamical Medium'. *ApJ* **146**, 831–837.

Miller, R. H.: 1971, 'Experimental Studies of the Numerical Stability of the Gravitational $n$-Body Problem'. *J. Comp. Phys.* **8**, 449–463.

Miller, R. H.: 1972, 'A Matrix Eigenvalue Problem'. Unpublished Report, Section B in ICR Quarterly Report No. 33.

Miller, R. H.: 1974, 'Numerical Difficulties with the Gravitational $n$-Body Problem'. In: D. G. Bettis (ed.): *Proceedings of the Conference on the Numerical Solution of Ordinary Differential Equations*. Berlin, pp. 260–275. Vol. 362 of Lecture Notes in Mathematics.

Quinlan, G. D. and S. Tremaine: 1992, 'On the reliability of gravitational $N$-body integrations'. *MNRAS* **259**, 505–518.

# DISTRIBUTION OF KOLMOGOROV-SINAÏ ENTROPY

## IN SELF-CONSISTENT MODELS OF

## BARRED GALAXIES

H. WOZNIAK

*IGRAP/Observatoire de Marseille, F-13248 Marseille cedex 4, France*
*E-mail: wozniak@observatoire.cnrs-mrs.fr*

and

D. PFENNIGER

*Observatoire de Genève, CH-1290 Sauverny, Switzerland,*
*E-mail: Daniel.Pfenniger@obs.unige.ch*

**Abstract.** The properties of chaos in 2D self-consistent models of barred galaxies are investigated using Kolmogorov-Sinaï entropy $h_{KS}$. These models are constructed with Schwarzschild's method which combines orbits as elementary building blocks.

Most models are dominated by chaos near the 2/3 of the length of the bar and close to corotation. These locations correspond to regions where star-forming HII regions are observed because gas clouds could shock, shrink and fragment such that star formation could be ignited.

The model the most similar to $N$-body models shows a peak of $h_{KS}$ between the corners of the rectangular-like $x_1$ orbits and the maximum extension points of the Lagrangian orbits. This emphasizes the role of Lagrangian orbits in the morphology of bars. Most models essentially contain 'semi-chaotic' orbits confined inside the corotation.

## 1. Introduction

The determination of the analytical distribution function of galaxies, in particular barred galaxies, remains a long standing problem. Freeman (1966) has been the first to propose an analytical formulation of the distribution function of a stellar bar with a model of an homogeneous triaxial ellipsoid for which the centrifugal and central forces are in equilibrium along the major-axis of the bar. This model has, however, unrealistic properties (e.g. the $x$-axis Lagrangian points extend over the whole major axis instead of being, as expected, located slightly beyond the end of the bar).

A numerical line of attack has been initiated by Schwarzschild (1979) to study triaxial elliptical galaxies. From a library of orbits in a triaxial mass model he determined numerically one of the all possible distribution functions reproducing the initial mass model.

With some improvements, Pfenniger (1984b; 1985) and Wozniak & Pfenniger (1996; 1997) used it to compute self-consistent models of barred galaxies. Apart from Zhao's study (1996) of the Milky Way bar, these studies remain the only attempts to apply Schwarzschild's method to fast rotating bars. These studies exploited more extensively the possibilities offered by Schwarzschild's technique. Instead of finding a single feasible kinematic model for a given mass model, the set of possible solutions was explored by looking at its extreme global properties

*Celestial Mechanics and Dynamical Astronomy* 73: 149–158, 1999.
© 1999 *Kluwer Academic Publishers.*

(energy, angular momentum, etc.) that a given mass distribution does allow while remaining self-consistent. The obtained distribution functions and linear superpositions of them are all possible self-consistent solutions satisfying the equilibrium condition. So the advantage of Schwarzschild's method over $N$-body models is that the whole solution set allowed by an arbitrary mass distribution can be explored, while its main drawback is that no stability information is provided. The stability of such models is best studied with a $N$-body code whose initial conditions of particles are drawn from a particular solution (cf. Zhao, 1996).

Chaotic motion is expected to form a significant part of the orbits populating stellar bars. However, because realistic galactic distribution functions are difficult to construct, the proportion of chaotic orbits is barely determined. Pfenniger (1984b) found between 10 and 30% of chaotic orbits in his models. In their $N$-body models, Sparke & Sellwood (1987) and Pfenniger & Friedli (1991) found a "hot" population which belong both to the bar and the stellar disc. The hot population may contribute up to 30% of the total mass. Kaufmann (1993) and Kaufmann & Contopoulos (1996) confirmed that between 5 and 14% of chaotic orbits populate the bar and as much belong to the "hot" population. However, chaotic orbits play a major role in the secular evolution of galaxies because they introduce irreversibility. A notion such as adiabatic invariance is based on the assumption that phase space is structured by isolating quasi-integrals, thus the absence of some of these means that much more freedom to morphology changes is left to slightly or slowly perturbed orbits. Thus, we decided to quantitatively determine the proportion of such orbits in Wozniak & Pfenniger's (1997) models of barred galaxies. The Kolmogorov-Sinai entropy was used as a tool to quantify the amount of orbital chaos in a stellar system by Udry & Pfenniger (1988).

We will describe Schwarzschild's technique used to built self-consistent models in Sect. 2, the properties of the models in Sect. 3, and how we compute the Kolmogorov-Sinai entropy in Sect. 4. In Sect. 5, we present our preliminary results. Finally, we give our conclusions and a few implications for the morphology of barred galaxies in the last section.

## 2. Schwarzschild's technique

In Schwarzschild's (1979) method, the configuration space is discretized into $N_{cell}$ compact cells. Phase space is discretized by $N_{orb}$ orbits that are computed in the gravitational potential generated by the chosen mass density. The fraction of time spent by each orbit in each cell $(B_{ij})$ is called the occupation. The mass $M_i$ inside any cell $i$ is thus a weighted sum of the $B_{ij}$. The $N_{orb}$ weights $X_j$ of the sum are the unknowns, and are constrained to be positive or zero (i.e. non-negative) to represent a physical mass. Equivalently, we can express this problem as a set of

$N_{cell}$ linear equations with $N_{orb}$ unknowns:

$$M_i = \sum_{j=1}^{N_{orb}} B_{ij} X_j \quad i = 1, \ldots N_{cell}, \tag{1}$$

$$X_j \geq 0, \qquad j = 1, \ldots N_{orb}. \tag{2}$$

This is a linear programming problem (cf. Chvátal, 1983). Of course, with the positivity constraint the number of orbits $N_{orb}$ must be larger than the number of cells $N_{cell}$ to have any possibility to find at least one solution.

As in Pfenniger (1984b) and Wozniak & Pfenniger (1997), instead of the traditional Simplex algorithm of linear programming, we used the NNLS (Non Negative Least Squares) algorithm (Lawson & Hanson, 1974, 1995) which finds a positive least squares solution of Eq. (1). The advantage of NNLS over the Simplex algorithm is to provide, in case no exact solution does exist, the nearest approximate solution in the least-squares sense, instead of nothing for the Simplex. But when a solution set does exist, it finds *one of the exact solutions*, the one which minimizes $\|X\|$. The ability of NNLS to find exact solutions when they exist has been sometimes overlooked. Indeed, when the residuals vanish the solution found by NNLS is exact. For all basic solutions computed by Wozniak & Pfenniger (1997), the relative error on the mass as computed with Eq. (1) is of the order $10^{-7}$, i.e., of the order of the round-off errors.

The minimization or maximization of basic solutions is obtained by adding a 'cost' function perturbing the set of equations (1) (see Pfenniger (see Pfenniger, 1984b for more details). The cost function, also called objective function, can be any linear function of the weights $X_j$. If $Z_j$ represents a physical quantity then the objective function is:

$$\sum_{j=1}^{N_{orb}} Z_j X_j \tag{3}$$

normalized so that $\max(Z_j) = 1$ and $\min(Z_j) = 0$. For a given orbit $j$, $Z_j$ can be the Jacobi constant $E_{Jj}$, the time-averaged $z$-angular momentum $\overline{L_{zj}}$, the absolute value of the time-averaged $z$-angular momentum $|\overline{L_z}|_j$, the energy $E_j = E_{Jj} + \Omega_p \overline{L_{zj}}$, where $\Omega_p$ is the rotation frequency of the bar pattern, or the Kolmogorov-Sinai entropy $h_{KS_j}$. It can also be any linear relation between different of the above quantities. Since the domain of feasible solutions is convex (cf. Pfenniger, 1984b) Fig. 5), we need only to compute 'extreme' models that serve to delimit the domain. Hereafter, we call $\min(Z)$ or $\max(Z)$ models which respectively minimize or maximize the objective function $Z$. Intermediate models can always be constructed by a linear superposition of basic models, while the converse is false.

### 3. The self-consistent models

We reuse Wozniak & Pfenniger's (1997) self-consistent models to allow some comparisons. We have also computed two new models which are extremum of the Kolmogorov-Sinai entropy (cf. next section). These models share the following properties:

1. The mass model consists of a Miyamoto disk and a $n = 2$ Ferrers (1877) bar (cf. Pfenniger, 1984a for more details). This mass model is a first order approximation of the real mass distribution in barred galaxies.
2. Inside the corotation radius, the computational polar grid contains 24 cells in $r$, 20 in $\theta$. The radial resolution is 4 cells per length unit, i.e. 250 pc. The central part ($r = 0$) of the grid consists of a single circular cell. We have used the symmetries of the mass density w.r.t. both axes to fold space onto the first quadrant.
3. The set of orbits which solves Eq. (1) belongs to a wide library of roughly 4000 orbits. Wozniak & Pfenniger (1997) carefully built this library keeping four properties under control: 1) the occupation of each individual orbit $B_{ij}$ is time independent in order to ensure the construction of a model to a pre-determined level of time independence, 2) the library includes trapped orbits around stable periodic orbits as well as chaotic orbits, 3) identical occupation numbers $B_{ij}$ for two orbits with distinct initial conditions imply that the $B$ matrix is degenerate and the problem becomes "ill-posed". Redundant orbits have thus been removed, 4) aliases between the time steps, total time integration and the orbit natural frequencies are avoided.

The only difference between the models is the kind of objective function which is minimized or maximized. The have retained 6 models from Wozniak & Pfenniger (1997), namely the $\min(|\overline{L_z}|)$, $\max(|\overline{L_z}|)$, $\min(\overline{L_z})$, $\max(\overline{L_z})$, $\min(E_J)$ and $\min(E)$ models. Moreover, we have computed two new models ($\min(h_{KS})$ and $\max(h_{KS})$) described in the next section.

### 4. Computation of Kolmogorov-Sinai entropy

As we would like to measure the level of chaos of each models, we have computed the Lyapunov exponents for each orbits of the library. We globally followed the same rules as Udry & Pfenniger (1988): we integrate the equations of motion simultaneously with the variational equations. The two pairs $(\lambda_k, -\lambda_k)$ of Lyapunov exponents are computed with a Gram-Schmidt orthogonalization. For each orbit, the computations are performed until the fluctuations of the space density become lower than a given threshold (0.5%). This ensures that each orbit is close to be time independent. The most chaotic orbits do require longer integration times. In WP97, integration times range between $2T_{bar}$ for the regular orbits closest to the centre and $4500\,T_{bar}$ for the most chaotic ones. Here, we have computed orbits during at least $T_{min} = 18\,T_{bar}$. We thus name 'regular' orbits with $h_{KS} < \log(T_{min})/T_{min} \approx 0.0038$.

Such long calculation times are justified on the ground that the exercise here is to mimic analytical models, which if integrable would correspond to an infinite integration time. Therefore the maximum retained integration time has been deliberately chosen, when necessary, much larger than a typical galaxy physical age.

The Kolmogorov-Sinai entropy $h_{KS}$ of an orbit can be viewed as the rate at which it looses information about its initial conditions. This quantity has been shown by Pesin (1977) to be equal to the sum of the positive Lyapunov exponents. For a given orbit $j$, a proper measure of this loss (or gain depending on the viewpoint) of information is:

$$h_{KS_j} = \sum_{\lambda_{kj}>0, k=1}^{4} \lambda_{kj} . \tag{4}$$

In Hamiltonian systems the entropy $h_{KS_j}$ vanishes only for regular orbits. Orbits with non-zero $h_{KS_j}$ have a sensitive dependence on initial conditions which is a possible criterion of chaos.

For the computation of the solutions for the $\min(h_{KS})$ and $\max(h_{KS})$ models, the objective function is:

$$h_{KS} = \sum_{j=1}^{N_{orb}} h_{KS_j} X_j \tag{5}$$

where $h_{KS}$ is the Kolmogorov-Sinai entropy of the whole system.

The spatial distribution of Kolmogorov-Sinai entropy ($h_{KS_i}$) is obtained with:

$$h_{KS_i} = \frac{1}{M_i} \cdot \sum_{j=1}^{N_{orb}} h_{KS_j} B_{ij} X_j \tag{6}$$

for each cell $i$.

## 5. Results

Figs. 1 and 2 show the spatial distribution of $h_{KS}$. Regions with $h_{KS} < 0.0038$ ('regular' orbits) are restricted to the inner part of the bar and to the minor-axis ($3 \lesssim x \lesssim 4$). The nucleus is always a global minimum because in this model near the centre the potential is nearly harmonic, i.e., integrable. This does not mean that 'regular' orbits are excluded from the other regions of the model but their contribution is negligible w.r.t. non-regular orbits.

The contribution of chaotic orbits leads to maxima in the $h_{KS}$ distribution in essentially two regions:

1. Along the corotation radius, especially near the $L_{1,2}$ Lagrangian points. Pfenniger (1984b) and Kaufmann & Contopoulos (1996) already claimed that chaotic orbits contribute essentially to the mass near $L_{1,2}$ points. Our study confirm this behavior.

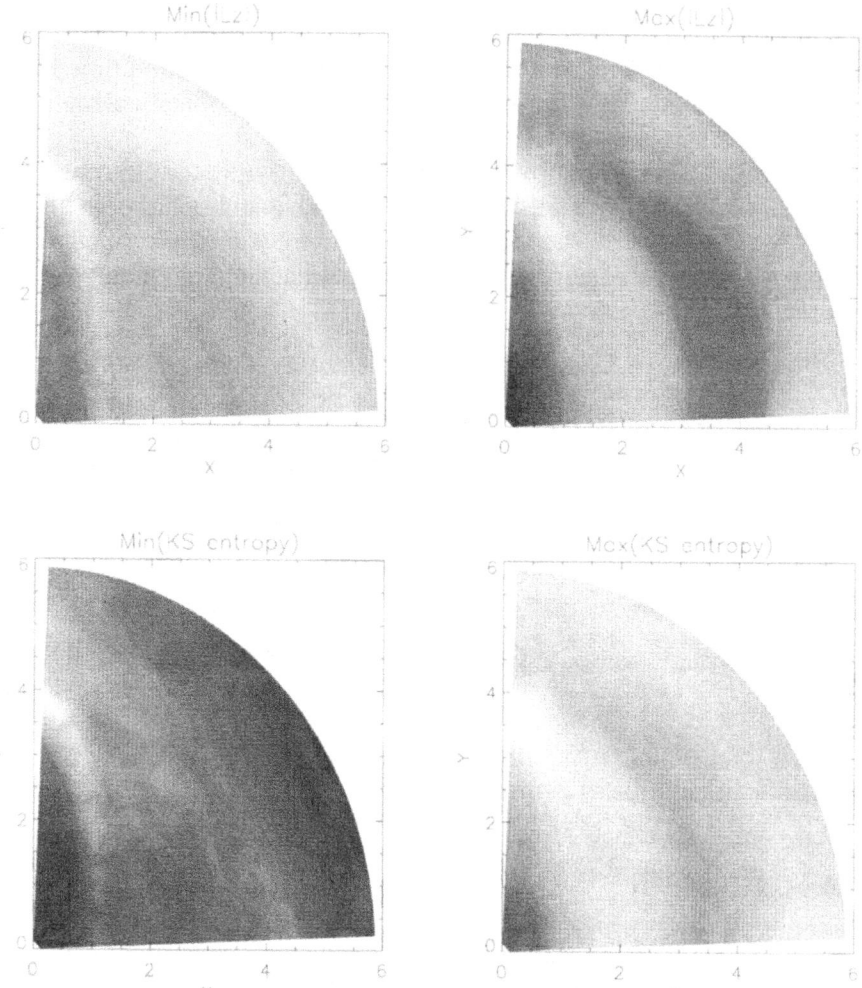

Fig. 1. Spatial KS entropy distribution $h_{KS_i}$ for the first quadrant of our mass model. Models $\min(E_J)$, $\min(E)$, $\min(\overline{L_z})$ and $\max(\overline{L_z})$ are displayed. The stellar bar (6 kpc long) is aligned with the $y$-axis. The highest values of $h_{KS_i}$ are white, the lowest are black

2. *Along the major-axis of the bar* ($3.5 \lesssim y \lesssim 4$), at the apocentre of the more elongated elliptical-like $x_1$ orbits. Indeed, the mass density response of an orbit trapped around the $x_1$ family is maximum at this location (cf. Fig. 3). Thus, most of the mass in this region is due to $x_1$-trapped orbits.

The most chaotic models are the $\min(\overline{L_z})$, $\min(|\overline{L_z}|)$ and, obviously $\max(h_{KS})$ models. Wozniak & Pfenniger (1997) noted that the $\min(|\overline{L_z}|)$ model shows a similar distribution function than an $N$-body model by Pfenniger & Friedli (1991). The $\min(|\overline{L_z}|)$ model maximizes the contribution of orbits elongated along the bar. It favors the orbits trapped around the $x_1$ family at the expense of the hot population

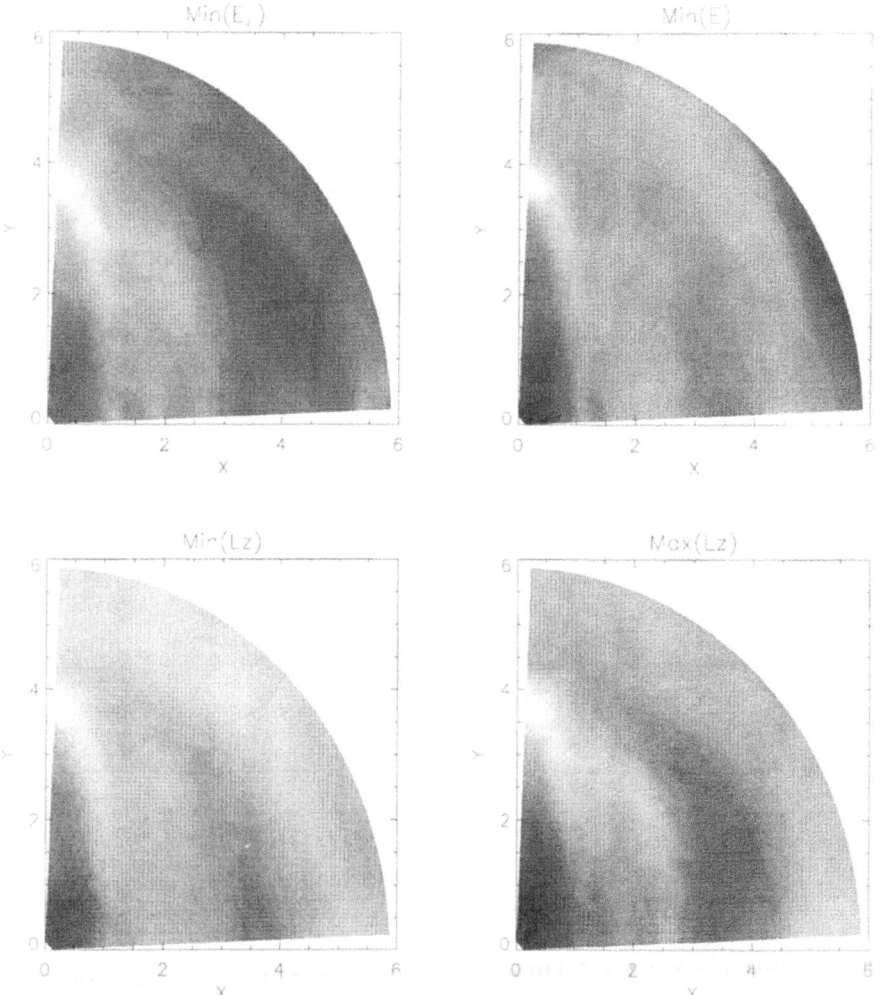

Fig. 2. As for Fig. 1, but for models $\min(|\overline{L_z}|)$, $\max(|\overline{L_z}|)$, $\min(h_{KS})$ and $\max(h_{KS})$

which populates mostly a region that includes the corotation. The maximum of $h_{KS}$ located at $x \approx 2.6$, $y \approx 4.7$ is mostly due to rectangular-like $x_1$-trapped orbits with Jacobi constants higher than the one at the ultra-harmonic 4/1 bifurcation. Moreover, the maximum extension of Lagrangian orbits which circulate around the $L_{4,5}$ Lagrangian points is close to this maximum of $h_{KS}$.

The less chaotic models are the $\min(E_J)$ and $\min(h_{KS})$ models. Pfenniger (1984b) already noted that the $\min(E_J)$ model has the simplest velocity field.

The mass distribution as a function of $h_{KS}$ (Fig. 4) shows that most of the mass is due to orbits with $h_{KS} \lesssim 0.008$. This means that a single population of 'regular' orbits cannot account for the mass distribution. Non-regular orbits are thus unavoidable. For comparison, the frequency $\Omega_p = 0.0547$, which shows that the

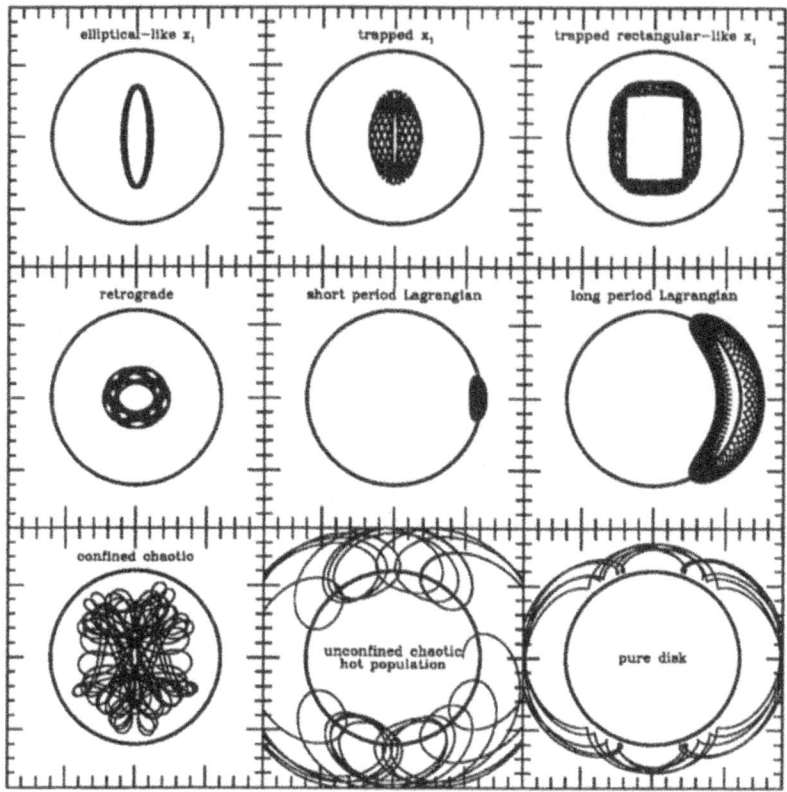

Fig. 3. Examples of typical orbits in the library

time-scale of chaos $h_{KS}^{-1} \gtrsim 125$ is similar to the dynamical time-scale $2\pi/\Omega_p \approx 115$.

One can also define another population in non-regular orbits which peaks at $h_{KS} \approx 0.01$. This component is clearly present in almost all models but $\min(E_J)$ and $\min(h_{KS})$. Orbits with $h_{KS} \gtrsim 0.013$ contribute less than a few percent to the total mass of the models. Globally, when chaotic orbits contribute to the mass density, they are only moderately chaotic orbits ($h_{KS} \lesssim 0.013$), even for the model that maximizes the Kolmogorov-Sinai entropy. These orbits remain confined inside the corotation (cf. Fig. 3). They are similar to the 'semi-chaotic' orbits of Wozniak (1994).

## 6. Astrophysical implications and conclusions

Most models are not only dominated by chaos close to corotation but also near the 2/3 of the length of the bar. This is of great importance for the gas flow along the bar. Indeed, this highly chaotic region should correspond to a region where the gas

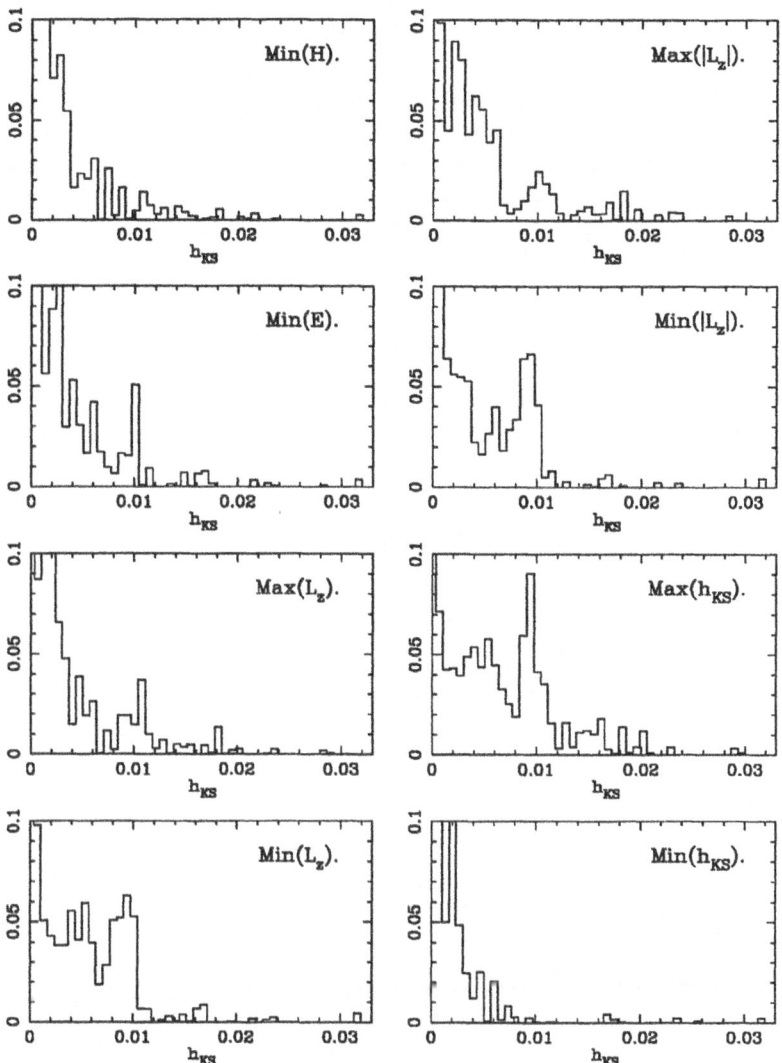

Fig. 4. Mass distribution (in fraction of mass inside the corotation radius) w.r.t. the KS entropy for the eight models

clouds could shock, shrink and fragment such that star formation could be ignited. In many barred galaxies one observes indeed HII regions located near the middle of the major-axis of the stellar bar.

The $\min(|\overline{L_z}|)$ model (the most similar to $N$-body models) shows a peak of $h_{KS_i}$ between the corners of the rectangular-like $x_1$ orbits and the maximum extension points of the Lagrangian orbits. This emphasizes the role of Lagrangian orbits in the morphology of bars. Moreover, although our mass model is not designed to study the spiral structure, it should be noted that the location of these maximums roughly corresponds to the starting points of the spiral structure of strong barred galaxies.

All the models, even those with high total KS entropy (e.g. $\min(\overline{L_z})$ and $\max(h_{KS})$), essentially contain orbits with $h_{KS_j} < 0.013$ which are not extremely chaotic orbits. Indeed, these orbits remain confined inside the corotation. They are in fact *semi-chaotic* orbits. These orbits have the same nature that the ones which seem to be responsible for the rectangular-like shape of strong bars (Wozniak, 1991), (Wozniak, 1994).

## Acknowledgements

All the computations were done on the IGRAP computer facilities.
This work has been supported by the Swiss National Science Foundation. We thank the anonymous referee for constructive remarks.

## References

Chvátal V.: 1983, Linear Programming, Freeman, New York
Ferrers N.M.: 1877, *Quart. J. Pure Appl. Math.* **14**, 1
Freeman K.C.: 1966, *Mon. Not. R. Astr. Soc.* **134**, 1
Kaufmann D.E.: 1993, Ph D. Thesis, Univ. of Florida
Kaufmann D.E., Contopoulos G.: 1996, *A&A* **309**, 381
Lawson C.L., Hanson R.J.: 1974, Solving Least Squares Problems, Prentice-Hall, Englewood Cliffs, New Jersey; 1995, Classic in Applied Mathematics 15, SIAM, Philadelphia (updated NNLS routines are available at http://www.netlib.org)
Pesin Y.B.: 1977, *Russ. Math. Surveys* **32**, 55
Pfenniger D.: 1984a, *A&A* **134**, 373
Pfenniger D.: 1984b, *A&A* **141**, 171
Pfenniger D.: 1985, Ph D Thesis, University of Geneva
Pfenniger D., Friedli D.: 1991, *A&A* **252**, 75
Schwarzschild M.: 1979, *ApJ* **232**, 236
Sparke L., Sellwood J.A., 1987, MNRAS 225, 653
Udry S., Pfenniger D.: 1988, *A&A* **198**, 135
Wozniak H.: 1991a, Ph.D. Thesis, University of Paris 7
Wozniak H.: 1994, in *Ergodic Concepts in Stellar Dynamics*, Gurzadyan V.G. & Pfenniger D. (eds.), Lecture Notes in Physics 430, Springer-Verlag, Heidelberg, p. 264
Wozniak H., Pfenniger D., 1996, in: Barred Galaxies, Buta R., Elmegreen B.G., Crocker D.A. (eds.), Proc. IAU Coll 157, ASP Conferences Series, p. 445
Wozniak H., Pfenniger D.: 1997, *A&A* **317**, 14
Zhao H.: 1996, *MNRAS*, **283**, 149

# REGULAR AND CHAOTIC MOTION IN GLOBULAR CLUSTERS

DANIEL D. CARPINTERO, JUAN C. MUZZIO and FELIPE C. WACHLIN

*Facultad de Ciencias Astronómicas y Geofísicas - UNLP and PROFOEG – CONICET*
*E-mail: ddc@fcaglp.unlp.edu.ar,jcmuzzio@fcaglp.unlp.edu.ar, fcw@fcaglp.edu.ar*

**Abstract.** As a first step towards a comprehensive investigation of stellar motions within globular clusters, we present here the results of a study of stellar orbits in a mildly triaxial globular cluster that follows a circular orbit inside a galaxy. The stellar orbits were classified using the frequency analysis code of Carpintero and Aguilar and, as a check, the Liapunov characteristic exponents were also computed in some cases.

The orbit families were obtained using different start spaces. Chaotic orbits turn out to be very common and while, as could be expected, they are particularly abundant in the outer parts of the cluster, they are still significant in the innermost regions. Their relevance for the structure of the cluster is discussed.

**Key words:** globular clusters – orbit classification – chaotic motion

## 1. Introduction

We tend to think of globular clusters as spherical stationary stellar systems that are well described by King's or Michie's models (see, e.g., (Binney and Tremaine, 1987)). Obviously, nobody in his right mind would search for chaotic motions in such systems, but the truth is that: a) Globular clusters are not spherical and exhibit different degrees of ellipticity (see, e.g., (Han and Ryden, 1994)); b) Globular clusters are not isolated systems and the motions of their stars are governed, not only by the cluster's field, but by the tidal forces of the galaxy where the cluster belongs as well. Thus, as neither angular momentum nor energy has to be conserved, it is very reasonable to expect to find chaotic motions in the stellar orbits within globular clusters.

The presence of significant chaotic motions would certainly have important consequences for the structure of the cluster and the models should take into account this fact. The present work is just a first step to show that, even under very simple hypotheses, chaotic orbits turn out to be very abundant in globular clusters, thus paving the way for future, more detailed, studies on this subject.

## 2. The Model

We wanted to begin our investigation of chaos in globular clusters with the simplest possible case, so that:
a)   We adopted a circular orbit for the motion of the globular cluster around the galaxy.
b)   We assumed that the cluster is deformed by the effect of the tidal forces only.
c)   We neglected the effects of stellar encounters within the cluster.

The adoption of more realistic conditions should increase chaos because: a) With elongated cluster orbits we lose Jacobi's integral; b) More triaxial potentials might enhance chaoticity; c) Impulsive forces will contribute to chaos.

*Celestial Mechanics and Dynamical Astronomy* 73: 159–168, 1999.
© 1999 *Kluwer Academic Publishers.*

The galaxy was represented by a spherically symmetrical logarithmic potential and the globular cluster with a modified Satoh (Satoh, 1980) distribution, whose potential is:

$$\Phi_S(x, y, z) = -\frac{GM}{\sqrt{x^2 + y^2 + z^2 + g(g + 2\sqrt{y^2 + (z/b)^2 + h^2})}}. \tag{1}$$

Here the origin of coordinates lies at the center of the globular cluster, the $x$ axis points in the direction opposite to the galactic center, the $y$ axis in the direction of motion of the cluster around the galaxy and the $z$ axis perpendicular to the orbital plane. Notice that we have interchanged $x^2 + y^2$ with $z^2$, in order to obtain a prolate (rather than Satoh's oblate) system, and that we have also divided $z$ by a parameter $b$, in order to get a triaxial system (tidal deformation yields the shortest axis perpendicular to the orbital plane).

The main advantage of this election is that isodensity surfaces increase their ellipticity as we move outwards, just as it should happen with a system that is tidally deformed, as shown by the full lines in Figure 1. We have also included in the figure (dashed lines) the effective equipotential curves (i.e., those that result from adding the centrifugal term and the galactic potential to the modified Satoh potential).

The equations of motion are:

$$\ddot{x} = -\frac{GMx}{S^3} - \omega^2 R^2 \frac{R + x}{(R + x)^2 + y^2 + z^2} + \omega^2(R + x) + 2\omega\dot{y}; \tag{2}$$

$$\ddot{y} = -\frac{GMy(1 + g/T)}{S^3} - \omega^2 R^2 \frac{y}{(R + x)^2 + y^2 + z^2} + \omega^2 y - 2\omega\dot{x}; \tag{3}$$

$$\ddot{z} = -\frac{GMz(1 + g/(b^2 T))}{S^3} - \omega^2 R^2 \frac{z}{(R + x)^2 + y^2 + z^2}, \tag{4}$$

where

$$S = \sqrt{x^2 + y^2 + z^2 + g(g + 2T)}, \quad T = \sqrt{y^2 + (z/b)^2 + h^2}, \tag{5}$$

and $M$ is the mass of the globular cluster, $R$ is the radius of its orbit, and $\omega$ is its angular velocity. The Jacobi integral is:

$$E_J = \frac{1}{2}(\dot{x}^2 + \dot{y}^2 + \dot{z}^2) - \frac{1}{2}\omega^2\left[(R + x)^2 + y^2\right] + \Phi(x, y, z), \tag{6}$$

where $\Phi$ is the sum of the potential of the globular cluster, $\Phi_S$, and that of the galaxy:

$$\Phi_G(x, y, z) = \frac{1}{2}\omega^2 R^2 \ln\left[(R + x)^2 + y^2 + z^2\right]. \tag{7}$$

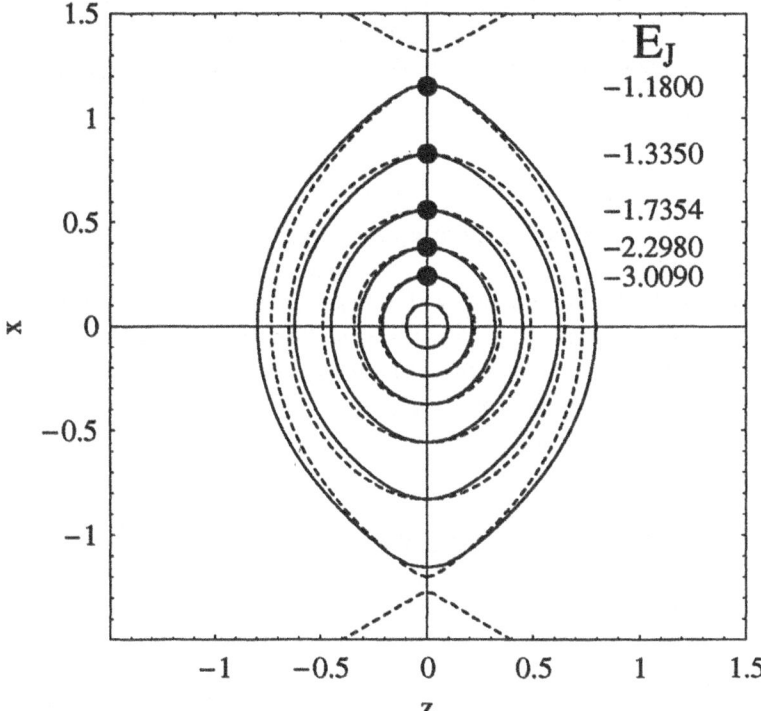

Fig. 1. Isodensity curves for the modified Satoh potential (full lines) in the $x$–$z$ plane. The equipotentials that result from adding the centrifugal term and the galactic potential are shown as dashed lines.

We adopted the following values: $b - 0.8$, $h = 0.5$, $g - 0.05$, $R = 100$, $\omega = 0.5$, which result in a tidal radius $r_t = x_t = 1.24$, and the half–mass radius is $r_h = 0.28$.

If, for example, we choose $R = 10$ kpc and the mass of the galaxy within that radius as $M_g = 1.25 \times 10^{11} M_\odot$, then we have a tidal radius of about $r_t = 120$ pc and a cluster mass of $M = 5 \times 10^5 M_\odot$, that is, reasonable values for a globular cluster.

## 3. Orbital Analysis

### 3.1. LIAPUNOV CHARACTERISTIC EXPONENTS

D. Pfenniger kindly let us use his LIAMAG routine that computes the six Liapunov exponents following Benettin's method.

We are not so much interested on whether a specific stellar orbit is chaotic or not, as in having statistical information on a large number of orbits. Therefore, we followed an approach similar to that of Merritt and Fridman (1996): 1) We integrated the orbits for about 10,000 orbital periods (rather than 100, as they did); 2) We used as estimator the sum of the three non–negative Liapunov exponents (also called Kolmogorov entropy); 3) We dubbed, rather arbitrarily, chaotic those orbits where:

$$\ln(s_1 + s_2 + s_3) > -5, \tag{8}$$

where the $s_i (i = 1, 2, 3)$ are the positive estimates of the Liapunov exponents after 10,000 orbital periods.

We used this method just as a check, because: 1) It only allows one to decide between regular and chaotic orbits, providing no further information on the kind of orbit one has; 2) It is very slow (about one day of computing on a Pentium Pro, 200 MHz, personal computer for 150 orbits).

### 3.2. FREQUENCY ANALYSIS

This technique was introduced by Binney and Spergel (1982), (1984) and extended, in a different form, by Laskar (1993). Carpintero and Aguilar (1998) refined the original method and prepared a FORTRAN code that allows one to automatically classify large numbers of orbits.

The basis of the method is that regular orbits move on a torus–like manifold and are quasi–periodic. Fourier spectra of the time series of the coordinates of a regular orbit consist of discrete lines whose frequencies are integer linear combinations of the frequencies of the angle variables. Thus, from the Fourier spectra one can classify the regular orbits. Besides, chaotic orbits yield continuous spectra and can be recognized too (see (Carpintero and Aguilar, 1998), for examples of different orbits). The main limitation of the method is the difficulty to recognize whether finite precision numbers have a rational quotient.

### 3.3. INITIAL CONDITIONS

We prepared sets of initial conditions for several values of the Jacobi integral (−3.0090, −1.7354, −1.3350 and −1.1800). For each value of the integral, we selected four sets: 1) Zero initial velocity; 2) $x–y$ plane and $\dot{z}$ initial velocity; 3) $x–z$ plane and $\dot{y}$ initial velocity; 4) $y–z$ plane and $\dot{x}$ initial velocity.

Schwarzschild (1993) proposed the use of the initial conditions 1 and 3 for non–rotating potentials. As we have a rotating potential, we preferred to add also initial conditions 2 and 4. We expect to have sampled the whole phase space with these sets, at the price of some possible overlap, as we will see later on.

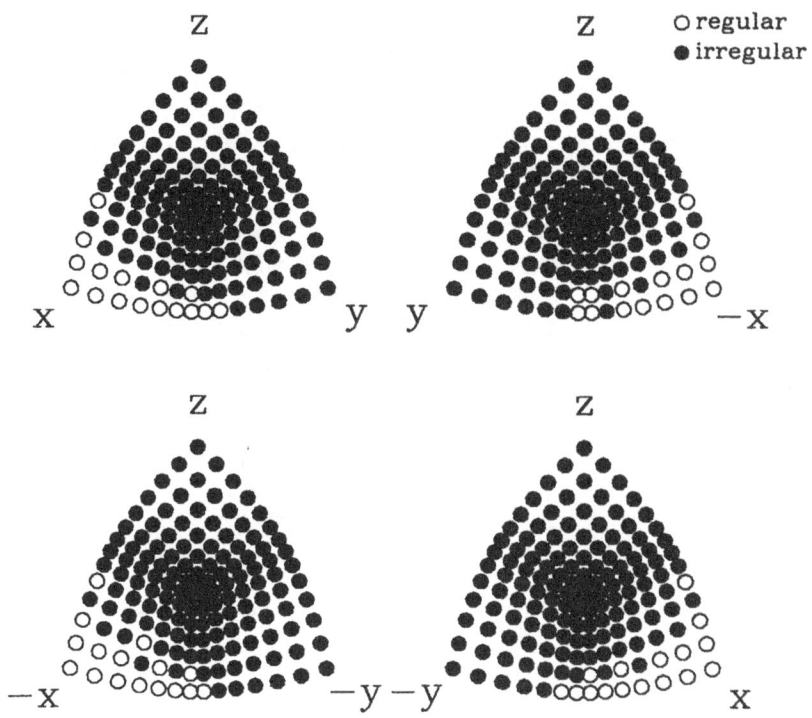

Fig. 2. Zero velocity start space for $E_J = -1.335$. The regular or irregular character of the stellar orbits was decided from the Liapunov exponents analysis.

## 4. Results

For every set of initial conditions we prepared colour plots showing, with different colours, the different types of stellar orbits that result from those initial conditions. Black and white examples are given in Figures 2 through 5, which correspond to a value of the Jacobi integral of $E_J = -1.335$. Figures 2 and 3 are results obtained from the Liapunov exponents analysis: Figure 2 corresponds to zero velocity initial conditions and Figure 3 to initial conditions in the $x$–$z$ plane. Figures 4 and 5 are results from the frequency analysis and correspond, respectively, to the same initial conditions of Figures 2 and 3. Notice that the density of points is two orders of magnitude larger for Figures 4 and 5, resulting in much better definition, thanks to the short computing times needed for the frequency analysis. There is a generally good agreement between the results of both methods, but there are a couple of

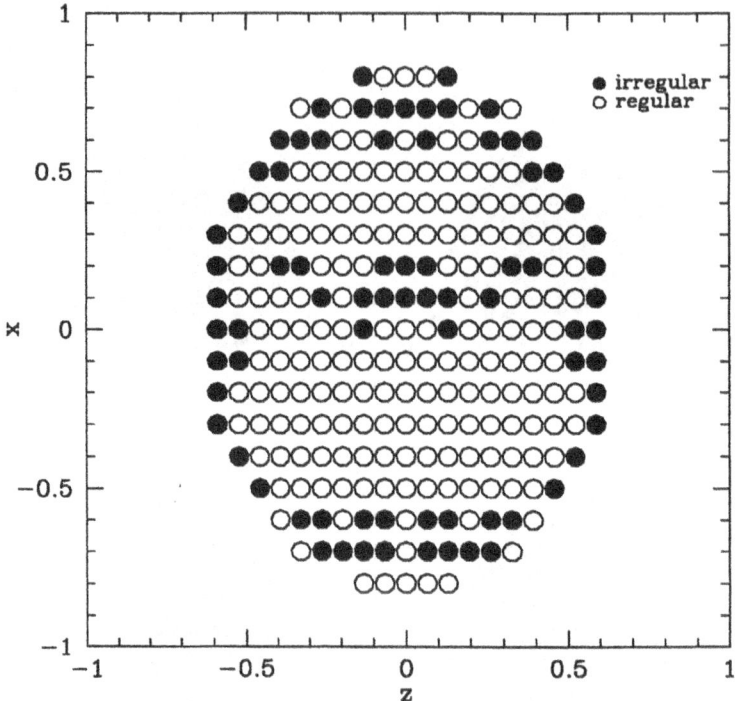

Fig. 3. $x$–$z$ start space for $E_J = -1.335$. The regular or irregular character of the stellar orbits was decided from the Liapunov exponents analysis.

caveats. First, a small fraction of the orbits (less than 10%) could not be classified with the frequency analysis code; from our experience with that code, we know that most of those orbits turn out to be chaotic on a more detailed analysis, so that we counted them as such. Second, the Liapunov exponents tend to give somewhat larger fractions (by about 10%) of chaotic orbits. The most likely explanation for this discrepancy comes from the very different integration times: between 100 and 200 periods for the frequency analysis and 10,000 periods for the Liapunov one; as a result, orbits that behave regularly most of the time, although they are truly chaotic, have a much larger chance of getting detected in the second case.

We noticed that the $x$–$z$ and $y$–$z$ initial conditions gave the same fractions for the different kinds of orbits, so that we suspect that with those sets of initial conditions we are sampling esentially the same parameter space. Therefore, we combined the results of both start spaces together in what follows.

Figures 6a, b and c give the fractions of the different kinds of stellar orbits, as a function of the Jacobi integral, for the zero initial velocity, $x$–$y$ and $x$–$z$ plus $y$–$z$ start spaces, respectively. Boxes and chaotic orbits clearly dominate for zero initial velocity conditions, while small–axis tubes are the most abundant orbits in

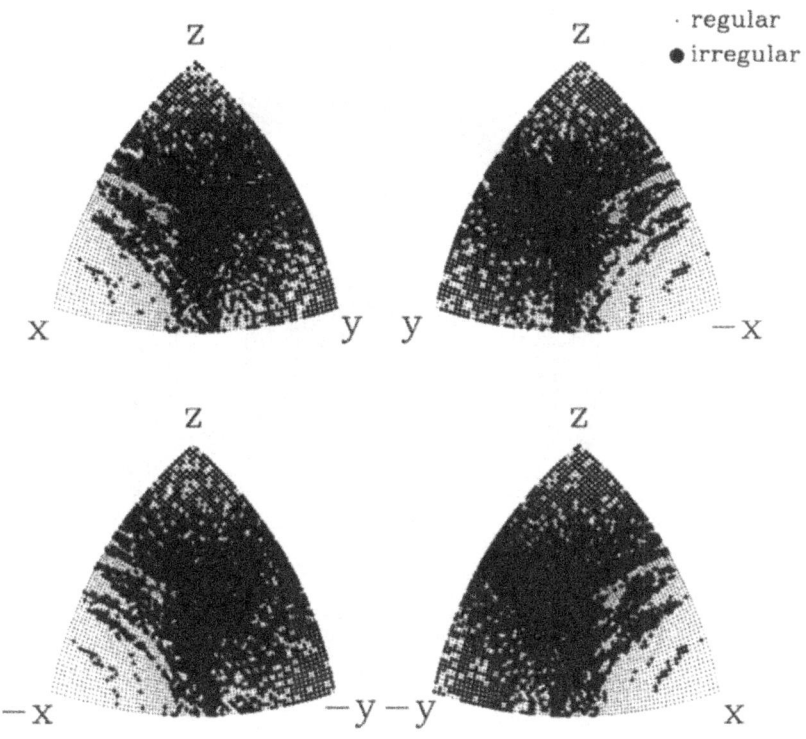

Fig. 4. Zero velocity start space for $E_J = -1.335$. The regular or irregular character of the stellar orbits was decided from the frequency analysis. The bulk of the regular orbits are long–axis boxes, as shown by their proximity to the $x$–axis.

the other cases. Long–axis tubes are almost non–existent. As expected, chaotic orbits predominate for low absolute values of the Jacobi integral, i.e., mainly in the outermost parts of the globular cluster. Nevertheless, chaos is still significant for $E_J = -3.009$, with 37% of the orbits with zero initial velocity, 35% of those on the $x$–$y$ start space and 21% of those on the $x$–$z$ and $y$–$z$ start spaces. These results are particularly important because the zero velocity $E_J = -3.009$ surface encloses about 50% of the total mass of the globular cluster, so that chaos is present well inside the cluster and, moreover, seems to dominate in its outer half.

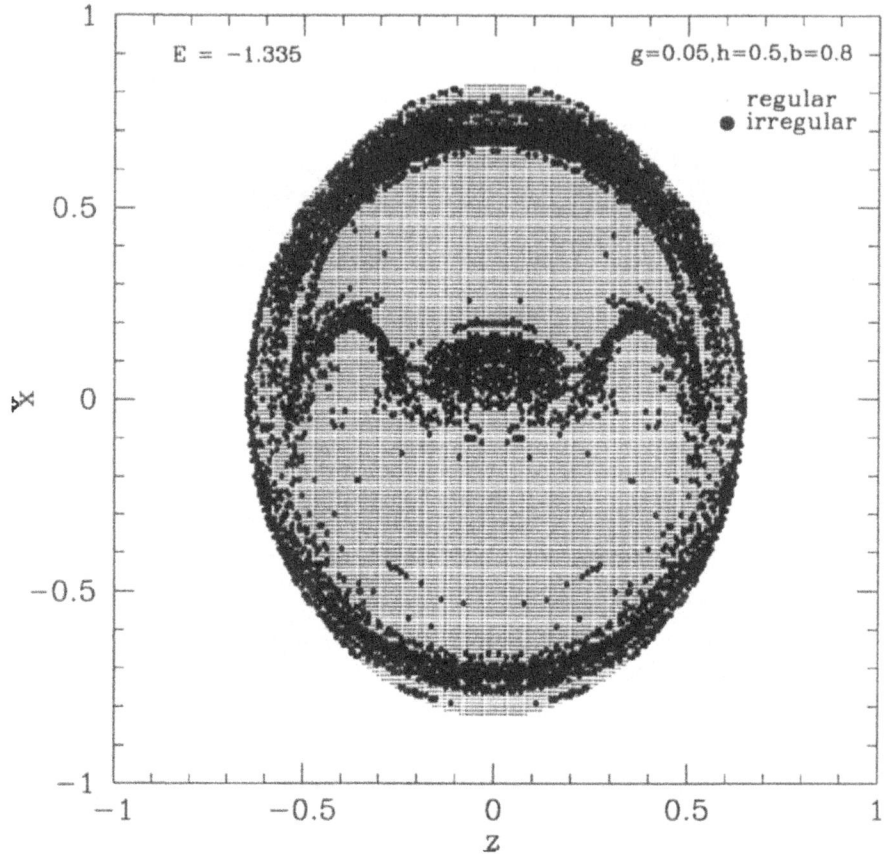

Fig. 5. $x$–$z$ start space for $E_J = -1.335$. The regular or irregular character of the stellar orbits was decided from the frequency analysis.

## 5. Conclusions

From a methodological point of view, we see that the results of frequency analysis are in generally good agreement with those from the Liapunov exponents. The advantages of frequency analysis over Liapunov exponents are that the former needs much less computing time and, in addition to decide between regular and chaotic motion, it also allows the classification of the regular orbits.

Stellar orbits within globular clusters are highly chaotic. For the stars that (barely) reach the tidal limiting surface, the fraction of chaotic orbits may lie somewhere in between 50% and 90%. Nevertheless, it is even more surprising that the innermost parts of the cluster are also affected, and that as much as about 30% of the orbits that reach the half mass limiting surface might be chaotic.

Moreover, the Liapunov times we obtained for $E_J = -1.335$ are surprisingly

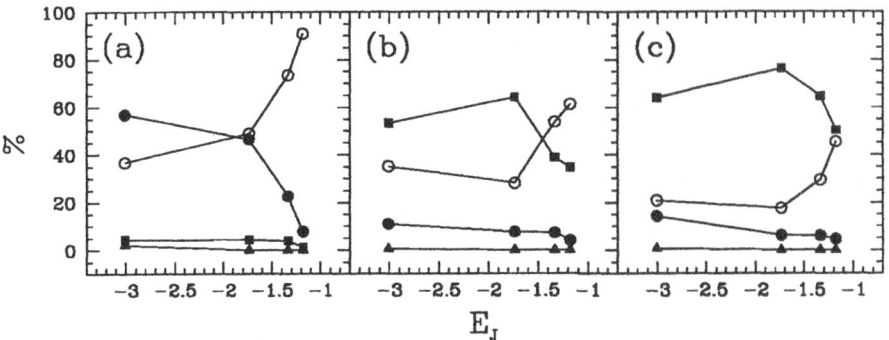

Fig. 6. Fractions of the different kinds of stellar orbits, for ($a$) the zero initial velocity, ($b$) $x$–$y$ and ($c$) $x$–$z$ plus $y$–$z$ start spaces. Filled circles: box orbits; filled triangles: long–axis tubes; filled squares: short–axis tubes; open circles: chaotic orbits.

short: they tend to crowd near 10 to 30 time units, while orbital periods are in the range between 1.4 and 7.5 time units. Not only is this short in terms of orbital periods but also in terms of cluster age: for the reasonable choice of units mentioned above, the cluster age would be about 50 to 100 time units, that is, longer than the Liapunov times. Evolution should thus be very fast, at least in the outermost parts of the cluster.

Long–axis tube orbits are very rare, even at the innermost parts of the cluster, while short–axis tubes are the most common orbits for non–zero initial velocity conditions. Box orbits only seem to dominate in the innermost parts, for zero velocity initial conditions. The scarcity of box orbits in the outermost, and most elongated, parts of the cluster may pose some problems to build self–consistent models. Such models should probably rely on the more abundant chaotic orbits but that might, in turn, complicate the building of stationary models, particularly considering the short Liapunov times detected here.

## Acknowledgements

We are very grateful to J. A. Núñez for very useful suggestions and advice, and to R. E. Martínez, H. R. Viturro, E. Suárez, M. C. Fanjul de Correbo and S. D. Abal de Rocha for their technical assistance. This investigation was supported with grants from the Universidad Nacional de La Plata and the Consejo Nacional de Investigaciones Científicas y Técnicas de la República Argentina.

## References

Binney, J. and Spergel D.: 1982, *ApJ*, **252**, 308
Binney, J. and Spergel D.: 1984, *MNRAS*, **206**, 159
Binney, J. and Tremaine, S.: 1987, *Galactic Dynamics*, Princeton University Press

Carpintero, D.D. and Aguilar, L.A.: 1998, *MNRAS*, **298**, 1
Han, C. and Ryden, B.S.: 1994, *ApJ*, **433**, 80
Laskar, J.: 1993, *Physica D*, **67**, 257
Merritt, D. and Fridman T.: 1996, *ApJ*, **460**, 136
Satoh, C.: 1980, *PASJ*, **32**, 41
Schwarzschild, M.: 1993, *ApJ*, **409**, 563

# PERIODIC ORBITS AROUND A MASSIVE STRAIGHT SEGMENT

ANDRÉS RIAGUAS*

*TERMA at European Space Operations Centre. 64293 Darmstadt. Germany*

ANTONIO ELIPE

*Grupo de Mecánica Espacial. Universidad de Zaragoza. 50009 Zaragoza. Spain*

and

MARTÍN LARA

*Real Observatorio de la Armada. 11110 San Fernando. Spain*

**Abstract.** In this paper, we consider the motion of a particle under the gravitational field of a massive straight segment. This model is used as an approximation to the gravitational field of irregular shaped bodies, such as asteroids, comet nuclei and planets's moons. For this potential, we find several families of periodic orbits and bifurcations.

## 1. Introduction

The interest for studying orbits around small celestial bodies such as asteroids and nuclei comets has grown considerably in the recent years. The most important space agencies have included missions to such small celestial bodies in their current programs, like the NEAR and ROSETTA missions (13; 17), just to mention but a few. These missions consider the flying of a spacecraft around an asteroid and even the landing on its surface (ROSETTA). On the other hand, in binary asteroids (e.g. Ida and Dactyl), the density of one asteroid may be inferred through the dynamics of its satellite (3).

The classical approach for handling the gravitational potential of a celestial body is to expand it in series of spherical harmonics (see e.g. (9), (8)); when the shape of the body is an spheroid, the convergence of the series is fast enough. However, when the shape is irregular, which happens in many of the celestial objects (asteroids, comet nuclei or planets' satellites like Phobos), these series hardly converge in the vicinity of the body. New models that fit better the main shape features of the body must be used instead.

When irregular shaped bodies are considered, such as the asteroids Eros, Ida, Amaltea (J5), etc., we found the body elongation as their main shape feature. This elongated shape makes pseudospherical approach to the gravitational field of this mass distribution far from the true effect. Indeed, the series expansion of the gravitational potential has its convergence guaranteed outside any sphere centered at the center of mass of the body and radius such that it encloses completely the mass of the body; hence, in the cases of elongated bodies there is a gap where the representation of the field force is uncertain.

* On leave from the Dpto. de Matemática Aplicada and Grupo de Mecánica Espacial. Universidad de Zaragoza. 50009 Zaragoza. Spain

*Celestial Mechanics and Dynamical Astronomy* **73**: 169–178, 1999.
©1999 *Kluwer Academic Publishers.*

Some alternative models to the expansion in spherical harmonics have already been proposed. For instance, Werner (18) and Scheeres *et al.* (15; 16; 19) use the potential and force of an homogeneous polyhedron close in shape to the asteroid. Prieto and Gómez–Tierno (14) model this type of bodies by a massive dipole; they find also that an axial symmetric body can be replaced by a massive wire lying in the axis of symmetry with appropriate mass distribution. Halamek (7) also studies the gravitational field of a massive straight segment.

In this paper, we consider the gravitational field originated by a massive straight segment that is fixed in the space. For this body, we express the potential function in closed form. Once the potential is obtained, we compute families of periodic orbits by means of a generalization of the method of numerically continuation of periodic orbits with respect to a parameter given by Deprit and Henrard (5). As an starter of the method, we compute Poincaré sections. We found several families and their bifurcations.

## 2. Potential of a Straight Segment

Let us consider a straight segment of length $2\ell$ and mass $M$. The gravitational potential per unit mass created by this *one* dimensional body at a certain point $P$ in space is given by the line integral

$$U(P) = -G \int_L \frac{dm}{r}.$$

where $G$ is the Gaussian constant. Assuming the linear mass density ($\sigma$) to be constant, this quadrature is not difficult to solve, and appears in textbooks on potential theory (9). Here we will use its expression in terms of intrinsic quantities such as is given in (2).

Let us consider an inertial reference frame $Oxyz$, and let $\boldsymbol{x}_1$, and $\boldsymbol{x}_2$ be the position vectors of the end points of the straight segment. The position vector of a point of the segment is

$$\boldsymbol{x} = \boldsymbol{x}_1 + \nu\,(\boldsymbol{x}_2 - \boldsymbol{x}_1) = \boldsymbol{x}_1 + \nu\,\boldsymbol{x}_{12} \quad \text{with} \quad 0 \le \nu \le 1,$$

hence, $r^2 = \boldsymbol{x} \cdot \boldsymbol{x} = 4\nu^2\ell^2 + \nu\,(r_2^2 - r_1^2 - 4\ell^2) + r_1^2$, and the length element is $dr = 2\ell\,d\nu$, where we denote $r_1 = \|\boldsymbol{x}_1\|$, $r_2 = \|\boldsymbol{x}_2\|$, and $2\ell = \|\boldsymbol{x}_{12}\|$.

Thus, the above quadrature is

$$U(P) = -2\,G\,\sigma\ell \int_0^1 \frac{d\nu}{r} = -GM \int_0^1 (4\nu^2\ell^2 + \nu(r_2^2 - r_1^2 - 4\ell^2) + r_1^2)^{-1/2}\,d\nu,$$

which results is

$$U(O) = -\frac{GM}{2\ell} \log\left(\frac{r_1 + r_2 + 2\ell}{r_1 + r_2 - 2\ell}\right), \tag{1}$$

equation that depends only on the distances ($2\ell$, $r_1$ and $r_2$).

As it was expected, equation (1) shows that the potential is symmetric with respect to the line containing the segment. This axial symmetry suggests the convenience of using cylindrical coordinates, for by virtue of Noether's theorem, the projection of the angular momentum on this axis (one of the conjugate moments in cylindrical coordinates) is an integral.

## 3. Equations of Motion

Let us assume now that the straight segment is fixed in the space. In that case, we may choose the origin to be at the center of mass of the segment and take the Cartesian axes in such a way that the segment lies along the $x$-axis.

Under these conditions, and denoting $\mu = GM$, $s = r_1 + r_2$ and $p = r_1 r_2$, the Lagrangian of the motion of a particle of unit mass in cylindrical coordinates $(\rho, \lambda, x)$ is

$$\mathcal{L} = \frac{1}{2} \left( \dot{\rho}^2 + \dot{x}^2 + \rho^2 \dot{\lambda}^2 \right) + \frac{\mu}{2\ell} \log \left( \frac{s + 2\ell}{s - 2\ell} \right), \tag{2}$$

where, now, $s$ and $p$ are

$$s = \sqrt{\rho^2 + (x - \ell)^2} + \sqrt{\rho^2 + (x + \ell)^2}, \quad p = \sqrt{\rho^2 + (x - \ell)^2}\sqrt{\rho^2 + (x + \ell)^2}.$$

Thus, since the coordinate $\lambda$ is cyclic, the conjugate moment $\Lambda = \rho^2 \dot{\lambda}$ is an integral, and the problem is reduced to only two degrees of freedom.

The equations of motion in cylindrical coordinates are

$$\ddot{\rho} = \frac{\Lambda^2}{\rho^3} - \frac{2\mu s \rho}{p(s^2 - 4\ell^2)}, \quad \ddot{x} = \frac{-2\mu x}{sp}, \quad \frac{d}{dt}(\rho^2 \dot{\lambda}) = 0. \tag{3}$$

Theoo arc, precisely, the equations ot motion we shall use to find periodic orbits.

Besides, from these equations, we can detect some particular motions. Indeed, when $\dot{\lambda} \equiv 0$, the plane of the motion contains the segment; when $x \equiv 0$, the motion takes place on this plane; and when $\rho \equiv 0$, the motion takes place along the $x$-axis (we do not deal here with collision problems, neither with the passing of the particle through the segment). These two last cases are one degree of freedom system (in $\rho$ and $x$ respectively), hence, integrable.

## 4. Periodic Orbits

To obtain information about the solutions and global behavior of the dynamical system we plot Poincaré sections; they will be of some help in finding initial conditions for periodic solutions.

Although the motion takes place in a 3-D space, the fact that the $\Lambda$ is an integral, allows to study the motion on a plane, namely the plane $Ox\rho$, that is rotating about the axis $Ox$ with angular velocity $\dot\lambda$. Besides, the system is autonomous (see Eq. (2)), hence the energy $h$

$$h = \frac{1}{2}\left(\dot\rho^2 + \dot x^2 + \frac{\Lambda^2}{\rho^2}\right) - \frac{\mu}{2\ell}\log\left(\frac{s+2\ell}{s-2\ell}\right). \tag{4}$$

is an integral.

With this, we define the surface of section as $h = $ constant and $x = 0$, $\dot x > 0$. On this 2-D surface of section, we plot points $(\rho, \dot\rho)$.

In the case here considered, we take the units of length, time and mass in such a way that $2\ell = 1$, $\mu = 1$ and the energy $h = -1/2$. For each plot, we compute typically about 20 orbits of about 100 revolutions. We made several surfaces sections for different values of the integral $\Lambda$.

From the expression of the energy (4), we find that the closed curve in the plane $(\rho, \dot\rho)$ that bounds the region where the motion is possible satisfies the equation

$$\dot\rho = \pm\sqrt{2h + \frac{\mu}{\ell}\log\left(\frac{s+2\ell}{s-2\ell}\right) - \frac{\Lambda^2}{\rho^2}},$$

and easily is checked that itself is solution of the differential system (3).

In Figure 1 we show a Poincaré section for $\Lambda = 0.5$. It is seen that the whole manifold of solutions for this case is dominated by a central fixed point, on the $\rho$-axis, of the stable elliptic type. This periodic orbit is surrounded by quasi periodic orbits lying on concentric tori. Besides, we detect several fixed points, some of then inside islands and others corresponding to hyperbolic type equilibria. The initial conditions for some of these orbits are listed on Table I.

For these orbits, we computed an index of stability, namely the trace $(\text{Tr}(T))$ of the matrizant of the associated Hill equation at the end of one period (5). As it is well known, if $|\text{Tr}(T)| > 2$, the characteristic exponents of the orbit are of the unstable type; if $|\text{Tr}(T)| < 2$, they are of the stable type; and $|\text{Tr}(T)| = 2$ represents a case of indifferent stability. This stability index appears on the fourth column of the table.

For these initial conditions, we compute the periodic orbits. Note that these orbits are periodic on the meridian plane $(\rho, x)$, but not necessarily in the Cartesian frame. The Figure 2 shows some of these 2D periodic orbits, both in the $\rho\,x$–plane and in the Cartesian system.

We compute Poincaré sections for several values of the parameter $\Lambda$. They are presented in Figure 3. For high values of $\Lambda$, the motion is strongly dominated by the central periodic orbit. Nevertheless, as $\Lambda$ decreases, several island and elliptic and hyperbolic points appear, such as the ones described above. For $\Lambda = 0.4$, the cross section is different; indeed, the central elliptic points has bifurcated, and now it is converted into hyperbolic, and two new elliptic equilibria appear on the $\rho$-axis. But this situation does not last much, since in the plot for $\Lambda = 0.35$, a new pitchfork bifurcation occurs, the unstable point being now stable, and two new symmetric

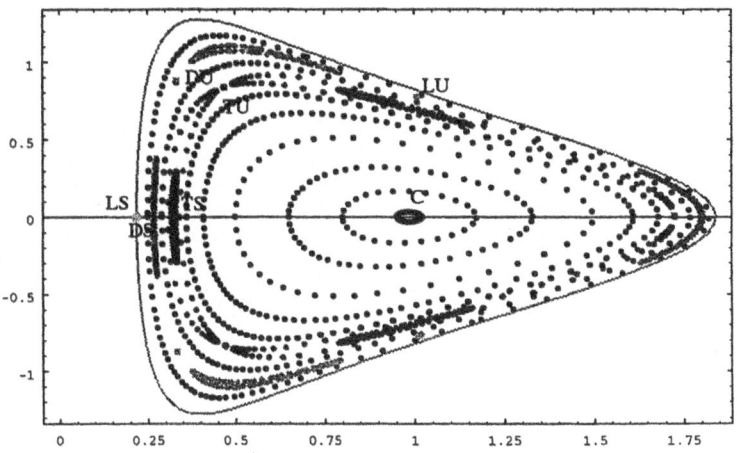

Fig. 1. Poincaré map for $\Lambda = 0.5$.

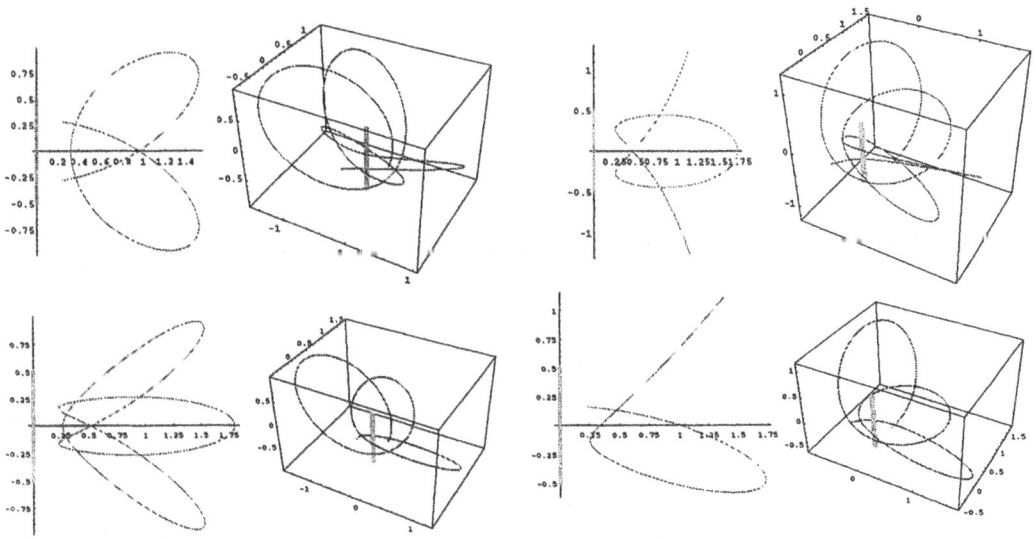

Fig. 2. Periodic orbits in the orbits $\rho x$–plane for $\Lambda = 0.5$. Top: TS (left) and TU (right). Bottom: DS (left) and DU (right). The $xyz$ plots are the same orbits in the Cartesian frame. Note that they are not 3D periodic orbits.

TABLE I

Initial conditions of some periodic orbits for $\Lambda = 0.5$ and $h = -0.5$ ($\lambda = x = 0$)

| Orbit | Stability | Period | $|\text{Tr}(T)|$ | $\rho$ | $d(\rho)/dt$ | $d(x)/dt$ | $d(\lambda)/dt$ |
|-------|-----------|--------|---------|--------|-----------|-----------|-----------|
| C     | S         | 6.4523  | 1.959   | 0.985  | 0.000     | 0.833     | 0.515     |
| LS    | S         | 12.8095 | 1.902   | 0.218  | 0.000     | 0.119     | 10.51     |
| LU    | U         | 6.4059  | 2.012   | 0.980  | 0.837     | 0.000     | 0.521     |
| DS    | S         | 19.0162 | 1.807   | 0.268  | 0.000     | 1.022     | 6.971     |
| DU    | U         | 19.0169 | 2.191   | 0.332  | 0.875     | 0.873     | 4.515     |
| TS    | S         | 25.1312 | 0.796   | 0.323  | 0.000     | 1.221     | 4.783     |
| TU    | U         | 25.1153 | 3.362   | 0.469  | 0.835     | 0.934     | 2.268     |

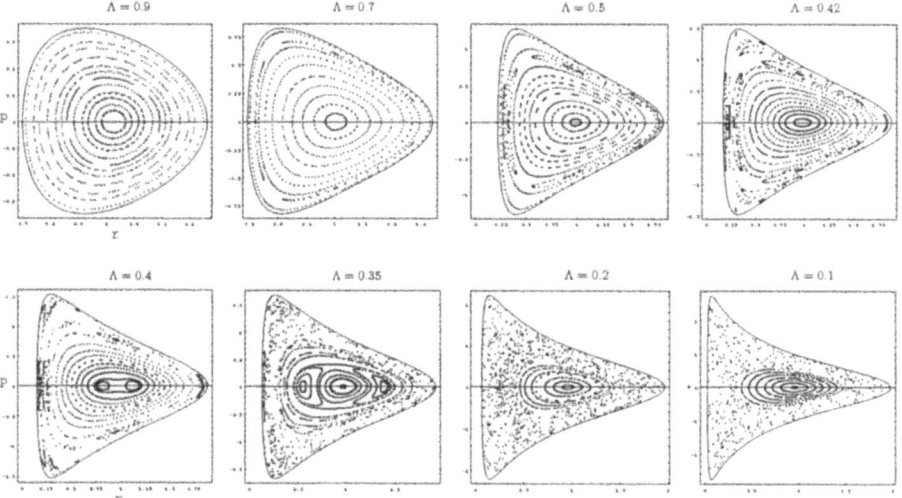

Fig. 3. Poincaré surfaces of section for different decreasing values of the parameter $\Lambda$. The central point corresponding to a periodic orbit can be easily identified.

unstable points appear outside the $\rho$-axis. However, this configuration disappears for $\Lambda = 0.2$, where the region of chaotic behavior is wider.

The change of the topology due to the change of stability of the central fixed point and the appearance of a small unstable region along the family of the central fixed points is quite common in other problems, like resonances in asteroids (see e.g. 6) or the critical inclination in artificial satellite theory (see e.g. 4; 1).

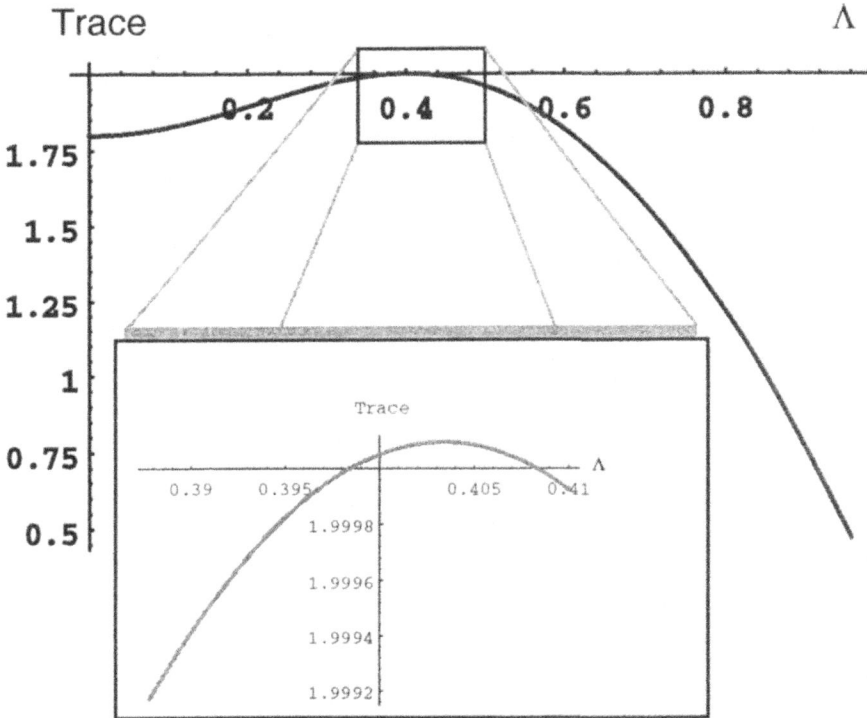

Fig. 4. Evolution of the trace of the matrizant versus $\Lambda$ for the family of circular orbits $(0 \leq \Lambda \leq 1)$. The family is unstable in the interval $0.39844 < \Lambda < 0.40830$.

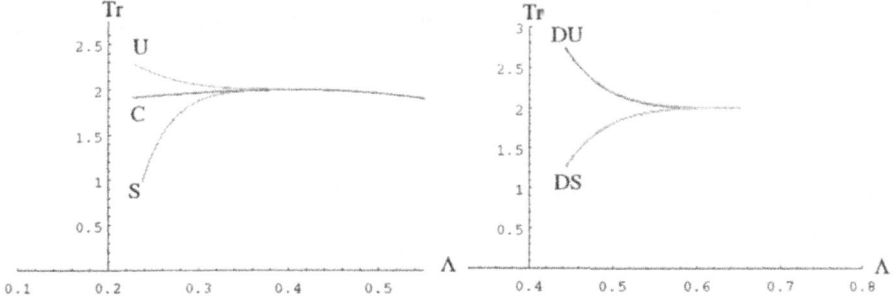

Fig. 5. Trace of the matrizant versus $\Lambda$. Left plot: family of circular orbits and the stable and unstable families that bifurcate from the circular family (the stable family S appears at $\Lambda = 0.40830$, and the unstable U at $\Lambda = 0.39844$). Right plot: families DU (unstable) and DS (stable); these families appear at $\Lambda = 0.6492$ and $\Lambda = 0.6498$ respectively, and they are bifurcations of another family.

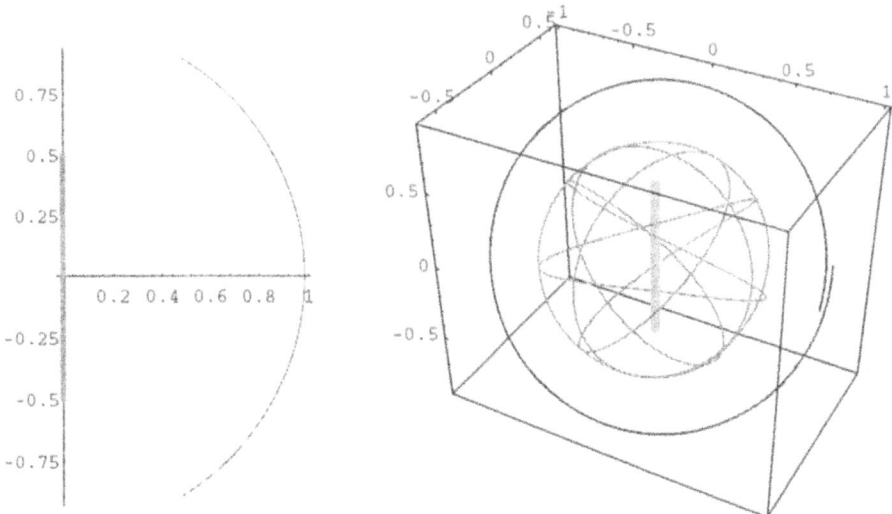

Fig. 6. Circular periodic orbit $\Lambda = 0.5$. Left figure (for $h = -0.5$) is the periodic orbit in the $\rho x$–plane, that is not a closed orbit in the 3D Cartesian frame (The outer orbit in the right plot). To have a 3D periodic orbit, we continue the family for variations of the energy. The inner orbit (right plot) is a closed orbit in the $xyz$–Cartesian frame and correspond to a value of the energy $h = -0.78728$

## 5. Families of Periodic Orbits

In order to determine the exact values of the parameter $\Lambda$ at which bifurcations occur, we use the method of numerical continuation with respect to a parameter of families of periodic orbits. The method is essentially the one given by Deprit and Henrard 5 with some additions made in Lara *et al.* 10; 11. The process addresses a Boundary Value Problem for the variational equations relative to conservative dynamical systems with two degrees of freedom.

Briefly, it consists in the following: starting with a set of initial conditions close to one periodic solution, we correct this initial set to obtain initial conditions for a true periodic orbit. Then, we vary the value of the parameter ($\Lambda$ in our case), and by calculating and refining a tangent prediction we obtain new initial conditions corresponding to a periodic orbit for the new value of the chosen parameter. In order to improve the prediction, we must numerically integrate the equations of motion and their tangent and normal variations, the variational equations associated with this solution. The main feature of this method, is that it splits the normal displacements along an orbit from the tangent ones: the latter, indeed, are secular in nature. For details, the reader is addressed to 5; 10.

We computed the family starting with the circular orbit C. For each periodic orbit belonging to this family, after one period $T$, we determine the trace ($|\text{Tr}(T)|$) of the associated Hill equation. The transition from $|\text{Tr}(T)| < 2$ to $|\text{Tr}(T)| > 2$ (and conversely) will correspond to the bifurcation. In Figure 4 we present the evolution of the trace $|\text{Tr}(T)|$ versus the parameter $\Lambda$. At the value $\Lambda = 0.40830$ we have the first pitchfork bifurcation; the elliptic point in the Poincaré section is split into three points, one hyperbolic and two of the elliptic type. The next pitchfork happens for $\Lambda = 0.39844$; the hyperbolic point is converted, again, into an elliptic point and two new hyperbolic points (with $\rho \neq 0$) appear. From the evolution of the trace, we can see the bifurcation and the stability of the three families (Figure 5, left).

Similar results are obtained for the families DS and DU. These families exist for $\Lambda < 0.6498$, and they are bifurcations of another family (Figure 5, right). For small values of the parameter $\Lambda$, the families enter into chaotic regions.

The periodic orbits just found are periodic in the $\rho\,x$–plane, but not necessarily in the Cartesian frame, since the angle $\lambda$ is moving at the $\dot\lambda$ rate. In order to have a 3D periodic orbit, it is necessary to impose a commensurately relation, namely, there must be two integers $m$ and $n$ such that

$$\lambda(t_0 + T) - \lambda(t_0) = \frac{m}{n}2\pi. \tag{5}$$

Thus, if we fix the two integers $m$ and $n$, we have to find the new period of the periodic orbit for a given value of the parameter $\Lambda$ in order to fulfill the periodicity condition (5). This is equivalent to find periodic orbits that repeat its ground track 12. Since the period varies with the altitude —that depends on the energy, and since our method for continuation of natural families is valid for whatever parameter, we compute the natural families of circular periodic orbits for variation of the energy $h$, that now is taken as a parameter. As an illustration, we compute the 3D circular orbit for $n = 5$, $m = 1$ and $\Lambda = 0.5$. As starter, we use the circular orbit found for $\Lambda = 0.5$ and $h = -0.5$ (see Table I). By varying the energy, we found that the repetition condition (5) is fulfilled for $h = -0.78728$, to which corresponds a circular orbit (see Figure 6) with initial conditions $\rho = 0.648485$, $x = 0.818833$, $\dot\rho = \dot x = 0$ and the period is $T = 3.885305$.

## Conclusions

The gravitational field of a very elongated celestial body is modeled by a massive straight segment fixed in the space. For this logarithmic potential we found families of periodic orbits and their bifurcations. Work for a rotating segment is in progress.

## Acknowledgements

Thanks are due to Professor Broucke for suggesting us this problem. Comments from an anonymous referee improved the final version of the manuscript. Partial support comes from the Spanish Ministry of Education (CICYT # PB95-0807).

# References

Broucke, R.: 1994, "Numerical Integration of Periodic Orbits in the Main Problem of Artificial Satellite Theory," *Celest. Mech. & Dynam. Astr.*, **58**, 99–123.

Broucke, R.: 1995, "Closed form expressions for some gravitational potentials: triangle, rectangle, pyramid and polyhedron." *AAS/AIAA Spaceflight Mechanics Meeting*, **AAS 95-190**, Albuquerque.

Byrnes, D.V., D'Amario, L.A.: 1995, "Dactyl orbit determination analysis," *AAS/AIAA Astrodynamics Specialist Conference*, **AAS 95-315**, Halifax.

Coffey, S. L., Deprit, A., Miller, B. R.: 1986, "The Critical Inclination in Artificial Satellite Theory," *Celest. Mech. & Dynam. Astr.*, **39**, 365–406.

Deprit, A. and Henrard, J.: 1967, "Natural families of periodic orbits," *Astronomical Journal* **72**, 158–172.

Hadjidemetriou, J.D.: 1993, "Asteroid Motion near the 3:1 Resonance," *Celest. Mech. & Dynam. Astr.*, **56**, 563–599.

Halamek, P.: 1988, *Motion in the potential of a thin bar.* Ph. D. dissertation. Univ. of Texas. Austin.

Heiskanen, W. A., Moritz, H.: 1967, *Physical Geodesy*, W. H. Freeman and Co., San Francisco.

Kellogg, O. D.: 1954, *Foundations of Potential Theory.* Dover Publications, Inc. New York.

Lara, M., Deprit, A., Elipe, A.: 1995, "Numerical continuation of frozen orbits for the zonal problem." *Celest. Mech. & Dynam. Astr.*, **62**, 167–181.

Lara, M.: 1996, "On numerical continuation of families of periodic orbits in a parametric potential." *Mechanics Research Communications*, **23**, 291–298.

Lara, M.: 1997, "On periodic polar orbits on the Artificial Satellite Problem." *Journal of the Astronautical Sciences*, **45**, 321–328.

"NEAR.- Special Issue on the NEAR Mission to 433 Eros": 1995, *Journal Astronautical Sciences* **43**.

Prieto-Llanos, T., Gómez-Tierno, M.A.: 1994, "Station keeping at libration points of natural elongated bodies," *Journal of Guidance Control and Dynamics*, **14**, 787–794.

Scheeres, D.J.: 1995, "Satellite dynamics about Eros," *AAS/AIAA Spaceflight Mechanics Meeting*, **AAS 95-110**, Albuquerque.

Scheeres, D.J., Ostro, S.J., Hudson, R. S., Werner, R. A.: 1996, "Orbits close to Asteroid 4769 Castalia," *Icarus*, **121**, 67–87.

Schwehm, G., Hechler, M.: 1994, "Rosetta - ESA's Planetary cornerstone mission," *ESA Bulletin*, **77**, 7–18.

Werner, R. A.: 1994, "The gravitational potential of a homogeneous polyhedron or don't cut corners," *Celest. Mech. & Dynam. Astr.*, **59**, 253–278.

Werner R.A., Scheeres D.J.: 1996 , "Exterior gravitation of a polyhedron derived and compared with harmonic and mascon gravitation representation of asteroid 4769 Castalia," *Celest. Mech. & Dynam. Astr.*, **65**, 313–344.

# ESCAPE FROM A CRISIS IN FOKKER-PLANCK MODELS

SIMON F. PORTEGIES ZWART* and KOJI TAKAHASHI

*Department of Earth Sciences and Astronomy, College of Arts and Sciences,
University of Tokyo, 3-8-1 Komaba, Meguro-ku, Tokyo 153-8902, Japan*

**Abstract.** Recent $N$-body simulations have shown that there is a serious discrepancy between the results of $N$-body simulations and the results of Fokker-Planck simulations for the evolution of globular and rich open clusters under the influence of the galactic tidal field. In some cases, the lifetime obtained from Fokker-Planck calculations is more than an order of magnitude smaller than those from $N$-body simulations. In this paper we show that the principal cause for this discrepancy is an over-simplified treatment of the tidal field used in previous Fokker-Planck simulations. We performed new Fokker-Planck calculations using a more appropriate implementation for the boundary condition of the tidal field. The implementation is only possible with *anisotropic* Fokker-Planck models, while all previous Fokker-Planck calculations rely on the assumption of isotropy. Our new Fokker-Planck results agree well with $N$-body results. Comparison of the two types of simulations gives a better understanding of the evolution of such clusters.

## 1. Introduction

Star clusters range in mass from a few hundred to several million solar-masses. In order to understand their formation and dynamical evolution, detailed numerical modeling is required. There are, however, many effects which complicate their evolution and numerical models of star clusters are just beginning to incorporate deviations from the ideal star cluster (see Vesperini and Heggie 1997; Portegies Zwart et al. 1998b).

Collisional $N$-body simulations are very expensive in terms of computer time. Even with super computers or special-purpose machines, it is impossible to do a simulation with the number of particles comparable to that of a real globular cluster. Therefore we are forced to rely on either $N$-body simulations with smaller number of particles or more approximate methods such as Fokker-Planck techniques. In theory, these two approaches should give identical results.

In order to check the reliability of the Fokker-Planck models with other models ($N$-body, gaseous, Monto Carlo, etc.), some authors compared the results of various types of numerical simulations (Aarseth et al. 1974; Giersz and Heggie 1994a, 1994b; Giersz and Spurzem 1994; Spurzem and Takahashi 1995). These comparisons demonstrate that for isolated clusters made of point masses the results of Fokker-Planck simulations are in good agreement with $N$-body computations.

Recent comparison between the same techniques for clusters in the galactic tidal field, however, gave a completely different view (Fukushige and Heggie 1995; Heggie et al. 1998, Aarseth and Heggie 1998); the result of the $N$-body simulations did not seem to converge to that of the Fokker-Planck simulations in the limit for $N \to \infty$, contrary to what was expected.

The disagreement between Fokker-Planck models and $N$-body models was even more clearly shown by Portegies Zwart et al. (1998b, PZHMM). They performed a

---

* Japan Society for the Promotion of Science Fellow

*Celestial Mechanics and Dynamical Astronomy* **73**: 179–186, 1999.
©1999 *Kluwer Academic Publishers.*

series of $N$-body simulations with up to 32768 stars with identical initial conditions as one of the Fokker-Planck simulations of Chernoff and Weinberg (1990, CW).

The results of the computations of PZHMM can be summarized as follows: 1) The $N$-body model with the largest number of particles has a lifetime more than an order of magnitude longer than that of the comparable model of CW. 2) The lifetime of the cluster depends on the number of stars in a rather complex way. Since the fundamental assumption of Fokker-Planck calculations is that the evolution is independent of the number of stars, the results of PZHMM might imply that the results of Fokker-Planck calculations for clusters in a tidal field and with stellar evolution are of questionable validity.

The purpose of this paper is to explore what caused this discrepancy between the $N$-body models of PZHMM and the Fokker-Planck models of CW.

## 2. The Models

### 2.1. THE $N$-BODY MODEL

The direct $N$-body integration program Kira (Hut 1994; Hut et al. 1995) is used in combination with the stellar evolution package SeBa (Portegies Zwart and Verbunt 1996; Portegies Zwart and Yungelson 1998). Both models are part of the Starlab software tool set (version 3.1, for the details of its implementation see PZHMM).

The numerical integration of the motion of the stars is performed using a fourth-order individual–time-step Hermite scheme (Makino and Aarseth 1992). For all $N$-body simulations we used GRAPE-4 (Makino et al. 1997).

### 2.2. THE FOKKER-PLANCK MODEL

The model used by CW is an orbit-averaged Fokker-Planck scheme in which the velocity distribution of the stars is assumed to be *isotropic*. In this paper we report the results of an *anisotropic* Fokker-Planck scheme in which the distribution function $f$ depends both on energy $E$ and angular momentum $J$. The two-dimensional orbit-averaged Fokker-Planck equation in $(E, J)$-space is solved numerically (see Cohn 1979; Takahashi 1995, 1997; Takahashi et al. 1997). Although anisotropy is usually unimportant in the central parts of the clusters, it is significant in the outer parts. Therefore we expect that the effects of anisotropy on the escape rate of stars from the clusters can be large. Furthermore we have to consider the $J$-dependence of the distribution function when we like to use a realistic escape criterion (see below).

In CW's isotropic model, a star is removed from the stellar system when its energy exceeds the potential energy at the tidal radius $r_t$ (which we will call the *energy criterion*). In an isotropic Fokker-Planck model, one has no choice but to use the energy as a criterion for escape. In the anisotropic model a more realistic escape condition is used: the *apocenter criterion* as is introduced by Takahashi et al. (1997). In the apocenter criterion, a particle is removed if its apocenter distance

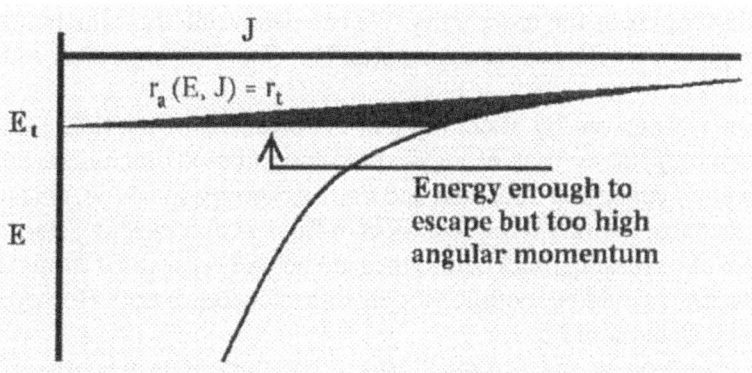

Fig. 1. Schematic diagram of the energy–angular momentum plane for a star cluster. Using the energy criterion all stars with energy $E$ greater than the energy at the tidal radius $E_t$ escape. If the apocenter criterion is applied pariceles in the darkest region are not in an escape orbit and therefore remain bound to the cluster.

$r_a$, which is a function of $E$ and $J$, exceeds the tidal radius (see Fig. 1). The energy criterion removes a larger number of stars than the apocenter criterion. For example: a star with energy equals to the tidal energy cannot reach the tidal radius, except if its orbit is purely-radial, i.e.; zero angular-momentum. This is illustrated in Fig. 1.

Both CW and Takahashi *et al.* (1997), removed particles from the cluster immediately after the escape criterion was satisfied. This assumption is justified if the orbital time scale at the tidal radius is negligible compared to the relaxation time. In real globular clusters this is generally the case, but in the small $N$-limit where $N$-body models operate, this criterion is violated and stars are usually removed from the stellar system too quickly.

Since a star has to move from one end of the cluster to the other end, it is important to account for the travel time of an escaping star. In our treatment an escaper time scale is introduced by applying the following formalism for the escapers (see Lee and Ostriker 1987):

$$\frac{df}{dt} = -\alpha_{esc} f \left[ 1 - \left( \frac{E}{E_t} \right)^3 \right]^{1/2} \frac{1}{2\pi} \sqrt{\frac{4\pi}{3} G \rho_t}. \tag{1}$$

Here $E_t$ is the potential energy at the tidal radius, $\rho_t$ is the mean mass density within the tidal radius, $G$ is the gravitational constant, and $\alpha_{esc}$ is a dimensionless constant which determines how quickly escapers leave the cluster. Note that there is a misprint (concerning the factor $2\pi$) in the original equation (Eq. 3.5) of Lee and Ostriker 1987. A star in an escaping orbit leaves the cluster within its orbital time scale, which is –on average– comparable to the crossing time for the tidal radius. The parameter $\alpha_{esc}$ relates the time scale on which escapers are removed from the cluster relative to the dynamical time scale of the cluster. It is therefore expected that $\alpha_{esc}$ is of the order unity. We can determine its value by calibrating the Fokker-Planck results to $N$-body results. The Coulomb logarithm was taken as $\log \Lambda = \log N$.

Equation 1 is derived by assuming that the presence of the tidal force for the escaping stars: $df/dt = 0$ at $E = E_t$. Our computations include a tidal cutoff rather than a self consistent tidal field and Eq. 1 is, strictly speaking, not applicable. However, the most important improvement of Eq. 1 is that escaping stars take time (of the order of a crossing time) before they are actually discarded from the cluster. In principle, Eq. 1 could be modified for anisotropic models also. However, we did not make any chances in Eq. 1.

Stellar evolution in the Fokker-Planck computations is performed with the same stellar evolution model as is used in the $N$-body computations. For a better comparison with CW's Fokker-Planck computations we performed a few runs with the same stellar evolution treatment as they adopted.

## 2.3. INITIAL CONDITIONS

All clusters have initially the same half-mass relaxation time as in the models IR of PZHMM, which is 2.9 Gyr. The other conditions are taken identical to that of CW's family 1. The dimensionless depth of the initial King model $W_o$ is 3. We use a mass function of the form $dN(m) \propto m^{-2.5}$ between $0.4 M_\odot$ and $15 M_\odot$. All clusters initially fill their Roche-lobe; the King radius equals the tidal radius. Note that Heggie and Ramamani (1995) developed new King models which account correctly for the tidal field of the parent galaxy and which would be more appropriate here. In the $N$-body model, stars that are outside the tidal radius are removed. This simple cutoff was chosen in order to facilitate direct comparison with the Fokker-Planck results.

Apart from testing the various escape treatments in the Fokker-Planck models the only parameter which we change is the number of stars (see PZHMM for more details).

# 3. Results

## 3.1. COMPARISON WITH CHERNOFF AND WEINBERG (1990)

Figure 2 shows the evolution of the total mass of the star cluster (normalized to its initial value) as a function of time. The result of CW's computation is presented as a dot in Fig. 2 (taken from their table 5). This is the end point of the simulation which CW regarded as the end of cluster lifetime, i.e.: the moment of disruption.

Our isotropic Fokker-Planck model (denoted as model $Ie$: $I$ for isotropic and $e$ for the energy criterion) is given as a dashed line. The same stellar evolution model and the same mass bins as CW are used for this model. Therefore the result should coincide with that of CW's run. The agreement, however, is not very good. Our run reaches CW's end mass almost 40% later. We repeated computations using several different sets of time steps and numbers of mass bins, but the result did not change

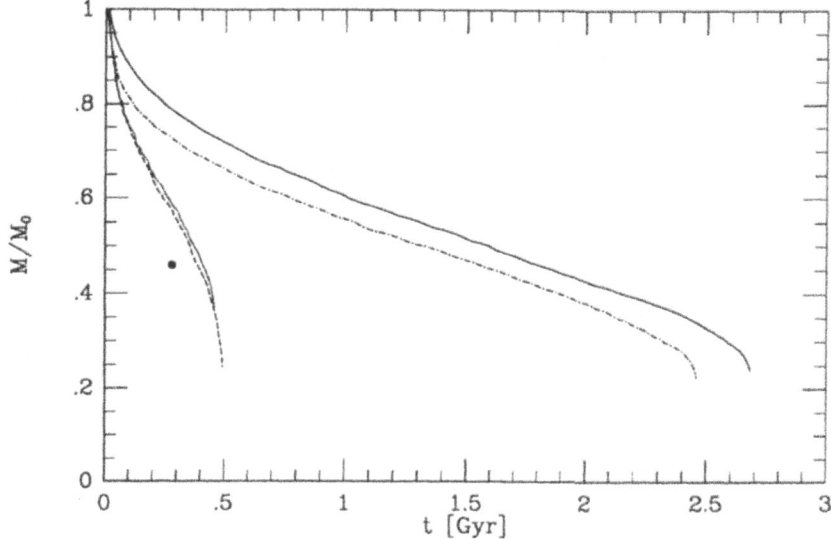

Fig. 2. The total mass of the simulated clusters (normalized to the initial mass) as a function of time for the Fokker-Planck models in which the number of particles is $\infty$ (by definition). The result of CW is presented as a $\bullet$ (to the left) at the mass and age of the system where they considered the cluster to cease to exist. The models in which the energy criterion is used are presented as the dashed line for the isotropic models $Ie$ and the dotted line for the anisotropic model $Ae$. The *two* lines to the right give the results of the anisotropic Fokker-Planck model in which the apocenter criterion is used (model $Aa$). The dash-dotted line uses the same stellar evolution model as is adopted by CW and for the solid line the stellar evolution program SeBa is used. All runs are stopped at the points where the self-consistent potential could not be found.

very much. A series of comparison runs with other initial conditions shows that there is a tendency that the agreement improves for models with a longer lifetime (see Takadashi and Portegies Zwart, in preparation). We did not investigate further the origins of this disagreement, but decided to choose the result of the $N$-body simulations as a base for our discussion.

A second run with the anisotropic Fokker-Planck model (denoted as $Ae$, where $A$ stands for anisotropic) is presented in Fig. 2 as a dotted line. The difference between the isotropic model ($Ie$) and the anisotropic model ($Ae$) is small (see Fig. 2). In both models the same stellar evolution prescription as adopted by CW was used.

The largest difference is between models with the energy criterion and the apocenter criterion (model $Aa$, where $a$ stands for apocenter criterion). The disruption time for model $Aa$ is about five times longer than that for the models $Ie$ and $Ae$. The evolution of models $Ie$ and $Ae$ are similar when the ratio of the tidal radius to the half-mass radius is small (Takahashi and Lee 1999, in preparation). This is because a strong tidal cutoff (as in a King model) suppresses the development of anisotropy in the halo. The apocenter criterion allows particles which would have escaped while using the energy criterion to stay in the cluster. The escape rate in models which use the apocenter criterion is therefore considerably lower than in

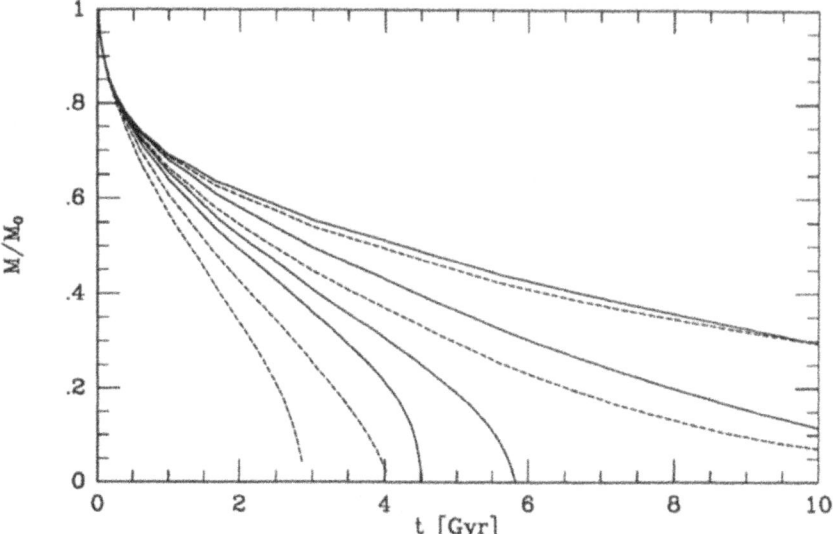

Fig. 3. Mass as a function of time for a number of Fokker-Planck models. The four solid lines represent the results of model Aa with 32k, 16k, 4k and 1k particles from left to right, respectively. Dotted curves present model *Ie* for the same numbers of particles as for model *Aa*. The time scale for escapers via Eq.(1) with $\alpha_{esc} = 2$ for all models.

models which use the energy criterion (see Fig. 1).

### 3.2. EFFECTS OF STELLAR EVOLUTION MODELS

The stellar evolution model used by PZHMM is different from that adopted by CW. In the computations of CW the post main-sequence evolution of the stars is neglected and stars in PZHMM's model live therefore somewhat longer. In fig. 2 the results of two models *Aa* are presented of which one is computed using the stellar evolution model of CW (dash-dotted line) and the other with PZHMM's model (solid line). The difference in the evolution of the mass of the star clusters is small, as excepted. The dissipation time of the two models differ by less than 10%.

### 3.3. $N$-DEPENDENCE OF THE CLUSTER EVOLUTION

For all computations in this section, we use the stellar evolution models according to the prescription in SeBa and employ Eq. 1 as escape condition.

Figure 3 shows the results of the calculations using $\alpha_{esc} = 2$ (see Eq. 1) in models *Ie* and *Aa*. The choice for $N$ at which we should calibrate $\alpha_{esc}$ is rather delicate. The results of the Fokker-Planck computation is more sensitive to $\alpha_{esc}$ for small $N$ than for large $N$. However, for a smaller number of particles the $N$-body results tend to become more noisy. We decided to use $N = 16384$ (16k) to calibrate $\alpha_{esc}$. It turns out that $\alpha_{esc} = 2$ gives good agreement between $N$-body and Fokker-Planck models (see however, Takahashi & Portegies Zwart 1999, in preparation, who prefered $\alpha_{esc} = 2.5$ which gives satisfactory results over a wider range of inital conditions).

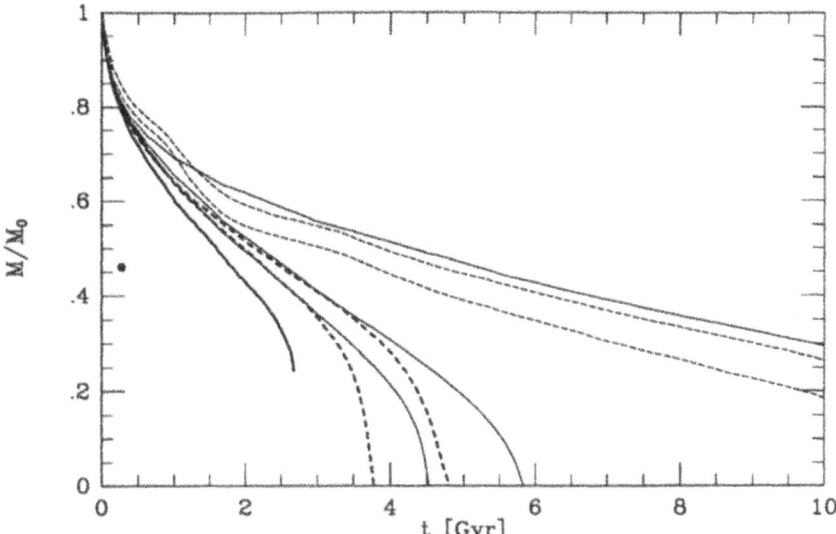

Fig. 4. Mass as a function of time for the $N$-body models (dotted lines) and Fokker-Planck models $Aa$ (solid lines). The **thick** solid line to the left is for $\infty$ stars, the three subsequent solids are with 32k, 16k and 1k stars, respectively. The left and right thick dashed lines give the results of the $N$-body simulations for 32k and 16k stars, respectively. The two *thin* dashed lines represent the $\frac{1}{2}\sigma$ deviation from the mean of the 10 performed runs with 1k stars. The result obtained by CW is presented as a •.

Figure 4 presents the results of a number of $N$-body computations and compares these with the results of the Fokker-Planck models $Aa$ with $\alpha_{esc} = 2$. In order to minimize the statistical fluctuations in the $N$-body results we performed 10 identical computations with $N = 1024$ (1k). For economic reasons we performed three runs with $N = 16k$ and a single run with 32k stars. Each of the 1k runs took about an order of magnitude less computer time on GRAPE-4 than one of the anisotropic Fokker-Planck computations on a fast workstation (the $N$-body with 32k stars took approximately two orders of magnitude longer than the Fokker-Planck models, i.e.: almost three weeks). However, even with the mean of 10 runs the noise in these 1k computations is rather large (see Fig. 4). The $N$-body computation with $N = 32k$ is, due to historical reasons, performed with an upper mass limit of 14 $M_{\odot}$ instead of 15 $M_{\odot}$. The lifetime of this model is therefore expected to be slightly longer than if 15 $M_{\odot}$ would have been used. However, the difference is small, which we tested by using different mass cut-offs in the Fokker-Planck model.

The agreement between the Fokker-Planck results and the $N$-body model is quite good although there are still some deviations. After about 70% of the mass is lost, the deviation becomes noticeable. This may be related to the disruption of the cluster on the dynamical time scale as discussed by CW, Fukushige and Heggie (1995) and PZHMM.

## 4. Conclusions

We have found the reason why the Fokker-Planck calculations of CW and the $N$-body calculations of Fukushige and Heggie (1995) and PZHMM gave very different results. The assumption of velocity isotropy and the over-simplified escape criterion (the energy condition and removing stars instantaneously) caused an enormous over estimate of the escape rate. By using an anisotropic Fokker-Planck model with an improved escape criterion: we have succeeded to achieve excellent agreement between Fokker-Planck and $N$-body results.

The dependence of the dissipation time on the number of particles is also understood. Stars need some time to travel away from the cluster in order to be gobbled up by the Galaxy. This time scale is of the order of a crossing time at the tidal radius. Therefore the escape rate depends on the ratio of the relaxation time to the dynamical time, i.e.; on the number of stars.

## Acknowledgements

We are grateful to Toshiyuki Fukushige, Douglas Heggie, Piet Hut, Junichiro Makino and Steve McMillan for many discussions and software development. This work is supported supported in part by the Research for the Future Program of Japan Society for the Promotion of Science (JSPS-RFTP97P01102).

## References

Aarseth, S.J., Heggie, D.C.: 1998, *MNRAS*, **297**, 794
Aarseth, S.J., Henon, M., Wielen, R.: 1974, *A&A*, **37**, 183
Chernoff, D., Weinberg, M.: 1990, *ApJ*, **351**, 121
Cohn, H.: 1979, *ApJ*, **234**, 1036
Fukushige, T., Heggie, D.C.: 1995, *MNRAS*, **276**, 206
Giersz, M., Heggie, D. C.: 1994a, *MNRAS*, **268**, 257
Giersz, M., Heggie, D.C.: 1994b, *MNRAS*, **270**, 29
Giersz, M., Spurzem, R.: 1994, *MNRAS*, **269**, 241
Heggie, D.C., Ramamani:, 1995, *MNRAS*, **272**, 317
Heggie, D.C., Giersz, M., Spurzem, R.: Takahashi, K.: 1998, To appear in *Highlights of Astronomy Vol. 11*, ed. Johannes Andersen, in press (astro-ph/9711191)
Hut, P.: 1994, IAU 165, in *Compact stars in binaries*, eds. J. van Paradijs, E.P.J. van den Heuvel and E. Kuulkers, Kluwer, p. 377
Hut, P., Makino, J. McMillan, S.: 1995, , *ApJ*, **443**, 93
Lee, H.M., Ostriker, J.P.: 1987, *ApJ*, **322**, 123
Makino, J., Aarseth, S.J.: 1992, *PASJ*, **44**, 141
Makino, J., Taiji, M., Ebisuzaki, T., Sugimoto, D.: 1997, , *ApJ*, **480**, 432
Portegies Zwart, S.F., Verbunt, F.: 1996, *A&A*, **309**, 179
Portegies Zwart, S.F., Yungelson, L.: 1998, *A&A*, **332**, 173
Portegies Zwart, S. F., Tout, Ch., Lee, H.M.: 1998a, To appear in*Highlights of Astronomy Vol. 11*, ed. Johannes Andersen, in press (astro-ph/9710209)
Portegies Zwart, S.F., Hut, P., Makino, J. and McMillan, S.L.W.: 1998b, *A&A*, **337**, 363
Spurzem, R., Takahashi, K.: 1995, *MNRAS*, **272**, 772
Takahashi, K.: 1995, *PASJ*, **47**, 561
Takahashi, K.: 1997, *PASJ*, **49**, 547
Takahashi, K., Lee, H.M., Inagaki, S.: 1997, *MNRAS*, **292**, 331
Vesperini, E., Heggie, D.C.: 1997, *MNRAS*, **289**, 898

# BUILDING GALAXY MODELS WITH SCHWARZSCHILD METHOD AND SPECTRAL DYNAMICS

HONGSHENG ZHAO

*Sterrewacht Leiden, Niels Bohrweg 2, 2333 CA, Leiden, The Netherlands,*
*(hsz@strw.LeidenUniv.nl)*

**Abstract.** Tremendous progress has been made recently in modelling the morphology and kinematics of centers of galaxies. Increasingly realistic models are built for central bar, bulge, nucleus and black hole of galaxies, including our own. The newly revived Schwarzschild method has played a central role in these theoretical modellings. Here I will highlight some recent work at Leiden on extending the Schwarzschild method in a few directions. After a brief discussion of (i) an analytical approach to include stochastic orbits (Zhao 1996), and (ii) the "pendulum effect" of loop and boxlet orbits (Zhao, Carollo, de Zeeuw 1999), I will concentrate on the very promising (iii) spectral dynamics method, with which not only can one obtain semi-analytically the actions of individual orbits as previously known, but also many other physical quantities, such as the density in configuration space and the line-of-sight velocity distribution of a superposition of orbits (Copin, Zhao & de Zeeuw 1999). The latter method also represents a drastic reduction of storage space for the orbit library and an increase in accuracy over the grid-based Schwarzschild method.

## 1. Introduction

One of the classical problems in galaxy dynamics is building equilibrium models for a galaxy with an observed light distribution. The basic process can be illustrated by the simpliest form of the problem, which is to construct a spherical, isotropic model with a constant mass-to-light ratio $M/L$ and a stellar distribution function $DF$ which fits the light profile. This has the well-known solution (Eddington 1916),

$$DF(E) \propto \int_E^0 \frac{d^2\rho}{d\phi^2} \frac{d\phi}{\sqrt{\phi - E}}, \tag{1}$$

which is a function of the energy $E$ only, where the potential $\phi$ comes from solving the Poisson equation, and the volume density $\rho(r)$ comes from deprojecting the light profile $\mu(R)$

$$\rho(r) \propto \frac{M}{L} \int_r^\infty \frac{d\mu(R)}{dR} \frac{dR}{\sqrt{R^2 - r^2}}. \tag{2}$$

In general deprojecting the light and getting the potential are relatively easier parts of the problem.

While a simple problem in concept, it is challenging to extend the mathematical and numerical machinery to cope with realistic systems. In particular, galaxies are almost always flattened, and sometimes triaxial. They are also anisotropic in velocity distribution due to dissipational and dissipationless processes in formation. By formation they are often dominated by dark matter at very small and very large radii (central black holes as indicated by nuclear activities in AGNs and outer dark halos as by flat HI rotation curves). In short, none of the three simplifying assumptions (constant $M/L$, isotropic and spherical) are generally valid.

While progress has been made in the analytical direction, the application is generally limited. The Hunter & Qian (1993) method, for example, can construct two-integral models – with a DF being function of energy and angular momentum azimuthal component $DF(E, L_z)$ – for axisymmetric galaxies and has been applied, e.g., in the case of the nucleus of M32 (Qian et al. 1995). Formulaism also exists for building anisotropic non-axisymmetric models as long as the potential remains in Stäckel form (Teuben 1987, Statler 1987, 1991, Arnold et al. 1994, Dejonghe et al. 1995), and in a few cases for tumbling models (e.g. Freeman 1966, Vandervoort 1980). As a side comment separability is no guarantee for self-consistency; for example, the recent non-axisymmetric disc potentials by Sridhar & Touma (1997) require unphysically negative DF (Syer & Zhao 1998).

At the other end of the spectrum of methods straight N-body simulations can deal with all geometries (Aarseth & Binney 1976, Wilkinson & James 1982, Barnes 1996), but their power is again limited when it comes to sculpturing a simulation to fit a set of observations in certain $\chi^2$ sense. It is generally very computationally expensive to cover enough degree of freedom to find the true best fitting model; nevertheless see Fux's (1997) models for the Milky Way.

The most promising approach so far is the so-called Schwarzschild (1979) method, after his pioneering efforts in this direction. Basically one tries to match the observed distribution of the light with typically a few hundred or thousand building blocks with each being one stellar orbit populated with certain amount of stars. One adjusts the mass assigned to each orbit until a best match is obtained (see reviews by Binney 1982, de Zeeuw & Franx 1991, de Zeeuw 1996, Merritt 1996, 1999).

## 2. Schwarzschild Method with Bells & Whistles

The Schwarzschild method has now been extensively applied to study nearby elliptical and S0 galaxies under the assumption of a static axisymmetric or triaxial potential (e.g., Richstone & Tremaine 1988, Merritt & Fridman 1996, Rix et al. 1997, van der Marel et al. 1998), and has also been applied to build 2-dimensional models of external bars (Pfenniger 1984, Wozniak & Pfenniger 1997) and 3-dimensional models of the tumbling bar of our own galaxy (Zhao 1996). These applications have also greatly generalized the original layout of the Schwarzschild method, and in particular, it is possible to match the orbits to a variety of kinematic data of gas and stars, and to derive a smooth physical solution (cf. the schematic Fig. 1). Nevertheless there are three main limitations of Schwarzschild approach and these are best overcome by joining force with the analytical and the N-body approaches.

Limitation A: Stability of a Schwarzschild model needs to be addressed by an N-body simulation. An interesting idea, due to Syer & Tremaine (1996) is to do the $\chi^2$ fitting and N-body simulation at the same time, adjusting the mass of each particle as the simulation evolves towards a best match of data with the distribution of the particles. A simpler, better understood approach is to design a Schwarzschild model first, then populate each library orbit with $N_A$ particles with random phase where $N_A$ is proportional to the weight assigned to the orbit with actions A, and

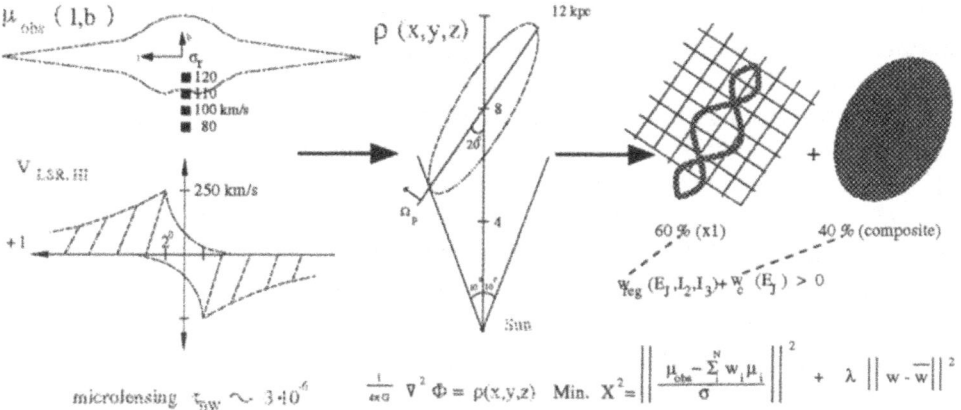

Fig. 1. An illustration of the steps in building a bar model for the Milky Way with a modified Schwarzschild method. We start from (1) a dust-corrected near infrared surface brightness map $\mu_{obs}(l, b)$ of the Galaxy from COBE/DIRBE observations, (2) the velocity dispersion $\sigma_r$ measurements on the minor axis of the Galaxy, (3) the atomic hydrogen intensity map in the longitude $l$ vs. line of sight velocity plane, which is basically a measure of the rotation curve $V_{LSR,HI}$ of the Galaxy, and (4) the probability (optical depth) of microlensing $\tau_{BW}$ towards the Baade Window ($l = 1^\circ$, $b = -4^\circ$). We then use these input quantities to build the potential $\Phi(x, y, z)$ of the bar via the Poisson equation, and constrain the axis ratio and the orientation of the bar. Finally we seek a positive DF of regular orbits ($x_1$, $x_2$, $x_4$ orbits etc.) of weights $w(E_J, I_2, I_3)$ plus super-orbits (composite orbits) of weights $w(E_J)$ such that the ensemble $\sum_i^N w_i \mu_i$ matches the observed distribution $\mu_{obs}$ in a set of rectangular grid cells. The last step is a $\chi^2$ minimization problem, regularized by slightly ($\lambda \sim 10^{-4}$) penalizing DF where adjacent orbits in phase space have wildly different weights $w_i$. We feed the final model to an N-body code to test stability.

finally feed these particles to an N-body simulation code to test stability. This has been applied successfully to the Galactic bar, which is found to be stable (Zhao 1996).

Limitation B: Stochastic orbits in a Schwarzschild model make the model evolve on time scales of the mixing time (several hundred dynamical time, cf. Merritt & Valluri 1996). Merritt & Fridman (1996) propose to average out this effect by explicitly summing up many stochastic orbits to achieve a good phase-mix. This is challenging because it means integrating a few hundred orbits for a few thousand of dynamical times to beat down the time-dependent fluctuations. An alternative approach has been used in the case of the Galactic bar (Zhao 1996). The hybrid model makes use of two types of building blocks for the Galaxy (cf. Fig. 1): a library of regular orbits obtained by direct integration for several hundred dynamical time, and a library of "super-orbits", which are nothing but many delta-like DFs $\sum_i N_i \delta(E_J - i\Delta)$, where the weighting $N_i$ are to be found by the same Non-Negative Least Square fitting code as with the weighting of the regular orbits. Each delta function includes all orbits with the same Jacobi integral $E_J \equiv E - \Omega J_z$

implicitly, where $\Omega$ is the tumbling speed of the bar and $E_J$ is the only analytical integral. Such a prescription naturally incorporates stochastic orbits in the model without explicitly making the division of the fraction of mass in stochastic orbits vs. regular ones. Variations of our analytical way of including stochastic orbits have now been developed to model axisymmetric systems and bars by the dynamics groups at Leiden (Cretton et al. 1998, private communication) and Oxford (Häfner et al. 1998, private communication).

Limitation C: A Schwarzschild model is cell-dependent. Checking self-consistency of the model involves computing the amount of time an orbit spends inside a cell and comparing it with the amount of mass prescribed in the same cell. However it is possible to make cell-independent modeling. For example, to keep a triaxial galaxy in equilibrium requires a healthy mix of shapes of its building blocks with some orbits more flattened than the potential, some less flattened. It is well-known that loop orbits cannot reproduce a self-consistent triaxial potential because they move too fast and spend too little time at the major axis to match the relatively (compared to, say, the minor axis) high model density there. We find that this problem is actually more general (Zhao, Carollo & de Zeeuw 1999): it is easy to prove analytically that any regular orbit will reach a local maximum for its angular momentum $|J(t)|$ at the major axis, because the torque of triaxial potential is always directed towards the major axis (cf. Fig. 2). So in this regard a loop orbit or a boxlet orbit (with the shape of a banana, fish, pretzel etc) behaves like a pendulum with a stretchable length. Since a pendulum tends to swing too fast and spend too little time at its symmetry axis, the "pendulum effect" generally prevents loops and boxlets from putting many stars at the major axis. This can be used as a cell-independent argument against making strongly flattened and triaxial galactic nuclei with bananas, fishes etc., consistent with previous authors (e.g., Gerhard & Binney 1985, Pfenniger & de Zeeuw 1989).

## 3. Spectral Dynamics Method

Another very promising cell-independent method of building galaxies is the spectral dynamics method. This method, as introduced by Binney & Spergel (1982), provides a conceptually simple representation of a regular orbit, by decomposing it into a truncated Fourier series involving three fundamental frequencies. The basic idea here is that a regular orbit in a 3D potential is simplest described in the action angle space since it satisfies periodic boundary conditions on the torus (cf. Fig. 3). Let an orbit be labeled by its three actions $\mathbf{A}$, then the phase space coordinates $[\mathbf{x_A}(t), \mathbf{v_A}(t)]$ are periodic with respect to the three action angles $\theta \equiv (\omega_1, \omega_2, \omega_3)t$, ie., we have the following truncated Fourier series

$$\mathbf{x_A}(t) = \sum_{\lambda \equiv (l,m,n)}^{L} X_\lambda \cos(\lambda \cdot \theta + \chi_\lambda), \quad \theta \equiv (\omega_1, \omega_2, \omega_3)t. \qquad (3)$$

where the $\omega$'s are the three basic frequencies, the coefficients $X_\lambda$ are the amplitudes of each frequency combination and $L$ is the highest order harmonics before

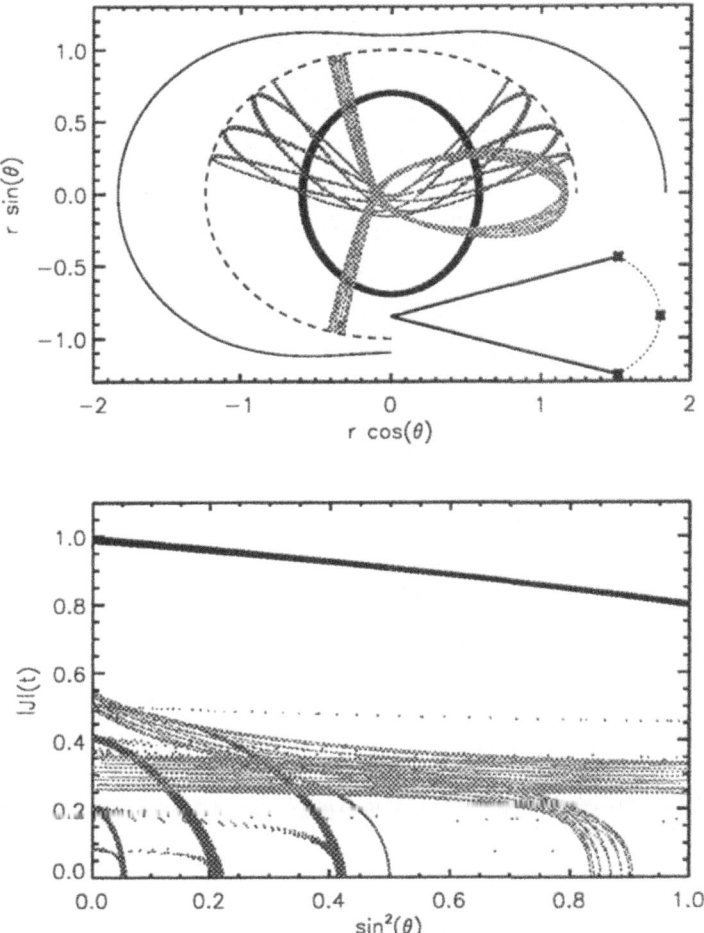

Fig. 2. the upper panel shows three types of centro-phobic orbits: a loop orbit (thick band), a banana orbit (dotted lines) and a fish orbit (solid thin lines) together with the iso-density and iso-potential contours (heavy solid and dashed lines respectively). The lower panel shows the angular momentum $|J(t)|$ of a star as a function of the angle from the major axis ($\theta = 0$) along the same three orbits, where each dot is one time step of the orbit. Note that $|J(t)|$ peaks on approaching the major axis for all boxlet orbits, like it does for the loops and the pendulum. A pendulum with a variable length is also sketched in the upper panel.

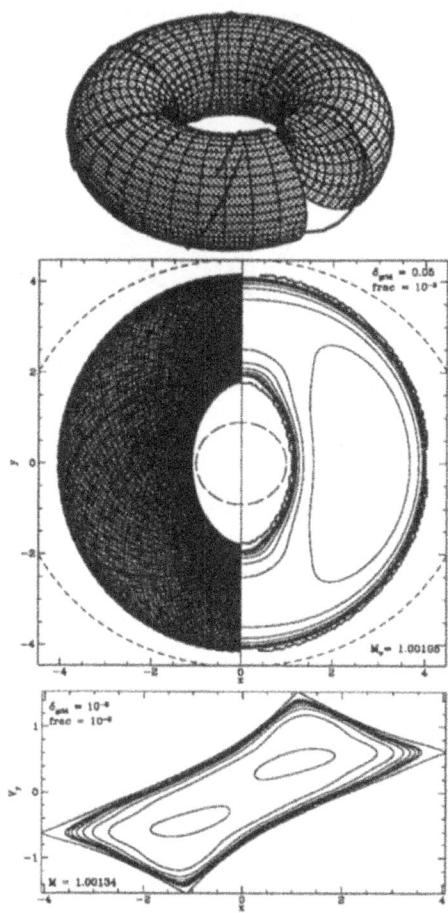

Fig. 3. A planar loop orbit in various cuts of the phase space. The top panel shows the phase space torus of the orbit, which will be populated uniformly after integrating the orbit for a long time. The orbit in the configuration space is related to the torus by eq. (3), and the uniformly populated torus projects to a non-uniform distribution in the configuration space via a Jacobian (cf. eq. 6). The middle panel shows the reconstructed orbit in the $xy$ plane, folded to the left half, and the density contour map of the orbit, folded to the right half. The bottom panel shows the line of sight velocity distribution of the orbit in the impact parameter $x$ vs. velocity $v_y$ plane, computed with eq. (7); the tiled parallelogram is indicative of the rotation of the loop (Copin et al. 1999).

truncation. Similarly the velocity of the orbit at any time, related to the position by a time derivative, can be written down as

$$\mathbf{v_A}(t) = - \sum_{\lambda \equiv (l,m,n)}^{L} \omega_\lambda X_\lambda \sin(\lambda \cdot \theta + \chi_\lambda), \quad \omega_\lambda = l\omega_1 + m\omega_2 + n\omega_3, \qquad (4)$$

It is easy to work out the actions $\mathbf{A}$ by integrating along one of the three action angles,

$$\mathbf{A} = \frac{1}{2} \sum_\lambda X_\lambda^2 (l\omega_1, m\omega_2, n\omega_3). \qquad (5)$$

This method actually goes back to many years ago (e.g., Ratcliff, Chang & Schwarzschild 1984), but recent work by the Oxford group (e.g., Kaasalainen & Binney 1994), but recent work by the Oxford group (e.g., Kaasalainen & Binney 1994), and by Papaphilipou & Laskar (1997) and Carpintero & Aguilar (1998) has made it possible to extract the basic frequencies numerically from the time series data of a regular orbit. Namely the step from $[\mathbf{x_A}(i\Delta t), \mathbf{v_A}(i\Delta t)]$ to $[\omega_\lambda, X_\lambda]$.

Most important to the Schwarzschild method is that we can compute the volume density of the orbit $\mathbf{A}$ at a given point $(x, y, z)$. Since we know that a regular orbit is uniformly distributed in its action angle space, the density in the real space is given by

$$\rho_\mathbf{A}(x, y, z) = \frac{1}{(2\pi)^3} \left| \frac{\partial(x, y, z)}{\partial(\theta_1, \theta_2, \theta_3)} \right|^{-1}, \qquad (6)$$

where the partial derivatives are simply the Jacobian for the transformation between the action angle space $(\theta_1, \theta_2, \theta_3)$ and the coordinate space $(x, y, z)$. Since the Jacobian can be evaluated analytically with eq. (3), we have derived a rigorous expression for the spatial distribution of an orbit. Likewise the line-of-sight velocity $(v_z)$ distribution of an orbit in the direction $(x, y)$ (cf. Fig. 3) is given by

$$LOSVD_\mathbf{A}(x, y, v_z) = \frac{1}{(2\pi)^3} \left| \frac{\partial(x, y, v_z)}{\partial(\theta_1, \theta_2, \theta_3)} \right|^{-1}. \qquad (7)$$

For details see Copin et al. (1999).

The beauty of this method is that the description of regular orbits is conceptually simple. The description is time-independent, and involves no gridding and binning. It is also easy to store and recover an orbit, thus saving the amount of disc space for storing orbit libraries. Typically the number of quantities to store is about 10 times the dimension of the problem; this includes the basic frequencies and the leading amplitudes.

To conclude we remark that the most promising method might be some kind of generalized Schwarzschild method or hybrid method, where the computationally intentive stochastic orbits are implicitly modelled by the analytical super-orbits, and the spatial and velocity distribution of the regular orbits are treated with spectral dynamics.

I thank Tim de Zeeuw for a critical reading of an earlier version and Danny Pronk for making the electronic version of Figure 1.

# References

Aarseth S. & Binney J. 1976: *MNRAS*, **185**, 227

Arnold R., de Zeeuw P.T., & Hunter C.: 1994, *MNRAS*, **271**, 924

Barnes J.: 1996, in *the formation of Galaxies*, Proc. 5th Carnary Islands Winter School of Astrophys., Munoz-Tunon ed., ( Cambridge U. Press), p 399

Binney J.: 1982,*ARAA*, **20**, 399

Binney J., Spergel, D.: 1982, *ApJ*, **252**, 308

Carpintero D.D., Aguilar L.A.: 1998, *MNRAS*, **298**, 1

Copin Y., Zhao H. & de Zeeuw P.T.: 1998, in preparation

Dejonghe H. et al.: 1995, *A&A*, **306**, 363

de Zeeuw P.T.: 1996, in *Gravitational Dynamics*, Proc. 36th Herstmonceux Conf., Lahav,Terlevich & Terlevich eds, Cambridge U. Press

Eddington A.S.: 1916, *MNRAS*, **76**, 572

Freeman K.: 1966, *MNRAS*, **134**, 15

Fux R.: 1997, *A&A*, **327**, 983

Gerhard O.: 1994, in *Galactic Dynamics & N-body Simulations*, Proc. 6th European Summer School, Contopoulos & Spyrou ed., (Springer), p191

Gerhard O.E., Binney J.J.: 1985, *MNRAS*, **216**, 467

Hunter C. & Qian E.: 1993, *MNRAS*,**262**, 401

Kaasalainen M. & Binney J.: 1994,*MNRAS*, **268**, 1033

Kormendy J. & Richstone D.: 1995, *ARA&A*, **33**, 581

Merritt D.: 1996, in *The Nature of Elliptical Galaxies*, 2nd Stromlo Symposium, Arnaboldi, Da Costa and Saha eds

Merritt D.: 1999, *PASP*, **111**, 129

Merritt D.R., Fridman T.: 1996, *ApJ*, **460**, 136

Merritt D.R., Valluri M.: 1996, *ApJ*, **471**, 82

Papaphilipou, Y., Laskar, J.: 1996, *A&A*, **307**, 427

Pfenniger D.: 1984, *A&A*, **141**, 171

Pfenniger D., de Zeeuw P.T.: 1989, in *Dynamics of Dense Stellar Systems*, D. Merritt ed., Cambridge Univ. Press, 81

Qian E.E. et al.: 1995, *MNRAS*, **274**, 602

Qian E.E., de Zeeuw P.T., van der Marel R.P. and Hunter C.: 1995, *MNRAS*, **274**, 602

Ratcliff S.J., Chang K.M. & Schwarzschild M.: 1984, *ApJ*, **279**, 610

Richstone, D. & Tremaine, S. 1988, *ApJ*, **327**, 82

Rix H.-W. et al. 1997, *ApJ*, **488**, 706

Schwarzschild M.: 1979, *ApJ*, **232**, 236

Schwarzschild M.: 1982, *ApJ*, **263**, 599

Sridhar S. and Touma J.: 1997, *MNRAS*, **287**, L1

Statler T. S.: 1987, *ApJ*, **321**, 113

Statler T. S.: 1991, *A.J.*, **102**, 882

Syer D., Zhao H.S.: 1998, *MNRAS*, **296**, 407

Teuben P.: 1987, *MNRAS*, **227**, 815

van der Marel R.P. et al.: 1998, *ApJ*, **493**, 613

Vandervoort P.O.: 1980, *ApJ*, **240**, 478

Wilkinson A. & James R.A.: 1982, *MNRAS*, **199**, 171

Wozniak H. & Pfenniger D.: 1997, *A&A*, **317**, 14

Zhao H.S.: 1996, *MNRAS*, **283**, 149

Zhao H.S., Carollo C.M. & de Zeeuw P.T.: 1999, *MNRAS*, in press

# CONDITIONAL ENTROPY

## A Tool to Explore the Phase Space

P. CINCOTTA*

*Facultad de Ciencias Astronómicas y Geofísicas, Universidad Nacional de La Plata, Paseo del Bosque, 1900 La Plata, Argentina*

and

C. SIMÓ

*Departament de Matemàtica Aplicada i Anàlisi, Universitat de Barcelona, Gran Via 585, 08007 Barcelona, Spain, e-mail: carles@maia.ub.es*

**Abstract.** In this paper we show that the Conditional Entropy of nearby orbits may be a useful tool to explore the phase space associated to a given Hamiltonian. The arc length parameter along the orbits, instead of the time, is used as a random variable to compute the entropy. In the first part of this work we summarise the main analytical results to support this tool while, in the second part, we present numerical evidence that this technique is able to localise (stable) periodic and quasiperiodic orbits, 'aperiodic' orbits (chaotic motion) and unstable periodic orbits (the 'source' of chaotic motion). Besides, we show that this technique provides a measure of chaos which is similar to that given by the largest Lyapunov Characteristic Number. It is important to remark that this method is very simple to compute and does not require long time integrations, just realistic physical times.

**Key words:** Chaos – Lyapunov Characteristic Number – Entropy

## 1. Introduction

It is well known that the largest Lyapunov Characteristic Number (LCN) provides a measure of stochasticity. By definition, the LCN is an asymptotic value for $t \to \infty$. For practical applications in Galactic Dynamics, good asymptotic results are obtained for motion times of the order of $10^5 - 10^6 T_D$, where $T_D$ is a characteristic period of motion of the system, while the Hubble time is in the order of $10^2 T_D$, $10^3 T_D$ as much. Therefore, the motion time needed to compute the LCN is very unrealistic for these applications (see, however, Merritt & Valluri, 1996 and Udry & Pfenniger 1988). Also, such long computing times turn this tool unsuitable to carry out a detailed study of the orbital structure of a given potential. Besides, since the LCN is a time–average magnitude for large $t$, (the mean rate of exponential divergence of nearby orbits), relevant information about the dynamics is missed. One alternative is the spectra of stretching numbers or Lyapunov numbers for finite times (see Contopoulos & Voglis 1996 and references there; Voglis et al., 1999). Another powerful techinque is the Frequency Map Analysis (Laskar, 1990, 1993; Papaphilippou & Laskar, 1998 and Wachlin & Ferraz–Mello, 1998 for applications to realistic galactic models). In this work we propose the 'Conditional Entropy' of nearby orbits as an effective tool to investigate the phase space structure of a

* Present address (until August 1999): Departament de Matemàtica Aplicada i Anàlisi, Universitat de Barcelona, Gran Via 585, 08007 Barcelona, Spain. E-mail: pablo@zeus.maia.ub.es – pmc@fcaglp.edu.ar

*Celestial Mechanics and Dynamical Astronomy* **73**: 195–209, 1999.
© 1999 *Kluwer Academic Publishers.*

given Hamiltonian in short motion times, i.e., $t \lesssim 10^3 T_D$. The latter concept, that was introduced first by Núñez et al. (1996), differs from the standard entropy in the fact that the arc length parameter of the orbit under consideration is used as a random variable. In that work the authors presented numerical evidence that this technique is efficient to separate ordered and stochastic regions of the phase space in relatively short times.

This work is divided in two parts. The first one deals with some theoretical considerations about the Conditional Entropy that lack in (Núñez et al., 1996). However a more detailed theory behind this method will be addressed in a separate paper. In the second part, we illustrate the use of this tool using a simple 2D system. As it was shown in (Núñez et al., 1996), this technique seems to work very well also in 3D systems.

## 2. Set Up, Definitions and Notation

Let us consider the Hamiltonian:

$$H(\mathbf{p}, \mathbf{q}) = \frac{\mathbf{p}^2}{2} + \phi(\mathbf{q}), \qquad \mathbf{p}, \mathbf{q} \in \mathbf{R}^N, \tag{1}$$

where $\phi$ is a smooth potential. For the sake of simplicity we write:

$$\mathbf{x} = (\mathbf{p}, \mathbf{q}) \in \mathbf{R}^{2N}, \qquad \mathbf{v} = (-\partial H/\partial \mathbf{q}, \; \partial H/\partial \mathbf{p}) = (-\nabla\phi, \mathbf{p}) \in \mathbf{R}^{2N}. \tag{2}$$

Then the equations of motion are:

$$\dot{\mathbf{x}} = \mathbf{v}(\mathbf{x}). \tag{3}$$

Let $M_h$ be the manifold (energy surface): $M_h = \{\mathbf{x} : H(\mathbf{p}, \mathbf{q}) = h\}$, so $\phi(\mathbf{q}) \leq h$. Throughout the present work we consider that the motion is bounded in phase space, that is, $M_h$ is compact. Let $\gamma \subset M_h$ be an arc of an orbit of $\mathbf{v}$:

$$\gamma = \{\mathbf{x}(t; \mathbf{x}_0) : \mathbf{x}_0 \in M_h, \; 0 \leq t \leq T < T_*\}, \tag{4}$$

where $T_*$ bounds the total motion time. Since $T$ is finite, the length of $\gamma$:

$$L_\gamma = \int_\gamma ds, \tag{5}$$

is a finite quantity. In (5), $s$ is the arc length parameter associated with the orbit $\gamma$. Clearly it depends on the metric; in this case $ds^2 = dx_1^2 + \ldots + dx_{2N}^2$.

Let $\psi(\mathbf{x})$ be any scalar function of the phase–space coordinates. We define the average value of $\psi$ along the orbit $\gamma$ as:

$$\langle \psi \rangle_\gamma \equiv \frac{1}{L_\gamma} \int_\gamma \psi(\mathbf{x}(s)) ds, \tag{6}$$

where $\chi\left(s(t)\right) = \mathbf{x}(t)$. Besides:

$$\int_\gamma \psi(\chi(s))ds = \int_0^T \psi(\mathbf{x}(t))|\dot{\mathbf{x}}(t)|dt = T\langle\psi|\mathbf{v}|\rangle_T, \tag{7}$$

where $\langle.\rangle_T$ denotes the time–average but over a finite time, $|.|$ is the usual Euclidean norm and we have made use of (3). From (5), (6) and (7) we readily see that:

$$L_\gamma = T\langle|\mathbf{v}|\rangle_T, \qquad \langle\psi\rangle_\gamma = \frac{\langle\psi|\mathbf{v}|\rangle_T}{\langle|\mathbf{v}|\rangle_T}. \tag{8}$$

Set $\psi = \ln|\mathbf{v}(\mathbf{x})|$. We call *entropy of the orbit* $\gamma$, $S(\gamma)$, the magnitude given by:

$$S(\gamma) = -\langle\ln|\mathbf{v}|\rangle_\gamma + \ln L_\gamma. \tag{9}$$

The term $\ln L_\gamma$ is introduced just for convenience. Indeed, from (8) and (9) it is not difficult to show that $S(\gamma)$ can be written in the form:

$$S(\gamma) = -\int_0^T \rho_\gamma(t)\ln\rho_\gamma(t)dt, \qquad \rho_\gamma(t) = \frac{|\mathbf{v}(\mathbf{x}(t))|}{L_\gamma} \geq 0. \tag{10}$$

From (10) we see that $\int_0^T \rho_\gamma dt = 1$. Formally, the definition of $S(\gamma)$ resembles the familiar continuous entropy:

$$S_c(X) = -\int_a^b \rho(x)\ln\rho(x)dx, \tag{11}$$

for the set $X = \{x \in \mathbf{R}, a < x < b\}$ and the distribution density $\rho(x)$. The latter may be considered as the continuous limit of:

$$S_d(X) = -\sum_{i=1}^m \mu(x_i)\ln\mu(x_i) \geq 0, \tag{12}$$

for the set $X = \{x_i, i = 1, ..., m, a < x_i < b\}$, where $\mu$ is the probability associated to $x_i$, i.e., the normalised measure of the elements of a given partition in $X$, $x_i$ being a point which represents an element of the partition and has assigned the corresponding measure (just as it is done in the Riemann sums). Therefore, we can follow Fraser & Swinney (1986). Let us summarise the main results.

Consider a measurable space $M$ provided with a normalised measure (probability) $\mu$. Let $A = \{a_i, i = 1, ..., n\} \subset M$. The entropy of the set $A$, $S(A)$, is defined as in (12). Consider another set $B = \{b_j, j = 1, ..., m\} \subset M$. The conditional entropy of $A$ relative to $B$ is defined by:

$$S(A|B) = \sum_{j=1}^m \mu(b_j)S(A|b_j) = -\sum_{j=1}^m \mu(b_j)\sum_{i=1}^n \mu(a_i|b_j)\ln\mu(a_i|b_j),$$

where $\mu(a_i|b_j)$ is the conditional probability of $a_i$ for a given $b_j$. Since $\mu(a_i|b_j) = \mu(a_i, b_j)/\mu(b_j)$, with $\mu(a_i, b_j)$ the joint probability of $a_i$ and $b_j$, then one readily finds:

$$S(A|B) = -\sum_{i,j} \mu(a_i, b_j) \ln \mu(a_i, b_j) + \sum_j \mu(b_j) \ln \mu(b_j) = S(A, B) - S(B),$$

where $S(A, B)$ is the joint entropy of both sets. Introducing $I(A, B)$ as the symmetric conditional entropy: $I(A, B) \equiv \frac{1}{2}[S(A|B) + S(B|A)]$ we get:

$$I(A, B) = S(A, B) - \frac{1}{2}[S(A) + S(B)]. \tag{13}$$

It is not difficult to verify that if, for example, $n = m$ and $\mu(a_i) = \mu(b_i)$ for all $i$, then $S(A, B) = S(A)$ and $I$ reduces to zero. On the other hand if $A$ and $B$ are independent sets (in a probabilistic sense), then $S(A, B) = S(A) + S(B)$ and $I = [S(A) + S(B)]/2 > 0$. In general, it can be proved that $I(A, B) \geq 0$ for any sets $A$ and $B$, while $I(A, B) = 0$ if and only if $\mu(a_i) = \mu(b_i)$ (see Arnold & Avez, 1989 for a formal presentation of the subject and proofs). Since $\mu(a_i|b_j)$ is a measure of the correlation between the sets $A$ and $B$ we can state the former mathematical results in 'statistical language' as follows: *if the sets $A$ and $B$ are strongly correlated, then $I(A, B) \approx 0$, while $I(A, B) \approx [S(A) + S(B)]/2$ whenever $A$ and $B$ are uncorrelated.*

When we deal with a continuous case, like in (11), we extend the former definitions but the probability, $\mu(a_i)$, has to be replaced by a continuous probability density $\rho(a)$, such that $d\mu(a) = \rho(a)da$ and $S(A) = -\int_A d\mu(a) \ln \rho(a)$. Actually it should be added a constant, say '$S_0 = \infty$'. So in the continuous limit, we lose the positivity of the entropy. Proceed in a similar way for $S(B)$ and $S(A, B)$. Hence, since $I$ is a relative entropy, we conserve the property $I \geq 0$ that holds in the discrete case (for further details see Katz, 1967).

Let us return to the Hamiltonian (1), the orbit $\gamma$ defined in (4) and identify $\gamma \to A$. Now consider another orbit $\gamma'$ similar to $\gamma$, on the same or a nearby, $M_{h'}$, level of energy but for a slightly different initial condition: $x_0' = x_0 + \delta_0$. Let us denote as $\dot{x}' = v(x') \equiv v'$ the associated vector field. In the same way as we did with $\gamma$, we can construct $\rho_{\gamma'}$ and $S(\gamma')$. Then we identify $\gamma' \to B$. To define the joint entropy of both, $\gamma$ and $\gamma'$, we proceed as follows. Let $\Gamma = \gamma \times \gamma' \subset M_h \times M_{h'}$ be the curve defined through the vector field $V = (v, v')$ such that $|V|^2 = |v|^2 + |v'|^2$. Then $\rho_\Gamma = |V(x(t), x'(t))|/L_\Gamma$. Thus we can define $I(\gamma, \gamma')$ in the same way as in (13):

$$I(\gamma, \gamma') = S(\Gamma) - \frac{1}{2}[S(\gamma) + S(\gamma')]. \tag{14}$$

Suppose that the Hamiltonian (1) is a near–integrable one, where the associated phase space is shared between regular and chaotic motion. Let us denote with $\Sigma_r, \Sigma_c \subset M_h$ those regions of the energy surface where the motion is regular and chaotic respectively. Among others, the main feature that distinguish both components is the following: if $x_0 \in \Sigma_r$, then $\gamma$ and $\gamma'$ will remain close one another, diverging in mean at a linear rate (in fact, any subexponential rate will do). On the other hand, if $x_0 \in \Sigma_c$, then $\gamma$ and $\gamma'$ will diverge at an exponential rate. Using the results given above –in the statistical sense– we can state: *Let $x_0 \in \Sigma_r$ and $T$ finite. Then $\gamma$ and $\gamma'$ will be strongly correlated and $I(\gamma, \gamma') \approx 0$; on the other hand, if $x_0 \in \Sigma_c$ then, for $T > T_c$, $\gamma$ and $\gamma'$ will be uncorrelated and $I(\gamma, \gamma') > 0$, where $T_c$ is some critical time.* In other words, for very short times ($T < T_c \ll T_*$) we expect the same behaviour of $I$ for both, regular and chaotic orbits ($I \approx 0$). This is just a consequence of the fact that, for this time interval, the divergence of initially nearby orbits is very small and thus $\gamma$ and $\gamma'$ are very similar. As the time ($T$) increases, the correlation between the two orbits evolves. But for $T > T_c$, this evolution will be rather different if $x_0$ belongs to a regular or to a chaotic region of the phase space. It is clear that $T_c$ can be associated to the time needed to pass close to an hyperbolic object with non coincident separatrices. If the initial condition lies in a regular region, we expect that $\gamma$ and $\gamma'$ will lose its correlation very slowly and then $I \ll 1$ for all $T$. On the other hand, if $x_0$ lies in a chaotic region, $\gamma$ and $\gamma'$ will lose its correlation very fast (due to the exponential divergence), and $I$ will take then a larger value. The 'distributions' $\rho(t)$ contain all the information about the flow and should reveal this rather different behaviour.

In the next paragraphs we shall summarise the main analytical results for $I$. By definition, for any $t$, we have $\dot{x}'(t) = \dot{x}(t) + \dot{\delta}(t)$. Denote with $v(t) \equiv |\dot{x}(t)|$, $\delta(t) \equiv |\delta(t)|$, $d(t) \equiv |\dot{\delta}(t)|$ and $\xi(t) = d(t)/v(t)$, which is assumed to be small. Then, up to second order in $\xi$, an elementary but tedious computation shows that:

$$\rho_{\gamma'} \approx \rho_\gamma (1 + a_1 + a_2 - k_1 a_1), \quad \rho_\Gamma \approx \rho_\gamma \left( 1 + \frac{a_1}{2} + \frac{a_2}{4} - k_1 \frac{a_1}{4} + a_3 \right), \quad (15)$$

where:

$$a_j(t) = \frac{\Delta_j(t)}{v(t)} - \frac{\langle \Delta_j \rangle_T}{\langle v \rangle_T} \quad j = 1, 2, 3, \qquad k_1 = \frac{\langle \Delta_1 \rangle_T}{\langle v \rangle_T}, \qquad (16)$$

with:

$$\Delta_1 \equiv \Delta v = \frac{\dot{\delta} \cdot \dot{x}}{v} = d \cos \zeta, \qquad \Delta_2 = \frac{1}{2} \frac{d^2}{v} \sin^2 \zeta, \qquad \Delta_3 = \frac{d^2}{8v}. \qquad (17)$$

In (17), $\zeta(t)$ is the angle between $\dot{x}(t)$ and $\dot{\delta}(t)$. From (15), it is straightforward to verify that $\rho_{\gamma'}$ and $\rho_\Gamma$ are normalised. As we see from (15)–(17), $\rho_{\gamma'}$ and $\rho_\Gamma$ differ from $\rho_\gamma$ –at $\mathcal{O}(\xi)$– by the instantaneous fluctuation of $\Delta v(t)$, the first variation of $|v(x(t))|$. After some algebra we obtain for $I$:

$$I(\gamma, \gamma') \approx \frac{\langle v a_1^2 \rangle_T}{8 \langle v \rangle_T}. \qquad (18)$$

For the moment being we shall assume that $\langle v \rangle_T$ depends *in a mild way* on $T$ (see, however, the remark at the end of this section). Then we write $\langle v \rangle_T \approx X$, independent of $T$, and we can approximate $I$ by:

$$I(\gamma, \gamma') \approx \frac{1}{8X^2} \left( \langle \Delta_1^2 \rangle_T - \langle \Delta_1 \rangle_T^2 \right) \tag{19}$$

As we note the dependence of $I$ on $T$ is given by the variance of $\Delta_1$ between $t = 0$ and $t = T$. From (17), $\Delta_1$ is expected to have a relatively small average since it depends linearly on $\cos \zeta$. However this is not the case for $\Delta_1^2$. Therefore, to obtain $I$ we must be able to compute the time–mean square value of $\Delta_1$. From the second equality in the first of (17), the second in (2) and recalling that:

$$\dot{\delta}^k = \frac{\partial v^k}{\partial x^l} \delta^l, \qquad v^k \equiv (\mathbf{v})^k, \tag{20}$$

(the sum over repeated indexes is understood) then it is straightforward to get:

$$\Delta_1 \approx \frac{1}{X}(\phi_k \phi_{kl} - \phi_l)\delta q^l = \frac{1}{X}\Phi_l \delta q^l, \tag{21}$$

where the subscripts in $\phi$ denote derivatives respect to the position, $\Phi_l = \phi_k \phi_{kl} - \phi_l$ and $\delta q$ is the displacement from $\mathbf{q}$ at the time $t$. The latter is given by the variational equations (20). However, unless we are near to an equilibrium point, these equations cannot be solved analytically. Therefore, to make some progress in the theoretical approach, we have to introduce further assumptions. Let us consider the simplest case: regular motion. In this case we know that $\delta q^l(t) \approx \delta_0 \lambda_l t$ is the expected law for the time evolution of $\delta$, where $\delta_0 \ll 1$ is the initial displacement and $\lambda_l$ is the time–rate of linear divergence in a neighbourhood of the initial point $(\mathbf{x}_0)$ in the $l$–direction. Let $\lambda(\mathbf{x}_0) = \max\{\lambda_l\}$, so $\lambda_l = \lambda b_l$, where $b_l \leq 1$. Then, introducing this in (21) we have:

$$\Delta_1(t) \approx \frac{\delta_0 \lambda t}{X} \Psi\left(\mathbf{q}(t)\right), \tag{22}$$

where $\Psi = \Phi_l b_l$ depends only on the position through the derivatives of the potential. If $\omega$ is the frequency vector associated with the invariant torus where the motion proceeds, then we can expand $\Psi\left(\mathbf{q}(t)\right)$ in Fourier series and, provided that $(\omega.\mathbf{k})T$ is large enough for all integer vectors $\mathbf{k} \neq 0$, then from (19) and (22) we finally obtain for $I(\gamma, \gamma')$:

$$I(\gamma, \gamma') \approx \frac{\delta_0^2 \lambda^2 A}{24X^4} T^2, \tag{23}$$

where $A$ is a positive constant (the average value of $\Psi^2$) which is almost independent of $T$ for $T \gg T_D$, the characteristic period of motion. As we note, $I$ depends on the initial condition mainly through $\lambda(\mathbf{x}_0)$, so we can write $I(\gamma, \gamma') \equiv I(T; \mathbf{x}_0)$. Therefore, for two different initial conditions in $\Sigma_r$ (and for the same values of $\delta_0$

and $T$), we have: $I_1/I_0 \approx \lambda_1^2/\lambda_0^2$, where the smaller $\lambda$ appears for $x_0$ in a neighbourhood of a stable periodic orbit (think about using a Normal Form around this orbit).

At this step, one has to keep in mind that (23) is valid for small $\xi(t)$. Using $d(t) \approx R\delta(t)$, $R \sim 1$ (these two quantities cannot differ too much since $\delta^2 = |\delta q|^2 + |\delta p|^2$ while $d^2 = |\delta p|^2 + |\delta \nabla \phi|^2$; in fact $d \approx R\delta$, with $R$ of bounded magnitude seems to be a reasonable approximation), then we can state that (23) is true for $\lambda T \ll X/\delta_0$. Therefore, $I \sim \mathcal{O}\left(\delta_0^2 T^2\right) \ll 1$. We see then that for regular motion, $I$ behaves in the expected way: $I$ takes a very small value due to the strong correlation between $\gamma$ and $\gamma'$.

Since $I$ depends on $T$, it will be interesting to know its time–rate in general. Using the analytic formula for the derivative of the entropy (see (28)) and up to second order in $\xi$ we get:

$$\frac{dI}{dT} \approx \frac{1}{8T} \left(a_1^2(T) - 8I(T)\right) \frac{v(T)}{\langle v \rangle_T}.$$

We introduce then the quantity $J(\gamma, \gamma') \equiv d \log I / d \log T$, which is given by:

$$J(\gamma, \gamma') = \frac{T}{I} \frac{dI}{dT} \approx \left(\frac{a_1^2(T)}{8I(T)} - 1\right) \frac{v(T)}{\langle v \rangle_T} \approx \frac{\Delta_1^2(T)}{\langle \Delta_1^2 \rangle_T} - 1, \qquad (24)$$

where the last follows for $v(t) \approx \langle v \rangle_T \approx X$ and $\langle \Delta_1 \rangle_T = 0$. This relation shows that $J$ seems to be a more interesting magnitude than $I$. For example, in the case of regular motion, we get:

$$J(\gamma, \gamma') \approx 2, \qquad T \gg T_D, \qquad (25)$$

independent of $\delta_0, T, \lambda$ and any other scale parameter. Besides, it is not difficult to see that $J$ only depends on $\delta$: that is, if $\delta$ grows with some power of $t$, say $t^r$, then $J \approx 2r$.

For the chaotic case, the analytical results are not so accurate. We shall say that if $\delta$ grows exponentially: $\delta(t) \approx \delta_0 e^{\sigma t}$, then the dominant terms in $I$ and $J$ are:

$$I(\gamma, \gamma') \sim \frac{\delta_0^2}{X^4} \frac{e^{2\sigma T}}{\sigma T}, \qquad J(\gamma, \gamma') \sim 2\sigma T - J_0, \qquad T \gg T_D. \qquad (26)$$

where $J_0 \sim 1$ is a constant independent of $\sigma(x_0)$, the mean–rate of exponential divergence in a neighbourhood of $x_0$.

As we have already mentioned, the approximate analytical results rest on the assumption $\langle v \rangle_T \approx X$. By definition

$$v^2 = 2[h - \phi(\mathbf{q})] + |\nabla \phi(\mathbf{q})|^2.$$

Since the motion is bounded in the phase space (confined to the energy surface) and if $\phi$ is smooth enough, then $\langle v \rangle_T$ is a bounded quantity of the order of the size of the system. In the simple case of an $N$–dimensional harmonic oscillator, $v(t) = \langle v \rangle_T = \sqrt{2h}$.

## 3. Numerical Examples

For the numerical study of this technique, we considered the Hénon–Heiles model (Hénon & Heiles 1964), for the energy level $h = 0.118$ ($T_D \lesssim 10$). Within this energy surface, we have restricted the analysis to the region shown in Fig. 1(a). Fixing $p_2 = 0$, for $0.3 \leq q_2 \lesssim 0.508$ we have a large regular component, where $q_2 \approx 0.305$ corresponds to the stable 1–periodic orbit. For $0.508 \lesssim q_2 \lesssim 0.57$ we observe a 5–periodic island with its stochastic layer around the separatrix and bounded by a KAM curve. This KAM curve, which can be seen in this figure, separates different stochastic regions and disappears for $h = 0.119$. For $q_2 \gtrsim 0.57$ we see a highly stochastic region. Certainly, within the whole domain, a large number of small islands is present. In Fig. 1(b) we show the computed LCN for 1000 initial conditions along the $q_2$ axis, $0.55 \leq q_{2_0} \leq 0.60$ and for $T = 2.5 \times 10^5$. This was done in order to have an estimation of the measure of chaos in the stochastic components. Even though the Hénon–Heiles is a very simple model, the structure of the region considered in Fig. 1(a) is always present in almost all near–integrable 2D Hamiltonian systems.

For the explicit computation of $I$ and $J$ we take advantage that from (10), $S(\gamma) \equiv S(T; \mathbf{x}_0)$ can be written in the form:

$$S(T; \mathbf{x}_0) = -\frac{1}{L(T; \mathbf{x}_0)} \int_0^T |\mathbf{v}(t; \mathbf{x}_0)| \ln |\mathbf{v}(t; \mathbf{x}_0)| dt + \ln L(T; \mathbf{x}_0), \qquad (27)$$

where $L(T; \mathbf{x}_0)$ is the time–integral of the vector field (3). $S(\gamma')$ and $S(\Gamma)$ satisfy similar formulae. Then we use (14) to obtain $I(T; \mathbf{x}_0)$. Besides, from (27) one readily finds:

$$\frac{dS}{dT} \equiv \dot{S}(T; \mathbf{x}_0) = \rho(T; \mathbf{x}_0)\left(1 - S(T; \mathbf{x}_0)\right) - \rho(T; \mathbf{x}_0) \ln \rho(T; \mathbf{x}_0). \qquad (28)$$

Then $\dot{I} = \dot{S}(\Gamma) - \frac{1}{2}[\dot{S}(\gamma) + \dot{S}(\gamma')]$ and finally, $J = T\dot{I}/I$. In all the numerical integrations we considered $T < T_* = 10^3 T_D \sim 10^4$.

The numerical calculations can be done in three different ways. The alternatives are: (A1) To take two nearby initial conditions, $\mathbf{x}_0$ and $\mathbf{x}_0'$, and to integrate the equations of motion (3) for both initial conditions to get explicitly $\gamma$ and $\gamma'$. Then with the help of (27) and (28), we obtain $I$ and $J$. (A2) To approximate $\mathbf{x}'(t) = \mathbf{x}(t) + \delta(t)$ where $\mathbf{x}_0' = \mathbf{x}_0 + \delta_0$ and $\delta(t)$ is the solution of the variational equations (20). Having computed $\gamma$ and $\gamma'$ we proceed then like in (A1). This approximation is justified by the fact that, even though $I$ depends on $\delta^2$, $I \propto \langle va_1^2 \rangle_T$, this term is the square of a term that comes from the first variational equations, while those quadratic terms that come from the second variationals cancel. (A3) To use the second order formulae given by (18) and by the first in (24). A1 was used in (Núñez et al, 1996) and has the restriction that $\delta(t) = |\mathbf{x}'(t) - \mathbf{x}(t)|$ reaches a saturation value given by the size of the system for the corresponding energy level. A2 is used here for the calculation of $\rho(T)$, with $\delta_0^k = \delta_0/2$, $\delta_0 = 10^{-6}$, $k = 1, ... 4$

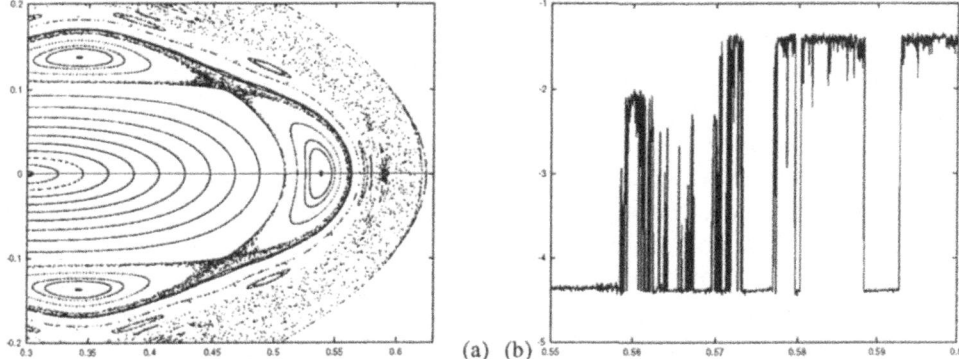

Fig. 1. (a) Surface of section $(q_2, p_2)$, $p_1 > 0$ for the Hénon–Heiles model with $h = 0.118$ and several initial conditions: $q_{1_0} = 0$ and $(q_2, p_2)_0$ taken on the line $p_{2_0} = 0$, with $0.3 \leq q_{2_0} \leq 0.62$. (b) log (LCN), after $T = 2.5 \times 10^5$, for $p_{2_0} = 0$ and 1000 values of $q_{2_0}$ in the range $0.55 \leq q_{2_0} \leq 0.60$.

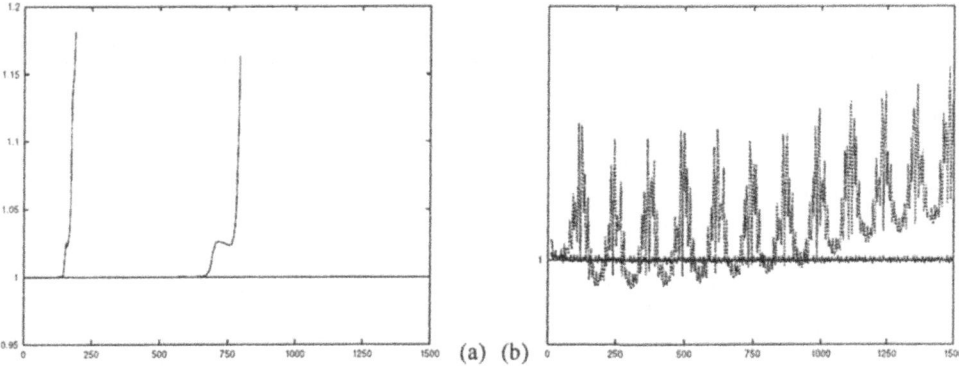

Fig. 2. (a) $\rho_{\gamma'}/\rho_\gamma$ vs. $T$, for $p_{2_0} = 0$, $q_{2_0} = 0.305$, 0.5, 0.5085 (these three curves look as the same, very close to 1), 0.509 (dotted line) and 0.6 (dotted–dashed line). (b) Zoom in the neighbourhood of $\rho_{\gamma'}/\rho_\gamma = 1$, for the first three curves in (a), window: [0.999995, 1.000015] (see text).

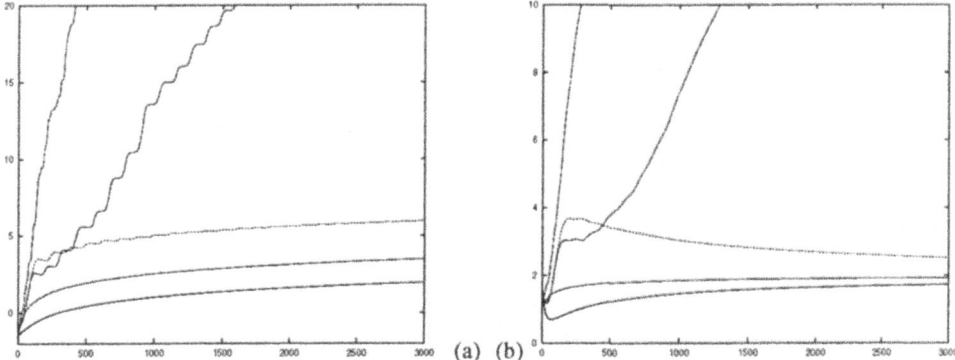

Fig. 3. (a) log $I$ vs. $T$; (b) $J^{(2)}$ vs. $T$ for the same orbits than in Fig. 2. Starting from the solid curve: $q_{2_0} = 0.305$, 0.5, 0.5085, 0, 509 and 0.6.

(note that $\gamma$ and $\gamma'$ are in different but very close energy levels). Even though with the A2 procedure $\delta(t)$ does not have a saturation level as in A1, this second alternative has some numerical limitations. Indeed, in both limits, $\delta_0 \to 0$ and/or $\delta$ large ($\sim X \sim$ the size of the system), the algorithm becomes numerically unstable. Therefore A2 has the restriction that $I$, and consequently $J$, depend on the selected value of $\delta_0$. However, within the range $10^{-7} \le \delta_0 \le 10^{-4}$, $J$ is almost independent of $\delta_0$ while for $I$, the latter behaves as an scale factor (see (23)). A3, that is used here for all the computations of $I$ and $J$, does not require the explicit calculation of $\gamma'$ and therefore, is numerically stable. In this case $\delta_0$ is always a scale factor that we set (arbitrarily) equal to one. Therefore, $I \sim \mathcal{O}\left(T^2\right) \gg 1$ for the regular component while $I \sim \mathcal{O}\left(\exp(\sigma T)\right) \ggg 1$ for the stochastic component.

In any case, since $J$ depends explicitly on the instantaneous value of $\rho(T)$ we averaged the computed $J$, just performing the sum of the successive values of $J(T_n)$ for $T_n \le T$ and dividing the sum by the integer time $n$. In this direction, if $T_n = T_0 + n\delta T$ ($\delta T$ is the time step) we denote then by $J^{(0)}(T_n) = J(T_n)$, $J^{(1)}(T_n) = \frac{1}{n}\sum_{k\le n} J(T_k)$ and by $J^{(2)}(T_n) = \frac{1}{n}\sum_{k\le n} J^{(1)}(T_k)$. This was done in order to lower as much as possible the effect of fast oscillations that do not affect the aperiodical changes. This is, in fact, a simple smoothing procedure.

Fig. 2(a) shows the time evolution of $\rho_{\gamma'}/\rho_\gamma(T)$ (averaged once in the sense mentioned above and using A2) for initial conditions in different zones along the $q_2$ axis: $q_{2_0} = 0.305$ (near to the stable 1–periodic orbit), $q_{2_0} = 0.5$ (quasiperiodic orbit associated with the latter periodic orbit), $q_{2_0} = 0.5085$ (similar to the former but closer to the separatrix), $q_{2_0} = 0.509$ (inside the stochastic layer), $q_{2_0} = 0.6$ (in the highly stochastic sea). We did not plot $\rho_\Gamma$ because its behaviour is almost the same than $\rho_{\gamma'}$. We see that for regular motion, $\rho_{\gamma'}$ and $\rho_\gamma$ seem to be identical, while for chaotic motion the behaviour is similar up to certain value of $T = T_c$. For $T > T_c$, $\rho_{\gamma'}$ diverges form $\rho_\gamma$ in an exponential way (see (15)–(17)). Note that, for the orbit in the highly stochastic sea, this divergence occurs after a few periods of motion while for that in the stochastic layer, this divergence occurs after several periods. Besides, the time–rate of exponential divergence seems to be different in both cases. A magnification around $\rho_{\gamma'}/\rho\gamma = 1$ (Fig. 2 (b) – note the size of the window in the vertical axis) reveals that, for the quasiperiodic orbit ($q_{2_0} = 0.5085$) which is very close to the separatrix, $\rho_{\gamma'}/\rho_\gamma$ exhibits small periodic pulses and a drift. However this effect is not observed for the other regular orbits. For $q_{2_0} = 0.5$, $\rho_{\gamma'}/\rho_\gamma$ oscillates very fast about 1 with a very small amplitude while for $q_{2_0} = 0.305$, $\rho_{\gamma'}/\rho_\gamma = 1$ for the resolution of the figure. This periodical behaviour for the outermost quasiperiodic orbit is due to the interaction between the latter and the 5–periodic hyperbolic orbit (in Fig. 1(a) we just see two of the five 'hyperbolic points'). Indeed, the motion in the vicinity of a hyperbolic point is mainly determined by its associated stable and unstable manifolds (or stable and unstable separatrices). Therefore, while the motion is confined to this small region of the phase space, regular nearby orbits diverge exponentially during certain interval $\Delta t$. The latter interval is the interaction time between both orbits.

The width of a pulse is then a measure of $\Delta t$. The drift is due to the cumulative changes produced by the latter interaction. It is important to remark that while $I$ is a time–averaged quantity, $J$ is not. Thus $I$ will not 'see' the periodical variations of $\rho_{\gamma'}$, but of course, the drift will do. On the other hand $J$ will be sensitive to both effects.

In Fig. 3 we show the time evolution of $I$ and $J^{(2)}$ for the orbits considered in Fig. 2 and using the A3 procedure (note the scale in $I$). We see that $\log I$ has a logarithmic dependence with $T$ (in fact, $I \propto T^2$) for $x_0 \in \Sigma_r$ (see (23)) while it varies nearly linear for $x_0 \in \Sigma_c$ (the first in (26)). As we mentioned above, the smallest value of $I$ corresponds to $x_0$ very close to the stable 1–periodic orbit. Also $J$ behaves in the expected way: for $x_0 \in \Sigma_r$ and $T \gtrsim 10^2 T_D$, $J^{(2)} \approx 2$ (see (25)) while it depends linearly with $T$ for $x_0 \in \Sigma_c$ and $T > T_c$ (the second in (26)). These figures confirm that the time–rate of exponential divergence is slower for $x_0$ in the stochastic layer. Besides, we see that for the quasiperiodic orbit close to the separatrix, $J^{(2)} \to 2$ in a different way than the other regular orbits. We observe that, in this case, $J^{(2)}$ reaches a maximum value and then it decreases asymptotically to 2. This is just a consequence of the effect observed in Fig. 2(b): we are seeing the pulses and the drift. The reason to observe just one instead of several peaks (as in Fig. 2(b)) is due to the fact that we have plotted $J^{(2)}$ (not $J^{(0)}$) and therefore, only the first peak, for small $T$, is significant in this case. Note that for $x_0 \in \Sigma_c$, $I$ and $J$ cover the whole time interval (both curves, $I(T)$ and $J^{(2)}(T)$ escape from Fig. 3 ) while in Fig. 2(a), these curves for $\rho_{\gamma'}/\rho_\gamma$ end at $T \lesssim 10^3$. Indeed, these figures clearly show the limitation of the A2 procedure (that was used to perform Fig. 2). In the latter case, the integration was stopped when $I$ became negative. As we have already mentioned, $I$ must be a positive quantity (see Section 2). This is true while the orbit $\gamma'$ is on an energy surface $M_{h'}$ such that $h' \approx h$ (recall that we took $\gamma'$ in a slightly different energy level than $\gamma$). Then for $x_0 \in \Sigma_c$ and after some time interval (that for which $\delta \gtrsim X \sim$ the size of the system), $\gamma'$ will be in an energy surface which is far from $M_h$ and then the procedure A2 leads $I$ to decrease, in almost all the cases, monotonically. However both, A2 and A3, provide the same results for regular motion for all $T$ (except for a scale factor in $I$) and, for stochastic motion, the results agree up to $T \lesssim 10^3$ and, in any case, for $\delta_0 \geq 10^{-7}$ when the A2 procedure is used.

In Fig. 4 we have plotted the final value of $\log I$ for $T = 4000$, and for $x_0$ along the $q_2$ axis for several values of $q_{2_0}$ within the 1–periodic island. From Fig. 4(a) we confirm once again that $I$ depends on the initial condition through $\lambda(x_0)$, its minimum corresponds to the stable 1–periodic orbit. Besides, we observe that $I$ increases very fast as $q_{2_0}$ approaches to the separatrix (recall the logarithmic scale). Fig. 4(b) shows that $J^{(2)} \approx 2$ along the 1–periodic island in accordance with the expected value for regular motion. Both, $I$ and $J$, are highly sensitive to the presence of small periodic islands. The 'discontinuities' observed in $I$ and $J^{(2)}$ reveal the existence of thin chaotic layers around the separatrices of high–order resonances for the corresponding values of $x_0$. But in fact it is the passage close to

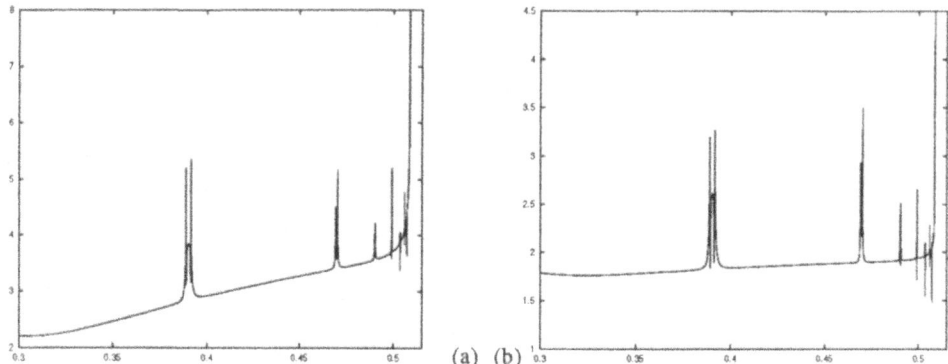

(a)  (b)

Fig. 4. (a) $\log I$ vs. $q_{2_0}$, (b) $J^{(2)}$ vs. $q_{2_0}$ for 3000 initial conditions within the 1–periodic island and $T = 4000$.

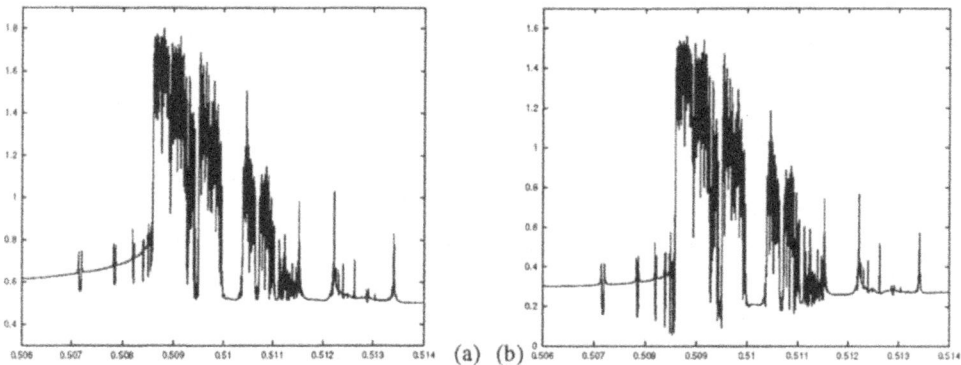

(a)  (b)

Fig. 5. The same as Fig. 4 but for $q_{2_0}$ in the neighbourhood of the separatrix of the 5–periodic island for 4000 initial conditions and $T = 4000$. (a) $\log(\log I)$ vs $q_{2_0}$; (b) $\log J^{(2)}$ vs. $q_{2_0}$, $\log 2 \approx 0.3$.

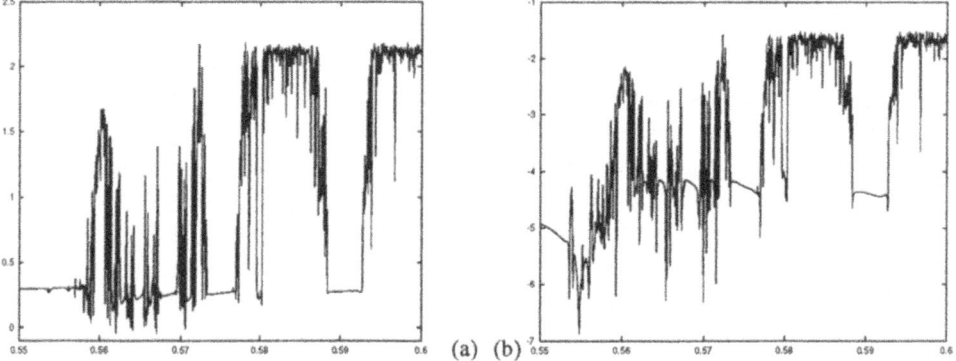

(a)  (b)

Fig. 6. (a) $\log J^{(2)}$ vs. $q_{2_0}$, (b) $\log \sigma_E$ vs. $q_{2_0}$ for the same set of initial conditions than Fig. 1(b) but $T = 7000$ (see text for details).

the hyperbolic periodic points which produces this effect on $I$ and $J$, that would be seen even for integrable systems. Fig. 5 is a continuation of Fig. 4 for $q_{2_0}$ across the separatrix of the 5–periodic island. It clearly shows the presence of the gross stochastic layer as well as the regular regions at both sides of the separatrix ($\log 2 \approx 0.3$). Note that the regular and the stochastic component are separated by several orders of magnitude (see the scale used in the $I$ axis), in accordance with the estimations given above. A simple inspection of (26) shows that Figs. 5(a) and 5(b) should be similar for the stochastic component and this is, in fact, observed.

In Fig. 6(a) we have plotted the final value of $\log J^{(2)}$ for the same set of initial conditions than that used for the computation of the LCN but for $T = 7000$. We see again that $J^{(2)} \approx 2$ in the ordered component while $J^{(2)}$ is clearly much larger that 2 in the stochastic regions. In fact, $J$ reaches higher values in the large stochastic sea than in the stochastic layer. Then, because we can separate different stochastic components and, obviously, the regular one, we can state that $J$ provides a measure of chaos. We can easily relate $J$ with the LCN. Indeed, recalling the second of (26) we see that $dJ/dT \sim LCN$. Therefore since $J^{(2)}$ (and also $J^{(1)}$) depends nearly linear with $T$ for $T > T_c$ (see Fig. 3(b)), we can fit by least squares the linear part of the curve $J^{(2)}(T)$. This was done for each initial condition of Fig. 6(a). The expected value for the slope is $\sigma \approx 0$ in the regular component (since $J^{(2)} \approx 2$) while $\sigma \sim LCN$ in the stochastic component. As we are computing $J^{(2)}$ instead of $J^{(0)}$, the factor 2 in front of the second of (26) compensates in part the averaging procedure. Fig. 6(b) shows the computed value of $\sigma$, $\sigma_E$. A comparison of this figure with that for the LCN (Fig. 1(b)) reveals a good agreement between both magnitudes. It is important to remark that while $\sigma_E$ was computed for $T = 7000$, the LCN was for $T = 2.5 \times 10^5$. Actually, $T = 5000$ is enough to get $\sigma_E$ while the total motion time used to compute the LCN is, perhaps, not sufficient to get a good asymptotic value. Besides, note that Fig. 6(b) shows the structure of the regular component while Fig. 1(b) does not provide any information about it (further details about these matters can be found in Cincotta & Simó, 1998).

Finally, let us consider the quasiperiodic orbits near the separatrix. As we have already shown in Fig. 3(b), $J^{(2)}$ presents a maximum and then it goes asymptotically to 2. In Fig. 7(a) we show the computed maximum, $J_{max}^{(2)}$, for several orbits in the vicinity of the separatrix. We see that while we move towards the separatrix from the regular side, $J_{max}^{(2)}$ is slightly sensitive to the presence of small periodic islands (compare with Fig. 5(b)). However the 'continuum' is smooth and 'diverges' on the separatrix. On the other side of the separatrix (stochastic), we can still observe nearly the same 'continuum' that in the regular side but the presence of many sharp 'lines', reveals the existance of the stochastic layer. This structure is similar to that observed in Fig. 5(b) for the borders of the small islands, but in that case the motion is almost all regular (stochastic motion, if it is present, is insignificant). We can use the regular part of this curve as a tool to find out the location of the 5–periodic hyperbolic orbit. Indeed, we have already mentioned that the existence

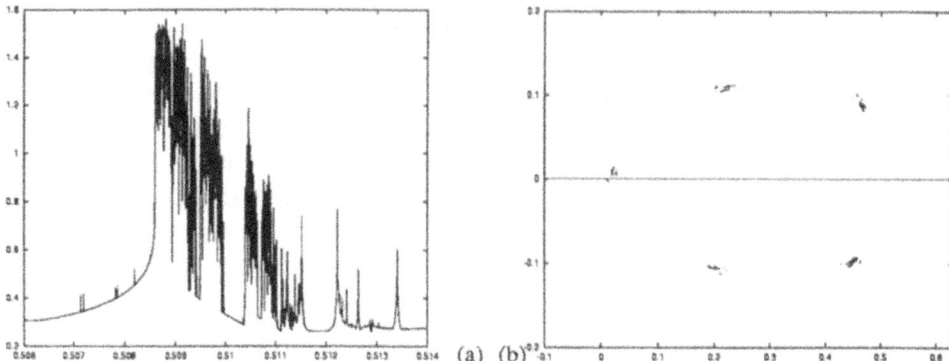

Fig. 7. (a) $\log J_{max}^{(2)}$ vs. $q_{2_0}$ in the neighbourhood of the separatrix for 4000 initial conditions. (b) 'Surface of section' $(q_2, p_2)_{T \approx T_m}$ for 1300 initial conditions in the interval $0.506 \leq q_{2_0} \leq 0.50859$, (see text for details).

of a maximum in $J^{(2)}$ is due to the presence, somewhere, of this unstable orbit. Therefore if we compute the time $T_m$ for which $J^{(2)}(T_m) = J_{max}^{(2)}$, then we can integrate each initial condition (in the regular part) up to $T \approx T_m$, and then plot the final value $(q_2, p_2)$ when $q_1 = 0$, $p_1 \geq 0$. This is shown in Fig. 7(b). A comparison of this figure with Fig. 1(a) confirms that the maximum value observed in $J^{(2)}$ is really due to the interaction between the quasiperiodic orbits and the unstable periodic orbit.

## 4. Conclusions

In this paper we showed that the Conditional Entropy, defined through the arc length parameter along the orbits, is an efficient tool to explore the phase space in short motion times. It is important to mention that this method provides an easy way to find out the location of unstable periodic orbits. The preliminary analytical results agree with that obtained by numerical simulations (at least for very simple systems). Besides, the mean rate of exponential divergence of nearby orbits can be estimated using this technique, but for motion times which are two or three orders of magnitude less than that for the computation of the LCN using the standard procedures. The numerical efficiency of this method, the true convergence of $J/2T$ to the LCN as well as a numerical analysis of the information about the dynamics contained in $J^{(0)}$ and $J^{(1)}$ were also studied. These results, however, are presented in a separate paper (Cincotta & Simó, 1998).

Finally we would like to remark that there are some topics that still remain to be done: i) a more complete theory; ii) numerical study of 3D models and iii) application to realistic dynamical systems. All of these issues will be addressed in next papers.

# Acknowledgements

The first author would like to acknowledge the Consejo Nacional de Investigaciones Científicas y Técnicas de la República Argentina who supports his visit to the University of Barcelona. The second author has been supported by DGICYT grant PB 94-0215 (Spain). Partial support from the EC grant ERBCHRXCT940460 and the catalan grant CIRIT 1996S0GR-00105 (moved to 1998S0GR-00042) is also acknowledged. Finally we would like to acknowledge the referee, I. Schevchenko, for a careful reading of the manuscript and his suggestions that helped us to improve it.

# References

Arnold, V. & Avez, A.: 1989, *Ergodic Problems of Classical Mechanics*, (New York: Addison–Wesley), 2nd. ed.
Cincotta, P. & Simó, C.: 1998, preprint
Contopoulos, G. & Voglis, N.: 1996, *Cel. Mech. & Dynam. Astron.*, **64**, 1
Fraser, A. & Swinney, H.: 1986, *Phys. Rev. A*, **33**, 1134
Hénon, M. & Heiles, C.: 1964, *AJ*, **69**, 73
Katz, A.: 1967, *Principles of Statistical Mechanics, The Information Theory Approach*, (San Francisco: W. H. Freeman & Co.)
Laskar, J.: 1990, *Icarus*, **88**, 266
Laskar, J.: 1993, *Physica D*, **67**, 257
Merritt, D. & Valluri, M.: 1996, *ApJ*, **471**, 82
Núñez, J., Cincotta, P. & Wachlin F.: 1996, *Cel. Mech. & Dynam. Astron.*, **64**, 43
Papaphilippou, Y. & Laskar, J.: 1998, *A&A*, **329**, 451
Udry, S. & Pfenniger, D.: 1988, *A&A*, **198**, 135
Voglis, N., Contopoulos, G. & Efthymiopoulos, C.: 1999, *Cel. Mech. & Dynam. Astron.*, in press.
Wachlin, F. & Ferraz–Mello, S.: 1998, *MNRAS*, **298**, 22

# DETECTION OF ORDERED AND CHAOTIC MOTION USING THE DYNAMICAL SPECTRA

N. VOGLIS[2], G. CONTOPOULOS[1,2] and C. EFTHYMIOPOULOS[1,2]

[1]*Research Center for Astronomy, Academy of Athens*
[2]*Department of Astronomy, University of Athens*

**Abstract.** Two simple and efficient numerical methods to explore the phase space structure are presented, based on the properties of the "dynamical spectra". 1) We calculate a "spectral distance" $D$ of the dynamical spectra for two different initial deviation vectors. $D \to 0$ in the case of chaotic orbits, while $D \to const \neq 0$ in the case of ordered orbits. This method is by orders of magnitude faster than the method of the Lyapunov Characteristic Number (LCN). 2) We define a sensitive indicator called ROTOR (ROtational TOri Recongnizer) for 2D maps. The ROTOR remains zero in time on a rotational torus, while it tends to infinity at a rate $\propto N$ = number of iterations, in any case other than a rotational torus. We use this method to locate the last KAM torus of an island of stability, as well as the most important cantori causing stickiness near it.

## 1. Introduction

Let $\mathbf{R} = (x_1, x_2, ..., x_n)$ be the position vector and $\boldsymbol{\xi} = (dx_1, dx_2, ..., dx_n)$ be the deviation vector in the n-dimensional phase space of a dynamical system. The evolution of these vectors can be given by the maps

$$\mathbf{R}' = \mathbf{F}(\mathbf{R}; K), \quad \boldsymbol{\xi}' = \frac{\partial \mathbf{F}}{\partial x} \boldsymbol{\xi} \qquad (1)$$

in time steps of $\Delta t$. A number of useful quantities can be defined in terms of these two vectors. Namely, we can define the 'stretching number'

$$a = \frac{1}{\Delta t} \lim \ln \frac{|\boldsymbol{\xi}'|}{|\boldsymbol{\xi}|}, \quad |\boldsymbol{\xi}| \to 0 \qquad (2)$$

(Nicolis 1983, Fujisaka 1983, Froeschlé et al 1993, Voglis and Contopoulos 1994) which gives the current rate of deviation of nearby orbits. If $\boldsymbol{\xi}_{ij} = (dx_i, dx_j)$ is the projection of $\boldsymbol{\xi}$ on the plane $(x_i, x_j)$ we can define the 'helicity angle' $\Phi_{ij}$ that gives the orientation of the vector $\boldsymbol{\xi}_{ij}$ with respect to the axis $x_i$, denoted by $\Phi_{ij} = ang(x_i, \boldsymbol{\xi}_{ij})$. The 'twist angle' $\phi_{ij} = \frac{1}{\Delta t} ang(\boldsymbol{\xi}_{ij}, \boldsymbol{\xi}'_{ij})$ gives the current angular frequency of rotation of the vector $\boldsymbol{\xi}_{ij}$. Similar quantities can be defined also for the position vector, e.g. the rotation angle $\theta_{ij} = \frac{1}{\Delta t} ang(\mathbf{R}_{ij}, \mathbf{R}'_{ij})$ gives the current angular frequency of rotation of the vector $\mathbf{R}_{ij}$.

If $Q$ is one of these quantities, i.e. $Q = a, \Phi_{ij}, \phi_{ij}, \theta_{ij}...$ at the $k^{th}$ iteration, we define as the 'dynamical spectrum of $Q$' the probability density of the values of $Q$ given by

$$S(Q) = \frac{dN(Q, Q + dQ)}{N \, dQ} \qquad (3)$$

where $N$ is the total number of iterations and $dN(Q, Q + dQ)$ is the number of values of $Q$ in the interval $(Q, Q + dQ)$.

*Celestial Mechanics and Dynamical Astronomy* **73**: 211–220, 1999.
© 1999 *Kluwer Academic Publishers.*

## 2. Detection of Ordered and Chaotic Orbits in 4D Maps

The dynamical spectra can be used very efficiently to distinguish between ordered and chaotic motion. The following properties of invariance of the spectra $S(Q)$ are known: (Voglis and Contopoulos 1994, Contopoulos and Voglis 1996, 1997, Voglis and Efthymiopoulos 1998, Voglis et al. 1998):

1)The spectra $S(Q)$ are invariant with respect to the initial conditions $R_0$ along an orbit (invariance in time).

2)The spectra $S(Q)$ are invariant with respect to $R_0$ and $\xi_0$ provided that $R_0$ belongs to a connected chaotic domain (invariant in space).

3) a) In 4D maps: If the motion is ordered, two different initial deviation vectors $\xi_{01}, \xi_{02}$ give different dynamical spectra, i.e.

$$S(Q, \xi_{01}) \neq S(Q, \xi_{02}), \tag{4}$$

while if the motion is chaotic they give the same spectrum

$$S(Q, \xi_{01}) = S(Q, \xi_{02}). \tag{5}$$

3) b) In 2D maps, the property 3a is reduced to a similar property for the spectrum of the helicity angles only. Namely, if the motion is ordered, an opposite initial deviation vector gives a helicity spectrum shifted by $\pi$

$$S(\Phi, \xi_0) = S(\Phi + \pi, -\xi_0), \tag{6}$$

while if the motion is chaotic the two helicity spectra are equal and $\pi$-symmetric.

$$S(\Phi, \xi_0) = S(\Phi, -\xi_0) = S(\Phi + \pi, \xi_0), \tag{7}$$

The properties 3a,b are explained as follows. In 4D maps the ordered motion occurs on a 2D torus. Any initial deviation vector $\xi_0$ becomes tangent to this torus after a short transient period. In general, two different initial vectors $\xi_{01}$ and $\xi_{02}$ become tangent to different directions on the torus. Thus they produce different sequences of vectors $\xi_1$ and $\xi_2$ and hence different spectra.

On the other hand, two initially different deviation vectors of a chaotic orbit tend to coincide, after a few transient periods, on the direction defined by a nearby most unstable manifold. Thus, they produce the same dynamical spectra.

The convergence of the two deviation vectors is much faster than the convergence of the spectra to their final form (which happens in the same time scale as the stabilisation of the LCN to its constant value). Thus, by measuring the *difference* of the spectra we can recognise a chaotic orbit much faster than by calculating the LCN (or the final spectra).

We demonstrate the eficiency of the above method using the model of two coupled standard maps, i.e.

$$
\begin{aligned}
x_1' &= x_1 + x_2', & x_2' &= x_2 + \frac{K}{2\pi}\sin 2\pi x_1 - \frac{\beta}{\pi}\sin 2\pi(x_3 - x_1) \\
x_3' &= x_3 + x_4', & x_4' &= x_4 + \frac{K}{2\pi}\sin 2\pi x_3 - \frac{\beta}{\pi}\sin 2\pi(x_1 - x_3)
\end{aligned}
\qquad (mod\,1) \;(8)
$$

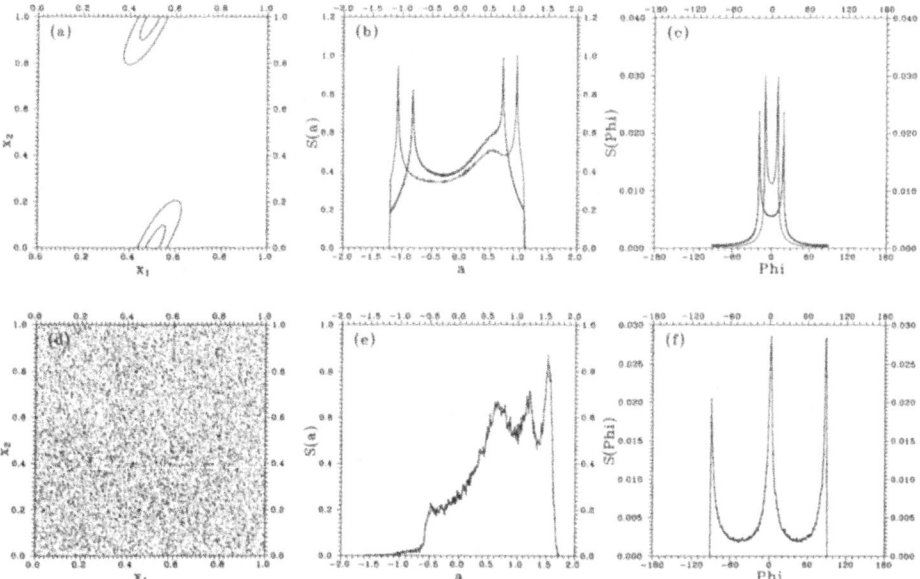

Fig. 1. Projections of the orbits (a) A1 and (d) B on the plane $(x_1, x_2)$. The respective spectra of (b,e) stretching numbers and (c,f) helicity angles.

For $K = 3$ each uncoupled map ($\beta = 0$) has a phase portrait with one main island embedded in a chaotic sea. We select the following initial conditions: $(x_1, x_2, x_3, x_4) = (0.55, 0.1, 0.62, 0.2)$ for three values of $\beta$, namely $\beta = 10^{-5}$ (orbit A1), $\beta = 10^{-1}$ (orbit A2), $\beta = 0.3051$ (orbit A3), and $(x_1, x_2, x_3, x_4) = (0.1, 0.5, 0.2, 0.6)$, for $\beta = 10^{-1}$ (orbit B). We iterate these orbits with two different initial deviation vectors, namely $\xi_1 = (1, 1, 1, 1)$, and $\xi_2 = (2, 2, 1, 1)$.

The projection of the orbit A1 on the plane $(x_1, x_2)$ gives two invariant curves (Fig.1a). This orbit, corresponding to two weakly coupled ordered 2D orbits, is ordered. On the other hand, orbit B is chaotic and its projection covers the plane $(x_1, x_2)$ uniformly (Fig. 1d).

The spectra of the stretching numbers $S(a)$ for the orbit A1 and the deviation vectors $\xi_1$ and $\xi_2$, after $10^5$ iterations, are shown in Fig. 1b. Both spectra are invariant in time, i.e. the next $10^5$ iterations produce the same spectra. However, the two spectra are clearly different from each other. The corresponding spectra of the helicity angles $S(\Phi)$ (Fig. 1c) are also different. In contrast, the spectra $S(a)$ and $S(\Phi)$ of the orbit B, and for the same two initial vectors $\xi_1$, $\xi_2$, coincide (Figs. 1e,f).

In order to quantify this behavior we define the distance $D$ between the two spectra of $Q$ ('spectral distance'), namely:

$$D^2 = \sum_{allQ} [S_1(Q) - S_2(Q)]^2 \tag{9}$$

In Fig. 2a the evolution of the finite time Lyapunov characteristic number

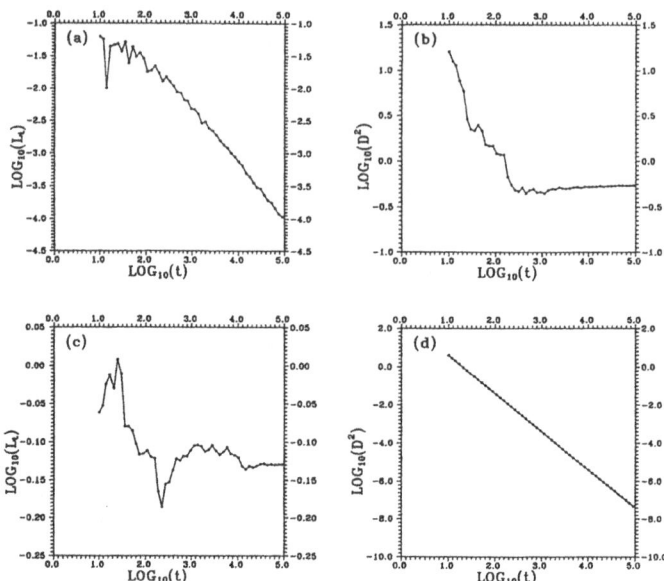

Fig. 2. The evolution of (a,c) the finite time Lyapunov characteristic number and (b,d) the spectral distance, for the orbits A1 and B.

$L_t$ is given for the orbit A1. As expected for regular orbits, $L_t$ decreases as $t^{-1}$, reaching the value $\approx 10^{-4}$ after $10^5$ iterations. The corresponding spectral distance of stretching numbers (fig. 2b), after a transient decrease for about $10^2$ periods, is stabilized near the value $D^2 \approx 0.54$, showing that the two spectra $S(a)$ have constant difference.

On the other hand, the value of $L_t$ of the chaotic orbit B is stabilized at the value $\approx 0.75$ after $10^5$ iterations (fig. 2c). The spectral distance $D$ (Fig. 2d) decreases at a constant logarithmic rate ($D^2 \propto t^{-2}$) and reaches the value of $10^{-2}$ after about 100 iterations, while it becomes smaller than $10^{-7}$ after $10^5$ iterations. Thus, the distinction between ordered and chaotic motion using $D^2$ is much faster than with the LCN criterion.

We apply now the method to the much more difficult case of the orbits A2 (regular) and A3 (chaotic with a very small LCN). The projections of these orbits on the plane ($x_1$, $x_2$) are shown in Figs.3a,d. The projections of both orbits give the impression of motion on a torus. In particular, both projections remain in the same limited area even after $10^9$ iterations.

Despite their visual similarity, the orbit A2 is regular while the orbit A3 is chaotic. The character of the two orbits is revealed very efficiently using the spectral distance $D$. Namely, the two spectra $S(a)$ and $S(\Phi)$ for the orbit A2 (and $\xi_1, \xi_2$ as above) are different (Figs.3b,c), while those of the orbit A3 are identical (Figs. 3e,f).

The evolution of the finite time LCN ($L_t$) for both orbits is shown in Fig. 4. For

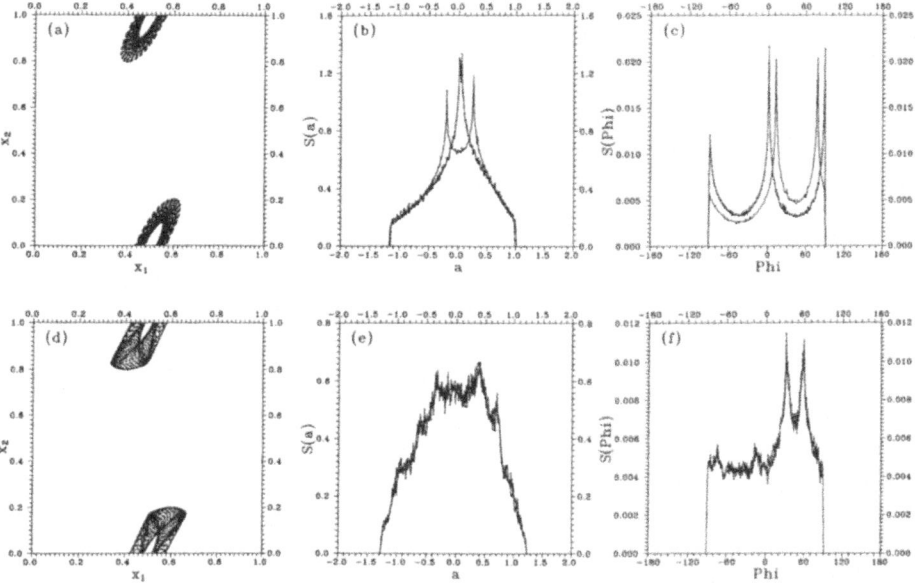

Fig. 3. Same as Fig.1 for the orbits A2 and A3.

the orbit A3 (solid line), $L_t$ converges to a very small positive value $4 \times 10^{-7}$ after $10^9$ iterations, while, for the orbit A2 (dashed line) $L_t$ continues falling as $t^{-1}$. In the same figure, the spectral distances $\log(D^2)$ for the two orbits (dashed for $A_2$, solid for $A_3$) after a transient phase of about $10^3$ iterations evolve in a different way. The spectral distance for the orbit A2 is stabilized to the value $D^2 \approx 0.4$, while the spectral distance for the orbit A3 keeps decreasing, reaching a level of $10^{-3}$ after about $10^5$ iterations. The distinction of the two orbits A2 and A3 using $D^2$ is again much faster than with the finite time LCN.

Similar results are found in 3D Hamiltonian systems by considering either their spectra of stretching numbers along the flow or their 4D Poincaré sections. In 2D maps, this method can be applied by calculating the spectra and spectral distances of the helicity angles.

## 3. Detection of KAM Tori

In this section we describe a method for 2D maps that detects KAM tori surrounding a central stable periodic orbit at a point $O(x_0, y_0)$ (rotational tori). The method is based on finding the average frequency of rotation of both the position vector $\mathbf{R}$ (with respect to $O$), and the deviation vector $\xi$ (Voglis 1996).

Two succesive vectors $(\mathbf{R}_i, \mathbf{R}_{i+1})$ and $(\xi_i, \xi_{i+1})$ define the *rotation angle* $\theta_i = ang(\mathbf{R}_i, \mathbf{R}_{i+1})$ and the *twist angle* $\phi_i = ang(\xi_i, \xi_{i+1})$. The angles $\theta_i$ and $\phi_i$ give

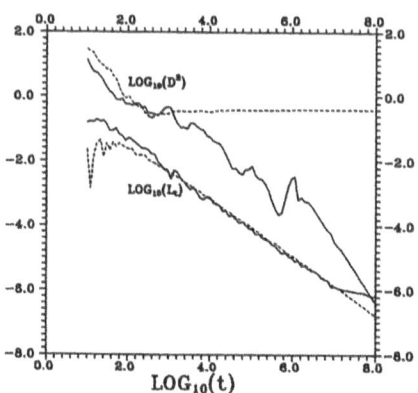

**Fig. 4.** The evolution of the finite LCN ($L_t$) and of the spectral distance $D^2$ for the orbit A2 (dashed) and A3 (solid).

the rotation number $\nu_\theta$ and the twist number $\nu_\phi$ defined respectively as

$$\nu_\theta = \lim \frac{\sum_{i=0}^{N} \theta_i}{2\pi N} , \quad \nu_\phi = \lim \frac{\sum_{i=0}^{N} \phi_i}{2\pi N} , \quad N \to \infty \tag{10}$$

where $N$ in the number of iterations. The rotation and twist numbers $\nu_\theta$ and $\nu_\phi$ give the average frequency of rotation of the vectors $\mathbf{R}$ and $\xi$ around $O$. The difference $\nu_\kappa = \nu_\phi - \nu_\theta$ (epicyclic frequency) gives the average frequency of relative rotation of the vector $\xi$ with respect to $\mathbf{R}$.

If the initial position vector $\mathbf{R}_0$ belongs to a KAM torus around $O$ the epicyclic frequency is by definition zero, while it is non zero in all other cases (Voglis and Efthymiopoulos 1998). In fact, even in the first case, the limit $\nu_\kappa = 0$ is reached only for $N \to \infty$, while the actual value of $\nu_\kappa$ for finite $N$ is given by

$$\nu_\kappa = 0 + \delta/N, \quad 0 < \delta < 1 . \tag{11}$$

We define the 'ROTOR' (acronym of 'ROtational TOri Recognizer') as the integer part of $N\nu_\kappa$, namely

$$ROTOR \equiv r = [N\nu_\kappa] = [N(\nu_\phi - \nu_\theta)] = [\frac{\sum_{i=0}^{N}(\phi_i - \theta_i)}{2\pi}] . \tag{12}$$

The ROTOR is a function $r(\mathbf{R}_0, N)$ of the initial position vector and the current number of iterations. If $\mathbf{R}_0$ belongs to a KAM torus, then, as $N \to \infty$, the value of $r$ is constantly zero. Otherwise, $r$ tends to infinity proportionally to $N$, i.e.

$$r(\mathbf{R}_0, N) = \begin{cases} 0 & \text{if } \mathbf{R}_0 \text{ belongs to a KAM torus} \\ \propto N \to \infty & \text{otherwise} \end{cases} \tag{13}$$

In practice, the correct evaluation of $\nu_\phi$ and $\nu_\theta$ as well as of the ROTOR is not a trivial task because of the multivalued character of the angles $\theta_i$ and $\phi_i$. This difficulty can be avoided if we define $\nu_\phi$ and $\nu_\theta$ as the moments of the angular dynamical spectra $S(\phi)$ and $S(\theta)$ (Voglis and Efthymiopoulos 1998), i.e.

$$\nu_\phi = \frac{1}{2\pi} \oint S(\phi)\phi d\phi, \qquad \nu_\theta = \frac{1}{2\pi} \oint S(\theta)\theta d\theta \qquad (14)$$

and select a proper interval of definition of $\theta$ or $\phi$ so that there is no discontinuity of the corresponding spectrum due to modulo terms introduced at $k\pi = 0, \pm\pi, \pm 2\pi, \dots$. For example, in the standard map

$$x' = x + K\sin(x+y), \qquad y' = x+y, \qquad (x, y \bmod 2\pi) \qquad (15)$$

for $K = 0.5 < K_c = 0.97...$, the correct twist number $\nu_\phi = 0$ for the rotational torus with initial conditions $x = 0.8, y = 0$ is obtained if $\phi$ is defined in the interval $(-\pi, \pi]$. In this interval the spectrum $S(\phi)$ (Fig.5a) is continuous. If, instead, we defined $\phi$ in the interval $(0, 2\pi]$, we would obtain a discontinuous spectrum $S(\phi)$ (Fig. 5b) and a wrong value of $\nu_\phi = 0.538157$. But for $K = -4.2$ and initial conditions $x = 0.05$, $y = 0$, the correct twist number $\nu_\phi = 0.434576$ is obtained in the interval $(0, 2\pi]$ where the corresponding spectrum $S(\phi)$ (Fig.5d) is continuous. If we adopt the interval $(-\pi, \pi]$ (Fig.5c) we obtain a wrong value $\nu_\phi = 0.042660$.

We apply now the ROTOR method in the standard map (Eqs. 16) for $K = -2.1$. In this case the phase portrait has one main island around the stable fixed point $(0, 0)$. This island is surrounded by a chaotic sea. We calculate the ROTOR for orbits near the outer limit of the island scanning outwards along the diagonal $(x = y)$ with a step $dx = 10^{-6}$ and $N_{max} = 10^5$ iterations for every orbit.

The dependence of the ROTOR on $x$ is shown in Fig. 6a. The right end of this figure (e.g. $x > 0.7440$) is inside the chaotic sea. In this region the ROTOR is systematically far from zero. All the inverse U-shapes in this figure indicate first order islands, while the smaller U-shapes inside the first order islands correspond to second order islands (Voglis and Efthymiopoulos 1998).

Moving to the left in this figure (towards the center of the main island) we find the first place where the ROTOR has zero value at $x = 0.743339$. The ROTOR maintains this zero value even after $10^9$ iterations. Thus, the ROTOR gives a good estimate of the position of the last KAM curve surrounding the main island.

In the region $0.7436 < x < 0.7438$ the ROTOR becomes small (near 1), but not zero. The closer approaches of the ROTOR to zero indicate orbits moving near cantori that are more efficient in producing stickiness. In Fig.6b the phase portrait near the last KAM curve is shown. There are only two orbits drawn in this figure: 1)the orbit with the above initial condition $x = y = 0.743339$ (last KAM curve) and 2) a nearby chaotic orbit ($x = y = 0.7436$), which is sticky for about $9 \times 10^8$ periods before escaping to the chaotic sea. The darker regions in Fig.6b correspond to the position of cantori that limit the chaotic diffusion. In these darker regions the ROTOR has a value close to 1. The positions of cantori can be located also in terms of their nearby periodic orbits (Efthymiopoulos et al. 1997). We find that

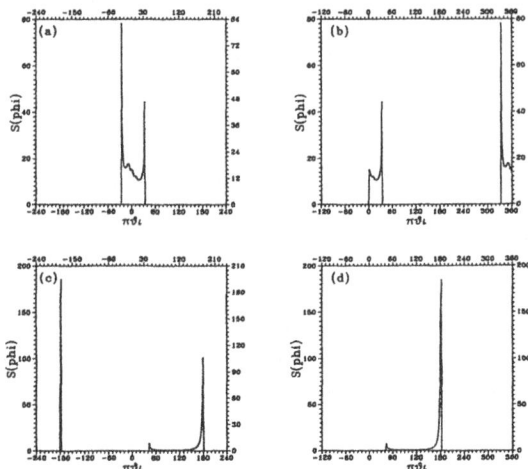

Fig. 5. The spectrum of twist angles $S(\phi)$ of the orbit with initial conditions $x = 0.8, y = 0$ in the standard map (Eq.16) for $K = 0.5$ in the interval a) $(-\pi, \pi]$ and b) $[0, 2\pi)$. c,d) similar as a,b for the orbit $x = 0.05, y = 0$ and for $K = -4.2$.

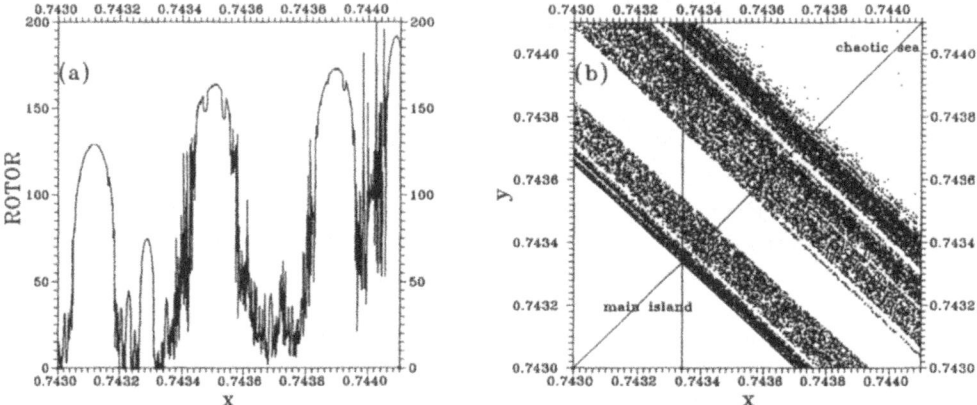

Fig. 6. a) The ROTOR as a function of $x$ in the standard map (Eq.16) along the scanning line x=y from $x = 0.743$ to $x = 0.7441$. b) The phase portrait in the same region. The vertical line gives the last position where the ROTOR is zero. The intersection of this line with the diagonal $y = x$ marks the position of the last KAM torus.

the values of the ROTOR near 1 indicate the positions of the cantori with good accuracy.

## 4. Discussion and Conclusions

In this paper we present two simple numerical methods, based on the properties of dynamical spectra, to investigate the structure of phase space in dynamical systems. The first method (section 2) distinguishes between ordered and chaotic orbits by measuring the spectral distance $D$ for two deviation vectors from the same orbit.

The spectral distance either becomes constant in time after a short transient period (ordered orbit), or decreases in time reaching a logarithmic slope $dlnD/dlnt = -1$ (chaotic orbit).

The second method (section 3) is based on a quantity called ROTOR, which is the integer part of the product of the epicyclic frequency $\nu_\kappa$ multiplied by the time $t$, i.e. $ROTOR = [t\nu_\kappa]$, where $t = N$, in a 2D map. If the ROTOR remains zero in time, the motion takes place on a KAM rotational torus. In the case of a secondary island the curve giving the ROTOR as a function of the initial conditions is U-shaped, or inverse U-shaped. In the case of chaos the curve is very irregular. This method is very sensitive because $t$ can be made arbitrarily large, thus the contrast between the values of ROTOR in rotational tori and in islands, or chaotic regions, can be made arbitrarily large.

In recent years various methods have been proposed for distinguishing between ordered and chaotic orbits. Such methods are based on the frequency analysis of the orbits, or on the deviations between nearby orbits.

Examples of methods of the first kind are:

1) the frequency analysis method of Laskar (1990) and Laskar et al. (1992)

2) the CLEANEST method of Foster (1995, 1996) and Gallardo and Ferraz-Mello (1997)

3) the "Hénon method" of continued fraction approximations (Lega and Froeschlé 1996), and

4) the wavelet transform method (Arneodo et al. 1988, Bendjoya and Slezak 1993, Michtchenko and Ferraz-Mello 1995, Michtchenko and Nesvorny 1996).

Examples of methods of the second kind are:

1) the fast Lyapunov indicators method of Froeschlé et al. (1996), and

2) the methods of dynamical spectra (Voglis and Contopoulos 1994, Contopoulos and Voglis 1996, 1997, Voglis and Efthymiopoulos 1998, Voglis et al. 1998).

Every method has its own advantages. In the present paper we stress only the advantages of our methods. In particular, the methods presented here combine:

a) Speed. The method of spectral distance described in section 2 can distinguish between ordered and chaotic motion without the need of evaluating the fundamental frequencies of the orbits. It is based on a property of the variational equations that lose the memory of initial conditions exponentially fast in the case of chaotic motions. This property makes the method faster than the frequency analysis methods in which the convergence of the frequencies follows a power law.

b) Simplicity. As all the methods of dynamical spectra, the two methods presented here, namely the method of spectral distance and the method of ROTOR, require only the integration of an orbit and its variational equations.

c) Sensitivity and information content. Both our methods can be applied equally well in regions of strong or weak chaos. The spectral distance method gives a very sensitive distinction of whether the motion is on a torus or not, for systems of any number of degrees of freedom. This avoids the need to visualize the orbits on either the phase space or the frequency space. On the other hand, in the case of 2D systems, the method of angular dynamical spectra gives also the fundamental frequencies (rotation and epicyclic frequency). In particular, the evaluation of the epicyclic frequency allows a sharp distinction between islands and KAM tori. This makes the method particularly suitable in exploring the resonant structure (i.e. location of KAM tori,islands, and cantori) near the border of transition to stochasticity.

## Acknowledgements

This research was supported in part by the Research Committee of the Academy of Athens (grant 200/409). C.E. was supported by the Greek Foundation of State Scholarships (IKY).

## References

Arneodo, A., Grasseau, G., Holschneider, M.: 1988, *Phys. Rev. Lett.*, **61**, 2281.
Bendjoya, P. and Slezak, E.: 1993, *Cel. Mech. Dyn. Astr.*, **56**, 231.
Contopoulos, G., and Voglis, N.: 1996, *Cel. Mech. Dyn. Astr.*, **64**, 1.
Contopoulos, G., and Voglis, N.: 1997, *Astron. Astrophys.*, **317**, 73.
Efthymiopoulos, C., Contopoulos, G., Voglis, N., and Dvorak, R.: 1997, *J. Phys.*, A **30**, 8167.
Foster, G.: 1995, *Astron. J.*, **109**, 1889.
Foster, G.: 1996, *Astron. J.*, **111**, 541.
Froeschlé, C., Froeschlé, Ch., and Lohinger, E.: 1993, *Cel. Mech. Dyn. Astron.*, **56**, 307.
Froeschlé, C., Lega, E. and Gonczi, R.: 1997, *Cel. Mech. Dyn. Astron.*, **67**, 41.
Fujisaka, H.: 1983, *Prog. Theor. Phys.*, **70**, 1264.
Gallardo, T. and Ferraz-Mello, S.: 1997, *Astron. J.*, **113**, 863.
Laskar, J.:1990, *Icarus* **88**, 266.
Laskar, J., Froeschlé, C. and Celletti, A.: 1993, *Physica*, **D56**, 253.
Lega, E. and Froeschlé, C.: 1996, *Physica*, **D95**, 97.
Michtchenko, T.A. and Ferraz-Mello, S.:1995 *Astron. Astrophys.*, **303**, 945.
Michtchenko, T.A. and Nesvorny, D.: 1996, *Astron. Astrophys.*, **313**, 674.
Nicolis, J.S., Meyer-Kress, G., and Haubs, G.: 1983, *Z.Naturfosch.*, **38a**, 1157.
Voglis, N.:1996, Human Capital and Mobility Workshop, Santorini, Greece (oral presentation).
Voglis, N., and Contopoulos, G.: 1994, *J. Phys.*, A **27**, 4899.
Voglis, N., and Efthymiopoulos, C.: 1998, *J. Phys.*, A **31**, 2913.
Voglis, N., Contopoulos, G., and Efthymiopoulos, C.: 1998, *Phys. Rev.* , E **57**, 372.

# CANTORI, ISLANDS AND ASYMPTOTIC CURVES IN THE STICKINESS REGION

C. EFTHYMIOPOULOS[1,2], G. CONTOPOULOS[1,2] and N. VOGLIS[2]
[1]*Research Center for Astronomy, Academy of Athens*
[2]*Department of Astronomy, University of Athens*

**Abstract.** The resonant structure near a noble cantorus is found. Islands of stability are located near the gaps of the cantorus. The crossing of the gaps of the cantorus by the asymptotic curves of unstable periodic orbits is shown numerically (non-schematically). We discuss how these structures influence stickiness.

## 1. Introduction

The phenomenon of "stickiness" occurs when a chaotic orbit starting near a last KAM torus remains close to this torus for long times before diffusing into a surrounding large chaotic sea. Among the first to give examples of stickiness were Contopoulos (1971), and Shirts and Reinhardt (1982).

A theoretical understanding of stickiness was advanced mainly for 2D systems (e.g. Greene 1979, MacKay 1983, MacKay et al. 1984; see Meiss 1992 for a review). It is now established that stickiness is due to the existence of *cantori* (Aubry 1978, Percival 1979, Mather 1982, Schmidt and Bialek 1982).

Cantori are the remnants of the KAM tori formed when the latter are destroyed by an increase of the perturbation. A cantorus forms a Cantor set of points on the surface of section. This set is invariant under the section mapping and has rotation number equal to the rotation number of its progenitor KAM curve.

A cantorus constitutes a partial barrier to *local* chaotic *diffusion*. Chaotic orbits can cross the cantorus only by passing through the gaps of the cantorus.

On the other hand, the *global* chaotic *transport* is limited by the existence of several consecutive cantori in the stickiness region. It is numerically observed that in many cases the holes of such cantori are delineated along "chimneys" (MacKay et al. 1984), where chaotic transport mainly takes place.

MacKay (1983) found numerically that the flux through gaps is minimal for those cantori having noble rotation numbers, i.e. numbers of which the continued fraction

$$a = [a_1, a_2, ...] \equiv 1/(a_1 + (1/a_2 + ...)) \tag{1}$$

(where $a_i$ are integers) has $a_i = 1$ for all $i$ above a certain order $N$. Thus, only noble cantori are important for the stickiness problem.

MacKay et al. (1984) modelled the flux through a gap of the cantorus as a "turnstile" and obtained estimates of transport rates by assuming a simple Markov chain model with given transition probabilities between any two successive cantori.

In order to construct a more accurate model of diffusion through cantori, one requires knowledge of the resonant phase space structure in the stickiness region.

*Celestial Mechanics and Dynamical Astronomy* **73**: 221–230, 1999.
©1999 *Kluwer Academic Publishers.*

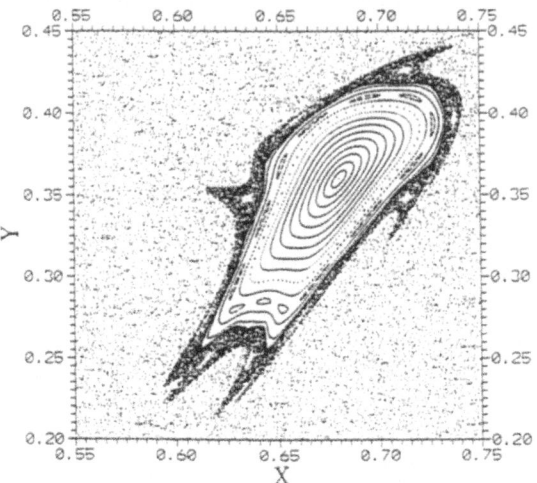

Fig. 1. The main island of stability and the surrounding stickiness region in the standard map for $K = 5$.

In particular, we find that the chaotic diffusion through a cantorus is substantially influenced by: a) the existence of islands of stability close to the gaps of a cantorus, and b) the form of the asymptotic curves (of unstable periodic orbits) crossing the gaps of the cantorus.

In this paper we summarize some of our results based on a numerical study of the standard map and some new results that give a better understanding of the stickiness problem. The earlier results are presented in greater detail elsewhere (Efthymiopoulos et al. 1997, Contopoulos et al. 1998).

## 2. The Stickiness Zone

We take as our model the 2D standard map

$$x_{i+1} = x_i + y_{i+1}, \qquad y_{i+1} = y_i + \frac{K}{2\pi} \sin(2\pi x_i) \qquad (mod\,1) \tag{2}$$

where $K$ is the nonlinearity parameter. When $K = 5$ the phase space (unit square) is mostly chaotic, but contains two symmetric islands of stability around the stable periodic orbit $x_c \approx 0.68$, $y_c \approx 0.36$ of period 2. Considering only every second iteration, the orbit $(x_c, y_c)$ can be considered of period 1. The island around this stable periodic orbit is shown in Fig. 1. This island is surrounded by a stickiness zone consisting of two parts. An inner thin zone (very dark) where the stickiness time is large, of order $10^4 - 10^7$ periods, and a more extended outer zone (less dark), where the stickiness time is smaller, of order $10^2$. Outside the stickiness region is the large chaotic sea.

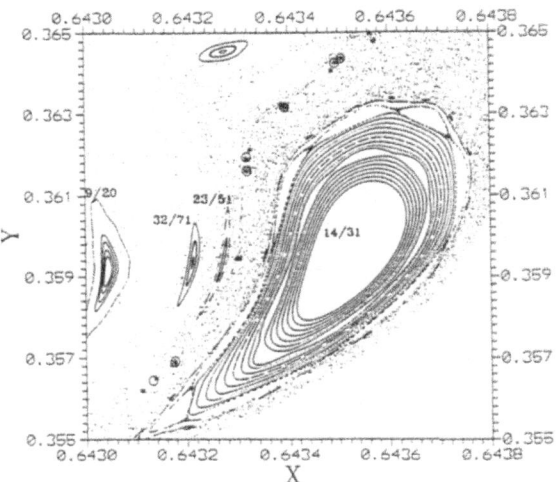

Fig. 2. A detail of the phase space near the border of the main island for $K = 5$. We distinguish the islands of stability $9/20, 14/31$, etc. and the unstable orbits $60/133$ (triangles), $97/215$ (squares) $157/348$ (circles) and $254/563$ (stars).

## 3. Cantorus and Islands

The most important cantorus in the stickiness region of Fig. 1 is the one with the noble rotation number $[2, 4, 1, 1, ...]$. This cantorus is formed at the critical value $K_c \approx 4.9974$ and surrounds the main island of stability.

The successive rational truncations of the noble number $[2, 4, 1, 1, ...]$ are:

$$\frac{1}{2}, \frac{4}{9}, \frac{5}{11}, \frac{9}{20}, \frac{14}{31}, \frac{23}{51}, \frac{37}{82}, \frac{60}{133}, \frac{97}{215}, \frac{157}{348}, \frac{254}{563}, \frac{411}{911} \cdots \quad (3)$$

and they form a Farey sequence. These rationals correspond to periodic orbits surrounding the main island, one inside (closer to the island) and one outside the cantorus. They approach the cantorus closer as the order increases.

Fig. 2 shows a detail of the phase space where several periodic orbits are marked. In particular, the periodic orbits $9/20$, $14/31$ and $23/51$ are stable and are surrounded by islands (the island $32/71$ does not belong to the sequence (3)), while the periodic orbits $97/215$, $254/563$ (inside the cantorus), and $60/133$, $157/348$ (outside the cantorus) are unstable. These unstable orbits are very close to each other and they define essentially the gaps of the cantorus.

In Fig. 2, the island $23/51$ is outside but very close to a cantorus gap. Such islands close to the gaps pose limits to chaotic diffusion.

The existence of islands close to the gaps, for $K$ slightly above $K_c$, is guaranteed by the fact that the critical value $K_c$ is the *minimum* of the critical values of destabilisation of the periodic orbits (3) (Contopoulos et al. 1987).

For every rational $n/m$ of the sequence (3) correspond two periodic orbits, one stable and one unstable, which bifurcate from the central periodic orbit (of period

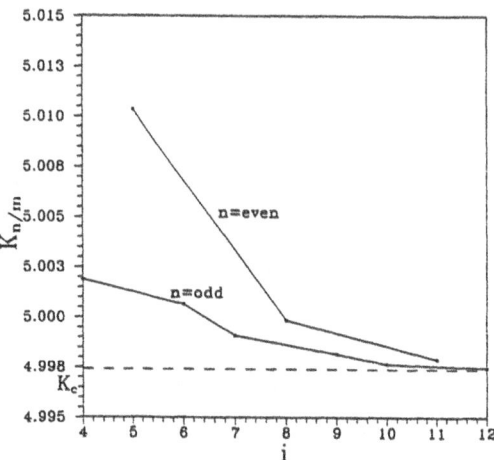

Fig. 3. The critical value $K_{n/m}$ of destabilisation of the orbit $n/m$ versus its corresponding order $i$ in the sequence (3).

1) at particular values of the perturbation $K$. As $K$ increases, both orbits move from the center outwards. At some particular perturbation value $K_{n/m}$, the stable periodic orbit $n/m$ becomes unstable by a single or double period bifurcation.

According to Greene's criterion (Greene 1979), the series of the critical values $K_{n/m}$ of the successive orbits $n/m$ has as its limit the critical value $K_c$ at which the noble torus $[2, 4, 1, 1, ..]$ turns to a cantorus.

In Fig. 3 we give the critical values $K_{n/m}$ versus the order of the orbits in the sequence (3). As the order increases, the values $K_{n/m}$ converge to a limit $K_c = 4.9974...$ (horizontal line). The convergence follows two different lines, when the numerator $n$ is even (upper line) or odd (lower line). Thus the deviations of $K_{n/m}$ from the limiting value $K_c$ are larger for $n = $ even than for $n = $ odd (for the same order of $n$). The value $K_c = 4.9974..$ is the critical value of destruction of the noble torus $[2, 4, 1, 1, ...]$. It is clear that the critical value $K_c$ is the *minimum* of the critical values $K_{n/m}$. Thus, for any particular orbit $n/m$, there is a finite interval of values $K_c < K < K_{n/m}$ for which the orbit is *stable* while the corresponding cantorus is already formed. Such stable orbits are surrounded by islands which limit diffusion through the gaps of the cantorus.

In Figs.4a,b the evolution of Hénon's stability index $a_{n/m}$ is given for each of the initially stable orbits $n/m$ of the sequence (3). For all orbits the stability index starts from $a_{n/m} = 1$ at the perturbation value $K$ of their generation from the center. As $K$ increases, the stability index $a_{n/m}$ initially decreases slowly and later abruptly. Then there are two possibilities: a) the stability index $a_{n/m}$ crosses the line $a = -1$ at the critical perturbation $K_{n/m}$ where the orbit becomes unstable. This happens when the numerator $n$ of the periodic orbit is odd. b) the stability curve $a_{n/m}$ becomes tangent to the line $a = -1$ and is reflected. Then the orbit becomes unstable at the value $K_{n/m}$ when $a_{n/m}$ crosses the line $a = +1$. This

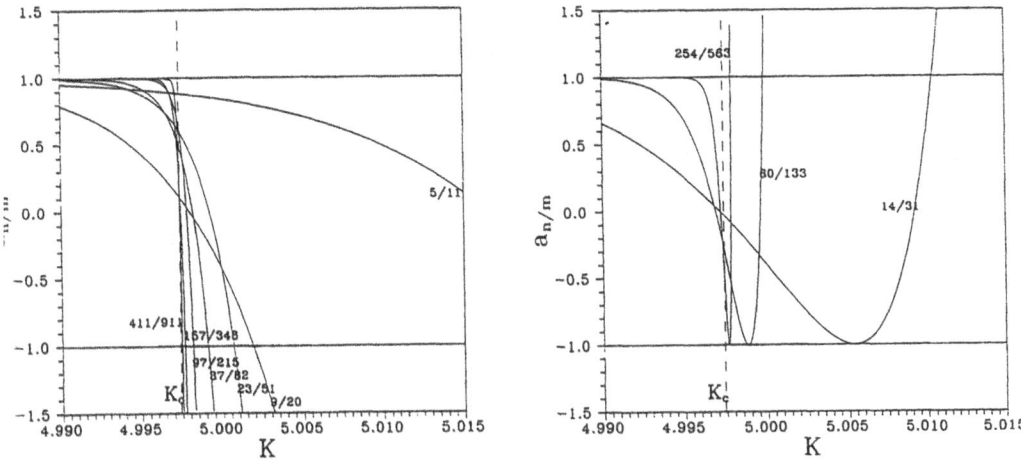

Fig. 4. The Hénon stability index as a function of $K$ for the orbits of the sequence (3). The vertical line is at the critical value $K_c$ of destabilisation of the noble torus $[2, 4, 1, 1, ...]$. a) Orbits with $n =$ odd. b) Orbits with $n =$ even.

happens when the numerator $n$ is even.

In any case, the transition to instability is more abrupt as the order of the periodic orbit increases. In Fig.4a, we show the evolution of the stability index of the orbits with odd numerator $5/11, 9/20, 23/51, 37/82, 97/215, 157/348$, and $411/911$ as a function of the perturbation $K$. The stability index curves, near destabilisation, are more abrupt for the higher order periodic orbits. The same phenomenon occurs for the periodic orbits with even numerator $14/31, 60/133$, and $254/563$ (Fig.4b). At the limit of infinite order (cantorus), the stability index curve takes the form of a step function changing slope from zero to $\infty$ abruptly at $K = K_c$.

The resonant structure near the noble torus or cantorus can be shown very efficiently by use of the recent method of angular dynamical spectra (Voglis and Efthymiopoulos 1998). Starting from the main center (period-1 orbit), we define the position vector $R_i$ of any point of an orbit around the island, and the infinitesimal deviation vector $\xi_i$ found by solving the variational equations of the map (2). Then we define the rotation angle $\theta_i$ as the angle between two successive vectors $R_i$ and $R_{i+1}$, and the twist angle $\phi_i$ as the angle between two successive vectors $\xi_i$ and $\xi_{i+1}$. Both angles are defined in appropriate intervals (see Voglis and Efthymiopoulos 1998 for details). Finally, we define the distributions of the rotation angles and twist angles (angular dynamical spectra $S(\theta)$ and $S(\phi)$). The angular moments are defined as:

$$\nu_\theta = \frac{1}{2\pi} \oint \theta S(\theta) d\theta, \qquad \nu_\phi = \frac{1}{2\pi} \oint \phi S(\phi) d\phi \quad , \tag{4}$$

and they correspond to the angular frequencies of rotation of the vectors $R$ and $\xi$.

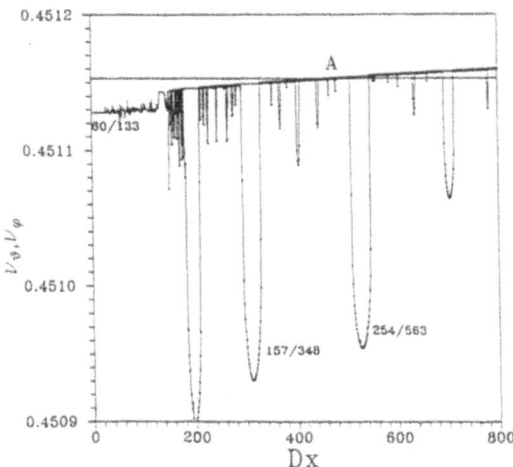

Fig. 5. The curves $\nu_\theta(x)$ (solid) and $\nu_\phi(x)$ (solid with dots) for $K = 4.997$ ($Dx$ is given in units of $3.75 \times 10^{-9}$, starting from $x = 0.64364$). The point A where the two curves coincide marks the position of the noble torus $[2, 4, 1, 1, ...]$.

Then the following relations hold (Voglis and Efthymiopoulos 1998):

$$\nu_\theta = \nu_\phi \qquad (5)$$

for a closed KAM curve and

$$\nu_\phi = \nu_\theta - \nu_k \qquad (6)$$

for higher order islands of stability, where $\nu_k$ is the mean angular frequency of epicyclic motion around the local center (stable periodic orbit) of the island.

In Fig. 5 we calculate the curves $\nu_\theta(x)$ and $\nu_\phi(x)$ for $K = 4.997 < K_c$. The U-shaped parts of the curve $\nu_\phi$ correspond to islands of stablility marked with their rotation numbers. The point A, where both curves $\nu_\theta(x)$ and $\nu_\phi(x)$ coincide on the horizontal line $\nu_\theta = \nu_\phi = [2, 4, 1, 1, ..]$, gives the position of the corresponding noble torus. The curves $\nu_\theta(x)$ and $\nu_\phi(x)$ coincide in a small segment around A, indicating that the noble torus $[2, 4, 1, 1, ..]$ and nearby tori still exist. The coincidence of the curves $\nu_\theta(x)$ and $\nu_\phi(x)$ in a *segment* around $[2, 4, 1, 1, ..]$ means that the volume of tori in the very close neighborhood of the noble torus tends to unity. This can be understood by a recent theorem of Morbidelli & Giorgilli (1995). Further away we distinguish many islands of stability. The islands 60/133, 157/348 and 254/563 belong to the sequence (3), while the rest of the islands belong to other Farey sequences.

The size of the islands for $K = 4.997$, as a function of their multiplicity, is given in Fig. 6. We see that the islands with $n =$ even are larger than those of the same order with $n =$ odd. The sizes of the islands for $n =$ even and $n =$ odd follow two parallel lines that correspond to the power law

$$DS = Am^{-2.75} \qquad (7)$$

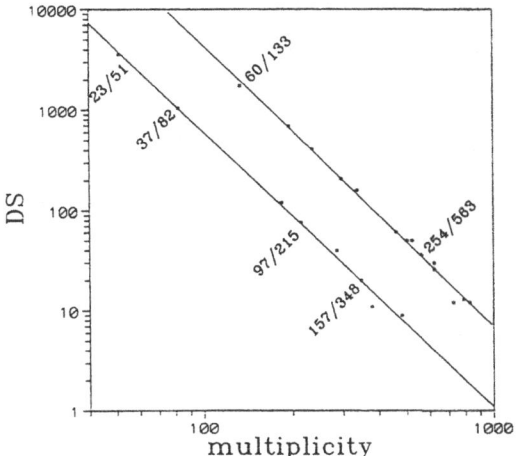

Fig. 6. The size of the islands as a function of the multiplicity $m$ for $K = 4.997$. The sizes $DS$ are measured along a line joining the periodic orbits 14/31 ($x = 0.6438327$, $y = 0.3593945$) and 23/51 ($x = 0.6437657$, $y = 0.3631837$), and are given in units of $1.267 \times 10^{-7}$. The sizes for $n =$ even and $n =$ odd follow the power law (7) with different constants $A$. The islands that belong to the Farey tree (3) are marked.

with $A = 2.8 \times 10^8$ for $n =$ odd and $A = 1.7 \times 10^9$ for $n =$ even.

When $K$ grows beyond $K_c$ the higher order periodic orbits become unstable and the corresponding islands are replaced by chaotic regions. This can be seen in Fig. 7 that gives the values of $\nu_\theta(x)$ and $\nu_\phi(x)$ for $K = 4.9975 > K_c$. In this case the noble torus $[2, 4, 1, 1, ...]$ has been destroyed and transformed into a cantorus. Then, the diference $\nu_\theta - \nu_\phi$ is no longer equal to zero, but it is minimum near the cantorus (Voglis et al. 1998). In Fig. 7 we see still some islands of lower order around the cantorus. These islands act as partial barriers for the diffusion of the orbits, and contribute to the stickiness in their neighbourhood. Even for $K = 5$ the low order orbits 4/9, 5/11, 9/20, 14/31 and 23/51 are still stable. Some of the corresponding islands of stability are shown in Fig. 2.

## 4. Stickiness and Asymptotic Curves

The asymptotic curves of unstable periodic orbits inside a cantorus cross the gaps of the cantorus and act also as barriers to chaotic diffusion. A numerical (non-schematic) example of such crossing of the cantorus by an ustable asymptotic curve is given in Fig. 8. The unstable asymptotic curve starting at the point $O$ of the unstable periodic orbit 87/215 (inside the cantorus $[2, 4, 1, 1, ..]$) is plotted until the time when it crosses the cantorus for the first time (Fig.8). This happens after 4 iterations of an initial segment of size $\Delta S_0 = 10^{-10}$ along the asymptotic curve near the unstable periodic orbit. The unstable asymptotic curve first moves downwards and makes four oscillations inside the cantorus. Then it moves upwards

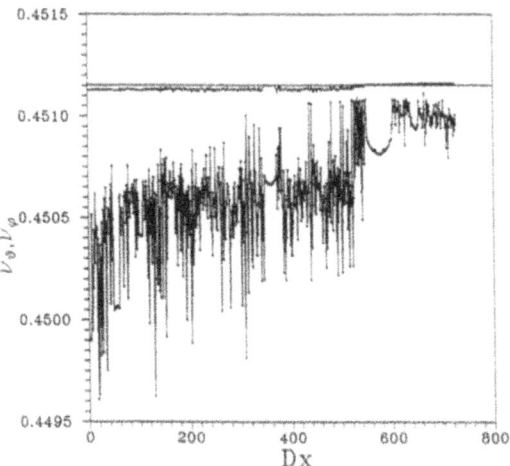

Fig. 7. Same as Fig.5 but for $K = 4.9975$ ($Dx$ starts at $x = 0.643583$). The two curves do not coincide anywhere, i.e. the noble torus no longer exists.

and crosses a gap of the cantorus. After its exit from the gap, the asymptotic curve moves almost parallel to the cantorus. If we iterate for longer times, we find that the asymptotic curve makes several oscillations crossing the cantorus successively inwards and outwards, before going to the large chaotic sea outside the island. The size of the asymptotic curve is approximately $\Delta S = \Delta S_0 |\lambda|^n$ at the nth iteration, where $\lambda$ is the larger eigenvalue of the periodic orbit. In the present case $\lambda \approx -190$ and the escape to the large chaotic sea occurs after $n = 6$ iterations.

The unstable asymptotic curve of Fig. 8 moves for some time almost parallel to the cantorus because the islands of stability close to the gaps of the cantorus (e.g. the islands 23/51) do not allow large excursions of the asymptotic curve in the transverse direction to the gaps. Furthermore, the asymptotic curve itself acts as a partial barrier to chaotic diffusion, since it cannot be intersected by the unstable asymptotic curve of any other unstable periodic orbit further inside the cantorus. Thus, for the latter to cross a gap of the cantorus, it necessarely has to move following a parallel path to the one of the unstable asymptotic curve of Fig. 8.

We conclude that the diffusion of any chaotic orbit inside the cantorus follows essentially the same path as defined by the unstable asymptotic curves which emanate from unstable periodic orbits inside the cantorus.

If we decrease the value of $K$ the stickiness time increases and becomes infinite when $K$ is smaller than the critical value $K_c$. In fact, as K decreases, the sizes of the gaps of the cantorus decrease and become zero for $K = K_c$. However the decrease of the size of the gaps is not the only reason for the increase of the stickiness time. In Fig. 9 we show part of the asymptotic curve of the unstable periodic orbit 254/563 which starts inside the cantorus $[2, 4, 1, 1, ...]$ for $K = 4.998$. This asymptotic curve passes through the gaps of the cantorus and finally reaches the large chaotic

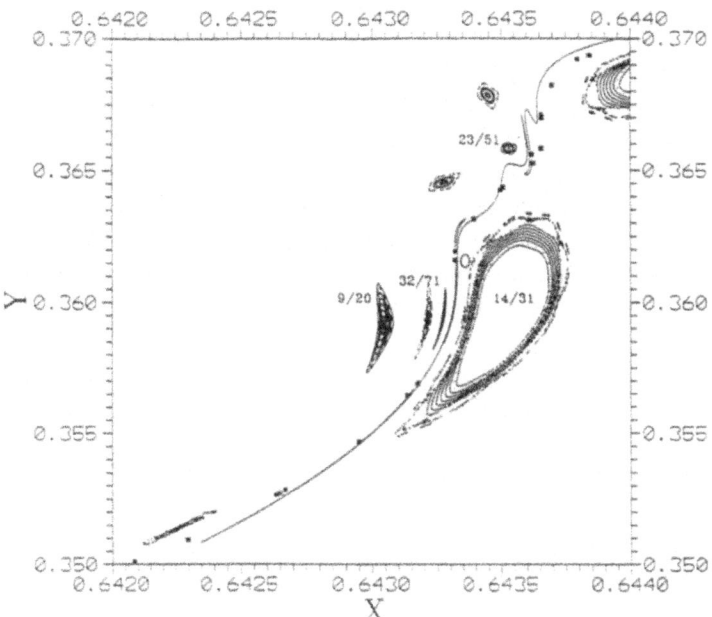

Fig. 8. Crossing of the cantorus $[2, 4, 1, 1, ...]$ by the unstable asymptotic curve of the orbit $97/215$ for $K = 5$. The stars and the squares represent the periodic orbits $97/215$ (inside the cantorus) and $157/348$ (outside the cantorus), and they define approximately the gaps of the cantorus.

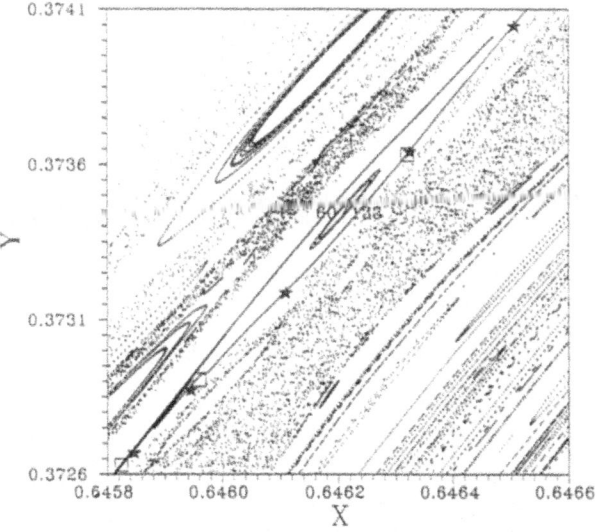

Fig. 9. The structure of the phase space near the cantorus $[2, 4, 1, 1, ...]$ for $K = 4.998$. The cantorus is approximated by the periodic orbits $157/348$ (stars, outside the cantorus) and $254/563$ (squares, inside the cantorus). A part of the asymptotic curve of the periodic orbit $254/563$ is also plotted (solid line), corresponding to the 11th iteration of an initial segment $\Delta S_0 = 10^{-10}$.

sea.

In this case the eigenvalue of the periodic orbit is about $\lambda \approx 8$. The first crossing of the cantorus occurs at the 9th iteration (starting with an initial segment $\Delta S_0 = 10^{-10}$). This is only a little over double the time required for crossing the same cantorus in the case $K = 5$. However for larger times the behaviour is very different. In Fig. 9 we see that after 11 iterations the asumptotic curve surrounds the islands 60/133 outside the cantorus, but there is no indication of a further diffusion outwards. In fact, we have found that an orbit requires a much longer time (of the order of $10^7$ iterations) in order to reach the outer chaotic sea.

We conclude:

(a) that the stickiness is much more pronounced in the case $K = 4.998$ than in the case $K = 5$. This is related to the fact that $K = 4.998$ is quite close to the limiting value $K_c = 4.9974$,

(b) that the stickiness is not due only to the cantorus $[2, 4, 1, 1, ...]$. Other cantori, outside the island 60/133, plus the higher order islands in this region produce an important barrier for the diffusion of the orbits in this neighboorhood. A further study of this phenomenon is in progress.

## Acknowledgements

This research was supported in part by the Research Committee of the Academy of Athens (grant 200/409). C.E. was supported by the Greek Foundation of State Scholarships (IKY)

## References

Aubry, S.: 1978, in Bishop, A.R. and Schneider, T.(eds) *"Solitons and Condensted Matter Physics"*, Springer, p.264.
Contopoulos, G.: 1971, *Astron.J.*, **76**, 147.
Contopoulos, G., Varvoglis, H. and Barbanis, B.: 1987, *Astron.Astrophys.*, **172**, 55.
Contopoulos, G., Harsoula, M., Voglis, N., and Dvorak, R.: 1998, *"Destruction of Islands of Stability"*, (in preparation).
Efthymiopoulos, C., Contopoulos, G., Voglis, N., and Dvorak, R.: 1997, *J. Phys. A:*, **30**, 8167.
Greene, J.M.: 1979, *J.Math.Phys.*, **20**, 1183.
MacKay, R.S.: 1983, *Physica*, **7D**, 283.
MacKay, R.S., Meiss J.D., and Percival I.C.: 1984, *Physica*, **13D**, 55.
Mather, J.N.: 1982, *Topology*, **21**, 457.
Meiss, J.D.: 1992, *Rev.Mod.Phys.*, **64**, 795.
Morbidelli, A., and Giorgilli, A.: 1995, *J. Stat. Phys.*, **78**, 1607.
Percival, I.C.: 1979, in Month, M. and Herrera, J.C. (eds) *"Nonlinear Dynamics and the Beam-Beam Interaction"*, Amer.Inst.Phys. p.302.
Schmidt, G., and Bialek, J.: 1982, *Physica*, **5D**, 397.
Shirts, R.B. and Reinhardt, W.P.: 1982, *J.Chem.Phys*, **77**, 5204.
Voglis, N., and Efthymiopoulos, C.: 1998, *J. Phys. A*, **31**, 2913.
Voglis, N., Contopoulos G., and Efthymiopoulos, C.: 1998, this volume.

# PARALLEL/VECTOR INTEGRATION METHODS
# FOR DYNAMICAL ASTRONOMY

TOSHIO FUKUSHIMA

*National Astronomical Observatory*
*2-21-1, Ohsawa, Mitaka, Tokyo 181-8588, Japan*
toshio@nao.ac.jp

**Abstract.** This paper reviews three recent[1] works on the numerical methods to integrate ordinary differential equations (ODE), which are specially designed for parallel, vector, and/or multi-processor-unit (PU) computers. The first is the Picard-Chebyshev method (Fukushima, 1997a). It obtains a global solution of ODE in the form of Chebyshev polynomial of large (> 1000) degree by applying the Picard iteration repeatedly. The iteration converges for smooth problems and/or perturbed dynamics. The method runs around 100-1000 times faster in the vector mode than in the scalar mode of a certain computer with vector processors (Fukushima, 1997b). The second is a parallelization of a symplectic integrator (Saha *et al.*, 1997). It regards the implicit midpoint rules covering thousands of timesteps as large-scale nonlinear equations and solves them by the fixed-point iteration. The method is applicable to Hamiltonian systems and is expected to lead an acceleration factor of around 50 in parallel computers with more than 1000 PUs. The last is a parallelization of the extrapolation method (Ito and Fukushima, 1997). It performs trial integrations in parallel. Also the trial integrations are further accelerated by balancing computational load among PUs by the technique of *folding*. The method is all-purpose and achieves an acceleration factor of around 3.5 by using several PUs. Finally, we give a perspective on the parallelization of some implicit integrators which require multiple corrections in solving implicit formulas like the implicit Hermitian integrators (Makino and Aarseth, 1992), (Hut *et al.*, 1995) or the implicit symmetric multistep methods (Fukushima, 1998), (Fukushima, 1999).

**Key words:** numerical integration – orbit – celestial mechanics

## 1. Introduction

It has been claimed that vector/parallel computation is not effective for the numerical integration of dynamics with small number of freedom such as the orbital and rotational motions of planets and satellites. This is due to the step-by-step nature of existing numerical integrators such as the Runge-Kutta methods, the linear multistep methods, and the extrapolation methods((Hairer *et al*, 1993). For example, the construction of Runge-Kutta methods takes the advantage of serial processing by assuming, in each phase, the availability of the result of all previous test integrations. The situation is unchanged in the symplectic methods (Kinoshita *et al.*, 1991), (Wisdom and Holman, 1991), which has been widely spread in the field of dynamical astronomy.

Recently, however, some papers appeared to destroy this barrier. They are

1. Picard-Chebyshev method (Fukushima, 1997a), (Fukushima, 1997b),
2. Parallel symplectic integrator (Saha *et al.*, 1997), and
3. Parallelized extrapolation method (Ito and Fukushima, 1997).

The first two seem[2] to be based on an idea to regard the ordinary differential equation as a one-dimensional partial differential equation. Parallel/vector com-

---

[1] All of the papers appeared in the Astronomical Journal in 1997.

[2] We are not sure whether Saha *et al.* had it in their minds when they wrote the paper.

*Celestial Mechanics and Dynamical Astronomy* **73**: 231–241, 1999.

puters are widely used in solving some partial differential equations. The key of parallelization is to rewrite the differential equations into a large set of nonlinear equations and to solve them by iterative procedures.

There are two ways to rewrite the differential equations into nonlinear equations. The one is to introduce the orthogonal function expansions in the expression of solutions. In this case, not the value of variables but the expansion coefficients are solved iteratively. The Picard-Chebyshev method is one of this family. It expands the solution in the form of Chebyshev polynomial. The other is the discretization. Saha *et al.* (1997) tactfully applies this idea to Hamiltonian systems and lead successfully a parallelization of a symplectic integration scheme, the implicit midpoint rule. On the other hand, the extrapolation method itself is easy to be adapted to parallel computation, since the method is based on the compilation of results of several (4-10) test integrations, which can be done in parallel. However, it was not experimented until the work of Ito and Fukushima (1997). The essence of their parallelization is the idea of folding, which makes the load balance among multiple processors almost equal.

In this short article, we will review these works. Before doing so, however, we make some remarks on the characteristics of these three new methods. First, apart from its parallel nature, the Picard-Chebyshev method is based on an approach being quite different from the existing integrators like Runge-Kutta methods, and other step-by-step integrators. Thus, we will discuss it in details. While, the last two methods are the parallelization of the existing methods; symplectic and extrapolation methods. The solution produced by these are just the same as those by the corresponding serial methods. Thus the readers can refer the literature of serial versions, say the textbook of Hairer *et al.* (1993), for the integration error and other properties except one thing; the speed-up by parallelization. Thus, we discuss only this factor.

Apart from the three parallel methods discussed, there remains a possibility to parallelize the multistep methods (Miranker and Liniger, 1967). The key idea is the concept of *pipeline*, which means the shift of timing between the predictor and the corrector(s). We will add a perspective on this approach.

## 2. Picard-Chebyshev Method

Consider solving a general first-order ordinary differential equation

$$\frac{dy}{dt} = f(y, t), \qquad y(t_0) = y_0 \tag{1}$$

One way is to start from an approximate solution $y^{(0)}(t)$ and to improve it iteratively. The series of the refined solutions, $y^{(n)}(t)$ for $n = 1, 2, \cdots$, are obtained successively by computing

$$y^{(n)}(t) = y_0 + \int_{t_0}^{t} f\left(y^{(n-1)}(s), s\right) ds \tag{2}$$

This is the Picard iteration method ((Hairer *et al.*, 1993), Section I.8).

We expand the solution $y(t)$ in a linear form of Chebyshev polynomials whose support is a certain long interval of period, say thousands of nominal orbital periods, for example. Then, each Picard iteration ((Fukushima, 1997a), Eq.(18)) is rewritten in a vector form mapping ((Fukushima, 1997b), Eq.(2)) as

$$\mathbf{Y}^{(n-1)} \to \mathbf{Y}^{(n)} \equiv \mathcal{G}\mathbf{f}\left(\mathcal{C}^T\mathbf{Y}^{(n-1)}, \mathbf{t}\right) \tag{3}$$

where $\mathbf{Y}^{(n)}$ is a column vector of the Chebyshev coefficients of the $n$-th approximate solution, $\mathcal{G}$ and $\mathcal{C}$ are certain matrices, t is a column vector of the zeros of the Chebyshev polynomial of the largest degree, and f means a vector notation of function evaluations. Of course, the Picard iteration method is not all-purpose. It works only if the iteration converges. When the perturbation is sufficiently small, (1) the zero polynomial is enough as a predictor, (2) the iteration converges rapidly, and (3) the integration interval for a single polynomial can be extended as long as hundreds of characteristic periods. As an example, Figure 1, which is taken from Figure 2 of (Fukushima, 1997a), shows the error distribution of intermediate solutions for a test problem integrated over 64 orbital periods;

$$\frac{dy}{dt} = \cos(t + \epsilon y) \tag{4}$$

where the perturbation parameter $\epsilon$ was set as $10^{-3}$.

In this figure, note that the final error is quite small, of the order of $10^{-13}$ or so. Next, remark that the final error distribution is quite different from those obtained by other type of integrators, namely being roughly uniform through the integration period. This comes from the almost mini-max nature of Chebyshev approximation. Further, if we inspect them in details, we will learn that the errors in the middle of the integration period are somewhat larger than those at the final epoch of the integration period. This owes to the fact that the distribution of the evaluation points, i.e. the zeros of a certain high-degree Chebyshev polynomial, is more sparse around the center than near the ends. Also this figure supports the expectation that the iteration converges linearly, although the speed of convergence is somewhat slower than the expected rate, $\epsilon$.

As for the computing speed, we remark that the function evaluation in the above mapping expression can be done in parallel, or more precisely speaking, can be vectorized easily. Figure 2, which is taken from Figure 1 of (Fukushima, 1997b), shows the comparison of the wall-clock time[3] for the Adams method in the scalar mode, the Picard-Chebyshev method in the scalar mode, and in the vector mode of the same computer, Fujitsu VX-1R. The curves in the figure are drawn as functions of the computational amount of function evaluations. Since the overhead of Picard-Chebyshev methods is larger than that of Adams method, the former outperforms the latter especially when the load of function evaluation is heavy.

---

[3] The usual CPU time is not appropriate in evaluating the performance of parallel/vector computers. Instead, used is the clock time in the real world, which is named *wall-clock* time.

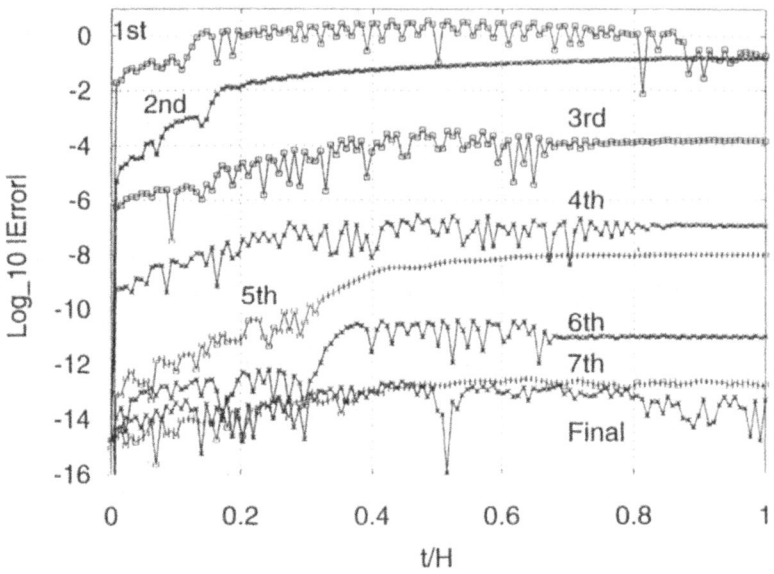

Fig. 1. Convergence of Picard-Chebyshev Method

Remark that the wall-clock time of Picard-Chebyshev method depends on the magnitude of perturbation parameter. In fact, if $\epsilon$ of Eq.(4) is zero, the problem reduces to the numerical quadrature of a known function of time, and therefore there is no need to repeat the Picard iteration. Then the total wall-clock time reduces by the factor of number of Picard iterations, 8 for $\epsilon = 10^{-3}$. Practically, there is a limitation of such reduction. At least two computation (i.e. one Picard iteration) is required to confirm the convergence. Thus the speed-up of factor 4 or 5 is expected for sufficiently small $\epsilon$. Figure 4 of Fukushima (1997a) supports this expectation.

As for the applicability to long integrations, the readers may refer Section 4.6 and Figure 5 of Fukushima (1997a). The error of the Picard-Chebyshev method increases in proportion to the 3/2 power of the integration period. Also the computational time does the same. This means that integrations over a very long period are not suitable. Rather, the method works best when integrating problems over a middle-size period, say the period of hundreds to thousands nominal revolutions. A typical problem would be the orbital improvement of Moon's orbit and rotation over 25 years of LLR observation, which includes some 300 revolutions.

In conclusion, the Picard-Chebyshev method directly provides the nearly minimax-approximated polynomials interpolating the solution almost uniformly within the whole integration interval. Therefore, the method is especially suitable for the orbit improvement where the previous ephemeris serves as a good approximation.

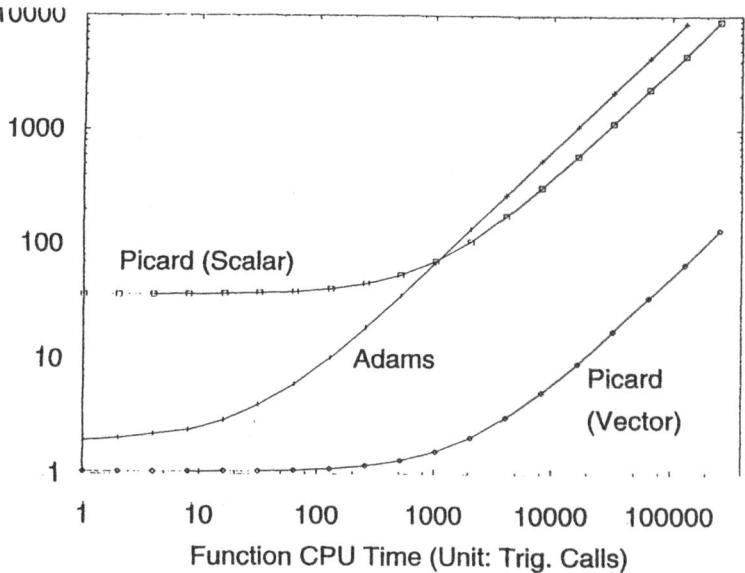

Fig. 2. Acceleration Factor of Picard-Chebyshev Method

Also the method is appropriate to solve perturbed dynamics where an approximate solution is known analytically. Typical examples would be the planetary motions, the satellite motions, the motions of comets and minor planets in Encke's method, and the rotational motions of the Earth and Moon.

## 3. Parallel Symplectic Integrator

The typical way to solve the initial value problem of ordinary differential equations for a long time span is to discretize the integration period into thousands of small time intervals and to go step by step from the initial epoch. To do this in parallel, let us rewrite the integral expression

$$y(t) = y_0 + \int_{t_0}^{t} f\left(y(s), s\right) ds \tag{5}$$

into a following discretized form by using the implicit midpoint rule;

$$y_m = y_0 + h \sum_{k=0}^{m-1} f\left(\frac{y_k + y_{k+1}}{2}\right) \tag{6}$$

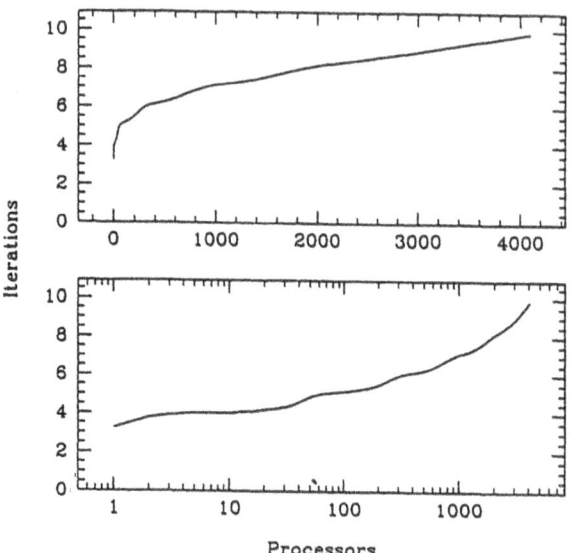

Fig. 3. Convergence of Parallel Symplectic Integrator

These can be regarded as a large set of nonlinear equations. Consider to solve them by a certain iterative procedure like

$$y_m^{(n-1)} \to y_m^{(n)} \equiv y_0 + h \sum_{k=0}^{m-1} f\left(\frac{y_k^{(n-1)} + y_{k+1}^{(n-1)}}{2}\right) \qquad (7)$$

Here it is easy to see that the function evaluation can be done in parallel.

Saha $et$ $al.$ (1997) applied this parallelization to integrate a perturbed Hamiltonian system;

$$H = H_0(p) + \epsilon H_1(p, q) \qquad (8)$$

The result is a following iterative scheme;

$$p_m^{(n)} \leftarrow p_0 + u_m\left(P^{(n-1)}, Q^{(n-1)}\right),$$

$$q_m^{(n)} \leftarrow q_0 + V_m\left(P^{(n)}\right) + v_m\left(P^{(n-1)}, Q^{(n-1)}\right) \qquad (9)$$

where

$$P = (p_0, p_1, \cdots), \qquad Q = (q_0, q_1, \cdots),$$

$$V_m(P) = h \sum_{k=0}^{m-1} \left(\frac{\partial H_0}{\partial p}\right)\left(\frac{p_k + p_{k+1}}{2}\right),$$

$$u_m(P, Q) = -\epsilon h \sum_{k=0}^{m-1} \left(\frac{\partial H_1}{\partial q}\right) \left(\frac{p_k + p_{k+1}}{2}, \frac{q_k + q_{k+1}}{2}\right),$$

$$v_m(P, Q) = \epsilon h \sum_{k=0}^{m-1} \left(\frac{\partial H_1}{\partial p}\right) \left(\frac{p_k + p_{k+1}}{2}, \frac{q_k + q_{k+1}}{2}\right) \quad (10)$$

Remark that this formulation is symplectic since the argument of $V_m$ is not $P^{(n-1)}$ but $P^{(n)}$, namely the momenta not before but after the kick. The local truncation error is of the order of $\epsilon h^2$, since the implicit midpoint rule is of the second order. The global truncation error is expected to grow linearly since the method is symplectic. The iteration converges linearly. Figure 3, which is taken from Figure 3 of Saha *et al.* (1997), shows the average number of iterations required for convergence as the number of simulated processors. Although they did not measure the performance by actual numerical experiments in parallel computers, based on this figure, Saha *et al.* (1997) gave an estimation of acceleration factor, i.e. the ratio of wall-clock time in parallel and serial computations, as much as around 50 for 1000 processors.

## 4. Parallelized Extrapolation Method

The extrapolation method is the direct application of Richardson's deferred extrapolation to the limit $h \to 0$ to the modified midpoint rule which assures the symmetric property and thus the $h^2$-expansion of the solution (Hairer *et al.*, 1993). Remark that all of the trial integrations, i.e. the integration by the modified midpoint rule with different $h$, can be done in parallel. Actually we prepare multiple processors sharing the memory and assign each processor unit (PU) to the test integration of different stepsizes like PU-1 for $h = H$, PU-2 for $h = H/2$, PU-3 for $h = H/3$, and so on[4]. Since the computational load of each test integration is inversely proportional to $h$, the above naive assignment allows an idling for the processors with larger $h$. In fact, in the above example, the PU-1 becomes idle just after one step integration, the PU-2 does after the second step, and so on. In order to avoid such an inefficiency, we introduce the concept of *folding*.

Imagine the case of 8-stage extrapolation method with the test stepsizes $h = H, H/2, \cdots, H/8$. Assume to prepare 4 PUs. If we assign processors with little care, like assigning PU-$n$ to the test integration of $h = H/(2n - 1)$ and $h = H/(2n)$, then the task of the PU-4, i.e. the test integrations of $h = H/7$ and $h = H/8$, becomes the bottle neck of the total procedures. In this case, the acceleration factor will be not so large as $(1 + 2 + \cdots + 8)/(7 + 8) = 2.4$. On the other hand, if we couple the test integration of $h = H/n$ with that of $h = H/(9-n)$, like $h = H$ and $h = H/8$, then the loads of 4 PUs become the same, and as a

---

[4] Here $H$ is the basic stepsize of the extrapolation method, namely the amount of time advanced after the extrapolation, and $h$ is the test stepsize, i.e. the stepsize of test integrations whose result are to be extrapolated.

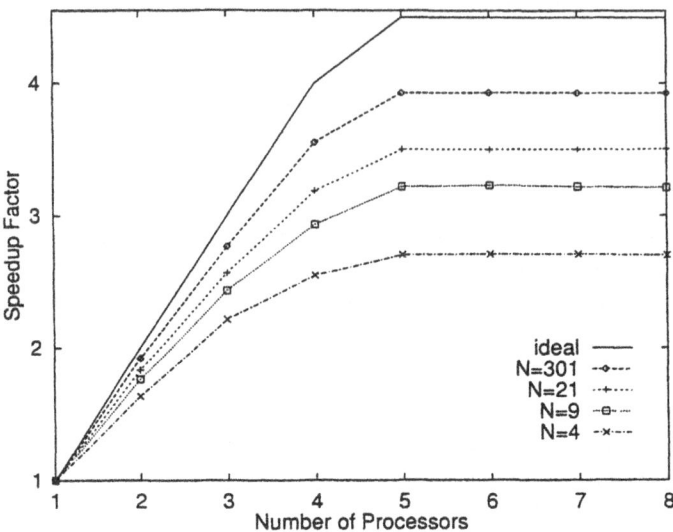

Fig. 4. Acceleration Factor of Parallelized Extrapolation Method

result, the acceleration factor becomes the same as the number of used processors as $(1 + 2 + \cdots + 8)/(1 + 8) = 4$. This is the technique of *folding*. In their Table 2, Ito and Fukushima (1997) provides the optimal foldings for the cases of up to 10 PUs. By using these, they developed a parallelization of the extrapolation method. Figure 4, which is taken from Figure 3 of Ito and Fukushima (1997), shows the acceleration factor as a function of number of processors. Even for a system of small freedom as 9-bodies problem, the acceleration factor of around 3.5 is achieved by using 4 or 5 PUs. It is remarkable that this situation is unchanged even when allowing the stepsize and/or the order variable.

In summary, the parallelized extrapolation method is all-purpose and leads to a speedup factor of 3-4 by using 4 PUs or so.

## 5. Pipelined Predictor-Corrector Method

More than three decades ago, Miranker and Liniger (1967) proposed a parallel method to accelerate the predictor-corrector methods. Their main idea is, in each process of prediction and correction, to advance not only a single step but also multiple steps together. Take an example of predictor phase. The ordinary predictor has a following general form (Hairer *et al.*, 1993);

$$y_{n+1} = -\sum_{j=0}^{J} \alpha_j y_{n-j} + h \sum_{j=0}^{J} \beta_j f_{n-j} \tag{11}$$

Miranker and Liniger (1967) predict not only this value at the $(n + 1)$-th step but also the values in future steps as

$$y_{n+k} = - \sum_{j=0}^{J} \alpha_j^{(k)} y_{n-j} + h \sum_{j=0}^{J} \beta_j^{(k)} f_{n-j}, \qquad (k = 2, \cdots) \qquad (12)$$

Since these predictions can be done in parallel, one can advance multiple integration steps at a single wall-clock step by using multiple processors. Unfortunately, this seems not practical. Usually the stepsize in multistep methods are taken as large as possible while the methods are numerically stable. This means that, in such extreme cases, the prediction/correction for doubly- and further advanced steps would cause instability.

In the author's viewpoint, rather the second idea of Miranker and Liniger (1967), which we call *pipeline*, seems effective especially when handling predictor-corrector methods requiring multiple correction stages[5]. Of course, it might be effective. However, are there such complicate methods worth to be applied? Yes, certainly. Good examples are the implicit time-symmetric methods such as the implicit Hermitian integrator (Makino and Aarseth, 1992), (Hut *et al.*, 1995) and the implicit symmetric multistep methods (Fukushima, 1998), (Fukushima, 1999). Now, the idea of pipeline is as follows. Imagine to perform a predictor-corrector method requiring $m$-correction stages such as PE(CE)$^m$ method[6]. Prepare $m + 1$ PUs and assign the PU-0 to the PE process of the $(n + 1)$-th step, the PU-1 to the first CE process of the $n$-th step, the PU-2 to the second CE process of the $(n - 1)$-th step, and so on;

$$\text{PU--0}: \quad y_{n+1}^{(0)} = - \sum_{j=0}^{J} \alpha_j^{(P)} y_{n-j}^{(j)} + h \sum_{j=0}^{J} \beta_j^{(P)} f_{n-j}^{(j)}, \qquad (13)$$

$$f_{n+1}^{(0)} = f\left( y_{n+1}^{(0)}, t_{n+1} \right) \qquad (14)$$

$$\text{PU--}k: \quad y_{n+1-k}^{(k)} = - \sum_{j=1}^{J} \alpha_j^{(C)} y_{n+1-(j+k)}^{(j+k)} + h \sum_{j=0}^{J} \beta_j^{(C)} f_{n+1-(j+k)}^{(j+k+1)}, \qquad (15)$$

$$f_{n+1-k}^{(k)} = f\left( y_{n+1-k}^{(k)}, t_{n+1-k} \right) \qquad (16)$$

where $y_n^{(k)}$ denotes the value at the $n$-th step after $k$ corrections and the superscripts of coefficients, (P) or (C), specifies the preditor or corrector, Remark that all these processes are independent with each other, and as a result, can be done in parallel.

---

[5] Miranker and Liniger (1967) applied the pipeline to the PECE method only, and therefore, did not stress so much about its applicability and effectiveness.

[6] Here P and C stand for the predictor and the corrector, respectively, while E does for the evaluator, i.e. the evaluation process of $f$.

TABLE I

Characteristics of Parallel/Vector Integration Methods

| Method | Suitable Problems | Acceleration Factor |
|---|---|---|
| Picard-Chebyshev | Perturbed Dynamics | 100-1000 |
| Parallel Symplectic | Hamiltonian Systems | ~ 50[7] |
| Parallel Extrapolation | General | 3-4 |
| Pipelined Multistep | Smooth | 3-4[8] |
| | [7] Expected | [8] Not yet experimented |

In other words, the pipelined predictor-corrector method can be done in the same wall-clock time as that of predictor-only formulas.

In general, the implicit methods have better properties than the explicit methods such as at the points of small error constants or of better numerical stability. Therefore, the technique of pipeline will enlighten the implicit symmetric methods.

## 6. Conclusion

We reviewed three numerical integrators designed for parallel/vector computers; the Picard-Chebyshev method (Fukushima, 1997a), (Fukushima, 1997b), the parallel symplectic integrator (Saha et al., 1997), and the parallelized extrapolation method (Ito and Fukushima, 1997). We also proposed the fourth scheme; the pipelined predictor-corrector method based on the idea of Miranker and Liniger (1967). Their characteristics are summarized in Table I although the acceleration factor for the last method was based on a rough estimation of the reduction of the number of function evaluations by means of parallelization. Also in listing the factor of Picard-Chebyshev method, we took the effect of magnitude of perturbation into account. In the case of parallel symplectic method, one may have an impression that the acceleration factor of 50 is a small gain at the cost of using 1000 processors. Further, the acceleration factor of 3 or 4 of the parallel extrapolation method and/or the pipelined multistep methods proposed here may be thought as a minor improvement. However, we stress that it is quite difficult to achieve a factor 2 in the parallelization of ordinary Runge-Kutta methods. Thus, we regard that even the factors 3 or 4 show good performance of the parallelizations.

Since the applicability of these methods are different one by one, it is difficult to recommend one of them. This comes from the fact that the comparison of serial integrators also depends on the problem to be solved. For example, in the case of orbital improvement, only one Picard iteration is enough. Thus the Picard-Chebyshev method would be the most appropriate. While, in doing long-time integration of Hamiltonian systems, symplectic methods are of the highest cost-performance. Thus, it would be worth to apply its parallelization to such problems. On the other hand, the extrapolation method is known to be tough, namely to be able to handle violent situations as close encounters. Thus the parallel extrapolation method would be recommended to integrate non-smooth problems like pure three body problems.

Finally let us make a comment about the way of comparisons presented here. As for the speed of these parallel methods, we have only considered the factor of acceleration, which was defined as the inverse ratio of total wall-clock time of the same processor used in serial and parallel/vector modes, respectively. Of course, the fastest serial processor would be faster than the serial usage of the fastest parallel/vector processor. Thus, the acceleration factors we presented do not mean the ratio of the fastest serial computation and the fastest parallel/vector computation. To evaluate the ratio, one needs a number of state-of-the-art serial and parallel/vector computers. It is beyond our ability. Rather, we think it is more appropriate to separate such machine-dependent effects with the machine-independent part like the acceleration factor.

## References

Fukushima, T.: 1997a, 'Picard Iteration Method, Chebyshev Polynomial Approximation, and Global Numerical Integration of Dynamical Motions', *Astron. J.*, **113**, 1909-1914.

Fukushima, T.: 1997b, 'Vector Integration of Dynamical Motions by the Picard-Chebyshev Method', *Astron. J.*, **113**, 2325-2328.

Fukushima, T.: 1998, 'Symmetric Multistep Methods Revisited', *in Proc. 30th Symp. on Cele. Mech. (Fukushima et al. eds)*, 229-247.

Fukushima, T.: 1999, 'Symmetric Multistep Methods Revisited II: Numerical Experiments', *in Proc. IAU Coll. No.173*, to be printed.

Hairer, E., Nørsett, S.P., and Wanner, G.: 1993, *Solving Ordinary Differential Equations I (2nd ed.)*, Springer-Verlag, Berlin.

Hut, P., Makino, J., and McMillan S.: 1995, *Astrophys. J. Lett.*, **443**, 93-.

Ito, T., and Fukushima, T.: 1997, 'Parallelized Extrapolation Method and Its Application to the Orbital Dynamics', *Astron. J.*, **114**, 1260-1267.

Kinoshita, H., Yoshida, H., and Nakai, H.: 1991, *Cele. Mech. and Dyn. Astr.*, **50**, 59.

Lambert, J.D., and Watson, I.A.: 1976, 'Symmetric Multistep Methods for Periodic Initial Value Problems', *J. Inst. Maths Applics*, **18**, 189-202.

Makino, J., and Aarseth S.J.: 1992, *Publ. Astron. Soc. Japan*, **44**, 141.

Miranker, W.L., and Liniger W.: 1967, 'Parallel Methods for the Numerical Integration of Ordinary Differential Equations', *Math. Comp.*, **21**, 303-320.

Quinlan, G.D., and Tremaine, S., 1990, 'Symmetric Multistep Methods for the Numerical Integration of Planetary Orbits', *Astron. J.*, **100**, 1694-1700.

Saha, P., Stadel, J., and Tremaine, S.: 1997, 'A Parallel Integration Method for Solar System Dynamics', *Astron. J.*, **114**, 409-415.

Wisdom, J., and Holman, M.: 1991, 'Symplectic Maps for the $N$-body Problem', *Astron. J.*, **102**, 1528-1538.

# DO AVERAGE HAMILTONIANS EXIST?

S. FERRAZ-MELLO

*Instituto Astronômico e Geofísico, Universidade de São Paulo, Caixa Postal 3386, São Paulo, SP, Brasil, E-mail: sylvio@usp.br*

**Abstract.** The word "average" and its variations became popular in the sixties and implicitly carried the idea that "averaging" methods lead to "average" Hamiltonians. However, given the Hamiltonian $H = H_0(J) + \epsilon R(\theta, J)$, ($\epsilon << 1$), the problem of transforming it into a new Hamiltonian $H^*(J^*)$ (dependent only on the new actions $J^*$), through a canonical transformation given by zero-average trigonometrical series has no general solution at orders higher than the first.

## 1. Introduction

Hamiltonian perturbation theories reached their apex in 1954, when one of their many versions was used by Kolmogorov for the construction of solutions of a perturbed Hamiltonian system. Hamiltonian perturbation theories are theories seeking for a canonical transformation able to transform the given Hamiltonian system into another one whose energy depends only on the new actions (i.e., a Hamiltonian independent of angles). They are used in Celestial Mechanics since the XIX$^{th}$ century. Examples are Delaunay's theory of the motion of the Moon, the theory called "Lindstedt method" by Poincaré, and the Lie-series methods, introduced by Born and Hori in the study of Quantum Mechanics and Celestial Mechanics, respectively (Delaunay, 1868; Poincaré, 1893; Charlier, 1907; Born, 1926; Kolmogorov, 1954; Brouwer, 1959; Hori, 1966; Deprit, 1969).

The more ancient methods look for a classical Jacobian generating function $S(\theta_i, J_i^*)$, of the old angles $\theta_i$ and new actions (or momenta) $J_i^*$, and the canonical transformation is written as

$$\theta_i^* = \frac{\partial S}{\partial J_i^*} \qquad J_i = \frac{\partial S}{\partial \theta_i} \qquad (i = 1, 2, \cdots, N) . \tag{1}$$

The more recent Lie-series methods look for a function $W(\theta_i^*, J_i^*)$, of the new angles and actions (the use of action-angle variables is not necessary, but we adopt them for sake of simplicity), and the canonical transformation is written, using Lie series, as

$$\theta_i = E_W \theta_i^* \qquad J_i = E_W J_i^* \qquad (\text{for each } i ) . \tag{2}$$

The Lie series $E_W \phi^*$ of a given function $\phi(\theta_i, J_i)$ is defined by

$$E_W \phi^* = \phi(\theta_i^*, J_i^*) + \{\phi, W\} + \frac{1}{2!}\{\{\phi, W\}, W\} + \frac{1}{3!}\{\{\{\phi, W\}, W\}, W\} + \cdots \tag{3}$$

where { , } denote Poisson brackets. In both cases, the generating functions ($S$ or $W$) are periodic functions of the angles $\theta_1, \theta_2, \cdots, \theta_N$ (or $\theta_1^*, \theta_2^*, \cdots, \theta_N^*$) and are seek as zero-average Fourier series in these angles with coefficients that depend

*Celestial Mechanics and Dynamical Astronomy* **73**: 243–248, 1999.
© 1999 *Kluwer Academic Publishers.*

only on the actions. One problem is considered as solved (at a given order) when a zero-average generating function ($S$ or $W$) and the resulting "average" Hamiltonian $H^*(J_i^*)$ are found.

The word "average", and its variations, became popular in the sixties and implicitly carried the idea that "averaging" methods lead to "average" Hamiltonians governing the secular variation of the given system. However, in at least one instance (Milani *et al.* 1987), the inverse transformation was explicitly calculated, to obtain the asymptotic (or formal) solutions of the given Hamiltonian and, for general disappointment, new non zero-average terms appeared in the solution! This is obvious when canonical transformations defined by generating functions are used. Indeed, in the case of methods using Lie Series, a glance at eqn. 3 is enough to see that even if $W$ is a pure zero-average trigonometric series, the series terms of order 2, and higher, will involve products of derivatives of $W$ among them, and constant terms will be generated. In the case of methods using Jacobian generating functions, the solutions given by eqns. 1 are in mixed form. To get them explicitly, say $\theta_i = \theta_i(\theta_i^*, J_i^*)$, $J_i = J_i(\theta_i^*, J_i^*)$, we have to use an inversion procedure, and any procedure will involve products of derivatives of $S$, thus leading to constant terms.

The above discussed drawbacks show that $H^*$ is not an "average". The actual solutions oscillate about the solutions of the Hamiltonian system defined by $H^*$, but with a non-zero average. This fact does not invalidate the classical perturbation theories (the zero average is not a necessary condition for their validity). It only conducts to some interpretation problems in Celestial Mechanics. The classical "secular theory" of Laplace and Lagrange is the construction of a first-order average Hamiltonian and the analysis of its solutions. The same is done for asteroids and serves to define "proper elements". However, proper elements are not "average" elements: second-order proper elements differ from mean elements by second-order quantitites.

In the following, we show that, in general, even if generating functions are not used, given the Hamiltonian $H = H_0(J) + \epsilon R(\theta, J)$, ($\epsilon << 1$), it is not possible to transform it into a new Hamiltonian $H^*(J^*)$ (dependent only on the new actions $J^*$), through a canonical transformation given by zero-average trigonometrical series.

## 2. Perturbation Theory with a Direct Canonical Transformation

Let us consider an N-degrees of freedom, non degenerate, integrable Hamiltonian $H_0(J)$, a perturbation $H_1 = \epsilon R(\theta, J)$, ($|\epsilon| << 1$), and one canonical transformation $(\theta, J) \Rightarrow (\theta^*, J^*)$ defined explicitly through

$$\theta_i = \theta_i^* + Q_1^i(\theta^*, J^*) + Q_2^i(\theta^*, J^*) + \cdots$$
$$J_i = J_i^* + P_1^i(\theta^*, J^*) + P_2^i(\theta^*, J^*) + \cdots, \tag{4}$$

where the functions $Q^i_k(\theta^*, J^*)$, $P^i_k(\theta^*, J^*)$ are zero-average Fourier series in $\theta$, of order $\mathcal{O}(\epsilon^k)$.

Let us adopt, for the canonical condition, the invariance of Poisson brackets: $\{x, y\} = \{x^*, y^*\}$, $x, y$ being any two of the canonical variables $\theta_i$, $J_i$. From eqns. 4, there follows:

$$\{x, y\} = \{x^*, y^*\} + \{x^*, Y_1\} + \{X_1, y^*\} + \{x^*, Y_2\} + \{X_1, Y_1\} + \{X_2, y^*\}$$

$$+\{x^*, Y_3\} + \{X_1, Y_2\} + \{X_2, Y_1\} + \{X_3, y^*\} + \cdots$$

where the letters $X$ and $Y$ were used instead of $P^i, Q^i$, since $x$ and $y$ can be any of the 2N canonical variables; their meanings are immediate. Then, because of the canonical condition,

$$0 = \{x^*, Y_1\} + \{X_1, y^*\} + \{x^*, Y_2\} + \{X_1, Y_1\} + \{X_2, y^*\}$$

$$+\{x^*, Y_3\} + \{X_1, Y_2\} + \{X_2, Y_1\} + \{X_3, y^*\} + \cdots. \tag{5}$$

If we assume that the above equation is satisfied identically in $\epsilon$, it decomposes itself into

$$
\begin{array}{ll}
\{x^*, Y_1\} + \{X_1, y^*\} = 0 & \mathcal{O}(\epsilon^1) \\
\{x^*, Y_2\} + \{X_1, Y_1\} + \{X_2, y^*\} = 0 & \mathcal{O}(\epsilon^2) \\
\{x^*, Y_3\} + \{X_1, Y_2\} + \{X_2, Y_1\} + \{X_3, y^*\} = 0 & \mathcal{O}(\epsilon^3)
\end{array} \tag{6}
$$

$\cdots\cdots$

The generic equation, at order $\mathcal{O}(\epsilon^k)$, is

$$\{X_k, y^*\} = \{Y_k, x^*\} - \Gamma_{k,x,y}, \tag{7}$$

where $\Gamma_{k,x,y}$ represents a known function of $X_1, \cdots, X_{k-1}, Y_1, \cdots, Y_{k-1}$.

Let us, now, write $H_0$ and $H_1$ in terms of the transformed variables. Limiting ourselves to the Taylor second-order terms, we have

$$H_0(J) = H_0(J^*) + \sum_i \nu_i P_1^i + \sum_i \nu_i P_2^i + \frac{1}{2} \sum_{i,j} \nu_{ij} P_1^i P_1^j + \mathcal{O}(\epsilon^3), \tag{8}$$

where

$$\nu_i = \frac{\partial H_0(J^*)}{\partial J_i^*}, \qquad \nu_{ij} = \frac{\partial^2 H_0(J^*)}{\partial J_i^* \partial J_j^*}; \tag{9}$$

and

$$H_1(\theta, J) = H_1(\theta^*, J^*) + \sum_i \frac{\partial H_1(\theta^*, J^*)}{\partial J_i^*} P_1^i + \sum_i \frac{\partial H_1(\theta^*, J^*)}{\partial \theta_i^*} Q_1^i + \mathcal{O}(\epsilon^3) \tag{10}$$

The sequence follows the same steps of other Hamiltonian perturbation theories. We substitute the above expansions in the law of conservation of the Hamiltonian under a time-independent canonical transformation:

$$H(\theta, J) = H^*(\theta^*, J^*) \equiv H_0^* + H_1^* + H_2^* + \cdots; \tag{11}$$

and identify in the powers of $\epsilon$. Then,

$$H_0^* = H_0(J^*),$$

$$H_1^* = \sum_i \nu_i P_1^i + H_1(\theta^*, J^*), \tag{12}$$

$$H_2^* = \sum_i \nu_i P_2^i + \frac{1}{2} \sum_{i,j} \nu_{ij} P_1^i P_1^j + \sum_i \frac{\partial H_1}{\partial J_i} P_1^i + \sum_i \frac{\partial H_1}{\partial \theta_i} Q_1^i,$$

etc.

The above equations may be compacted in the *homological* **equation**

$$\sum_i \nu_i P_k^i = H_k^* - \Psi_k(\theta^*, J^*) \tag{13}$$

for all $k \geq 1$. In all cases, the function $\Psi_k(\theta^*, J^*)$ is independent of $P_k$ and is known if the equations for the previous subscripts were solved.

The homological equation has $N$ unknowns $P_k^i$. Its indeterminacy is, however, only apparent, since the $P_i$ must obey at the corresponding canonical condition given by eqn. 7. Let us transform eqn. 13 by composing it with the $2N$ canonical variables, in Poisson brackets:

$$\begin{aligned} \sum_i \{\nu_i P_k^i, J_j^*\} &= \{H_k^*, J_j^*\} - \{\Psi_k, J_j^*\} \\ \sum_i \{\nu_i P_k^i, \theta_j^*\} &= \{H_k^*, \theta_j^*\} - \{\Psi_k, \theta_j^*\} \end{aligned} \tag{14}$$

or, decomposing the left-hand sides brackets,

$$\begin{aligned} \sum_i \nu_i \{P_k^i, J_j^*\} &= \{H_k^*, J_j^*\} - \{\Psi_k, J_j^*\} \\ \sum_i \nu_i \{P_k^i, \theta_j^*\} + \sum_i P_k^i \{\nu_i, \theta_j^*\} &= \{H_k^*, \theta_j^*\} - \{\Psi_k, \theta_j^*\} \end{aligned} \tag{15}$$

where we did take into account that $\{\nu_i, J_j^*\} = 0$ because $\nu_i$ is independent on the angles. We may, now, use eqn. 7, and transform the above set into

$$\begin{aligned} \sum_i \nu_i \{P_k^j, J_i^*\} + \sum_i \nu_i \Gamma_{k, J_i, J_j} &= \{H_k^*, J_j^*\} - \{\Psi_k, J_j^*\} \\ \sum_i \nu_i \{Q_k^j, J_i^*\} + \sum_i \nu_i \Gamma_{k, J_i, \theta_j} + \sum_i P_k^i \{\nu_i, \theta_j^*\} &= \{H_k^*, \theta_j^*\} - \{\Psi_k, \theta_j^*\}. \end{aligned} \tag{16}$$

After the computation of some elementary brackets, we obtain the homological system of equations:

$$\sum_i \nu_i \frac{\partial P_k^j}{\partial \theta_i^*} = \frac{\partial H_k^*}{\partial \theta_j^*} - \frac{\partial \Psi_k}{\partial \theta_j^*} - \sum_i \nu_i \Gamma_{k, J_i, J_j}, \tag{17}$$

$$\sum_i \nu_i \frac{\partial Q_k^j}{\partial \theta_i^*} = \sum_i \nu_{ij} P_k^i - \frac{\partial H_k^*}{\partial J_j^*} + \frac{\partial \Psi_k}{\partial J_j^*} - \sum_i \nu_i \Gamma_{k, J_i, \theta_j}. \tag{18}$$

### 3. Non-Existence of Average Hamiltonians

Let us consider, in this section, the question title of this article, and search for solutions of eqns. 17 and 18 such that $< P_k^i >=< Q_k^i >= 0$. The condition for the existence of such solutions is that the right-hand sides of the two equations have zero averages, that is,

$$< \frac{\partial H_k^*}{\partial \theta_j^*} > - \sum_i \nu_i < \Gamma_{k,J_i,J_j} >= 0 \tag{19}$$

and

$$- < \frac{\partial H_k^*}{\partial J_j^*} > + \frac{\partial < \Psi_k >}{\partial J_j^*} - \sum_i \nu_i < \Gamma_{k,J_i,\theta_j} >= 0 \tag{20}$$

These equations show that, for $k > 1$, it is not possible to find a solution of the given problem such that we have simultaneously (for all $i$) $< P_k^i >= 0$, $< Q_k^i >= 0$, and $H_k^*$ independent of $\theta^*$. Indeed, in this case, $< \frac{\partial H_k^*}{\partial \theta_j^*} >= 0$, and eqn. 19 can only be generally satisfied when $< \Gamma_{k,J_i,J_j} >= 0$, what is true, in general, only if $k = 1$.

#### 3.1. FIRST-ORDER AVERAGE HAMILTONIAN

For $k = 1$, since $\Gamma_{1,J_i,J_j} = 0$, eqns. 19 and 20 become

$$< \frac{\partial H_1^*}{\partial \theta_j^*} >= 0 \tag{21}$$

and

$$- < \frac{\partial H_1^*}{\partial J_j^*} > + \frac{\partial < H_1 >}{\partial J_j^*} = 0 \tag{22}$$

which have the trivial solution $H_1^* =< H_1 >$. It is worth recalling that, to this order, the methods using the generating functions ($S$ or $W$) also give this same result.

The calculations were done following a constructive scheme, but it is easy to make a verification using a reversed reasoning and prove that, indeed, this solution satisfies the condition given by the first of eqns. 6, for the components of $P_1$.

This result means that it is possible to obtain a first-order "average" Hamiltonian. This fact certainly played a role in the introduction of the word "average" and its variations in the study of the construction of asymptotic (or formal) solutions of perturbed systems.

## 4. Conclusion

The conclusion of the above sections is the following: Given the Hamiltonian $H = H_0(J) + \epsilon R(\theta, J)$, ($\epsilon << 1$), the problem of transforming it into a new Hamiltonian $H^*(J^*)$ (dependent only on the new actions $J^*$), through a canonical transformation given by zero-average trigonometrical series has no general solution at orders higher than the first. It is worth mentioning that a general solution cannot be found even in the particular case, usual in Celestial Mechanics, in which the disturbing potential $R(\theta, J)$ is a cosine series.

## Acknowledgements

This investigation was supported by CNPq.

## References

Born, M.: 1926, *Problems of Atomic Dynamics*, M.I.T., Cambridge (US).

Brouwer, D.: 1959, Solution of the problem of artificial satellite theory without drag, *Astron. J.*, **64**, 378-397.

Charlier, C.V.L.: 1907, *Die Mechanik des Himmels*, De Gruyter, Leipzig, Vol. II.

Delaunay, C.: 1868, *Mémoire sur la Théorie de la Lune*, Acad. Sc., Paris.

Deprit. A.: 1969, Canonical transformation depending on a small parameter, *Cel. Mech. & Dyn. Astr.*, **1**, 12-30.

Hori, G.-I.: 1966, Theory of General Perturbations with Unspecified Canonical Variables, *Publ. Astron. Soc. Japan*, **18**, 287-296.

Kolmogorov, A.N.: 1954, Preservation of conditionally periodic movements with small change in Hamiltonian function, *Dokl. Akad. Nauk*, **98**, 527-530.

Milani, A.; Nobili, A.M. and Carpino. M.: 1987, Secular variations of the semimajor axes: theory and experiments, *Astron. Astrophys.*, **172**, 265-279.

Poincaré, H.: 1893, *Les Méthodes Nouvelles de la Mécanique Celeste*, Gauthier-Villars, Paris, Vol. II.

# DIFFUSION CHARACTER IN FOUR-DIMENSIONAL
# VOLUME-PRESERVING MAP

YI-SUI SUN

*Department of Astronomy, Nanjing University, Nanjing 210093, P.R. China*

and

YAN-NING FU

*Purple Mountain Observatory, Academia Sinica, Nanjing 210008, P.R. China*

**Abstract.** Due to the existence of invariant tori, chaotic sea and hyperbolic structures in higher dimensional phase space of a volume-preserving map, the diffusion route of chaotic orbits will be complicated. The velocity of diffusion will be very slow if the orbits are near an invariant torus. In order to realize this complicated diffusion phenomenon, in this paper we study the diffusion characters in the different regions, i.e., chaotic, hyperbolic and invariant tori's regions. We find that for the three different regions, the diffusion velocities are different. The diffusion velocity in the vicinity of an invariant torus is the slowest one.

**Key words:** Volume-Preserving map – Invariant tori – Diffusion

## 1. Introduction

According to the KAM theorem for Hamiltonian systems with two degrees of freedom, the two dimensional invariant tori will prevent the escape of orbits on a three-dimensional energy surface. When the degrees of freedom, $n$, exceeds two, the n-dimensional invariant tori cannot divide the $(2n-1)$-dimensional energy surface into two disconnected parts, so the escape will appear. Because of the existence of n-dimensional invariant tori, the escape route across a net of invariant tori will be complicated and the velocity of escape will be very slow, this is called Arnold diffusion. This kind of diffusion can hardly be detected by numerical methods, as pointed out by Laskar with a four-dimensional symplectic map similar to the so-called Froeschlé map (Laskar 1993). Efthymiopoulos et al. (1998) also conclude that the diffusion can be practically ignored.

The problems related to invariant manifolds and diffusion orbits in four-dimensional symplectic maps have been carefully studied (e.g. Froeschlé 1971, Froeschlé 1972, Ding et al. 1990, Laskar 1993, Efthymiopoulos et al. 1998, etc.). In this paper, we study a similar problem in a more general kind of maps, i.e., four-dimensional volume-preserving maps. In order to realize the complicated diffusion phenomenon, it's worth to study at first the diffusion characters in chaotic, hyperbolic and invariant tori's regions, respectively. We investigate the above problem using a four-dimensional volume-preserving map obtained by coupling two area-preserving maps with a perturbation parameter, which possesses the hyperbolic and parabolic fixed point, respectively. In terms of the results on the three-dimensional volume-preserving map, the ordered region near the unstable fixed points will be changed into a chaotic one by perturbation, but for the ordered region distant from it the invariant tori will survive (Sun et al. 1988, Zhang et al. 1989). We suspect that we might find some invariant tori, chains of islands and hyperbolic structure in the above four-dimensional volume-preserving map.

## 2. The Map

We study the following volume-preserving map with four dimension.

$$
T_{hp} \begin{cases} x_{n+1}= s \left( x_n \cos \varphi_n - y_n \sin \varphi_n \right), \\ y_{n+1}= s^{-1} \left( x_n \sin \varphi_n + y_n \cos \varphi_n \right) + c \sin(x_{n+1} + z_{n+1}), \\ z_{n+1}= z_n - b \, t_n^3, \\ t_{n+1}= z_n + t_n - b \, t_n^3 + c \sin(x_{n+1} + z_{n+1}), \end{cases} \tag{1}
$$

$$
\varphi_n = (x_n^2 + y_n^2)^k. \tag{2}
$$

The map $T_{hp}$ is the coupling of the following two area-preserving maps

$$
T_h \begin{cases} x_{n+1}= s \left( x_n \cos \varphi_n - y_n \sin \varphi_n \right), \\ y_{n+1}= s^{-1} \left( x_n \sin \varphi_n + y_n \cos \varphi_n \right), \end{cases} \tag{3}
$$

$$
\varphi_n = (x_n^2 + y_n^2)^k. \tag{4}
$$

and

$$
T_p \begin{cases} z_{n+1}= z_n - b \, t_n^3, \\ t_{n+1}= z_n + t_n - b \, t_n^3. \end{cases} \tag{5}
$$

Here $b, s, k$ are parameters ($s \neq 1$) and $c$ the perturbation parameter ( coupling parameter ). We take $s = 1.05, k = -1.5, b = 1.5$ and $c = 0.03$, in which case the map $T_{hp}$ is not defined at the origin. Obviously, the map $T_{hp}$ is symmetric with respect to the origin. As can be easily verified, the map $T_{hp}$ is volume-preserving but not symplectic.

At first, we explore the structure of phase space for the map $T_{hp}$, i.e., to look for ordered and chaotic regions and the hyperbolic structure. We have investigated the structure of phase space of the map $T_{hp}$ by computing the LCIs (Lyapunov Characteristic Indicators, the approximate values of Lyapunov Characteristic Exponents up to the finite time of computation ) and exploring fixed points and checking their stability. Fig.1(a) and Fig.1(b) are the projections of some invariant tori (roughly approximated by quasi-periodic orbits) and chaotic sea onto the planes $(x, y)$ and $(z, t)$, respectively. These figures display the global structure of phase space. We know that the dimensionality of the invariant tori is two and these tori can not divide the four-dimensional phase space into two disconnected parts. As a result, the chaotic orbits diffuse around the invariant tori, as easily seen in the Fig.1(a) and Fig.1(b).

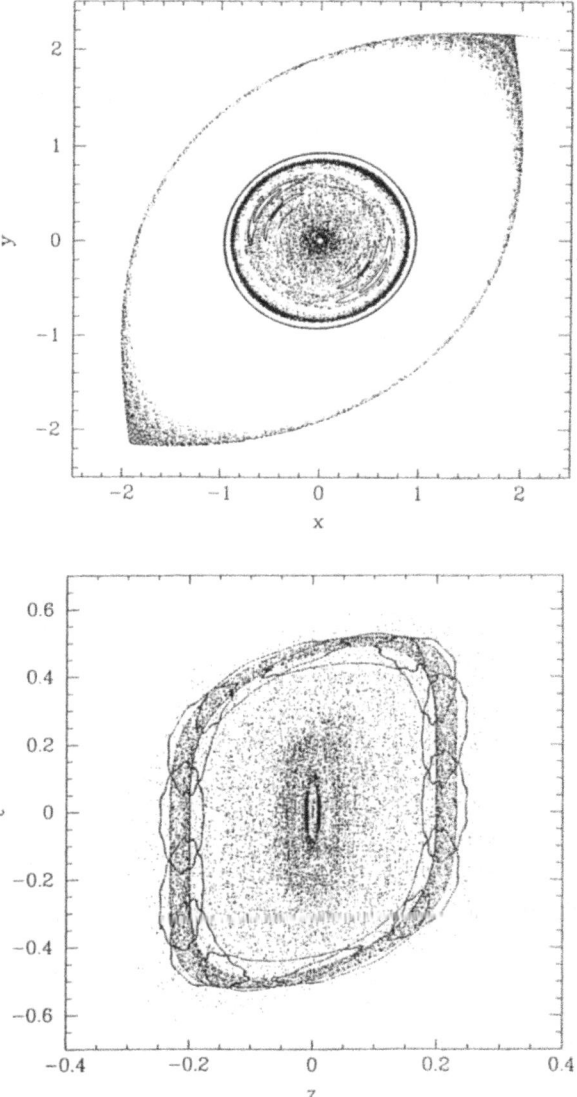

Fig. 1. The projections on the two coordinate planes of the quasi-periodic orbits starting from the following initial points: (0.376, 0.582, -0.0002, 0.008), (0.232, 0.658, 0.00019, 0.0038), (0.68, 0.68, 0.2125, 0.4250), (0.356, -0.246, -0.012, 0.003), (0.392, -0.268, 0, 0.0058), (0.2, -0.5, -0.0127, 0.0016), (0.40, -0.37, -0.0127 ,0.0016), (0.16, -0.65 ,0, 0.009), (0.40 ,-0.55, 0 ,0.009), (0.61 ,0.61, 0.190625 ,0.38125) and their symmetric points with respect to the origin for the map $T_{hp}$, and the chaotic orbits starting from the following initial points: (0.261101993556, 0.341296119830, -0.007526015390, 0.001646639192), (1.918131071904, 2.156373097694 -0.028487926171, -0.000415023486), (0.68 ,0.53 ,0.190625, 0.381250). ($s = 1.05, k = -1.5, b = 1.5$ and $c = 0.03$).

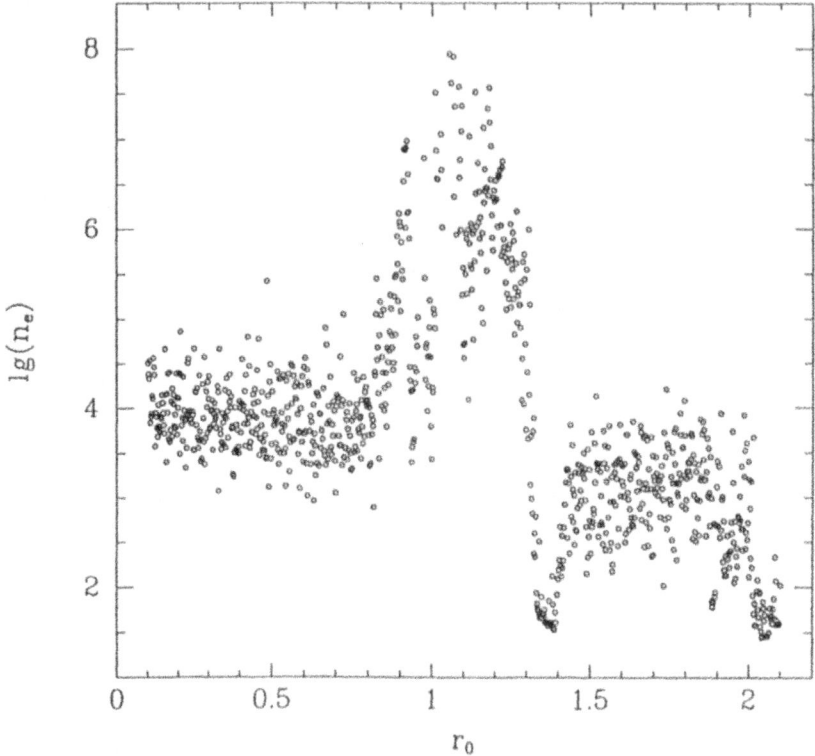

Fig. 2. Distribution of escape time of orbits with respect to the distance $d_0$ along the line (6).

## 3. Diffusion Characters in Different Regions

In order to clarify the diffusion of orbits in the phase space globally, we have made a transversal exploration of escape time along the line passing through the origin and the point of which the projections on the (x,y) and (z,t) planes are, respectively, the farthest boundary point of the ordered region for the maps $T_h$ and $T_p$. The equations of this line is as follows

$$5x = 5y = 16z = 8t. \tag{6}$$

We take 1001 initial points on the line (6), of which the distances to the origin are as follows

$$r_0 = 0.1 + 0.002i, \quad i = 0, 1, ..., 1000. \tag{7}$$

Defining the escape of orbits as $r = \sqrt{x^2 + y^2 + z^2 + t^2} > 2.5$, we calculate the escape time for each orbit. Fig.2 shows the result of this exploration (not including the orbits on the invariant tori). We find a slow escape region near $r = 1.05$, where exist most of the explored initial points on invariant tori. It implies that the escape

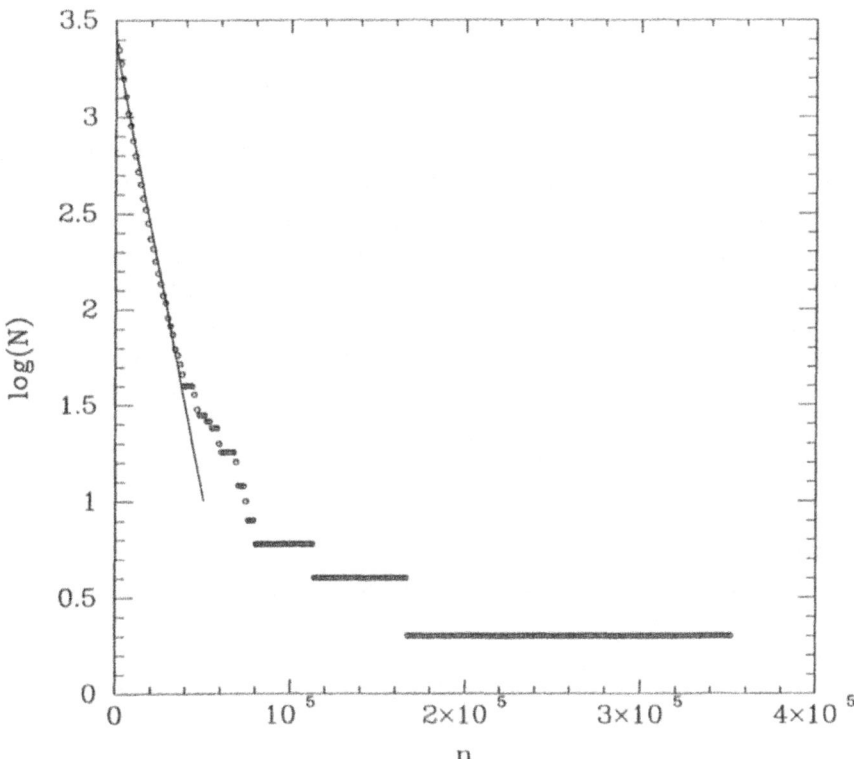

Fig. 3. Diagrams of the number of orbits having not escaped before the iteration number $n$, for the case (a). circle: computational values, curve: fitted values.

velocity of orbits near the invariant tori is much slower than that in the chaotic region, and this is due to the "stickiness" effect of invariant tori as can be seen below. From Fig.2 we can also see that even in the "dense" region of invariant tori, there exist chaotic zones with faster escape. In the following we will investigate the diffusion characters in different regions.

(a) Chaotic sea

We choose 2352 initial points $(x_0, y_0, z_0, t_0)$ in the chaotic region near the origin as follows

$$(x_0, y_0, z_0, t_0) = (i, j, l, m)s,$$
$$s = 0.1, i, j, l, m = 0, \pm 1, \pm 1.5, \pm 2 \ (excluding \ i = j = 0) \tag{8}$$

Defining the escape of orbits as above, i.e., $r = \sqrt{x^2 + y^2 + z^2 + t^2} > 2.5$, we count the number, $N$, of orbits which have not escaped yet before the number of iterations $n$. In Fig.3 the circles stand for the numerical results of $\log(N)$ versus $n$. Because there are positive LCEs for the chaotic orbits, we try to fit the numerical results by an exponential law, i.e.,

$$N = N_0 \times 10^{-\alpha n} \tag{9}$$

However, this law can not be applied for large $n$, as can be easily seen from Fig.3. Actually, there should exist some "island" tori which exert the "stickiness" effect on the orbits near them (Meiss and Ott 1985, Meiss and Ott 1986, Lee 1988, Ding et al. 1990, Lai 1992). Therefore, we fit the numerical results only for $n < 40000$ (including about 99% of the explored orbits) and obtain an analytic curve with $N_0 = 2532$ and $a = 4.735163 \times 10^{-5}$ (see Fig.3). This analytic curve is in good agreement with the numerical values for $n < 40000$. Accordingly, we suspect that the diffusion of orbits in a "complete" chaotic sea, i.e., the region where the area occupied by islands can be neglected, should possess the exponential law. As it is difficult to find out small tori in four-dimensional phase space, the above point is just a conjecture.

(b) Vicinity of invariant torus

There exists a slow escape region near the point on the line (6), of which the distance to the origin is $r = 1.05$ (see Fig.2).

By computing the LCIs and drawing the projective figures of the orbit starting from this point, we find that the point is on an invariant torus. In the following, we study the diffusion character of orbits in the vicinity of this torus by taking 2401 initial points $(x_0, y_0, z_0, t_0)$ as follows

$$\begin{aligned}
&(x_0, y_0, z_0, t_0) = (x_c, y_c, z_c, t_c) + (i, j, l, m)s, \\
&s = 0.005, i, j, l, m = 0, \pm 1, \pm 2, \pm 3.
\end{aligned} \tag{10}$$

At first, we look for a region in which the orbits will stay for quite a long time ($n \geq 4096$) and we choose this region as small as possible so that it contains as less points far from the torus as possible. According to our testing calculation, we define the escape of orbits as they leave the following zone,

$$\begin{aligned}
G = \{ \ &(x, y, z, t): \\
&x \in (-0.98, 0.98), y \in (-0.94, 0.94), z \in (-0.26, 0.26), t \in (-0.55, 0.55), \\
&\sqrt{x^2 + y^2} \in (0.86, 0.99) \}
\end{aligned} \tag{11}$$

(Note: $\sqrt{x^2 + y^2}$ is an invariant of map $T_{hp}$ with $s = 1$ and $c = 0$.)

Fig.4 indicates the diagram of $\log(N)$ versus $n$ with 605 orbits escaped before $n = 10^8$ iterations, where $N$ is the number of orbits in the region $G$ at the iteration number $n$. We note that only a small fraction of the orbits escape before $n = 10^7$ iterations, which implies that the diffusion in the vicinity of invariant tori is very slow. This is due to the "stickiness" effect of invariant tori, but here it is different from the usual one which has an algebraic decay law of diffusion (Meiss and Ott 1985, Ding et al. 1990). According to the method used to study the usual "stickiness" effect in related papers, we suspect that the algebraic decay law

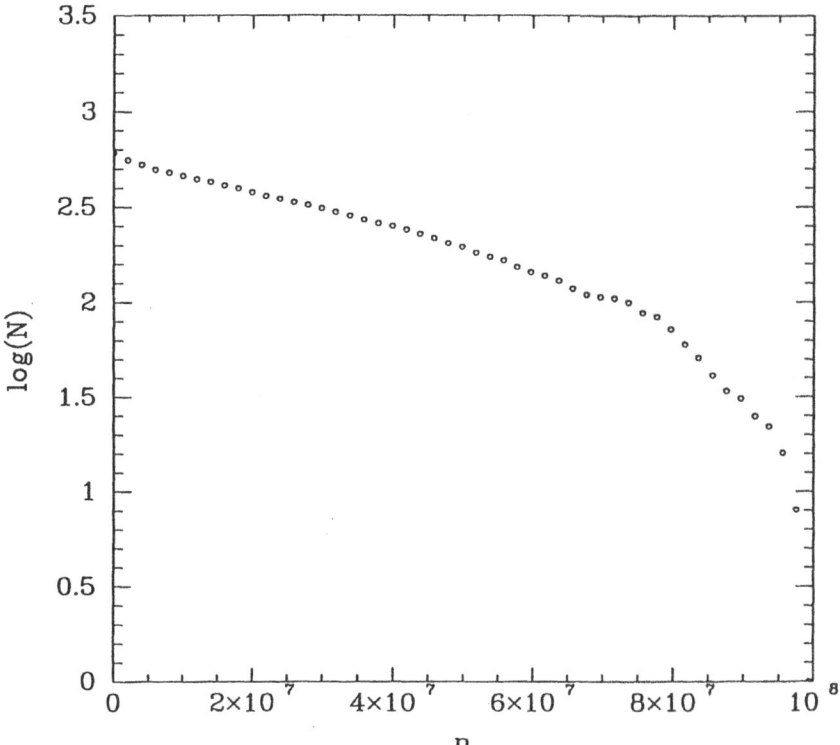

Fig. 4. Diagram of the number of orbits having not escaped before the iteration number $n$, for the case (b).

should be the "averaged" law of diffusion over the chaotic region with invariant islands and cantori. The orbits diffuse sometimes "freely" through chaotic sea, and sometimes by crossing cantori and passing around islands, so the usual "stickiness" effect of invariant tori is actually modified by the diffusion process in chaotic sea. If a region of higher dimensional phase space is filled "densely" by invariant tori, the "stickiness" effect would appear to be that in our case.

(c) Vicinity of hyperbolic-elliptic fixed point
According to the exploration about the fixed points, we find that

$$(x_h, y_h, z_h, t_h) = (1.918131071904, 2.156373097694, -0.028487926171, -0.000415023486)$$

is a hyperbolic-elliptic fixed point. Choosing 2401 initial points $(x_0, y_0, z_0, t_0)$ in its vicinity as follows

$$(x_0, y_0, z_0, t_0) = (x_h, y_h, z_h, t_h) + (i, j, l, m)s$$
$$s = 10^{-5}, i, j, l, m = 0, \pm 1, \pm 2, \pm 3, \tag{12}$$

we study the diffusion character in the vicinity of the above fixed point. The orbit with initial point sufficiently close to the hyperbolic-elliptic fixed point should

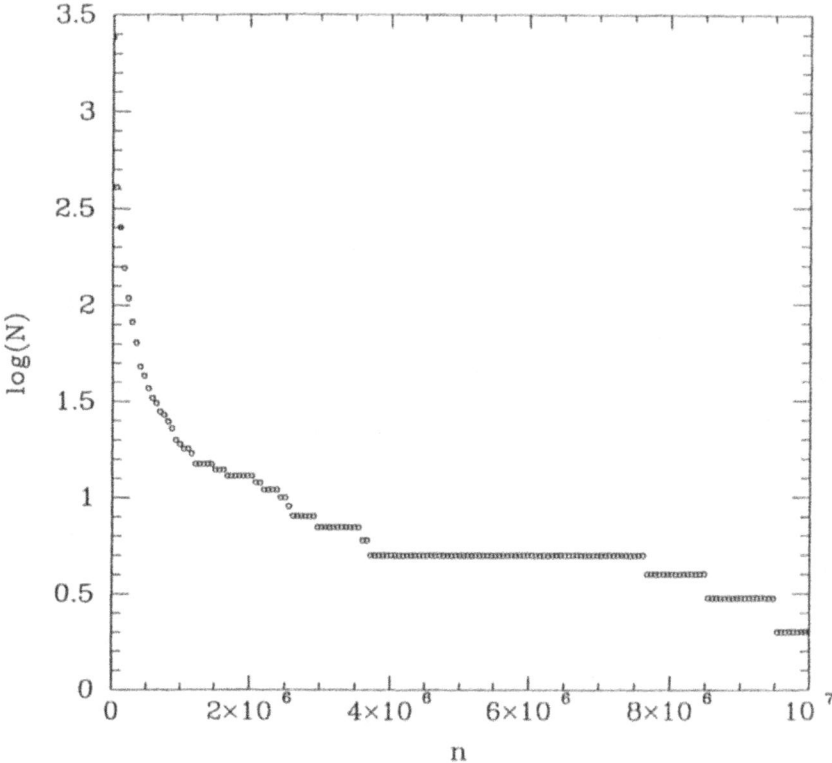

Fig. 5. Diagram of the number of orbits $N$ having not escaped before the iteration number $n$, for the case (c).

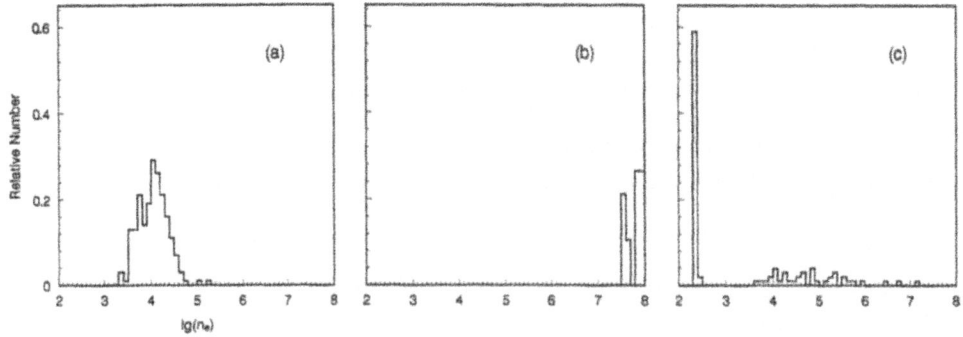

Fig. 6. Distribution of the relative number of orbits escaping at the iteration number $n$ for (a) chaotic sea, (b) vicinity of invariant torus and (c) vicinity of hyperbolic-elliptic fixed point.

wander near the hyperbolic structure on the plane $(x, y)$ and near the origin on the plane $(z, t)$ for quite a long time. After testing calculation, we find that, as soon as the distance $d$ between the fixed point and the returning point of orbit to the vicinity of the fixed point exceeds $\bar{d} = 0.670213829709$ or the distance $d'$ of the orbit to the origin exceeds 10, the orbit will diffuse to very distant point from the origin $(d' > 10^5)$ with no return, so we regard $d > \bar{d}$ or $d' > 10$ as escape. Fig.5 shows the variation of $N$, number of the orbits having not escaped yet, with the number of iteration $n$. We find that this result is similar to that in chaotic sea, except that there are some orbits with much slower diffusion, of which the initial points are very close to the hyperbolic-elliptic fixed point. This implies that the hyperbolic-elliptic fixed point also has "stickiness" effect on some nearby orbits. Another interesting feature is the appearance of fast escape orbits starting near the hyperbolic-elliptic fixed point, which causes fast decreasing of $N$ for small $n$ (about 49% of the explored orbits escaping before $n = 145$). Contopoulos et al. (1997) have found the "hole" of escape embedded in stickiness region near an invariant curve of two-dimensional area-preserving map. The above-mentioned feature shows that the "holes" of escape also exist in the vicinity of the hyperbolic-elliptic fixed point of four-dimensional volume-preserving map.

In the above studies, we discuss separately the diffusion characters in three different regions, and now we will compare the diffusion velocities for the different cases with the same definition of escape. For this purpose, we choose randomly 250 initial points in each region as above, and regard an orbit as escape if the distance of an orbit point to the origin exceeds 100. In the chaotic region and the vicinity of hyperbolic-elliptic fixed point, all of the 250 orbits escape, but in the vicinity of invariant torus, there are only 45 orbits escaping before $n = 10^8$ iterations. Fig.6 exhibits the results. We can see that the diffusion velocities in the different region are different. The diffusion in chaotic sea is the fastest, while the slowest one is in the vicinity of the invariant torus. In the vicinity of hyperbolic- elliptic fixed point, the diffusion velocity is between the above two cases, because of the existence of narrow weak "stickiness" zone nearby the perturbed hyperbolic-elliptic structure.

## 4. Conclusions

In a higher dimensional volume-preserving map, due to the existence of invariant tori, chaotic sea and hyperbolic structure, the diffusion of orbits will be very complicated. In this paper, we study the diffusion characters in the different regions, respectively, and compare the diffusion velocities with each other. This kind of work is the base of realizing the global diffusion in phase space. We find that for the three explored regions, the diffusion velocities are different and have different characters. The diffusion velocity in the vicinity of an invariant torus is the slowest one, the reason for this is the "stickiness" effect. We believe that the "stickiness" effect would take the main responsibility for very slow diffusion in higher dimensional phase space.

## Acknowledgements

We thank Drs. C. Froeschlé, N. Voglis and S. Ferraz-Mello for their valuable comments on the manuscript. This work is supported by the National Natural Science Foundation of China.

## References

Contopoulos, G., Voglis, N., Efthymiopoulos, C., Froeschlé, C., Gonczi R., Lega E., Dvorak R. and Lohinger E.: 1997, Transition spectra of dynamical systems, *Celest. Mech. and Dyn. Astro.*, **67**, 293

Ding, M.Z., Bountis, T. and Ott, E.: 1990, Algebraic escape in higher dimensional Hamiltonian systems, *Physics Letter A*, **151**, 395

Efthymiopoulos, C., Voglis, N. and Contopoulos, G.: 1998, Diffusion and transient spectra in a 4-d symplectic mapping, in "Advances in Discrete Mathematics and Applications, Volume 1, Analysis and Modelling of Discrete Dynamical Systems", D. Benest and C. Froeschlé (eds), Gordon and Breach Science Publishers, 91

Froeschlé, C.: 1971, On the number of isolating integrals in systems with three degrees of freedom, *Astrophys. Space Sci.*, **14**, 110

Froeschlé, C.: 1972, Numerical study of a four-dimensional mapping, *Astron. Astrophys.*, **16**, 172

Lai, Y.C., Ding. M.C. and Blümel, R.: 1992, Algebraic decay and fluctuations of the decay exponent in Hamiltonian systems, *Phy. Rev. A*, **46**, 4661

Laskar, J.: 1993, Frequency analysis for multi-dimensional systems. Global dynamics and diffusion, *Physica D*, **67**, 257

Lee, K.C.: 1988, Long-time tails in a chaotic system, *Phy. Rev. Lett.*, **60**, 1991

Meiss, J.D. and Ott, E.: 1985, Markov-tree model of intrinsic transport in Hamiltonian systems, *Phy. Rev. Lett.*, **55**, 2741

Meiss, J.D and Ott, E.: 1986, Markov-tree model of transport in area-preserving maps, *Physica D*, **20**, 387

Morbidelli, A. and Giorgilli, A.: 1995, Superexponential stability of KAM tori, *J. Stat. Phys*, **78**, 1607

Sun, Y.S. and Yan, Z.M.: 1988, A perturbed extension of hyperbolic twist mapping, *Celest. Mech. and Dyn. Astro.*, **42**, 369

Zhang, T.L. and Sun, Y.S.: 1989, Behaviour of a class of perturbed measure-preserving mappings. *J. Nanjing Uni.*, **25**, 187 (In Chinese).

# THE SEPARATRIX ALGORITHMIC MAP:
# APPLICATION TO THE SPIN-ORBIT MOTION

IVAN I. SHEVCHENKO

*Pulkovo Observatory, Russian Academy of Sciences, Pulkovskoje ave. 65/1, St.Petersburg 196140, Russia; E-mail: iis@gao.spb.ru*

**Abstract.** The planar rotational motion of a non-symmetric satellite in an elliptic orbit is considered. A two-dimensional map is constructed, describing the motion in a vicinity of the separatrix of the synchronous spin-orbit resonance. This map is a generalization of Chirikov's separatrix map, in the sense that the asymmetry of perturbation is taken into account. Phase portraits of the generalized map perfectly reproduce well-known examples of surfaces of section (first computed by Wisdom et al. (Wisdom *et al.*, 1984), Wisdom (Wisdom, 1987)) of the phase space of spin-orbit coupling for non-symmetric natural satellites. Moreover, it provides a straightforward analytical description of the phase space: analysis of properties of the map allows one to precalculate, by means of compact analytical relations, the locations of resonances and chaos borders, the emergence of marginal resonances, and even to describe bifurcations of the synchronous resonance's center, though far from the separatrix.

## 1. Introduction

Equations of the nonlinear pendulum with periodic perturbations constitute an important paradigm in various fields of modern physics and mechanics. Consider the Hamiltonian

$$H = \frac{\mathcal{G}p^2}{2} - \mathcal{F}\cos\varphi + a\cos(\varphi - \tau) + b\cos(\varphi + \tau), \tag{1}$$

where $\tau = \Omega t + \tau_0$. The first two terms represent the Hamiltonian of the pendulum, while the two remaining ones the periodic perturbations. The variable $\varphi$ is the pendulum angle (this angle measures deviation of the pendulum from the position of equilibrium), and $\tau$ is the phase angle of perturbation. The quantity $\Omega$ is the perturbation frequency and $\tau_0$ is the initial phase of perturbation; $p$ is the momentum; $\mathcal{F}, \mathcal{G}, a, b$ are constants. In what follows, the Hamiltonian of the unperturbed pendulum is designated as $H_0$, i.e. $H_0 = \frac{\mathcal{G}p^2}{2} - \mathcal{F}\cos\varphi$.

The motion in a vicinity of the separatrix of the Hamiltonian (1) in the symmetric case $a = b$ was considered by Chirikov (Chirikov, 1979). He showed that it is efficiently described by a map, which he called the whisker map. Now the term "separatrix map", hereafter SM, is customary for such kinds of maps. The SM in Chirikov's (Chirikov, 1979) form is a two-dimensional area-preserving map

$$
\begin{aligned}
w_{n+1} &= w_n - W\sin\tau_n, \\
\tau_{n+1} &= \tau_n + \lambda\ln\frac{32}{|w_{n+1}|} \quad (\mathrm{mod}\,2\pi),
\end{aligned}
\tag{2}
$$

where $w$ denotes the relative (with respect to the separatrix value) pendulum energy $w = \frac{H_0}{\mathcal{F}} - 1$, and $\tau$ is the phase of perturbation. Constants $\lambda$ and $W$ are parameters:

$\lambda$ is the ratio of $\Omega$, the perturbation frequency, to $\omega_0 = (\mathcal{F}\mathcal{G})^{1/2}$, the frequency of the small-amplitude pendulum oscillations; and

$$W = \frac{a}{\mathcal{F}}\lambda(A_2(\lambda) + A_2(-\lambda)) = 4\pi\frac{a}{\mathcal{F}}\lambda^2 \text{csch}\frac{\pi\lambda}{2}, \tag{3}$$

where

$$A_2(\lambda) = 4\pi\lambda\frac{\exp\frac{\pi\lambda}{2}}{\sinh(\pi\lambda)} \tag{4}$$

is the Melnikov–Arnold integral as defined in Refs. (Chirikov, 1979; Lichtenberg and Lieberman, 1987; Shevchenko, 1998). One iteration of the SM (2) corresponds to one period of the pendulum rotation, or a half-period of its libration.

In this paper, the separatrix algorithmic map (SAM) is constructed for the motion in a vicinity of the separatrix of the Hamiltonian (1). This map is a generalization of Chirikov's SM (2), in the sense that the asymmetry of perturbation is taken into account. It constitutes an algorithm containing conditional transfer statements. Besides, a regular projection algorithm (RPA) is proposed, which allows one to construct phase portraits of the motion at customary sections of phase space.

The Hamiltonian (1) emerges in a number of mechanical and physical problems. The reason for such universality is that the nonlinear pendulum provides a model of the nonlinear resonance under very general conditions (Chirikov, 1979; Lichtenberg and Lieberman, 1987). In this general setting, the phase angle of the model pendulum has the meaning of the resonant phase angle.

The SAM is applied here to the problem of the resonant rotational motion of a non-symmetric satellite in an elliptic orbit. Performance of the SAM and RPA is checked in comparisons with direct numeric integrations of the original system (1). It is shown that the theory of the SAM provides a straightforward analytical description of the structure of phase space of the near-separatrix motion.

## 2. The Separatrix Algorithmic Map

The separatrix map in the case of asymmetric perturbation, i.e. $a \neq b$ in Eq. (1), is different from that in the symmetric case, since the energy increments are different for prograde and retrograde motions of the model pendulum. (Henceforth the motion is called "prograde" or "retrograde" if the increase in $\varphi$ with time is respectively positive or negative.) Taking this difference into account, let us write down the following algorithm

$$\begin{aligned}
&\text{if } w_n < 0 \text{ and } W = W^- \text{ then } W = W^+,\\
&\text{if } w_n < 0 \text{ and } W = W^+ \text{ then } W = W^-;\\
&w_{n+1} = w_n - W\sin\tau_n,\\
&\tau_{n+1} = \tau_n + \Delta_{n+1}\tau \pmod{2\pi};
\end{aligned} \tag{5}$$

where $\Delta_{n+1}\tau$, which is approximately equal to $\lambda \ln \frac{32}{|w_{n+1}|}$ (as it is adopted for the ordinary SM (2)), is defined in a more exact sense below, and $W^+$, $W^-$ denote the values of the parameter $W$ for the prograde and retrograde motions respectively. In case of asymmetric perturbation these values are different.

The algorithm (5) constitutes the separatrix algorithmic map (SAM). Its essence is in taking into account alternations of the parameter $W$. The latter one alternates when the direction of motion alternates. This takes place when rotation changes to libration and when the motion is librational. The algorithm (5) does not contain conditions with $w_n > 0$, because the direction of motion does not change when they hold.

An important property of the original SM (2) is that it maps the motion asynchronously: the relative energy $w$ is mapped at $\varphi = \pm\pi$, while the phase angle of perturbation $\tau$ is mapped at $\varphi = 0$. The SAM (5) retains this asynchronism. The procedure of synchronization of the SM (2) to the unified surface of section $\varphi = 0$ is described in Ref. (Shevchenko, 1998). An analogous procedure of synchronization of the SAM (5) can be also derived. It is more complicated. However, it is not needed in what follows and is not presented here.

In order to find expressions for $W^+$, $W^-$, one should integrate the increment of energy per one iteration of the map, following the usual procedure (Chirikov, 1979), but making it separately for prograde and retrograde directions of motion. This gives

$$W^+(\lambda, \eta) = \frac{a}{\mathcal{F}}\lambda \left( A_2(\lambda) + \eta A_2(-\lambda) \right), \tag{6}$$

$$W^-(\lambda, \eta) = \frac{a}{\mathcal{F}}\lambda \left( \eta A_2(\lambda) + A_2(-\lambda) \right), \tag{7}$$

where $\eta = \frac{b}{a}$; the function $A_2(\lambda)$ is given by Eq. (4).

One should be aware that the expression for the increment of the phase $\tau$ in the ordinary SM (2) is an approximation. It is valid for a low strength of perturbation, i.e. for $W \ll 1$. According to Ref. (Shevchenko, 1998), if the perturbation is not weak, one can improve the performance of the map by means of replacing the logarithmic approximation of the phase increment by its exact value, which depends on what side of the line of the unperturbed separatrix the motion takes place. Making this replacement, one obtains the expression for the exact increment:

$$\Delta_{n+1}\tau = \begin{cases} 2\lambda K\left( (1 + \frac{w_{n+1}}{2})^{1/2} \right), & \text{if } w_{n+1} < 0; \\ 2\lambda(1 + \frac{w_{n+1}}{2})^{-1/2}K\left( (1 + \frac{w_{n+1}}{2})^{-1/2} \right), & \text{if } w_{n+1} > 0; \end{cases} \tag{8}$$

where $K(k)$ is the elliptic integral of the first kind. The first line in Eq. (8) corresponds to libration of the model pendulum, while the second one to its rotation. In the algorithm (5), henceforth the exact expression (8) is used.

## 3. The Regular Projection Algorithm

The SAM maps the motion of the system (1) on the plane $(\tau, w)$ at fixed values of the resonant phase angle $\varphi$ equal to 0 and $\pm\pi$ (as noted in Section 2, the mapping is asynchronous). When Poincaré sections are constructed numerically in applied problems, it is customary to use another plane, namely the plane $(\varphi, p)$ taken at a fixed value of the phase angle of perturbation, e.g. $\tau = 0 \pmod{2\pi}$.

Consider the problem how the section of the second kind can be found with the help of the separatrix map. In order to characterize the current state of the system, described by the SAM, introduce temporary designations $w = w_n$, $\tau = \tau_n$; and $\Delta\tau = \tau_n - \tau_{n-1}$. The value of $W$, which specifies the current prograde/retrograde state of the system, is taken at the next iteration $n + 1$. The matter is that the values of the variables $w_n$ and $\tau_n$, given by the SAM (5), as well as by the usual SM (2), correspond to different instants of time (the property of asynchronism, noted already in Section 2). The phase $\tau_n$ is mapped with a delay in relation to the relative energy $w_n$. The delay is equal to a half-period of rotation, or a quarter-period of libration of the model pendulum. Due to this delay, the value of $W$ in the formula for $w_{n+1}$ in the SAM (5) specifies the prograde/retrograde state of the system at the preceding instant $\tau_n$; or, inversely, the state of the system at $\tau_n$ is determined by the value of $W$ in the formula for $w_{n+1}$, i.e. at the next iteration.

Let us find the phase point which is connected with the position of the system at the current instant $\tau$ (corresponding to $\varphi = 0$) by a trajectory backwards in time, and which is situated at the nearest surface $\tau = 0 \pmod{2\pi}$. The trajectory is assumed to be regular and possessing energy $w$. For convenience, the current instant $\tau$ is taken moduli $2\pi$, while the increment $\Delta\tau$ is not. If $\Delta\tau \leq \tau$, there are no intersections with the plane of interest on the open interval of time backwards in relation to the current state, and no projection is made therefore at such an iteration of the map. Otherwise, the projection to the nearest surface $\tau = 0 \pmod{2\pi}$ is given by the formulae

$$\varphi = \begin{cases} \varphi\left(t = -\frac{\tau}{\Omega}\right), & \text{if } W = W^+ \text{ (prograde)}, \\ -\varphi\left(t = -\frac{\tau}{\Omega}\right), & \text{if } W = W^- \text{ (retrograde)}, \end{cases} \tag{9}$$

$$p = \begin{cases} p\left(t = -\frac{\tau}{\Omega}\right), & \text{if } W = W^+ \text{ (prograde)}, \\ -p\left(t = -\frac{\tau}{\Omega}\right), & \text{if } W = W^- \text{ (retrograde)}, \end{cases} \tag{10}$$

where $\varphi(t)$ and $p(t)$ represent the explicit solution of equations of the unperturbed nonlinear pendulum (cf. e.g. (Wisdom, 1985)):

$$\cos\varphi(t) = 1 - 2k^2 \text{sn}^2(\omega_0 t),$$
$$\sin\varphi(t) = 2k \, \text{sn}^2(\omega_0 t)(1 - k^2\text{sn}^2(\omega_0 t))^{1/2},$$
$$p(t) = \frac{2\omega_0 k}{G} \, \text{cn}(\omega_0 t), \tag{11}$$

for libration, and

$$\cos\varphi(t) = \mathrm{cn}^2(\omega_r t) - \mathrm{sn}^2(\omega_r t),$$
$$\sin\varphi(t) = 2\,\mathrm{sn}(\omega_r t)\,\mathrm{cn}(\omega_r t),$$
$$p(t) = \frac{2\omega_r}{\mathcal{G}}\,\mathrm{dn}(\omega_r t), \tag{12}$$

for rotation; sn, cn, dn are Jacobi elliptic functions, $\omega_r = \frac{\omega_0}{k}$,

$$k = \begin{cases} \left(1 + \frac{w}{2}\right)^{1/2}, & \text{if } w < 0 \text{ (libration)}, \\ \left(1 + \frac{w}{2}\right)^{-1/2}, & \text{if } w > 0 \text{ (rotation)}, \end{cases} \tag{13}$$

is the elliptic modulus (compare with Eq. (8)).

One iteration of the SAM can produce several (or even many) projected points. To find all projected points for a given iteration, one should employ the following algorithm

> while $\Delta\tau > \tau$ do
> evaluate $\varphi$, $p$ by Eqs. (9, 10)
> $\tau := \tau + 2\pi$
> end do

Note that the input value of $\tau$ is taken moduli $2\pi$, while the increment $\Delta\tau$ and consequent values of $\tau$ are not. In words, the contents of the algorithm are as follows. It is verified whether the intersection condition $\Delta\tau > \tau$ is valid, and if yes, a projection is made. Then the interval $\tau$ is incremented by $2\pi$ and it is verified whether the intersection condition is still valid. If yes, the projection is accomplished once more with the new value of $\tau$, and one more phase point on the plane $(\varphi, p)$, $\tau = 0 \pmod{2\pi}$, is found. The cycle is repeated until $\Delta\tau \leq \tau$.

The described procedure is applied at each iteration of the SAM. Since it is based on the regular approximation of the motion on small time scales, it is called henceforth the regular projection algorithm (RPA).

## 4. The Spin-Orbit Resonance Problem

One of straightforward applications of the SAM concerns the planar resonant rotational motion of a non-symmetric satellite in an elliptic orbit. In this Section, it is shown how the SAM can be used to describe the motion in a vicinity of the separatrix of the synchronous spin-orbit resonance.

The orbit of the satellite is assumed to be a fixed ellipse. The vector of the angular momentum coincides with the axis of the maximum moment of inertia and is perpendicular to the orbit plane (the rotational motion is planar). The Hamiltonian

TABLE I

Satellites. Inertial, orbital and SAM parameters

| Satellite | $\omega_0$ | $e$ | Ref. | $\lambda$ | $W^+$ | $W^-$ |
|---|---|---|---|---|---|---|
| Phobos | 0.86 | 0.015 | Wisdom, 1987 | 1.163 | −0.286 | 0.0336 |
| Deimos | 0.81 | 0.0005 | Wisdom, 1987 | 1.235 | −0.00962 | 0.00118 |
| Amalthea | 1.14 | 0.003 | Wisdom, 1987 | 0.877 | −0.0509 | 0.00408 |
| Janus | $0.14^{1/2}$ | 0.009 | Goźd., 1997 | 2.673 | −0.0850 | 0.0121 |
| Epimetheus | 0.87 | 0.007 | Goźd., 1997 | 1.149 | −0.133 | 0.0155 |
| Pandora | 0.93 | 0.004 | Goźd. & Macie. 1995 | 1.075 | −0.0749 | 0.00818 |
| Prometheus | 1.17 | 0.004 | Goźd. & Macie. 1995 | 0.855 | −0.0668 | 0.00503 |
| Bif. case | 0.50 | 0.01 | − | 2 | −0.152 | 0.0214 |

of the problem (Wisdom *et al.*, 1984; Celletti, 1990), if the terms beyond first order in the eccentricity are ignored, has the form

$$H = \frac{y^2}{2} - \frac{\omega_0^2}{4}\cos(2x - 2t) - \frac{7e\omega_0^2}{8}\cos(2x - 3t) + \frac{e\omega_0^2}{8}\cos(2x - t), \quad (14)$$

where $x$ is the orientation of the satellite, i.e. the angle between the axis of the minimum moment of inertia and the line of apsides; $y = \dot{x} \equiv \frac{dx}{dt}$, and $t$ is time. The parameters are: the inertial parameter $\omega_0 = \left(\frac{3(B-A)}{C}\right)^{1/2}$, where $A < B < C$ are the principal moments of inertia of the satellite; and the eccentricity $e$ of the satellite's orbit. The time unit is equal to $\frac{1}{2\pi}$ of the orbital period.

By means of the canonical transformation $p = \frac{y-1}{2}$, $\varphi = 2(x - t)$, the Hamiltonian (14) is reducible to the paradigm (1) of the perturbed pendulum:

$$H = 2p^2 - \frac{\omega_0^2}{4}\cos\varphi - \frac{7e\omega_0^2}{8}\cos(\varphi - t) + \frac{e\omega_0^2}{8}\cos(\varphi + t). \quad (15)$$

The motion in a vicinity of the separatrix of the Hamiltonian (15) is described by the SAM (5). Comparing Eqs. (1) and (15), one has $\mathcal{F} = \frac{\omega_0^2}{4}$, $\mathcal{G} = 4$, $\Omega = 1$, $a = -\frac{7e\omega_0^2}{8}$, $b = \frac{e\omega_0^2}{8}$. Thus the SAM parameters $\lambda = \frac{1}{\omega_0}$ and $W^+$, $W^-$, given by Eqs. (6, 7), are determined by the inertial parameter $\omega_0$ and by the eccentricity $e$. Since $a = -7b$ (i.e. $\eta = -1/7$) the asymmetry of perturbation is intermediate; if $\lambda > 1$ then $W^+ \approx -7W^-$.

Available data on the inertial and orbital parameters of natural satellites of planets, as well as the calculated values of the SAM parameters, are presented in Table 1. In Fig. 1, the main chaotic layer computed by the SAM–RPA is shown on the plane $(x, y)$, $t = 0 \pmod{2\pi}$, for the case of Phobos. Note that the transformation from

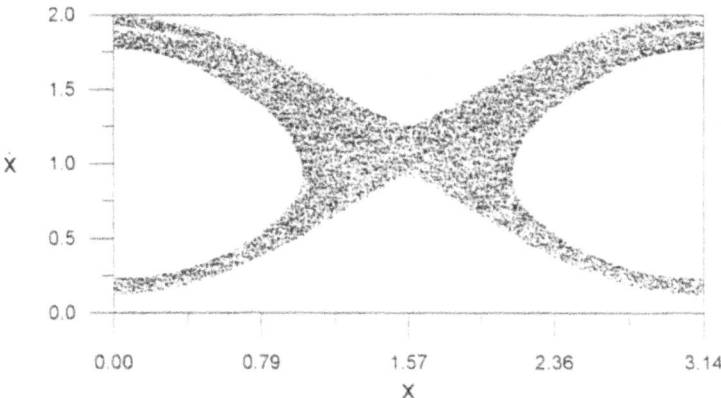

Fig. 1. Phobos. The main chaotic layer on the phase plane $(x, y)$, $t = 0$ (mod $2\pi$). Computed by means of the SAM–RPA.

$\varphi, p$ to $x$, $y$ at this plane is $x = \frac{\varphi}{2}$, $y = 1 + 2p$. In order to check the performance of the SAM–RPA, I constructed the corresponding section of phase space by means of a direct numeric integration and compared it to the phase portrait in Fig. 1. The integration was performed by the 8th order Dormand–Prince technique (Hairer *et al.*, 1987) with the local tolerance set to $10^{-10}$. Close, almost exact agreement between the SAM–RPA and the integration was observed. What is more, there is also close agreement with the corresponding surface of section obtained by a direct integration by Wisdom (Wisdom, 1987), cf. Fig. 1 of Ref. (Wisdom, 1987). Note that the SAM–RPA is a hundred times faster than usual numeric integrations.

Close agreement between the SAM–RPA and numeric integrations was observed for all satellites in Table 1. In Figs. 2, 3, the phase portraits are presented for Deimos and Janus, for which the surfaces of section were originally obtained by means of direct integrations by Wisdom (Wisdom, 1987) and Goździewski (1997); see Fig. 3 of Ref. (Wisdom, 1987) (Deimos) and Fig. 5a of Ref. (Goździewski, 1997) (Janus; note that the coordinate system in the latter figure is different).

Apart from fast construction of phase portraits, the SAM provides useful theoretical advantages in description of the phase space structure. Consider the location of spin-orbit resonances, e.g. the half-integer resonances 1/2 and 3/2. One has for the time averaged derivatives: $\langle \dot{\varphi} \rangle = 2 \langle y \rangle - 2$; therefore, these resonances both correspond to an integer resonance of the SAM. The resonance 1/2 corresponds to the retrograde, while 3/2 to the prograde resonant rotation of the model pendulum. This resonant rotation has winding number $Q = 1/\langle \dot{\varphi} \rangle = \pm 1$. If, say, $\langle y \rangle = 5/2$ then $Q = 1/3$.

The elliptic modulus $k^{(Q)}$ of the motion in the center of a resonance with winding number $Q$ can be found by means of numeric solution of the equation

$$\lambda k^{(Q)} K \left( k^{(Q)} \right) = \pi |Q|, \tag{16}$$

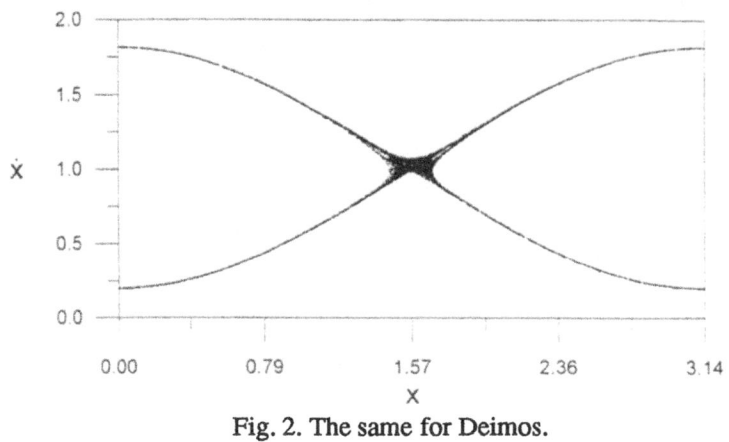

Fig. 2. The same for Deimos.

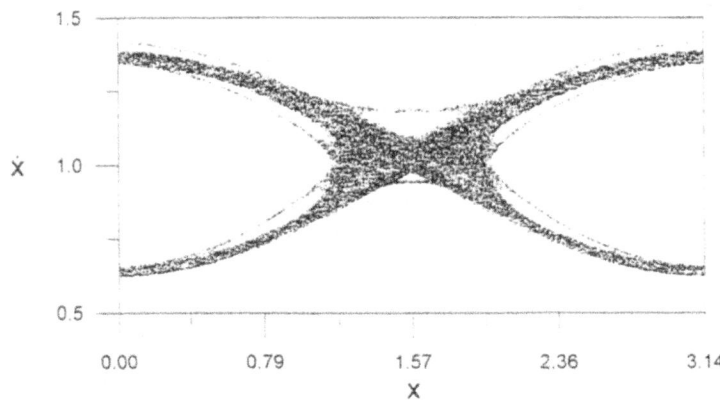

Fig. 3. The same for Janus.

following from Eq. (8); $\lambda = \frac{1}{\omega_0}$. It is efficiently solved by iterations. Then the energy deviation is $w^{(Q)} = 2\left(\left(k^{(Q)}\right)^{-2} - 1\right)$. On the other hand, the expression for $H_0$ (cf. Eq. (1)) is reducible to the relation

$$y = 1 \pm \frac{\omega_0}{2^{1/2}}(1 + w + \cos\varphi)^{1/2}. \tag{17}$$

The sign plus or minus refers to the prograde and retrograde motion respectively. In the coordinate $x$, the centers of resonances are located at $x = \pi/2$ (resonance 1/2) and at $x = 0$ (resonance 3/2), both mod $\pi$. Since $\varphi = 2x$ at $t = 0$, one has, via Eq. (17), the locations of centers of resonances $1/2$ and $3/2$ in the coordinate $y$:

$$y_{1/2} = 1 - \frac{\omega_0}{2^{1/2}}\left(w^{(1)}\right)^{1/2},$$
$$y_{3/2} = 1 + \frac{\omega_0}{2^{1/2}}\left(2 + w^{(1)}\right)^{1/2}. \tag{18}$$

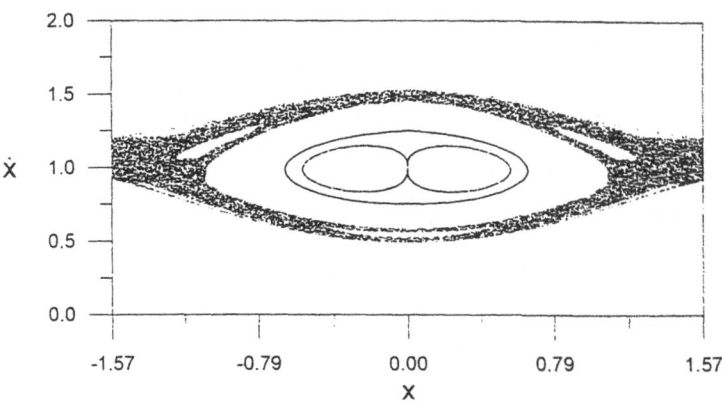

Fig. 4. A phase portrait in case of the resonance $\lambda = 2$ (see Table 1). Computed by means of the SAM–RPA. The period-doubling bifurcation of the synchronous resonance (the "∞" librational invariant curve in the center) is evident. Note that the coordinate $x$ is shifted for convenience.

Evaluations of $y_{1/2}$, $y_{3/2}$ by Eqs. (16, 18) can be compared with results (Wisdom *et al.*, 1984, Fig. 3), obtained by localization of the resonances by a direct numeric integration. Good correspondence is observed.

The localization of resonances gives just one example of the SAM theory applications. Besides, this theory allows one to precalculate locations of chaos borders on the phase plane; to predict the emergence of marginal resonances (i.e. prominent intermittent behaviour, see Ref. (Shevchenko, 1998)). Consider the external borders of the chaotic layer. The separatrix map can be linearized in $w$ to give the standard map; then the locations of the borders follow from conditions on the criticality of the parameter of the standard map (Chirikov, 1979). These considerations, as well as conditions for the emergence of marginal resonances (Shevchenko, 1998), are straightforwardly generalized to the case of the SAM.

The SAM allows one even to describe bifurcations of the synchronous resonance's center. In Fig. 4, a phase portrait in case of the resonance $\lambda = 2$ between the perturbation frequency and the frequency of the small-amplitude pendulum oscillations is presented. The relevant data of the model are in Table 1. The main chaotic layer and librational invariant curves near the center of the synchronous resonance are shown. The period-doubling bifurcation of the resonance's center clearly manifests itself, though the SAM is presumed to describe behaviour in a vicinity of the separatrix.

## 5. Conclusions

In this paper, the separatrix map (SM) in Chirikov's (Chirikov, 1979) form, describing the motion in a vicinity of the separatrix of the nonlinear resonance, is generalized to the case of asymmetric perturbation. The new map (5) is an algorithm containing conditional transfer statements.

Introduction of this separatrix algorithmic map (SAM) expands the area of applications of the SM. In particular, the SAM is directly applicable to the problem of the planar rotational dynamics of a non-symmetric satellite in an elliptic orbit. The motion in a neighbourhood of the separatrix of the synchronous spin-orbit resonance is reducible to the SAM. This neighbourhood is not at all a narrow region. It is usually large enough to engulf most important resonances other than the synchronous one.

The SAM, together with the regular projection algorithm (RPA), reproduces phase portraits of the motion which are in close agreement with well-known surfaces of section obtained by Wisdom et al. (Wisdom *et al.*, 1984), Wisdom (Wisdom, 1987), Goździewski (Goździewski, 1997) and others by means of numeric integration of equations of the rotational motion of non-symmetric natural satellites. The application of the SAM–RPA provides a hundred times advantage in the computation speed. The SAM theory allows one to analytically precalculate locations of resonances and chaos borders.

## Acknowledgements

The author is grateful to Christos Efthymiopoulos for useful remarks and comments on the manuscript. This work was partially supported by the Russian Foundation of Fundamental Research under Grant 97-01-01176.

## References

A. Celletti: 1990, Analysis of resonances in the spin-orbit problem in Celestial Mechanics: The synchronous resonance (Part I), *ZAMP*, **41**, 174–204.

B. V. Chirikov: 1979, A universal instability of many-dimensional oscillator systems, *Phys. Reports*, **52**, 263–379.

K. Goździewski: 1997, Rotational dynamics of Janus and Epimetheus, in: *Dynamics and Astrometry of Natural and Artificial Celestial Bodies*, I. M. Wytrzyszczak et al., eds, Kluwer Academic Publishers, Dordrecht, 269–274.

K. Goździewski and A. J. Maciejewski: 1995, On the gravitational fields of Pandora and Prometheus, *Earth, Moon and Planets*, **69**, 25–50.

E. Hairer, S. P. Nørsett and G. Wanner: 1987, *Solving Ordinary Differential Equations I: Nonstiff Problems*, Springer-Verlag, Berlin.

A. J. Lichtenberg and M. A. Lieberman: 1992, *Regular and Chaotic Dynamics*, Springer-Verlag, New York.

I. I. Shevchenko: 1998, Marginal resonances and intermittent behaviour in the motion in the vicinity of a separatrix, *Physica Scripta*, **57**, 185–191.

J. Wisdom: 1985, A perturbative treatment of motion near the 3/1 commensurability, *Icarus*, **63**, 272–289.

J. Wisdom: 1987, Rotational dynamics of irregularly shaped natural satellites, *Astron. J.*, **94**, 1350–1360.

J. Wisdom, S.J. Peale and F. Mignard: 1984, The chaotic rotation of Hyperion, *Icarus*, **58**, 137–152.

# CANONICAL VARIABLES OF THE SECOND KIND

## AND THE REDUCTION OF THE N-BODY PROBLEM

JOHN G. BRYANT

*47, avenue Felix Faure, 75015 Paris, France*

**Abstract.** We introduce a new kind of canonical variables that prove very useful when the order of a Hamiltonian system can be reduced by one, as in the case of isoenergetic reduction, and of what we call homogeneous reduction. The Kepler Problem, Geometrical Optics and McGehee Blow-up are discussed as examples. Finally we carry out the isoenergetic reduction of the general $N$-Body Problem using the new variables, and briefly discuss its application to the problem of collision.

## 1. The Definition of Canonical Variables of the Second Kind

When we consider a Hamiltonian function written in the standard form $H(p_i, q_i) = T(p_i) - U(q_i)$ (where the $(p_i, q_i)$ are a set of $2n$ canonical variables), and we want to reduce the corresponding Hamiltonian system thanks to the energy integral $H =$ constant $= h$, the question arises as to which variable we should choose to eliminate. The usual procedure is to eliminate one of the $p_i$, and the reduced system in the remaining variables is then a non-autonomous Hamiltonian system (with the role of the "time" variable now being played by the corresponding $q_i$). The drawback is that the individual $p_i$ as a rule do not have any intrinsic physical meaning (since they usually depend on the choice of reference frame). Instead of eliminating one of the $p_i$, it would clearly be more useful to eliminate the function $T(p_i)$ itself, considered as a single variable, and which generally does have a definite physical meaning that is independent of the choice of reference frame. This is especially true in problems like the $N$-Body Problem, where $T(p_i)$ becomes infinite every time the system approaches a collision (since then $U(q_i)$ becomes infinite).

In the second part of this paper, we will show how the elimination of $T(p_i)$ in the $N$-Body Problem can be accomplished. First of all we will consider the somewhat simpler case where $T(p_i)$ is written

$$T(p_i) = \frac{p^2}{2} \text{ where } p = \sqrt{\sum_{i=1}^{n} p_i^2}$$

i.e. we consider the hypothetical motion of a particle with mass 1 in $n$-dimensional space.

The basic idea is to incorporate $p$ in a new set of variables that, while no longer canonical in the traditional sense, still presents some remarkable properties. More precisely, we want to define a transformation

$$(p_i, q_i) \rightarrow (p, u, v, v_\lambda, w, w_\lambda) ; (\lambda = 1, \cdots, n-2)$$

*Celestial Mechanics and Dynamical Astronomy* **73**: 269–280, 1999.
© 1999 *Kluwer Academic Publishers.*

characterized by the property that the so-called Liouville 1-form (Libermann and Marie, 1987), $\pi = \sum_{i=1}^{n} p_i dq_i$ is written in the new variables

$$\pi = p\left(du + vdw + \sum_{\lambda=1}^{n-2} v_\lambda dw_\lambda\right)$$

where

$$p = \sqrt{\sum_{i=1}^{n} p_i^2} \; ; \; u = \frac{1}{p} \sum_{i=1}^{n} p_i q_i \; ; \; v = \frac{1}{p} \sqrt{\sum_{\substack{i,j=1 \\ i<j}}^{n} (p_j q_i - q_j p_i)^2}$$

Note that we have the relation $u^2 + v^2 = \sum_{i=1}^{n} q_i^2$.

We call the new set of variables Canonical Variables of the Second Kind (CVSK). Since the above transformation is not a canonical transformation, the Hamiltonian system no longer has its usual form in the new variables. As we will see however, the new system can be easily computed, and will still be designated as Hamiltonian.

We start with the simplest case $n = 2$. We want to define a transformation $(p_1, p_2, q_1, q_2) \rightarrow (p, u, v, w)$ such that the 1-form $\pi = p_1 dq_1 + p_2 dq_2$ can be written $\pi = p(du - vdw)$. The obvious choice is to define "polar coordinates" for the $p_i$, i.e. we define $w$ so that

$$p_1 = p \cos w \; ; \; p_2 = p \sin w$$

We can then write

$$\begin{aligned} \pi &= p_1 dq_1 + p_2 dq_2 \\ &= p(\cos w dq_1 + \sin w dq_2) \\ &= p[d(q_1 \cos w + q_2 \sin w) + (q_1 \sin w - q_2 \cos w)dw] \end{aligned}$$

Setting $u = q_1 \cos w + q_2 \sin w$ ; $v = q_1 \sin w - q_2 \cos w$ the transformation $(p_1, p_2, q_1, q_2) \rightarrow (p, u, v, w)$ is defined by the formulas

$$p = \sqrt{p_1^2 + p_2^2} \; ; \; \tan w = \frac{p_2}{p_1}$$

$$u = \frac{1}{p}(p_1 q_1 + p_2 q_2) \; ; \; v = \frac{1}{p}(p_2 q_1 - p_1 q_2)$$

Note that it is always possible to have $v > 0$, if need be by changing the order of the $(p_1, q_1)$ and the $(p_2, q_2)$. To write the Hamiltonian equations in the new variables, we make use of the fact that, although the above transformation is not a canonical transformation, the transformation $(p_1, p_2, q_1, q_2) \rightarrow (p, p' = pv, u, w)$ is canonical, so that the equations in these variables do have standard form. We can thus easily deduce from the Hamiltonian system defined by the Hamiltonian $H' = H'(p, p', u, w)$

$$\frac{dp}{dt} = -\frac{\partial H'}{\partial u} \quad ; \quad \frac{du}{dt} = \frac{\partial H'}{\partial p} \quad ; \quad \frac{dp'}{dt} = -\frac{\partial H'}{\partial w} \quad ; \quad \frac{dw}{dt} = \frac{\partial H'}{\partial p'}$$

the corresponding system in the $(p, u, v, w)$:

$$\frac{dp}{dt} = -\frac{\partial H}{\partial u} \qquad\qquad ; \qquad \frac{du}{dt} = \frac{\partial H}{\partial p} - \frac{v}{p}\frac{\partial H}{\partial v}$$

$$\frac{dv}{dt} = -\frac{1}{p}\left(\frac{\partial H}{\partial w} - v\frac{\partial H}{\partial u}\right) \quad ; \quad \frac{dw}{dt} = \frac{1}{p}\frac{\partial H}{\partial v}$$

where we set $H(p, u, v, w) = H'(p, pv, u, w)$. Although the above system no longer has standard form we still call it Hamiltonian. Note that when $\partial H/\partial w = 0$, $dp'/dt = d(pv)/dt = 0$ and $pv = $ constant.

The introduction of CVSK in the case $n = 2$ has several immediate applications.

*a) The Kepler Problem :* In variables $(p_1, p_2, q_1, q_2)$, the Hamiltonian of the Kepler Problem can be written

$$H(p_1, p_2, q_1, q_2) = \frac{p_1^2 + p_2^2}{2} - \frac{\alpha}{\sqrt{q_1^2 + q_2^2}} \qquad (\alpha = \text{ constant})$$

In variables $(p, u, v, w)$, $H$ is written

$$H(p, u, v, w) = \frac{p^2}{2} - \frac{\alpha}{\sqrt{u^2 + v^2}}$$

Since $H$ does not depend on $w$, we have the first integral $pv = $ constant $= C$. Setting $H = $ constant $= h$, and replacing $p$ by $C/v$, we have the following relation between $u$ and $v$

$$\frac{C^2}{2v^2} - \frac{\alpha}{\sqrt{u^2 + v^2}} = h$$

For a given value of $h$, we have a one-parameter flow in the $(u, v)$-space (the parameter being $C$). In the case $h < 0$, this gives the following figure

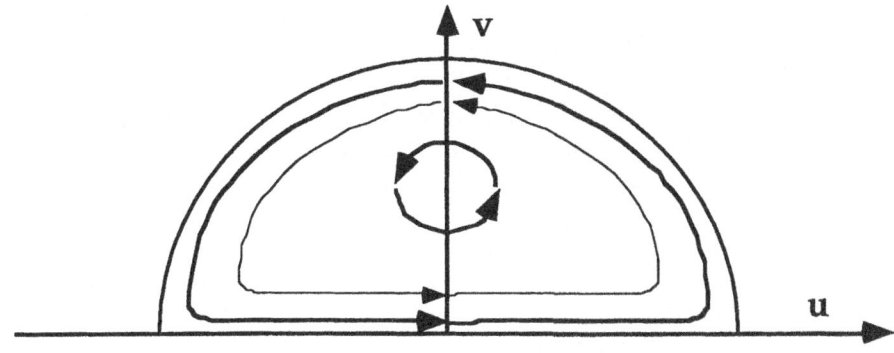

The flow along the $u$-axis, corresponding to $C = 0$, represents rectilinear motion. We see that the flow near the origin ($u = 0, v = 0$), which corresponds to collision, is parallel to the $u$-axis, i.e., the flow is regular. The introduction of CVSK leads to "instant" regularization.

*b) Geometrical Optics :* The propagation of a light ray in a medium with refractive index $n(q_1, q_2)$ is governed by Fermat's well-known Principle of Least Time. This is a variational principle and the paths of the ray are obtained from a Lagrangian system. It is not generally known that the paths of the light ray (as well as the motion in time) can also be obtained from the Hamiltonian system associated with the following *homogeneous* Hamiltonian

$$H(p_1, p_2, q_1, q_2) = \frac{c\sqrt{p_1^2 + p_2^2}}{n(q_1, q_2)}$$

where $c$ is the velocity of light in a vacuum. This is shown simply by comparing the two sets of Lagrangian and Hamiltonian equations. Introducing the CVSK and using the formulas $q_1 = u \cos w + v \sin w$ ; $q_2 = u \sin w - v \cos w$, we set $n(q_1, q_2) = \tilde{n}(u, v, w)$ and $H$ is written

$$\tilde{H}(p, u, v, w) = cp/\tilde{n}(u, v, w)$$

The corresponding Hamiltonian system is written

$$\frac{dp}{dt} = \frac{cp}{\tilde{n}^2}\frac{\partial \tilde{n}}{\partial u} \qquad ; \qquad \frac{du}{dt} = \frac{c}{\tilde{n}} + \frac{cv}{\tilde{n}^2}\frac{\partial \tilde{n}}{\partial v}$$

$$\frac{dv}{dt} = \frac{c}{\tilde{n}^2}\left(\frac{\partial \tilde{n}}{\partial w} - v\frac{\partial \tilde{n}}{\partial u}\right) ; \qquad \frac{dw}{dt} = -\frac{c}{\tilde{n}^2}\frac{\partial \tilde{n}}{\partial v}$$

We see that the last three equations do not contain $p$, i.e. the system *separates*. The equations in the $(u, v, w)$ determine what is known as a $3^{rd}$-order contact system (Arnold, 1978). It has the form

$$\frac{du}{dt} = K - v\frac{\partial K}{\partial v} ; \quad \frac{dv}{dt} = -\frac{\partial K}{\partial w} + v\frac{\partial K}{\partial u} ; \quad \frac{dw}{dt} = \frac{\partial K}{\partial v}$$

where $K(u, v, w) = \frac{c}{\tilde{n}(u,v,w)}$ is known as the Contact Hamiltonian, and is equal in this case to the velocity of the light ray. This reduction by an order of one of a homogeneous Hamiltonian system is a general property (Bryant, 1983) that becomes immediately apparent when we use CVSK. Note that this type of reduction, which we call *homogeneous* reduction is distinct from the standard isoenergetic reduction and does not involve any new first integral of the motion.

*c) McGehee Blow-up :* This is a more elaborate case of homogeneous reduction. The starting point is the Generalized Kepler Problem. In a system of extended polar coordinates $(p_r, p_\varphi, r, \varphi)$ defined by the formulas

$$r = \sqrt{q_1^2 + q_2^2} \; ; \; p_r = \frac{1}{r}(p_1 q_1 + p_2 q_2) \; ; \; p_\varphi = p_2 q_1 - p_1 q_2 \; ; \; \tan\varphi = \frac{q_2}{q_1}$$

we consider the Hamiltonian

$$H(p_r, p_\varphi, r, \varphi) = \frac{1}{2}\left(p_r^2 + \frac{p_\varphi^2}{r^2}\right) - \frac{\mu(\varphi)}{r}$$

Depending on the form of the function $\mu(\varphi)$, the above Hamiltonian can represent the anisotropic Kepler problem, the rectilinear 3-Body Problem, the Planar Isoceles 3-Body Problem, and, when $\mu = $ constant, the ordinary Kepler Problem. The homogeneous nature of the potential function can be interpreted geometrically thanks to the vector field

$$Y' = -p_r \frac{\partial}{\partial p_r} + 2r \frac{\partial}{\partial r} + p_\varphi \frac{\partial}{\partial p_\varphi}$$

for which we have

$$Y'.H = -p_r \frac{\partial H}{\partial p_r} + 2r \frac{\partial H}{\partial r} + p_\varphi \frac{\partial H}{\partial p_\varphi} = -2H \; .$$

Associated with $Y'$ is the $1-$form

$$\pi' = -p_r dr - 2r dp_r + p_\varphi d\varphi$$

which, like the standard Liouville $1-$form (whose expression in polar coordinates is $\pi = p_r dr + p_\varphi d\varphi$), verifies $d\pi' = \omega$, where $\omega = dp_r \wedge dr + dp_\varphi \wedge d\varphi$ is the symplectic 2-form. The idea is to have $\pi'$ play the role that $\pi$ has played up till now. The first step is to choose a set of variables $(p_1', p_2', q_1', q_2')$ such that $\pi'$ has the form $\pi' = p_1' dq_1' + p_2' dq_2'$. We can choose (Bryant, 1980)

$$p_1' = 2\sqrt{r} \; ; \; p_2' = p_\varphi \; ; \; q_1' = -p_r \sqrt{r} \; ; \; q_2' = \varphi$$

In these variables $Y'$ is written $Y' = p_1' \frac{\partial}{\partial p_1'} + p_2' \frac{\partial}{\partial p_2'}$ and $Y'.H = p_1' \frac{\partial H}{\partial p_1'} + p_2' \frac{\partial H}{\partial p_2'} = -2H$ means that $H$ is homogeneous of degree -2 in the $(p_1', p_2')$. The second step is to define CVSK for $\pi'$, and for this we set

$$p' = p_1' = 2\sqrt{r} \; ; \; u' = q_1' = -p_r \sqrt{r} \; ; \; v' = \frac{p_2'}{p_1'} = \frac{p_\varphi}{2\sqrt{r}} \; ; \; w' = q_2' = \varphi$$

so that $\pi' = p'(du' + v' dw')$. Note that in these variables, $Y' = p' \frac{\partial}{\partial p'}$, and $Y'.H = p' \frac{\partial H}{\partial p'} = -2H$ means that $H$ is homogeneous of degree -2 in $p'$. We now in fact have

$$H(p', u', v', w') = \frac{4}{p'^2}\left(\frac{u'^2}{2} + 2v'^2 - \mu(w')\right)$$

The Hamiltonian equations are written

$$\frac{dp'}{dt} = -\frac{\partial H}{\partial u'} = -\frac{4u'}{p'^2} \; ; \; \frac{du'}{dt} = \frac{\partial H}{\partial p'} - \frac{v'}{p'}\frac{\partial H}{\partial v'} = -\frac{4}{p'^3}(u'^2 + 8v'^2 - 2\mu(w')) \; ;$$

$$\frac{dv'}{dt} = -\frac{1}{p'}\left(\frac{\partial H}{\partial w'} - v'\frac{\partial H}{\partial u'}\right) = \frac{4}{p'^3}\left(\frac{d\mu}{dw'} + u'v'\right) \; ; \; \frac{dw'}{dt} = \frac{1}{p'}\frac{\partial H}{\partial v'} = \frac{16v'}{p'^3}$$

After the time change $dt' = 8p'^{-3}dt(= r^{-3/2}dt)$, the equations in $u', v', w'$, which no longer depend on $p'$, are equivalent to McGehee's equations (McGehee, 1974). We see that they result from the homogeneous reduction of the Generalized Kepler Problem, obtained by the introduction of a system of CVSK adapted to the 1-form $\pi'$.

In the case of the ordinary Kepler Problem ($\mu =$ constant), an interesting comparison can be made with the isoenergetic reduction given beforehand. As before, the problem can be reduced to a 2-dimensional flow, this time in the $(u', v')$ space. Using the first integrals $H = h$ ; $p'v' = C$ to eliminate $p'$, we obtain a 1-parameter family of curves given by the equation $v'^2(2u'^2 + 8v'^2 - 4\mu) = k(= hC^2)$. When $k < 0$, we have the following figure

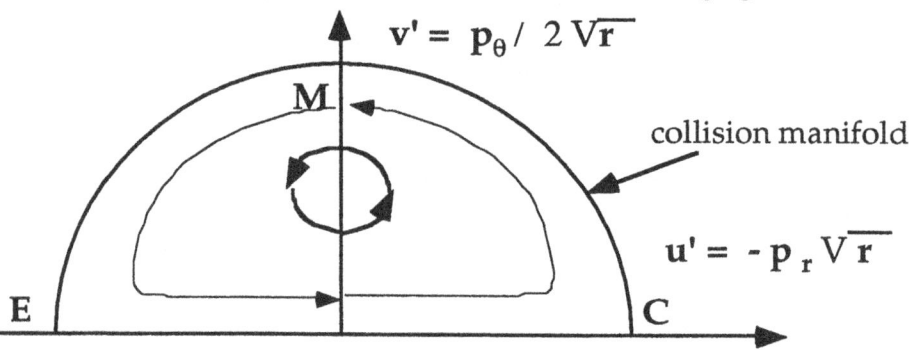

A near-collision orbit remains close to the $u'$-axis until it is near point $C$ when it follows the "collision manifold", i.e. the limiting curve $2u'^2 + 8v'^2 - 4\mu = 0$, and reaches closest approach at point $M$ on the $v'$-axis. McGehee variables entail a "blow-up" of the origin of physical space and allow a detailed description of near-collision orbits. They do not however supply us with a "natural" extension of actual collision orbits, which correspond to the flow on the $u'$-axis. Note that the flow near the origin ($u' = 0$ ; $v' = 0$) now corresponds to the maximum value of $r$, and no longer has anything to do with collision.

We now consider the case $n = 3$. In order to define a system of CVSK, we make use of an intermediate set of canonical variables. These are obtained thanks to the well-known Andoyer transformation (Boigey, 1981). This transformation goes from the variables $(p_1, p_2, p_3, q_1, q_2, q_3)$ to the Andoyer variables

$(p'_1, p'_2, p'_\beta, q'_1, q'_2, \beta')$ and is defined by the formulas

$$p_1 = p'_1 \cos \beta' - p'_2 \sin \beta' \cos i' \quad q_1 = q'_1 \cos \beta' - q'_2 \sin \beta' \cos i'$$
$$p_2 = p'_1 \sin \beta' + p'_2 \cos \beta' \cos i' \quad q_2 = q'_1 \sin \beta' + q'_2 \cos \beta' \cos 1'$$
$$p_3 = p'_2 \sin i' \qquad\qquad q_3 = q'_2 \sin i'$$

where by definition $\cos i' = p'_\beta / (p'_2 q'_1 - p'_1 q'_2)$. It is easily checked that the transformation verifies $\pi = p_1 dq_1 + p_2 dq_2 + p_3 dq_3 = p'_1 dq'_1 + p'_2 dq'_2 + p'_\beta d\beta'$, i.e. it is a canonical transformation. We have the following geometrical interpretation

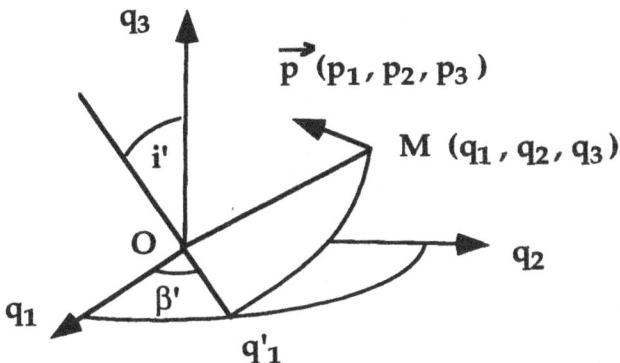

The $Oq'_1$-axis lies on the intersection of the horizontal $Oq_1 q_2$-plane with the so-called plane of instantaneous motion determined by $\alpha OM(q_1, q_2, q_3)$ and $\mathbf{p}(p_1, p_2, p_3)$. The $Oq'_2$-axis is perpendicular to the $Oq'_1$-axis in the plane of instantaneous motion. We immediately verify that

$$\|\mathbf{p}\|^2 = p^2 = \sum_{i=1}^3 p_i^2 = p_1'^2 + p_2'^2 \; ; \; \mathbf{OM.p} = \sum_{i=1}^3 p_i q_i = p'_1 q'_1 + p'_2 q'_2$$

$$\|\mathbf{OM}\|^2 = \sum_{i=1}^3 q_i^2 = q_1'^2 + q_2'^2 \; ; \; \|\mathbf{OM} \times \mathbf{p}\| = \sqrt{\sum_{\substack{i,j=1 \\ i<j}}^3 (p_j q_i - q_j p_i)^2} = p'_2 q'_1 - p'_1 q'_2$$

All we now have to do is apply to the $(p'_1, p'_2, q'_1, q'_2)$ the transformation defined in the case $n = 2$, i.e.

$$p = \sqrt{p_1'^2 + p_2'^2} \; ; \; \tan w = \frac{p'_2}{p'_1} \; ; \; u = \frac{1}{p}(p'_1 q'_1 + p'_2 q'_2) \; ; \; v = \frac{1}{p}(p'_2 q'_1 - p'_1 q'_2)$$

The $(p, u, v, w)$ along with $v_1 = \frac{p'_\beta}{p}$ and $w_1 = \beta'$ verify

$$\pi = p'_1 dq'_1 + p'_2 dq'_2 + p'_\beta d\beta' = p(du + vdw + v_1 dw_1)$$

as well as the conditions on $p, u$ and $v$ given at the beginning of this paper, and therefore determine a system of CVSK.

For each successive value of $n$, we must first of all define an intermediate set of variables which are in fact generalized Andoyer variables. These are obtained by iteration of the standard Andoyer transformation. In the case $n = 4$ for example, we define a transformation $(p_1, p_2, p_3, p_4, q_1, q_2, q_3, q_4) \rightarrow (p_1'', p_2'', p_\beta'', p_\beta'', q_1'', q_2'', \beta', \beta'')$ by two iterations of the Andoyer transformation. The first one, which leaves $p_4$ and $q_4$ invariant and goes from the $(p_1, p_2, p_3, q_1, q_2, q_3)$ to the $(p_1', p_2', p_\beta', q_1', q_2', \beta')$, is the transformation given previously. The second Andoyer transformation, which leaves $p_\beta'$ and $\beta'$ invariant, goes from the $(p_1', p_2', p_4, q_1', q_2', q_4)$ to the $(p_1'', p_2'', p_\beta'', q_1'',q_2'', \beta'')$, and is defined by the formulas

$$p_1' = p_1'' \cos \beta'' - p_2'' \sin \beta'' \cos i'' \qquad q_1' = q_1'' \cos \beta'' - q_2'' \sin \beta'' \cos i''$$
$$p_2' = p_1'' \sin \beta'' + p_2'' \cos \beta'' \cos i'' \qquad q_2' = q_1'' \sin \beta'' + q_2'' \cos \beta'' \cos i''$$
$$p_4 = p_2'' \sin i'' \qquad\qquad q_4 = q_2'' \sin i''$$

where by definition $\cos i'' = p_\beta'' / (p_2'' q_1'' - p_1'' q_2'')$. It is easily checked that the total transformation is canonical, since

$$\pi = (p_1 dq_1 + p_2 dq_2 + p_3 dq_3) + p_4 dq_4 = (p_1' dq_1' + p_2' dq_2' + p_\beta' d\beta') + p_4 dq_4$$
$$= (p_1' dq_1' + p_2' dq_2' + p_4 dq_4) + p_\beta' d\beta' = (p_1'' dq_1'' + p_2'' dq_2'' + p_\beta'' d\beta'') + p_\beta' d\beta'$$

and that the following relations hold

$$p^2 = \sum_{i=1}^{4} p_i^2 = p_1'^2 + p_2'^2 + p_4^2 = p_1''^2 + p_2''^2 \,,$$

$$\sum_{i=1}^{4} q_i^2 = q_1'^2 + q_2'^2 + q_4^2 = q_1''^2 + q_2''^2$$

$$\sum_{i=1}^{4} p_i q_i = p_1' q_1' + p_2' q_2' + p_4 q_4 = p_1'' q_1'' + p_2'' q_2''$$

$$\sum_{\substack{i,j=1 \\ i<j}}^{4} (p_j q_i - q_j p_i)^2 = (p_2' q_1' - p_1' q_2')^2 + (p_4 q_2' - p_2' q_4)^2 + (p_4 q_1' - p_1' q_4)^2$$

$$= (p_2'' q_1'' - p_1'' q_2'')^2$$

To determine a set of CVSK, we simply apply to the $(p_1'', p_2'', q_1'', q_2'')$ the same transformation as in the case $n = 2$, i.e.

$$p = \sqrt{p_1''^2 + p_2''^2} \; ; \quad \tan w = \frac{p_2''}{p_1''} \; ; \quad u = \frac{1}{p}(p_1'' q_1'' + p_2'' q_2'') \; ; \quad v = \frac{1}{p}(p_2'' q_1'' - p_1'' q_2'')$$

and add the variables $v_1 = p'_\beta/p$ ; $v_2 = p''_\beta/p$ ; $w_1 = \beta'$ ; $w_2 = \beta''$, so that $\pi = p(du + v\,dw + v_1\,dw_1 + v_2\,dw_2)$. The Hamiltonian system can be easily computed, and is written in the new variables

$$\frac{dp}{dt} = -\frac{\partial H}{\partial u} \; ; \; \frac{du}{dt} = \frac{\partial H}{\partial p} - \frac{v}{p}\frac{\partial H}{\partial v} - \frac{v_1}{p}\frac{\partial H}{\partial v_1} - \frac{v_2}{p}\frac{\partial H}{\partial v_2} \; ;$$

$$\frac{dv}{dt} = -\frac{1}{p}\left(\frac{\partial H}{\partial w} - v\frac{\partial H}{\partial u}\right) \; ; \; \frac{dv_1}{dt} = -\frac{1}{p}\left(\frac{\partial H}{\partial w_1} - v_1\frac{\partial H}{\partial u}\right)$$

$$\frac{dv_2}{dt} = -\frac{1}{p}\left(\frac{\partial H}{\partial w_2} - v_2\frac{\partial H}{\partial u}\right)$$

$$\frac{dw}{dt} = \frac{1}{p}\frac{\partial H}{\partial v} \; ; \; \frac{dw_1}{dt} = \frac{1}{p}\frac{\partial H}{\partial v_1} \; ; \; \frac{dw_2}{dt} = \frac{1}{p}\frac{\partial H}{\partial v_2}$$

Although the transformation from the $(p_1, p_2, p_3, p_4, q_1, q_2, q_3, q_4)$ to the generalized Andoyer variables $(p''_1, p''_2, p'_\beta, p''_\beta, q_1, q_2, \beta', \beta'')$ is not easy to write explicitly, it does receive a simple geometrical interpretation, thanks to the following figure

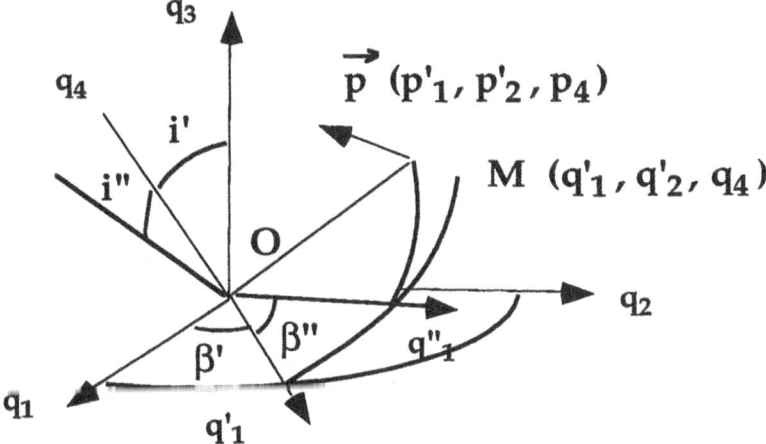

The $0q_4$-axis is perpendicular to the $0q'_1q'_2$ plane. The $0q''_1$-axis lies on the intersection of the $0q'_1q'_2$-plane with the "pseudo plane of instantaneous motion" determined by $\alpha OM(q'_1, q'_2, q_4)$ and $\mathbf{p}(p'_1, p'_2, p_4)$. The $0q''_2$-axis is perpendicular to the $0q''_1$-axis in the pseudo plane of instantaneous motion.

The generalization of the above transformation to arbitrary $n$ is accomplished by the iteration of the Andoyer transformation the number of times necessary, and then applying the transformation given at the beginning of this paper to the generalized Andoyer variables in order to obtain our system of CVSK.

## 2. The Reduction of the $N$-Body Problem

We take as our starting point the $N$-Body Problem formulated in Jacobi variables, for which the elimination of the center of mass motion is immediate, and allows us to interpret the reduced problem as the motion of $N-1$ fictitious bodies with Jacobi reduced masses in 3-dimensional space (Whittaker, 1927). Designating the Jacobi coordinates of each body and their conjugate momenta by $q_{\alpha i}, p_{\alpha i} (\alpha = 1, \cdots, N-1 ; i = 1, 2, 3)$ and the corresponding reduced mass by $m_\alpha$, the Hamiltonian function has the form

$$H(p_{\alpha i}, q_{\alpha i}) = T(p_{\alpha i}) - U(q_{\alpha i}) \text{ where } T(p_{\alpha i}) = \sum_{\alpha=1}^{N-1} \frac{p_\alpha^2}{2m_\alpha} \text{ and } p_\alpha^2 = \sum_{i=1}^{3} p_{\alpha i}^2$$

The Liouville 1-form is written $\pi = \sum_{\alpha=1}^{N-1} \pi_\alpha$ with $\pi_\alpha = \sum_{i=1}^{3} p_{\alpha i} dq_{\alpha i}$. We can eliminate the masses from the problem by "normalizing" the $(p_{\alpha i}, q_{\alpha i})$, i.e. by replacing them with $\tilde{p}_{\alpha i} = (1/\sqrt{m_\alpha})p_{\alpha i}$ ; $\tilde{q}_{\alpha i} = \sqrt{m_\alpha} q_{\alpha i}$. This simplifies the expression of $T \left(= \frac{1}{2} \sum_\alpha \tilde{p}_\alpha^2\right)$ and preserves the form of the $\pi_\alpha \left(= \sum_i \tilde{p}_{\alpha i} d\tilde{q}_{\alpha i}\right)$. (So as to not overburden the notation, we will keep on using $(p_{\alpha i}, q_{\alpha i})$ and assume normalization has already been carried out).

The definition of a system of CVSK is a two-step process. This is so we can take account of the fact that the variables come in $N-1$ groups, and that the problem is invariant by rotation (i.e. there exists an angular momentum integral). The first step is to define CVSK for *each one* of the $N-1$ sets of variables $(p_{\alpha i}, q_{\alpha i})$. This means replacing them by $N-1$ sets $(p_\alpha, u_\alpha, v_\alpha, v_{1\alpha}, w_\alpha, w_{1\alpha})$ according to the transformation corresponding to the case $n = 3$. We therefore have

$$\pi_\alpha = \sum_{i=1}^{3} p_{\alpha i} dq_{\alpha i} = p_\alpha(du_\alpha + v_\alpha dw_\alpha + v_{1\alpha} dw_{1\alpha})$$

along with the usual relations

$$p_\alpha^2 = \sum_{i=1}^{3} p_{\alpha i}^2 \qquad ; \quad u_\alpha = \frac{1}{p_\alpha} \sum_{i=1}^{3} p_{\alpha i} q_{\alpha i} ,$$

$$v_\alpha = \frac{1}{p_\alpha} \sqrt{\sum_{\substack{i,j=1\\i<j}}^{3} (p_{\alpha j} q_{\alpha i} - q_{\alpha j} p_{\alpha i})^2} \quad ; \quad u_\alpha^2 + v_\alpha^2 = \sum_{i=1}^{3} q_{\alpha i}^2$$

We can write

$$\pi = \sum_{\alpha=1}^{N-1} \pi_\alpha = \sum_{\alpha=1}^{N-1} p_\alpha du_\alpha + \sum_{\alpha=1}^{N-1} (p_\alpha v_\alpha) dw_\alpha + \sum_{\alpha=1}^{N-1} (p_\alpha v_{1\alpha}) dw_{1\alpha}$$

The second step is to define a system of CVSK *for the* $(p_\alpha, u_\alpha)$, i.e. we must define a system of $2(N-1)$ variables $(p, \overline{u}, \overline{v}, \overline{v}_\lambda, \overline{w}, \overline{w}_\lambda), (\lambda = 1, \cdots, N-3)$, such that

$$\pi = \sum_{\alpha=1}^{N-1} p_\alpha du_\alpha = p(d\bar{u} + \bar{v}d\bar{w} + \sum_{\lambda=1}^{N-3} \bar{v}_\lambda d\bar{w}_\lambda)$$

$$p^2 = \sum_{\alpha=1}^{N-1} p_\alpha^2 \; ; \; \bar{u} = \frac{1}{p} \sum_{\alpha=1}^{N-1} p_\alpha u_\alpha$$

$$\bar{v} = \frac{1}{p} \sqrt{\sum_{\substack{\alpha,\beta=1 \\ \alpha<\beta}}^{N-1} (p_\beta u_\alpha - u_\beta p_\alpha)^2} \; ; \; \bar{u}^2 + \bar{v}^2 = \sum_{\alpha=1}^{N-1} u_\alpha^2$$

To accomplish this, we simply make use of the general transformation with the $u_\alpha$ in the place of the $q_i$, and where $n = N - 1$. Having thus defined the $(p, \bar{u}, \bar{v}, \bar{v}_\lambda, \bar{w}, \bar{w}_\lambda)$, we add the variables $\tilde{v}_\alpha = p_\alpha v_\alpha/p$ ; $\tilde{v}_{1\alpha} = p_\alpha v_{1\alpha}/p$, and the Liouville 1-form $\pi$ takes on the desired form

$$\pi = p \left( d\bar{u} + \bar{v}d\bar{w} + \sum_{\lambda=1}^{N-3} \bar{v}_\lambda d\bar{w}_\lambda + \sum_{\alpha=1}^{N-1} \tilde{v}_\alpha dw_\alpha + \sum_{\alpha=1}^{N-1} \tilde{v}_{1\alpha} dw_{1\alpha} \right)$$

Note that it is possible to give a simple expression to the length $C$ of the angular momentum vector. For this we must assume that this vector lies along the vertical axis. We then have the classic result that $C$ is the sum of the projections of the angular momentum of each of the $N - 1$ bodies on the vertical axis, i.e. $C = \sum_\alpha p_\alpha v_{1\alpha} = p \sum_\alpha \tilde{v}_{1\alpha}$.

The $(q_{\alpha i})$, as well as the potential function $U(q_{\alpha i})$, when expressed in terms of the new variables, do not depend explicitly on $p$. To see this, we consider the so-called homothetical vector field associated with the Liouville 1-form, which has the following expression in terms of the $(p_{\alpha i}, q_{\alpha i})$

$$Y = \sum_{\alpha=1}^{N-1} \sum_{i=1}^{3} p_{\alpha i} \frac{\partial}{\partial p_{\alpha i}}$$

We therefore have $Y.q_{\alpha i} = 0$. Expressed in the new variables, $Y$ is written simply $Y = p\frac{\partial}{\partial p}$, and the relation $Y.q_{\alpha i} = p\frac{\partial q_{\alpha i}}{\partial p} = 0$ means that $q_{\alpha i}$ and hence $U(q_{\alpha i})$ do not depend on $p$. It follows that since $T(p_{\alpha i}) = \frac{1}{2}p^2$, the energy integral $H = T - U = \text{constant} = h$ can be easily solved for $p$, i.e.

$$p = \sqrt{2(U + h)}$$

Replacing $p$ by the above expression in the Hamiltonian equations effectively carries out the isoenergetic reduction of the system.

When the motion approaches collision, $U(q_{\alpha i})$ becomes infinite, and, because of the above relation, so does $p$. A remarkable feature of the CVSK defined for the problem is that all of the other variables remain finite. This is a direct consequence of the method we used to construct the system of CVSK. Therefore, once $p$ has been eliminated, collision and near-collision orbits in the reduced problem belong to a finite domain in the vicinity of collision.

The above property leads to a "practical" method for extending total collision orbits. A classical property of total collision orbits is that the length $C$ of the angular momentum is zero. The idea for extending such orbits through collision is to incorporate them in a one-parameter continuous family of orbits, the parameter being $C$. The only total collision orbit in the family is therefore the one where $C = 0$, which appears as a limiting case since we assume $C \geq 0$. If none of the orbits of the family exhibits a lower-order collision, all the orbits except the total collision one are well-defined and finite. A limiting orbit for the family therefore always exists when $C \to 0$, and this is the orbit we adopt as the extension of the collision orbit. Note that this method only extends the geometrical orbits through collision, but not the actual Hamiltonian system (i.e. it has not been regularized).

In a subsequent paper we will describe the above method for extending total collision orbits in more detail, and give several concrete examples. We will also show how to handle lower-order collisions.

## References

Arnold, V.I.:1978, *Mathematical methods of classical mechanics*, Graduate texts in mathematics, 60, Springer Verlag.

Boigey, F.: 1981, Transformations canoniques a variables imposées,. Applications a la mecanique celeste, These, Universite P. et M. Curie (Paris), .

Bryant, J.: 1983, Le formalisme de contact en mecanique classique et relativiste, *Ann. Inst. Henri Poincare, Physique Theorique*, **38**, 121–152

Bryant, J.: 1980, Sur une reduction du probleme des $N$-corps mettant en evidence une structure de contact, *C.R. Acad. Sc. Paris*, **291 A**, 205–208

Libermann, P. and Marie, C-M.: 1987, *Symplectic geometry and analytical mechanics*, Kluwer Academic, .

McGehee, R.: 1974, Triple collision in the collinear three-body problem, *Inv. Math*, **27**, 191–227

Whittaker, E.T.: 1927, *A treatise on the analytical dynamics of particles and rigid bodies*, Cambridge University Press

# MOTION NEAR THE UNIT CIRCLE IN THE THREE-BODY PROBLEM

ROGER A. BROUCKE

*Department of Aerospace Engineering, University of Texas, Austin, TX 78712*

## 1. Introduction

Many of the important applications of the circular planar restricted problem of three bodies involve motion in the vicinity of the unit circle, (as defined in canonical units). It is then of interest to develop simplified models which are valid in this region. These models preserve the gross characteristics of the original system but they possess simpler equations of motion.

We will also show that several simplified models can be seen as a perturbation of a very well known simple linear system: the Clohessy-Wiltshire equations used by NASA in all their rendezvous operations. These are actually very close to the well-known Hill problem. We will thus consider the Restricted problem as a perturbed Hill or Clohessy-Wiltshire problem. We also introduce the Clohessy-Wiltshire Lagrangian in polar coordinates.

Actually we use polar coordinates throughout, in a rotating frame with canonical units of length and time. In this frame, the larger primary is at the origin, while the smaller primary is at (1.0). We are especially interested in the motion near the unit circle, $(r = 1)$. This type of situation has numerous applications, such as for instance the Trojan Asteroids and the classification of orbits near the triangular points (such as "tadpoles" and "horseshoes"; see Dermott and Murray, 1981a). Many extensive numerical investigations of the manifolds of periodic orbits in the vicinity of the triangular points were made in the sixties by Deprit (1966) and Deprit and Henrard (1969) as well as Deprit, Henrard, Palmore and Price (1969).

Another motivation that prompted us to perform a systematic study of simplified models of three-body problem is the work of Garfinkel (1975, 1977, 1978) on the Ideal Resonance problem; (see also Giacaglia, 1976).

One of the newer applications is in the dynamics of the coorbital satellites of Saturn, discovered by the Voyager Spacecraft. The theory of the coorbitals was studied extensively, for instance by Dermott and Murray (1981b), by Harrington and Seidelmann (1981), by Yoder and Colombo as well as by Salo and Yoder (1988). The dynamics of the coorbitals is without doubt a unit-circle phenomenon, although the igeneral problemi is a better model than the Restricted problem of three bodies, (Konopliv, 1986).

One of the main results of the present paper is that we discovered a new symmetry property that exists in some of our simplified models; the symmetry with respect to the unit-circle. Using this symmetry theorem in the model described at the end of section 8, with the potential (8.6), we were able to compute horseshoe periodic orbits as well as the long and short-period periodic orbits that exist around $L_4$ and $L_5$ as well as the short period orbits around $L_3$. It is actually an amazing

*Celestial Mechanics and Dynamical Astronomy* **73**: 281–290, 1999.
© 1999 *Kluwer Academic Publishers.*

result that many of our rather strongly mutilated models of the restricted problem still have three classical libration points $L_3, L_4$ and $L_5$. Some of these models should be investigated for the existence of Brownis conjectures.

## 2. The Restricted Problem of Three Bodies

The equations will be derived using polar coordinates in which r is measured from the larger primary and $\theta$ is an angle in the rotating $xy$ system.

The Lagrangian per unit mass is

$$L = \frac{1}{2}(\dot{r}^2 + r^2\dot{\theta}^2) + r^2\dot{\theta} + \frac{r^2}{2} + \frac{1}{r} + \mu \left[\frac{1}{d} - r\cos\theta - \frac{1}{r}\right] . \tag{1}$$

where

$$d^2 = 1 + r^2 - 2r\cos\theta . \tag{2}$$

The above Lagrangian has been written in the form of two parts, the second part being proportional to the mass $\mu$ of the smaller primary (located at 1,0) which will often be considered as a small quantity:

If $\mu$ is set equal to zero, we obtain the Lagrangian of the classical Kepler problem, referred to a uniformly rotating frame of reference. So the Lagrangian made up of the first five terms will be called the Keplerian Lagrangian.

## 3. Motion near theUnit Circle; the First Approximate Model

In order to investigate motion in the vicinity of the unit circle, we will let $r = 1 + \varepsilon$ where $\varepsilon \ll 1$. In what follows we will thus expand the Lagrangian of the restricted problem, to different orders in $\varepsilon$. In some instances we will expand the 5 Keplerian terms to a different order from the 3 remaining perturbation terms in $\mu$. This will allow us to obtain a large variety of Lagrangians which may not be very accurate representations of the true restricted problem, but which may show us some important connections with other well-known dynamical systems, such as the simple pendulum or the Clohessy-Wiltshire system for instance.

In the expansions that will be made the three quantities $\varepsilon, \dot{\varepsilon}$ and $\dot{\theta}$ will usually be considered as small quantities of the same order of magnitude. The angle $\theta$ will be considered as a large unconstrained number. Let us first consider the expansion of the five Keplerian terms, say up to order 4 in the three variables $\varepsilon, \dot{\varepsilon}$ and $\dot{\theta}$. We have first of all

$$= 1 - \varepsilon + \varepsilon^2 - \varepsilon^3 + \varepsilon^4 . \tag{3}$$

The expansion of the Keplerian Lagrangian is thus

$$L = \left[\frac{1}{2}(\dot{\varepsilon}^2 + \dot{\theta}^2) + 2\varepsilon\dot{\theta} + \frac{3}{2}\varepsilon^2\right] + [\varepsilon\dot{\theta}^2 + \varepsilon^2\dot{\theta} - \varepsilon^3]\left[\frac{1}{2}\varepsilon^2\dot{\theta}^2 + \varepsilon^4\right] , \tag{4}$$

where the different brackets contain respectively the terms of order 2, order 3 and order 4. Note that we have dropped a term $\dot{\theta}$ which is an exact derivative; and that the linear terms in $\varepsilon$ canceled each other out. The Lagrangian consisting of the first four quadratic terms will be called the Clohessy-Wiltshire Lagrangian. It will be studied in more detail later.

The expression for the distance $d$ between the smallest primary (of mass $\mu$) and the massless satellite is given by equation (2.2).

Assuming that $r = 1 + \varepsilon$, it is easy to expand $1/d$ in powers of $\varepsilon$. We start from the exact formula

$$d^2 = 2(1 - \cos\theta)\left(1 + \varepsilon + \frac{\varepsilon^2}{2(1 - \cos\theta)}\right) . \tag{5}$$

The expression in the last pair of parentheses can now be expanded in $\varepsilon$, with the use of the binomial formula (to the second order in $\varepsilon$):

$$\frac{1}{d} = \frac{1}{\sqrt{2(1 - \cos\theta)}} \cdot \left[1 - \frac{\varepsilon}{2} + \frac{\varepsilon^2}{8}\left(3 - \frac{2}{1 - \cos\theta}\right)\right] . \tag{6}$$

### 4. The Unperturbed Clohessy-Wiltshire Lagrangian

A very interesting point in our present formulation of the restricted three-body problem in polar coordinates is its connection with the so-called Clohessy-Wiltshire equations which are well known in the space program, in relation with circular relative motions and rendezvous operations. If in the previous Lagrangians for the restricted problem, we neglect all the perturbation terms in $\mu$ and we keep only the terms up to order 2 in $\varepsilon$ and $\dot{\theta}$, we obtain

$$L = L_{CW} = \frac{1}{2}(\dot{\varepsilon}^2 + \dot{\theta}^2) + 2\varepsilon\dot{\theta} + \frac{3}{2}\varepsilon^2 , \tag{7}$$

or equivalently

$$L = L_{CW} = \frac{1}{2}(\dot{\varepsilon}^2 + \dot{\theta}^2) + (\varepsilon\dot{\theta} - \dot{\varepsilon}\theta) + \frac{3}{2}\varepsilon^2 , \tag{8}$$

the two expressions differing only by the exact derivative $(\varepsilon\theta)^{\bullet}$. The corresponding equations of motion are the well known linear Clohessy-Wiltshire equations:

$$\begin{aligned}\ddot{\varepsilon} &= +2\dot{\theta} + 3\varepsilon , \\ \ddot{\theta} &= -2\dot{\varepsilon}\end{aligned} . \tag{9}$$

We note that if we trace the origin of the term $3\varepsilon^2/2$, we find that two-thirds of it is due to gravity and one third to centripetal acceleration. This is because of the expression

$$\frac{r^2}{2} + \frac{1}{r} = \frac{1}{2}(1 + 2\varepsilon + \varepsilon^2) + (1 - \varepsilon + \varepsilon^2) = \frac{3}{2}(1 + \varepsilon^2) . \tag{10}$$

The term $r^2/2$ gives the centripetal acceleration, whereas the term $1/r$ gives the inverse-square-law gravity term.

## 5. The Perturbed Clohessy-Wiltshire Lagrangian

It is apparent, from what has been seen in the previous sections that the perturbed Clohessy-Wiltshire Lagrangian

$$L = \frac{1}{2}(\dot{\varepsilon}^2 + \dot{\theta}^2) + (\varepsilon\dot{\theta} - \dot{\varepsilon}\theta) + \frac{3}{2}\varepsilon^2 + \mu U(\varepsilon, \theta) \tag{11}$$

could form a very interesting dynamical system for the study of motions near the unit-circle in a rotating frame, at least if the proper function $U(\varepsilon, \theta)$ is selected, ($\mu$ being a small parameter, as usual). The equations of motion corresponding to (5.1) are

$$\begin{cases} \ddot{\varepsilon} = 2\dot{\theta} + 3\varepsilon + "U_\varepsilon \, , \\ \ddot{\theta} = -2\dot{\varepsilon} + \mu U_\theta \, . \end{cases} \tag{12}$$

If we make no approximations at all in the potential function $U$ for the restricted three-body problem we have the following expression with three terms:

$$U(\varepsilon, \theta) = \frac{1}{d} - (1 + \varepsilon)\cos\theta - \frac{1}{1 + \varepsilon} \, . \tag{13}$$

The two last terms, expanded to the second-order in $\varepsilon$ give us:

$$-\left(r\cos\theta + \frac{1}{r}\right) = (1 + \varepsilon)(1 - \cos\theta) - \varepsilon^2 + 2 \, . \tag{14}$$

An application of this type of formulation is the Lagrangian

$$L = L_{CW} + \mu\left[(1 + \varepsilon)(1 - \cos\theta) + \frac{1}{d}\right] \, , \tag{15}$$

which contains only one (first-degree) $\varepsilon$-term in the perturbation. The corresponding equations of motion are

$$\begin{aligned} \ddot{\varepsilon} &= +2\dot{\theta} + 3\varepsilon + \mu(1 - \cos\theta) \, , \\ \ddot{\theta} &= -2\dot{\varepsilon} + \mu\sin\theta. \left[(1 + \varepsilon) - \frac{1}{d^3}\right] \, . \end{aligned} \tag{16}$$

We assumed here that $d^2 = 2(1 - \cos\theta)$.
A slightly more complicated Lagrangian would be

$$L = L_{CW} + \mu\left[(1 + \varepsilon)(1 - \cos\theta) - \varepsilon^2 + \frac{1}{d}\right] \, , \tag{17}$$

where

$$d^2 = 1 + (1+\varepsilon)^2 - 2(1+\varepsilon)\cos\theta = 2(1+\varepsilon)(1-\cos\theta) + \varepsilon^2 ,$$
$$= 4(1+\varepsilon)\sin^2\frac{\theta}{2} + \varepsilon^2 , \tag{18}$$

or

$$d^2 = 2(1+\varepsilon)(1-\cos\theta) = 4(1+\varepsilon)\sin^2\frac{\theta}{2} . \tag{19}$$

or

$$d^2 = 2(1-\cos\theta)4\sin^2\frac{\theta}{2} \tag{20}$$

according to the degree of precision that is required in $\varepsilon$. In the last formula, $d$ can vary from 0 to 2.

If the second approximation for $d^2$ is used, the above Lagrangian (5.7) can also be written as

$$L = L_{CW} + \mu\left[\frac{d^2}{2} + \frac{1}{d}\right] . \tag{21}$$

In general the equations of motion for the last Lagrangian can be written as

$$\ddot{\varepsilon} = +2\dot{\theta} + 3\varepsilon + \mu\left(d - \frac{1}{d^2}\right)\frac{\partial d}{\partial\varepsilon} ,$$
$$\ddot{\theta} = -2\dot{\varepsilon} + \mu\left(d - \frac{1}{d^2}\right)\frac{\partial d}{\partial\varepsilon} . \tag{22}$$

For instance, if $d^2 = 2(1+\varepsilon)(1-\cos\theta)$, we have

$$\frac{\partial d}{\partial\varepsilon} = \frac{1-\cos\theta}{d} ; \quad \frac{\partial d}{\partial\theta} = \frac{(1+\varepsilon)\sin\theta}{d} . \tag{23}$$

## 6. A Typical Simplified Lagrangian $L_{CW} + \mu(\varepsilon, \theta)$

As an illustration, we will now consider the approximation of the restricted problem, where the Keplerian terms as well as $U$ have all been expanded to order two only in the three variables, $\varepsilon, \dot{\varepsilon}, \dot{\theta}$, while no assumption is made on $\mu$. We also expanded $1/d$ to order two in $\varepsilon$. We obtain then the following Lagrangian:

$$L = \tfrac{1}{2}(\dot{\varepsilon}^2 + \dot{\theta}^2) + 2\varepsilon\dot{\theta} + \tfrac{3\varepsilon^2}{2} + \mu\{-\varepsilon^2 + (1+\varepsilon)(1-\cos\theta)$$
$$+ \frac{\sqrt{2}}{2(1-\cos\theta)^{1/2}}\left[1 - \frac{\varepsilon}{2} + \frac{\varepsilon^2}{8}\left(3 - \frac{2}{1-\cos\theta}\right)\right]\} \tag{24}$$

which is clearly of the form of a perturbed Clohessy-Wiltshire Lagrangian:

$$L = \frac{1}{2}(\dot{\varepsilon}^2 + \dot{\theta}^2) + 2\varepsilon\dot{\theta} + \frac{3\varepsilon^2}{2} + \mu U(\varepsilon, \theta) \tag{25}$$

The equations of motion for the new Lagrangian system are:

$$\ddot{\varepsilon} = 2\dot{\theta} + 3\varepsilon + \mu\left\{-2\varepsilon + (1 - \cos\theta) - \frac{1}{\sqrt{2(1-\cos\theta)}}\left[\frac{1}{2} - \frac{\varepsilon}{4}\left(3 - \frac{2}{1-\cos\theta}\right)\right]\right\}$$
$$\ddot{\theta} = -2\dot{\varepsilon} + \mu\left\{(1 + \varepsilon)\sin\theta - \frac{\sqrt{2}\sin\theta}{4(1-\cos\theta)^{\frac{3}{2}}}\left[1 - \frac{\varepsilon}{2} + \frac{3\varepsilon^2}{8}\left(1 - \frac{2}{1-\cos\theta}\right)\right]\right\}. \tag{26}$$

It is not difficult to show, by letting $\dot{\varepsilon} = \ddot{\varepsilon} = \dot{\theta} = \ddot{\theta} = 0$, that the Lagrange points $L_3, L_4$, and $L_5$ still exist with the coordinates:

$$
\begin{aligned}
L_3 &: \varepsilon = \frac{-7\mu}{12-7\mu} \ , \quad \theta = 180^0 \ , \\
L_4 &: \varepsilon = 0 \qquad\quad , \quad \theta = +60^0 \ , \\
L_5 &: \varepsilon = 0 \qquad\quad , \quad \theta = -60^0 \ .
\end{aligned}
\tag{27}
$$

## 7. The Second Approximate Model

For many situations $\mu \ll 1$, so the additional assumption of neglecting terms of order 3 and higher in $\mu, \varepsilon$ and $\dot{\theta}$ will be made. This simplifies the Lagrangian to

$$L = \frac{1}{2}(\dot{\varepsilon}^2 + \dot{\theta}^2) + 2\varepsilon\dot{\theta} + \frac{3\varepsilon^2}{2} + \mu(1+\varepsilon)(1-\cos\theta) + \frac{\mu}{[2(1-\cos\theta)]^{\frac{1}{2}}}\left(1 - \frac{\varepsilon}{2}\right) \tag{28}$$

and the equations of motion:

$$\ddot{\varepsilon} = 2\dot{\theta} + 3\varepsilon + \mu(1 - \cos\theta)\left[1 - \frac{\sqrt{2}}{4(1-\cos\theta)^{\frac{3}{2}}}\right] \ ,$$
$$\ddot{\theta} = -2\dot{\varepsilon} + \mu\left[(1 + \varepsilon)\sin\theta - \frac{\sqrt{2}\sin\theta}{4(1-\cos\theta)^{\frac{3}{2}}}\left(1 - \frac{\varepsilon}{2}\right)\right] \ . \tag{29}$$

The Lagrange points are as before.

## 8. The Unit-Circle Symmetry in Polar Coordinates

We can easily show that the perturbed Clohessy-Wiltshire equations

$$\ddot{\varepsilon} = +2\dot{\theta} + 3\varepsilon + \mu U_\varepsilon(\varepsilon, \theta)$$
$$\ddot{\theta} = -2\dot{\varepsilon} + \mu U_\theta(\varepsilon, \theta) \tag{30}$$

exhibit a remarkable symmetry condition, under some special constraints on the potential function $U(\varepsilon, \theta)$. In particular, if we assume that $U_\varepsilon(\varepsilon, \theta)$ is an odd function of $\varepsilon$ and $U_\theta(\varepsilon, \theta)$ an even function of $\varepsilon$, then, the substitutions

$$\varepsilon \to -\varepsilon \ ; \ \theta \to \theta \ ; \ \dot{\varepsilon} \to \dot{\varepsilon} \ ; \ \dot{\theta} \to \dot{\theta} \ ; \ \ddot{\varepsilon} \to \ddot{\varepsilon} \ ; \ \ddot{\theta} \to \ddot{\theta} \tag{31}$$

leave the above equations of motion invariant. Indeed, we see that this substitution changes all the signs of the terms of the first equation, while it produces no change at all to the second equation of motion. This results in a mirror-image property: to every solution corresponds a mirror image solution relative to the horizontal axis in the $(\theta, \varepsilon)$-plane, (the axis $\varepsilon = 0$). In other words, to every arc of solution outside the unit-circle corresponds another arc of solution inside the unit circle. This property is thus very similar to the mirror-image theorem which is well known in the restricted problem, relative to the syzygy-axis. This symmetry property can also be used to find a special kind of periodic solutions: A sufficient condition of periodicity is that the solution must have <u>two</u> orthogonal intersections with the unit-circle. In other words the solution should have two points with initial and final conditions of the form

$$(0, \theta_0, \dot{\varepsilon}_0, 0) , \tag{32}$$

these two points being separated by a half-period. Another fact that can be noted in relation to these properties is that we could use the symmetry line $\varepsilon = 0$, (the unit-circle) as the line of sections for the construction of Poincare maps in the $(\theta, \dot{\theta})$-plane.

As for the potential function $U(\varepsilon, \theta)$, we see that it clearly must satisfy the general condition of being an <u>even</u> function of the variable $\varepsilon$. Let us mention that this symmetry property <u>does</u> not hold for the true untruncated restricted problem. We also note that the condition for the unit-circle symmetry does hold for the simplified systems of the form

$$\begin{aligned} \ddot{\varepsilon} &= +2\dot{\theta} + 3\varepsilon , \\ \ddot{\theta} &= -2\dot{\varepsilon} + B(\theta, \varepsilon^2) \end{aligned} \tag{33}$$

where $B$ is an arbitrary function not depending on $\varepsilon$ or even in $\dot{\varepsilon}$.

The above considerations are completely independent of the $\theta$-dependence of the potential $U(\varepsilon, \theta)$. However, the standard mirror-image theorem with the syzygy axis always holds: orbits with initial values $\theta_0 = 0$ or $\pi$ and $\dot{\varepsilon}_0 = 0$ are symmetric with respect to the syzygy-line (which joins the two primaries).

As for the distance $d$ between the second primary and the satellite, the two following approximations

$$d^2 = \varepsilon^2 + 2(1 - \cos\theta) \; ; \; d^2 = 2(1 - \cos\theta) \tag{34}$$

both satisfy the unit-circle symmetry criterion.

As an example of a potential function which would be a reasonable approximation for the restricted three-body problem and which at the same time satisfies the unit-circle symmetry condition, we propose

$$U(\varepsilon, \theta) = -(1 + \cos\theta) - \varepsilon^2 + \frac{1}{\sqrt{\varepsilon^2 + 2(1 - \cos\theta)}} , \tag{35}$$

which has the derivatives

$$U_\varepsilon = -\varepsilon \left(2 + \frac{1}{d^3}\right) \; ; \; U_\theta = +\sin\theta \left(1 - \frac{1}{d^3}\right) , \tag{36}$$

where $d$ represents the first of the two distance approximations (8.5) given above.

## 9. The Case where $U$ depends on $\theta$ only

An interesting special case corresponds to the situation where the function $U$, present in the perturbed Clohessy-Wiltshire Lagrangian contains explicitly only the variable $\theta$. We have then the equations of motion

$$\ddot{\varepsilon} = +2\dot{\theta} + 3\varepsilon \; ; \; \ddot{\theta} = -2\dot{\varepsilon} + \mu U_\theta , \tag{37}$$

so that the perturbation acts only on the theta-equation.

As an illustration, we mention the Lagrangian

$$L = L_{CW} + \mu \left[(1 - \cos\theta) + \frac{1}{\sqrt{2(1 - \cos\theta)}}\right] = L_{CW} + \mu \left[2\sin^2\frac{\theta}{2} + \frac{1}{2\sin\frac{\theta}{2}}\right] \tag{38}$$

The equations of motion derived from this Lagrangian are

$$\ddot{\varepsilon} = +2\dot{\theta} + 3\varepsilon \; ; \; \ddot{\theta} = -2\dot{\varepsilon} + \mu \left(1 - \frac{1}{8\sin^3(\theta/2)}\right)\sin\theta \tag{39}$$

In the case where we would have only motions which are extremely close to the unit circle, we could assume that $\ddot{\varepsilon} = 0$, so that, by differentiation of the right-hand side of the first equation $\dot{\varepsilon} = -2\dot{\theta}/3$, (Yoder *et al.*, 1983). The second equation becomes thus

$$\ddot{\theta} = \frac{+4\ddot{\theta}}{3} + \mu U_\theta , \tag{40}$$

which is equivalent to

$$\ddot{\theta} = -3\mu U_\theta . \tag{41}$$

Note that this equation has the Lagrangian

$$L = \frac{1}{2}\dot{\theta}^2 - 3\mu U(\theta) . \tag{42}$$

We believe that this Lagrangian is somewhat more realistic than the Lagrangian which is obtained by setting $\varepsilon = \dot{\varepsilon} = 0$ in (5.1):

$$L = \frac{1}{2}\dot{\theta}^2 + \mu U(\theta) . \tag{43}$$

As an illustration we mention again the case where

$$U(\theta) = (1 - \cos\theta) + (2(1 - \cos\theta))^{-1/2} = (1 - \cos\theta) + \tfrac{1}{d}$$
$$= 2\sin^2\tfrac{\theta}{2} + \left(2\sin\tfrac{\theta}{2}\right)^{-1} \tag{44}$$

which will clearly result in a perturbed pendulum problem.
The Lagrangian

$$L = \frac{1}{2}\dot{\theta}^2 - 3\mu U(\theta) \tag{45}$$

gives us the equation of motion

$$\ddot{\theta} = -3\mu\left(1 - \frac{1}{8\sin^3(\theta/2)}\right)\sin\theta \tag{46}$$

or

$$\ddot{\theta} = -3\mu\left[1 - \frac{2}{\sqrt{8(1 - \cos\theta)^3}}\right]\sin\theta \tag{47}$$

which has two important terms: the first term $(-3\mu\sin\theta)$ gives the usual type of pendulum motion. However, the second term has the opposite sign. In these formulations, the radial variable $\varepsilon$ can be recovered by using Yoderis equation $\varepsilon = 2\dot{\theta}/3$.

We note that Yoder et al. (1983) has used an equation of the general form of a perturbed pendulum

$$\ddot{\theta} = -3\mu\sin\theta + \cdots \tag{48}$$

in his work on the co-orbital satellites of Saturn.
The above dynamical system being conservative, it has the Energy equation:

$$E = \frac{1}{2}\dot{\theta}^2 + 3\mu\left[2\sin^2\frac{\theta}{2} + \frac{1}{2\sin(\theta/2)}\right] = \text{Constant.} \tag{49}$$

It will be convenient to define the Jacobi constant $C$ for this problem, with the equation

$$E = \frac{3}{2}\mu C. \tag{50}$$

It is clear that the present dynamical system is integrable. In fact the above energy equation gives the solution ($\dot{\theta}$ as a function of $\theta$, for instance).

The most important property of the present system is the existence of three equilibrium solutions. Their existence can be justified analytically, by finding the roots of the equation

$$\sin\theta.\left[1 - 1/\left(8\sin^3\frac{\theta}{2}\right)\right] = 0 . \tag{51}$$

The first root corresponds to $\theta = 0$. However this corresponds to the singularity at the center of the smaller primary at $(1, 0)$. By writing the equation of motion as

$$\ddot{\theta} = -6\mu\cos\frac{\theta}{2}\left[\sin\frac{\theta}{2} - 1/\left(8\sin^2\frac{\theta}{2}\right)\right] \tag{52}$$

we see that, at $\theta = 0, \ddot{\theta} = 6\mu\left(0 - \frac{1}{0}\right) \simeq +\infty$. This point $\theta = 0$ is thus a repulsive point. The other root is $\theta = \pi$, which is an unstable equilibrium point: $L_3$, one of the three collinear Lagrange points.

Finally we have the two roots $\sin\frac{\theta}{2} = \frac{1}{2}$, which are thus $\theta = \pm 60^0$, the two triangular Lagrangian points $L_4$ and $L_5$. It is remarkable that in this dynamical system, which is an extreme simplification of the restricted problem, three of the five Lagrange points still survive !

## References

Deprit, A. and Henrard, J.: 1978, "A Manifold of Periodic Orbits", *Advances in Astronomy and Astrophysics*, Vol. 6, Academic Press, New York.

Deprit, A.: 1966, "Motion in the Vicinity of the Triangular Libration Centers", *Lectures in Applied Mathematics*, Vol. 6, American Mathematical Society.

Deprit, A., Henrard, J., Palmore, J. and Price, J F.: 1969, "The Trojan Manifold in the System Earth-Moon", *Mon. Not. R. Astr. Soc.*, **137**, 311-335.

Dermott, S.F. and Murray, C.D.: 1981, "The Dynamics of Tadpole and Horseshoe Orbits: I. Theory", *Icarus*, **48**, 1-11.

Dermott, S. F. and Murray, C.D.: 1981, "The Dynamics of Tadpole and Horseshoe Orbits: II. The Coorbital Satellites of Saturn", *Icarus*, **48**, 12-22.

Garfinkel, B.: 1977, "Theory of the Trojan Asteroids, Part I", *Astron. J.*, **82**, 368-379.

Garfinkel, B.: 1975, "An Extended Ideal Resonance Problem", *Celestial Mechanics*, **12**, 203-214.

Garfinkel, B.: 1978, "Theory of the Trojan Asteroids. Part II", *Celest. Mech.*, **18**, 259-275.

Giacaglia, G. E. O.: 1976, comments on the paper by Boris Garfinkel: "An Extended Ideal Resonance Problem", *Celestial Mechanics*, **13**, 515-516.

Harrington, R.S. and Seidelmann P.K.: 1981, "The Dynamics of the Saturnian Satellites 1980S1 and 1980S3", *Icarus*, **47** , 97-99

Konopliv, A. S.: 1986, "Theory of Coorbital Motion", Ph.D. dissertation, Univ. of Texas at Austin.

Salo, H. and Yoder C F.: 1988, "The Dynamics of Coorbital Satellite Systems", *A. & A.*, **205**, 309-327

Yoder, C. F., G. Colombo, S. P. Synnott, and K. A. Yoder, 1983, "Theory of Motion of Saturn's Coorbiting Satellites", *Icarus*, **53**, 431-443.

# A GLOBAL ANALYSIS OF THE GENERALIZED SITNIKOV PROBLEM

STEVEN R. CHESLEY

*Dipartimento di Matematica, Universitá di Pisa, 56127 Pisa, Italy;*
*E-mail: chesley@dm.unipi.it*

**Abstract.** The isosceles three-body problem with Sitnikov-type symmetry has been reduced to a two-dimensional area-preserving Poincaré map depending on two parameters: the mass ratio, and the total angular momentum. The entire parameter space is explored, contrasting new results with ones obtained previously in the planar (zero angular momentum) case. The region of allowable motion is divided into subregions according to a symbolic dynamics representation. This enables a geometric description of the system based on the intersection of the images of the subregions with the preimages. The paper also describes the regions of allowable motion and bounded motion, and discusses the stability of the dominant periodic orbit.

## 1. Introduction

The isosceles three-body problem has been studied for a variety of reasons and for more than one hundred years. Early researches concerned the possible isosceles configurations (Fransén, 1895), the necessary conditions for the isosceles symmetry to be preserved (Wilczynski, 1913), and the integrability of certain cases (Macmillan, 1913). Later Sitnikov (1961) used a special case of this problem to prove the existence of so-called oscillatory motion—unbounded oscillations with no escape. In recent decades considerable attention has focused on the triple collision manifold of the planar isosceles problem (e.g., Simó and Martínez, 1988).

In the present work the isosceles problem is used as an example to demonstrate a method of global analysis based on a Poincaré map. In this method one uses the mapping to investigate the evolution of entire areas under a single iteration, rather than examining the behavior of a few points on the surface of section under hundreds or thousands of iterations as is typically done. This new approach permits insight into the structure and evolution of the *chaotic regions* (seas) of the phase space. On the other hand the traditional approach shows very clearly the structure and extent of the *stable regions* (islands). In this sense the new method is very much complementary to the traditional approach.

The isosceles problem is obtained by applying certain special symmetries to the initial conditions. The present formulation permits the representation of both the Sitnikov and planar configurations, which are usually modelled separately. In this formulation there are two free parameters which govern the dynamics, a mass ratio $\alpha$ and the angular momentum of the binary $c$. The Sitnikov configuration consists of a *symmetric* binary system with *nonzero* angular momentum and a secondary mass moving for all time along the unique line that is perpendicular to the plane of the binary and passes through the center of mass of the binary. In this configuration, the three masses always form an isosceles triangle, and the triangle is always rotating about its axis of symmetry. (See Fig. 1.) If the angular momentum of the binary is zero then the triangle is confined to an invariable plane. This is the planar isosceles configuration.

*Celestial Mechanics and Dynamical Astronomy* **73**: 291–302, 1999.
©1999 *Kluwer Academic Publishers*.

These two configurations are most importantly distinguished by the presence or absence of collisions. In the planar case the binary motion is rectilinear since $c = 0$, thus there is a binary collision at every pericenter. If the binary collision is coincident with a crossing of the syzygy line by the third mass then we have a triple collision. In contrast, the Sitnikov case has no possibility of binary collision, hence triple collision is clearly also excluded.

In a previous paper (hereafter Paper I) Zare and Chesley (1998) studied carefully the case of planar motion with three equal masses. In that paper the tools and principles used in the present study are described in much greater detail than is possible here. Later (Paper II) Chesley and Zare (1998) expanded the study to include all possible mass ratios, while maintaining the planar condition ($c = 0$). Finally, in this paper, the entire parameter space is explored, considering all possible values of angular momentum and mass ratio. New results are put into contrast with properties obtained in the previous papers. This paper also presents some more general results on the problem, including a discussion of the allowable motions, and a global analysis of the stability properties of the dominant periodic orbit.

## 2. The General Isosceles Problem

A full development of the system equations can be found in (Chesley, 1998), but for brevity let us begin with the Hamiltonian in physical variables

$$\mathcal{H} = \frac{1}{2}\left(p_1^2 + \frac{p_2^2}{\alpha}\right) - \frac{Gm}{4q_1} - \frac{Gm_3}{\sqrt{q_1^2 + q_2^2}} + \frac{c^2}{2q_1^2}. \tag{1}$$

Here $(q_1, q_2)$ are the distances of a binary element and the secondary, respectively, from the binary mass center (see Fig. 1), and $(p_1, p_2)$ are the corresponding conjugate momenta. The constant of gravitation is given by $G$, while $m$ is the mass of each of the binary components and $m_3$ is the mass of the secondary or "third mass". We also have the binary angular momentum $c$, and the ratio of the secondary mass to the total mass $\alpha = m_3/(2m + m_3)$, $0 < \alpha \le 1$. We will use the value of the Hamiltonian to represent the system energy $h = \mathcal{H}$.

**Remark**: The Hamiltonian formulation does not permit the analysis of the classical Sitnikov problem, which is the restricted case ($\alpha = 0$). This fact can be clearly seen from Eq. (1).

For reasons which will become apparent later, we shall apply the following normalization of units throughout

$$G = 1, \qquad m + 4m_3 = 1/2, \qquad h = -1/8. \tag{2}$$

There is no loss of generality with this normalization. The system fundamentally has the following parameters: the constant of gravity $G$, the energy $h$, the angular momentum $c$, and two mass parameters $m$ and $m_3$. Selecting the units of mass,

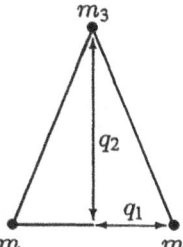

Fig. 1. Geometry of the isosceles problem. Note that the binary rotates about the axis of symmetry if $c > 0$.

length, and time according to the normalization given above fixes $G$, $h$, and one of the mass parameters. This leaves the angular momentum $c$, and the other mass parameter, which we may choose to be $\alpha$, as the only free parameters. Selection of $\alpha$ within this normalization has the effect of determining the mass values according to the identities $m = (1 - \alpha) / (1 + 7\alpha) / 2$ and $m_3 = \alpha / (1 + 7\alpha)$.

**Note**: This Hamiltonian has been scaled from the usual three-body energy expression by a factor $2m$ to allow a simpler presentation. This means that the usual system energy and usual binary angular momentum are given by $2mh$ and $2mc$, respectively.

## 2.1. REGULARIZATION

If $c = 0$ the equations of motion stemming from Eq. (1) are singular at the binary collision ($q_1 = 0$, $q_2 \neq 0$) and at the triple collision ($q_1 = q_2 = 0$). (If $c$ is small but nonzero the equations are near-singular at binary pericenter.) The singularity at or near the binary collision may be removed by the classical Levi-Civita regularization. In order to preserve the Hamiltonian form, we shall adapt the regularization to the extended phase space. First, we introduce the generating function $W = p_1 Q_1^2 + p_2 Q_2$, and the associated canonical transformation

$$q_i = \frac{\partial W}{\partial p_i}, \qquad P_i = \frac{\partial W}{\partial Q_i}, \qquad i = 1, 2,$$

leading to the Hamiltonian

$$\mathcal{H} = \frac{1}{2} \left( \frac{P_1^2}{4Q_1^2} + \frac{P_2^2}{\alpha} \right) - \frac{Gm}{4Q_1^2} - \frac{Gm_3}{\sqrt{Q_1^4 + Q_2^2}} + \frac{c^2}{2Q_1^4}.$$

Under these new variables one may construe $(Q_1, P_1)$ as the state vector of the binary, and $(Q_2, P_2)$ as the state vector of the third mass.

Introduction of the auxiliary variables $Q_0 = t$, $P_0 = -h$, and $d\tau = dt/Q_1^2$ provides the new equations of motion

$$\frac{dQ_i}{d\tau} = \frac{\partial \Gamma}{\partial P_i}, \qquad \frac{dP_i}{d\tau} = -\frac{\partial \Gamma}{\partial Q_i}, \qquad i = 0, 1, 2, \tag{3}$$

where the new Hamiltonian function in the extended phase space $\Gamma = Q_1^2 (P_0 + \mathcal{H})$ is given by

$$\Gamma = \frac{1}{2} \left( \frac{P_1^2}{4} + \frac{P_2^2 Q_1^2}{\alpha} \right) - \frac{Gm}{4} - \frac{Gm_3 Q_1^2}{\sqrt{Q_1^4 + Q_2^2}} + P_0 Q_1^2 + \frac{c^2}{2Q_1^2}.$$

On every trajectory we have the energy integral $\Gamma \equiv 0$, and the equations of motion are now regular at the binary collision ($Q_1 = 0$, $Q_2 \neq 0$).

## 2.2. REGION OF POSSIBLE MOTION IN THE CONFIGURATION SPACE

A general theory has been developed by Zare (1976) to obtain the regions of possible motion in the configuration space, and in particular the bifurcation sets of their topological classification (Zare, 1977). For the present problem we will follow the theory to obtain the maximum possible angular momentum for a given energy and to identify regions of bounded motion in the parameter space.

Starting with Eq. (1) we obtain the inequality

$$h \geq -\frac{Gm}{4q_1} - \frac{Gm_3}{q_1 \sqrt{1 + q_2^2/q_1^2}} + \frac{c^2}{2q_1^2}.$$

By introducing the new variable

$$\theta = \arctan \frac{q_2}{q_1}, \qquad -\frac{\pi}{2} \leq \theta \leq \frac{\pi}{2},$$

this inequality can be recast as

$$\psi(q_1, \theta) = 4 |h| q_1^2 - G(m + 4m_3 \cos \theta) q_1 + 2c^2 \leq 0, \tag{4}$$

where we have assumed $h < 0$. The variables here represent the configuration ($\theta$) and the scale ($q_1$) of the triangle formed by the three bodies. Note that $\theta = 0$ implies syzygy, and $\theta = \pm \pi/2$ implies either binary collision or the escape of the third mass. From Eq. (4) it is immediate that if $c \neq 0$ then $q_1 > 0$ for all time, hence the binary collision is forbidden. The possible range for $q_1$ is obtained from the quadratic equation where the discriminant is given by

$$\Delta(\theta) = G^2 (m + 4m_3 \cos \theta)^2 - 32c^2 |h| \geq 0. \tag{5}$$

Equation (5) provides the totality of possible configurations, independent of scale. Notice that $c^2 |h|$, sometimes referred to as the Zare integral, appears as an essential parameter. It is easy to show that $\Delta(\theta)$ has a maximum at $\theta = 0$ with maximal value

$$\Delta(0) = G^2 (m + 4m_3)^2 - 32c^2 |h| \geq 0.$$

From this we obtain the critical value

$$c^2 |h| \leq \left( c^2 |h| \right)_1 = G^2 (m + 4m_3)^2 / 32.$$

If we apply the normalization of units given by Eq. (2) then we obtain the permissible range of angular momentum: $0 \leq c \leq \frac{1}{4}$. (We shall assume $c \geq 0$, the direction of rotation being irrelevant to the dynamics.)

**Remark:** The maximum value of $c$ reflects an important orbit. Note that this value requires $p_1 = p_2 \equiv 0$ and $\theta \equiv 0$, which implies $q_2 \equiv 0$. In this case we have $q_1 \equiv \frac{1}{2}$ under our normalization according to Eq. (4). So for $c = \frac{1}{4}$ the motion is limited to a single point in the phase space. This orbit is the so-called Euler solution. Here the binaries rotate on circular orbits with radius $q_1 = \frac{1}{2}$ and the third mass remains forever fixed at the binary center of mass.

### 2.2.1. The Region of Bounded Motion

If the angular momentum is large enough the third mass cannot escape to infinity, while the binary remains bounded according to Eq. (4). This implies bounded motion for all particles for all time. To determine the smallest value of angular momentum where bounded motion is assured, let us continue by recasting Eq. (5) as

$$\cos \theta \geq \frac{\sqrt{32c^2 |h|} - Gm}{4Gm_3}. \tag{6}$$

From this we can see that for $c^2 |h| > (c^2 |h|)_1$ there exist no possible configurations, and for $c^2 |h| = (c^2 |h|)_1$ only $\theta = 0$ is possible. But for $c^2 |h| < (c^2 |h|)_1$ the possible configurations are $|\theta| \leq \theta_{max}$, where $\theta_{max}$ is obtained from Eq. (6). It is clear that $\theta_{max} = \pi/2$ at the critical value

$$\left(c^2 h\right)_2 = G^2 m^2 / 32.$$

Recall that $\theta = \pm \pi/2$ corresponds to escape of $m_3$ if $c > 0$. Therefore the third mass may go to infinity for $c^2 |h| \leq (c^2 |h|)_2$ since the possible configurations are $-\pi/2 \leq \theta \leq \pi/2$. But for $c^2 |h| > (c^2 |h|)_2$ we have $|\theta| < \pi/2$, and escape is impossible. After normalization of units according to Eq. (2) we have the final result

$$c \geq \frac{1 - \alpha}{4(1 + 7\alpha)} \implies \text{bounded motion.}$$

The corresponding region of the parameter space is plotted in Fig. 2.

### 2.3. THE BOUNDARIES OF THE PARAMETER SPACE

Each of the boundaries for the space of parameters $(\alpha, c) \in (0, 1] \times \left[0, \frac{1}{4}\right]$ has a special significance. (See Fig. 2.)

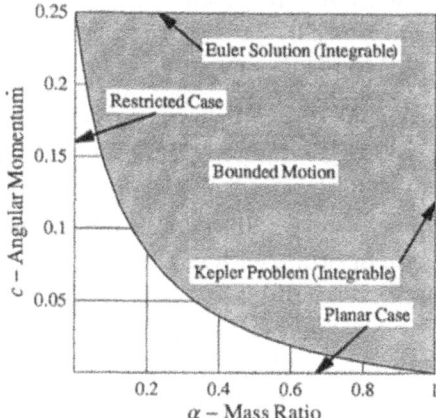

Fig. 2. Region of bounded motion in the $(\alpha, c)$-space of parameters. The special significance of each boundary is also depicted.

1. $c = 0$. This is the planar isosceles problem. This boundary was the focus of Paper II, and the point on this boundary at $(\alpha, c) = \left(\frac{1}{3}, 0\right)$ was studied in detail in Paper I.

2. $c = \frac{1}{4}$. The motion here corresponds to an integrable solution of the general three-body problem—the Euler solution discussed above. The motion is reduced to a single fixed point in the phase space.

3. $\alpha = 0$. Here the third mass is zero—the classical (restricted) Sitnikov problem. This is the only boundary that cannot be treated by the present formulation of the equations of motion. As one approaches this boundary, hyperbolic escape of $m_3$ approaches full measure. (See Paper II).

4. $\alpha = 1$. On this boundary all of the mass is concentrated in the third body. This leads to Keplerian motion, so here again the solution is integrable. This is the double Kepler problem described in Paper II. Here every point is a fixed point under the Poincaré mapping described in the next section.

## 3. Reduction to a Poincaré Mapping

The properties of the Poincaré map in the planar isosceles problem have been discussed extensively in Papers I and II. The primary aim of the present paper is to describe certain new properties which have appeared after extending the investigation to include nonzero angular momentum. For this reason, and for reasons of space, only a cursory introduction to the methods employed previously is presented here. For a comprehensive discussion, the interested reader is referred to Paper I, Paper II, and (Chesley, 1998).

To arrive at the desired surface of section we eliminate $P_2$ using the integral of energy, and take as a section the plane $Q_2 = 0$ (syzygy crossing). Thus one obtains a mapping in the $(Q_1, P_1)$-plane. This is an excellent surface of section

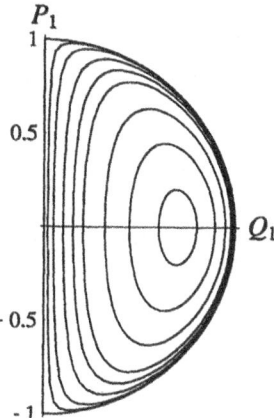

Fig. 3. Boundary curves for the regions of possible motion $A$ in the Poincaré section for varying angular momentum. The depicted curves correspond to the following values of $c$: $\{0, 0.01, 0.025, 0.05, 0.075, 0.10, 0.15, 0.20, 0.24\}$.

because almost every isosceles trajectory intersects it, with the only exceptions being certain isolated triple collision orbits. The region of allowable motion $A$ depends only upon the angular momentum $c$ according to

$$A = \left\{ (Q_1, P_1) \mid Q_1^2 + P_1^2 + \frac{4c^2}{Q_1^2} < 1, \ Q_1 > 0 \right\}.$$

We specify $Q_1 > 0$ without loss of generality since $q_1 = Q_1^2 \geq 0$, and on this section $Q_1 = 0$ implies a triple collision condition. Every point in $A$ represents a unique initial condition with $Q_2 = 0$ and $P_2 \geq 0$ computed from the energy integral. The bounding curves for varying levels of $c$ are depicted in Fig. 3.

For $c = 0$ (the planar case) $A$ reaches its greatest extent forming a semicircular region of unit radius. (In fact, the somewhat curious normalization of units in Eq. (2) has been selected to this end.) In this case the straight boundary of $A$ (on which $Q_1 = 0$) corresponds to the triple collision state, while the semicircular boundary of $A$ corresponds to the collinear homothetic solution. For $c > 0$ the boundary of $A$ corresponds to the Euler (collinear) solution.

The selection of this surface of section permits a useful physical interpretation of the mapping since $Q_2 = 0$ represents the syzygy configuration. Every trajectory started from a point in $A$ must return to another point in $A$ at the next syzygy crossing unless $m_3$ escapes or goes to triple collision. This may be viewed as a two-dimensional map $f : \bar{A} \rightarrow A$, where $\bar{A} \subset A$ is the set of points which neither escape nor lead to triple collision. Thus, for a given binary state $p_0 = (Q_1, P_1)_0 \in \bar{A}$, we define $f(p_0)$ to be the binary state $p_1 = (Q_1, P_1)_1 \in A$ taken at the next syzygy crossing.

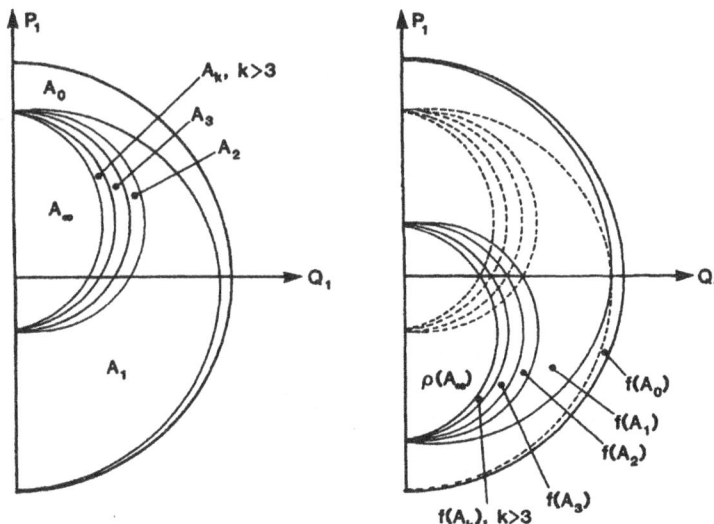

Fig. 4. The geometrical description of $f$ with $\alpha = \frac{1}{3}$ (equal masses) and $c = 0$ (planar). The left diagram depicts the partitioning of $A$ into $A_k$, $(k = 0, 1, 2, \ldots)$. The right diagram depicts the images of the $A_k$. Here the preimages are shown with dashed lines to emphasize the intersections discussed in the text.

## 3.1. PROPERTIES OF THE MAPPING

The global description of $f$, which is given in greater detail in Paper I, proceeds as follows. Since each point in $\bar{A}$ must return for a subsequent syzygy crossing, we may assign to it a non-negative integer $k$ representing the number of binary pericenters which ensue before the next syzygy crossing. This will partition $A$ into open subsets $A_k$ where all points in each subset have the same assigned $k$ (see Fig. 4). Additionally, we denote the points which lead to hyperbolic escape of $m_3$ by $A_\infty$. Separating $A_k$ and $A_{k+1}$ is a locus of points $B_k$ which lead to a binary pericenter at the next syzygy (triple collision in the planar problem) after $k$ binary collisions. The boundary of $A_\infty$, which we denote by $B_\infty$, corresponds to a set of points leading to parabolic escape of $m_3$.

Due to the conservative nature of this system and the associated time reversibility we have a useful symmetry. Consider the reflection $\rho$ about the $Q_1$-axis in the $(Q_1, P_1)$-plane defined by

$$(Q_1, P_1) \xrightarrow{\rho} (Q_1, -P_1).$$

Now $\rho$ can be considered as a reversal of the velocity, or, alternatively, a reversal of the arrow of time. Thus the backward orbit of the reflection of $p$ is the same as the reflection of the forward orbit of $p$ and it has the same number of binary collisions as the forward orbit of $p$. This leads to the identity $\rho f = f^{-1}\rho$. It also follows that for any $p \in A_k$, we have $\rho f(p) \in A_k$ or $f(p) \in \rho(A_k)$. That is

$$f(A_k) = \rho(A_k). \tag{7}$$

Furthermore, if the orbit of $p$ corresponds to the escape of $m_3$ the orbit of $\rho(p)$ corresponds to the capture of $m_3$. If the orbit of $p$ terminates with a triple collision the orbit of $\rho(p)$ initiates with a triple ejection.

The identity (7) holds only for the $A_k$ and is not true for other sets in general. However, it provides a convenient means of obtaining the forward images $f(A_k)$. The asymmetry and the intersections of these subregions $A_k$ with the $Q_1$-axis are the necessary ingredients for the intersections $A_j \cap \rho(A_k)$ illustrated in Fig. 4. These intersections allow one to ascertain the existence and amount of communication between the $A_k$. They are also a necessary condition for the existence of period-one and -two orbits under $f$ as discussed in the next section.

This partitioning of $A$ into $A_k$ and $B_k$ is particularly conducive to a symbolic sequence representation of an orbit. We define the $n$th character in the symbolic sequence to be $k$ if $f^n(p) \in A_k$, for all integer $n$. Thus a sequence of $k$'s corresponds to the sequence of $A_k$'s visited under $f^n$ (in both directions of time). If $f^n(p)$ is not defined due to hyperbolic escape (capture) then we terminate the sequence on the right (left) with the symbol "$\infty$". If $f^n(p) \in B_k$ we may terminate the sequence with some appropriate symbol, say "$*_k$". These sequences and subsequences are very valuable in any effort to qualitatively categorize and characterize orbits.

### 3.2. NEW RESULTS FOR $c > 0$

For $c = 0$ the regions $A_k$ for $k \geq 2$ were crescent shaped (Fig. 4), but for $c > 0$ they are segments of a spiral structure. Fig. 5 provides a particularly clear example of the spiral shape. In this figure there are two distinct curves along which the value of $k$ changes (the shading in the figure changes). The curve with an obvious spiral shape is the set of points for which the orbit *returns* to syzygy at pericenter. This spiral comprises the $B_k$, which were the orbits that returned to triple collision in the planar problem. The other, relatively straight, curve in Fig. 5 is the set of points for which the orbit *begins* at pericenter. In the planar case these were the points corresponding to triple collision. Geometrically what is happening is that, as the angular momentum increases above zero, the tips of the crescent-shaped $A_k$ of Fig. 4 bend inward to touch across this curve. On the other hand, as $c$ decreases to zero the curve diminishes to a segment on the $Q_1$-axis—the segment forming the border with $A_\infty$ in Fig. 4.

As described above, for large enough values of $c$ there is no possibility of escape. However, being below that curve is only a necessary condition for escape; it is not sufficient. There must, however, be a curve—on or below the one providing the necessary condition—that gives the sufficient condition. This curve can be defined as the locus of parameter values for which the only escape orbits are parabolic. Above this curve $A_\infty$ has been destroyed. As the mass ratio or angular momentum continues to increase each $A_k$ is destroyed sequentially until only $A_0$ and $A_1$ survive. So for each $k \geq 2$ there is a curve in the parameter space marking the destruction of the corresponding $A_k$. This is quite intuitive because as $c$ increases,

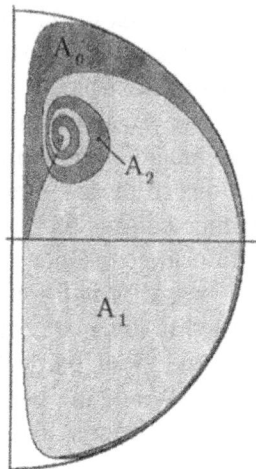

Fig. 5. Partitioning of $A_k$ at $\alpha = 0.67$, $c = 0.025$, showing the spiral effect described in the text. The axes are the same as Fig. 3.

the maximum ejections of the third body become more and more limited until $c = \frac{1}{4}$, at which point $Q_2(t) \equiv 0$. An approximate plot of the first and last curve is given in Fig. 6. The lower curve marks where $A_\infty$ is destroyed. The upper curve is where $A_2$—indeed all $A_k$, for $k \geq 2$—have been destroyed, and above this only $A_0$ and $A_1$ survive.

In terms of the symbolic dynamics, the destruction of $A_\infty$ is very important because it implies that all sequences are bi-infinite (except for a zero measure set of triple collision orbits). Furthermore, the symbolic alphabet becomes finite. These factors make the symbolic dynamics more tractable, and potentially more useful.

In Paper II we discussed a curious global bifurcation in the planar problem that occurred when, at a particular value of $\alpha$, all of the $A_k$ for $k \geq 2$ moved above the $Q_1$-axis. This "instantaneous" event, where an infinity of periodic and triple ejection-collision orbits are simultaneously destroyed, no longer occurs for $c > 0$. The reason is that the $A_k$ are no longer crescent shaped, but rather they have the spiral shape discussed above. This means that the $A_k$ move above the $Q_1$-axis sequentially rather than simultaneously. Thus the intersections illustrated in Fig. 4 and the associated periodic orbits (as opposed to the regions themselves) are *gradually* destroyed as $c$ or $\alpha$ increase until all of the $A_k$, $k \geq 2$, are above the $Q_1$-axis as in the example of Fig. 5. The dashed curves in Fig. 6 reflect this movement, the global bifurcation occurring where they intersect at $c = 0$.

## 4. Stability of the Main Periodic Orbit

In Paper I considerable attention was given to the *main periodic orbit* and the surrounding stable (quasi-periodic) region, where it was shown that this orbit has an important impact on the global dynamics. The orbit corresponds to the

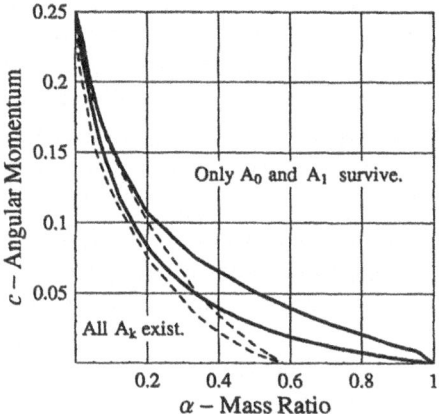

Fig. 6. Regions where particular $A_k$ survive. Below the lowest solid curve, all of the $A_k$ persist, including $A_\infty$. Above the upper solid curve, only $A_0$ and $A_1$ exist. The sequential elimination of the $A_k$ takes place between these two curves. The dashed curves depict the region where the $A_k$ move above the $Q_1$-axis, also sequentially.

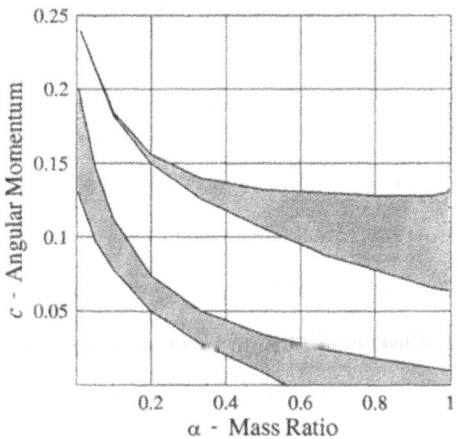

Fig. 7. Stability of the main periodic orbit. Unstable regions are shaded.

case when syzygy crossing occurs at binary apocenter, and there is exactly one binary pericenter between syzygy crossings. This means that triple interactions are minimized on this orbit, and the corresponding symbolic sequence for this orbit takes the form "...1, 1, 1...". In Paper II we showed that the main periodic orbit for the planar problem ($c = 0$) is stable for all values of $\alpha < 0.563...$ Now, for completeness, let us briefly expand that discussion to the entire parameter space.

The author has identified through computer experiments two bands in ($\alpha, c$)-space where the main periodic orbit is unstable. (See Fig. 7.) Each of these bands is bounded above and below by a bifurcation curve. The uppermost bifurcation curve is of the ordinary period doubling type, while the three lowest curves represent inverse period doubling bifurcations. The path of the curves very close (within 0.005) to the limiting values of $\alpha$ is not known at present. An analytic treatment of these situations would likely yield definitive answers, but this must be left for future study. Recently, Dvorak and Sun (1997) have obtained similar results, although they have studied only the equal mass ($\alpha = \frac{1}{3}$) case, and they have not computed orbits below approximately $c = 0.05$ since their equations of motion are unregularized.

An investigation is underway on the stability of the main periodic orbit within the framework of the general three-body problem. The important question is whether there exist regions of quasi-periodic motion in the full phase space, implying at least the possibility of the existence of near-isosceles triple stellar systems.

## Acknowledgements

The valuable contributions of Dr. K. Zare are gratefully acknowledged.

## References

Chesley, S. and K. Zare: 1998, 'Bifurcations in the Mass Ratio of the Planar Isosceles Three-Body Problem'. In: A. E. Roy and B. A. Steves (eds.): *Dynamics of Small Bodies in the Solar System*.

Chesley, S. R.: 1998, 'The Isosceles Three-Body Problem: A Global Geometric Analysis'. Ph.D. thesis, University of Texas at Austin.

Dvorak, R. and Y. S. Sun: 1997, 'The Phase Space Structure of the Extended Sitnikov Problem'. *Celest. Mech. Dyn. Astron.* **67**, 87–106.

Fransén, A. E.: 1895, 'Ett Specialfall Af Tre-Koppars-Problemet: Tva Himlakroppar Röra Sig Pa Lika Stora Afstand Fran Den Tredje'. *Öfversigt af Kongl. Vetenskaps-Akademiens Förhandlingar* **52**, 783–805.

MacMillan, W. D.: 1913, 'An Integrable Case in the Restricted Problem of Three Bodies'. *Astron. J.* **27**, 11–13.

Simó, C. and R. Martínez: 1988, 'Qualitative study of the planar isosceles three-body problem'. *Celest. Mech.* **41**, 179–251.

Sitnikov, K.: 1961, 'The Existence of Oscillatory Motions in the Three-Body Problem'. *Soviet Phys. Dokl.* **5**(4), 647–650.

Wilczynski, E. J.: 1913, 'Ricerche Geometriche Intorno al Problema Dei Tre Corpi'. *Annali di Matematica, Ser. 3* **21**, 1–31.

Zare, K.: 1976, 'The Effect of Integrals on the Totality of Solutions of Dynamical Systems'. *Celest. Mech.* **14**, 73–83.

Zare, K.: 1977, 'Bifurcation Points in the Planar Problem of Three Bodies'. *Celest. Mech.* **16**, 35–38.

Zare, K. and S. Chesley: 1998, 'Order and Chaos in the Planar Isosceles Three-Body Problem'. *Chaos* **8**(2), 475–494.

# THE HAMILTONIAN DYNAMICS
# OF THE TWO GYROSTATS PROBLEM

F. MONDEJAR, AND A. VIGUERAS

*Departamento de Matemática Aplicada, E. T. S.I.I., Universidad de Murcia, Paseo
Alfonso XIII, 30203 Cartagena (Murcia), Spain,
E-mail: fmalacid@plc.um.es and vigueras@plc.um.es*

**Abstract.** The problem of two gyrostats in a central force field is considered. We prove that the
Newton-Euler equations of motion are Hamiltonian with respect to a certain non-canonical structure.
The system posseses symmetries. Using them we perform the reduction of the number of degrees of
freedom. We show that at every stage of the reduction process, equations of motion are Hamiltonian
and give explicit forms corresponding to non-canonical Poissson brackets. Finally, we study the case
where one of the gyrostats has null gyrostatic momentum and we study the zero and the second order
approximation, showing that all equilibria are unstable in the zero order approximation.

## 1. Introduction

The problem of roto-traslatory motion of n-rigid bodies has been studied, amongst
other authors by Duboshin (1972), Aboelnaga (1979), Barkin (1980), Wang (1990;
1992), and by Maciejewski (1995). They considered the mutual interactions between orbital and rotational motion for artificial and natural bodies in the solar
system. But the model of a rigid body to represent celestial bodies implies the
absence of internal or relative motions. This is not always suitable, as was shown
by Volterra in the study of variation of latitude on the Earth's surface. He explained
the anomalies of the free rotation by means of internal or relative motions which
do not modify the distribution of masses. A gyrostat is a mechanical system S
composed of a rigid body S' and other bodies S'' connected to it; these other bodies
are either deformable or rigid, but their motion relative to S' does not alter the
distribution of masses of the system S.

In this paper the problem of roto-translatory motion of two gyrostats in a central
force field is considered following the papers of Maciejewski (7) and Wang (5). We
prove that the Newton-Euler equations of motion are Hamiltonian with respect to a
certain non-canonical structure. The system possesses symmetries and using them
we perform the reduction of the number of degrees of freedom. We show that on
every stage of this reduction process the equations of motion are Hamiltonian and
we give the explicit form of corresponding non-canonical Poisson bracket. Finally,
we restrict to the case where one of the gyrostats has null gyrostatic momentum and
we study the zero and second order approximation. In the first case it is shown that
only the Lagrangian equilibria can be obtained and furthermore those equilibria
are unstable. In the second case we identify the Lagrangian equilibria. For the non-
Lagrangian case we find a rather intricate non-linear systems of equations, where
solutions only can be obtained by numerical or perturbative methods in particular
cases. We show that apart from the Lagrangian equilibria, it can be found non-

*Celestial Mechanics and Dynamical Astronomy* 73: 303–312, 1999.
© 1999 *Kluwer Academic Publishers.*

Lagrangian configurations in contrast with the assert of Maciejweski in the second case (7).

## 2. Two gyrostats problem

### 2.1. NOTATION

We will denote by bold italic letters vectors in $\mathbf{R}^3$ as geometrical objects, $\mathbf{a}, \mathbf{b}, \mathbf{c}$. When a reference frame is fixed then the set of components of a vector $\mathbf{x}$ will be denoted by the corresponding bold Roman letter $\mathbf{x}$, and will be considered as one column matrix, it is, $\mathbf{x} = [x_1, x_2, x_3]^T$, where $\mathbf{A}^T$ denotes the transposition of a matrix $\mathbf{A}$. We use subscripts for numbering coordinates of a vector, and superscript for distinguishing vectors of different bodies. We will not distinguish the difference between $\mathbf{x}$ and $\mathbf{x}$ calling the last object coordinates vector. The standard scalar and vector products in $\mathbf{R}^3$, and the length of a vector will be denoted respectively by $< \mathbf{x}, \mathbf{y} >$ , $(\mathbf{x} \times \mathbf{y}$ and $\|\mathbf{x}\| = \sqrt{< \mathbf{x}, \mathbf{x} >}$. Gyrostatic motion is described in a chosen reference frame. The reference frames used in this paper are right handed and orthonormal. Here the special orthogonal group $SO(3)$ will be identified with all $3 \times 3$ real orthogonal matrices with determinant $+1$. Its Lie algebra, denoted by $so(3)$, is identified with all $3 \times 3$ anti-symmetric matrices.

For a vector $\mathbf{x}$ we denote by $\hat{\mathbf{x}}$ the image of $\mathbf{x}$ by the standard isomorphism between the Lie algebra $\mathbf{R}^3$ ( with the vector product as the algebra multiplication) and $so(3)$.

### 2.2. EQUATIONS OF MOTION

Let us consider two gyrostats $S_1$ and $S_2$ that interact mutually according to the universal law of gravitation. The quantities related to gyrostat $(i)$ will be denoted, in coordinates, by a superscript i, and in vector notation by subscript $i$. We describe the motion of gyrostats with respect to an arbitrary inertial frame $R = \{u_1, u_2, u_3\}$. The body fixed frame in each gyrostat $R_1$ and $R_2$ is attached to its mass center $C_i$, $i = 1, 2$ and the first frame coincide with the principal axes of inertia of $S_1$. We denote the versors of the inertial reference frame in the systems $R_i$ by $\alpha_i$, $\beta_i$ and $\gamma_i$ respectively. Then the matrices

$$\mathbf{A}_i = \begin{bmatrix} \alpha_i^T \\ \beta_i^T \\ \gamma_i^T \end{bmatrix} = \begin{bmatrix} \alpha_1^i & \alpha_2^i & \alpha_3^i \\ \beta_1^i & \beta_2^i & \beta_3^i \\ \gamma_1^i & \gamma_2^i & \gamma_3^i \end{bmatrix} \in SO(3) \quad i = 1, 2, \tag{1}$$

are called the attitude matrices or the direction cosines matrices.

We have the following equations for the time dependence of the attitude matrices

$$\dot{\mathbf{A}}_i = \mathbf{A}_i \widehat{\boldsymbol{\Omega}}_i \quad i = 1, 2, \tag{2}$$

where $\Omega_i$ is the body angular velocity of $S_i$. Coordinates of this vector in the inertial frame $R$ are $\omega_i = [\omega_1^i, \omega_2^i, \omega_3^i]^T = A_i \Omega_i$, $i = 1, 2$ and following equation is satisfied

$$A_i = \hat{\omega}_i A_i. \tag{3}$$

We will denote by $R_i$, $i = 1, 2$ the position vector of each mass center $C_i$ in the frame $R$. If $Q_i = [Q_1^i, Q_2^i, Q_3^i]^T$ are the coordinates of a generic point of $S_i$ in the inertial frame $R$ and $q_i = [q_1^i, q_2^i, q_3^i]^T$ are the coordinates in the frames $R_i$, $i = 1, 2$, then the relations $Q_i = R_i + A_i q_i$, $i = 1, 2$ are satisfied.

The linear and the total angular momenta of the gyrostats are

$$P_i = \int_{S_i} \frac{d}{dt}\{R_i + A_i q_i\}\, dm_i = m_i\, \dot{R}_i, \tag{4}$$

$$L_i = \int_{S_i} \{R_i + A_i q_i\} \times \frac{d}{dt}\{R_i + A_i q_i\}\, dm_i = R_i \times P_i + \pi_i + l_{r_i}, \quad i = 1, 2, \tag{5}$$

where

$$m_i = \int_{S_i} dm_i, \quad \pi_i = A_i \Pi_i, \quad \Pi_i = I_i \Omega_i,$$

$$I_i = \int_{S_i} \hat{q}_i \hat{q}_i^T\, dm_i, \quad l_{r_i} = A_i L_{r_i}, \quad L_{r_i} = \int_{S_i} (q_i \times \dot{q}_i)\, dm_i, \tag{6}$$

are the masses of the gyrostats, the rotational angular momenta of the gyrostats considered as rigid bodies in the frame $R$ and $R_i$, $i = 1, 2$, the tensors of inertia and the gyrostatic momenta of the gyrostats in the frame $R$ and $R_i$ (i.e. the relative angular momenta of the mobile parts of the gyrostats $S_i$), respectively. In the following we will assume that the $L_{r_i}$ are known constants.

The kinetic energy of the system is the sum of the kinetic energies of the individual gyrostats

$$T = T_1 + T_2,$$

$$T_i = \frac{1}{2}\int_{S_i} \left\| \dot{q}_i \right\|^2 dm_i = T_{1i} + T_{2i} + T_{ri},$$

$$T_{1i} = \langle L_{r_i}, I_i^{-1}\Pi_i \rangle, \quad T_{2i} = \frac{\|P_i\|}{2m_i} + \frac{1}{2}\langle \Pi_i, I_i^{-1}\Pi_i \rangle, \tag{7}$$

where $T_{ri}$ are the relative kinetic energies of the gyrostats $S_i$ and we will assume that $T_{ri}$ are known functions of time.

The gravitational potential energy of the system is given by

$$U(R_1, A_1; R_2, A_2) = -G\int_{S_1}\int_{S_2} \frac{dm_1\, dm_2}{\|R_1 + A_1 q_1 - R_2 - A_2 q_2\|}, \tag{8}$$

where **G** is the universal gravitational constant, see (Cid and Vigueras, 1995).

In the Newtonian description of motion, we choose as independent variables, the inertial coordinates of radii vectors of mass centers $\mathbf{R}_i$, the linear momenta $\mathbf{P}_i$, the attitude matrices $\mathbf{A}_i$ for $i = 1, 2$, and describe the rotational motions with respect to each gyrostats reference frames using the rotational angular momenta $\Pi_i$. We denote by $\mathbf{f}_{ij}$, $\mathbf{m}_{ij}$ $i, j = 1, 2$ the gravitational forces and the resultant torques acting on the gyrostat $S_i$ expressed in the inertial frame $R$. Newton equations have the following form

$$\dot{\mathbf{R}}_i = \frac{\mathbf{P}_i}{m_i}, \quad \dot{\mathbf{P}}_i = \mathbf{f}_{ij}, \quad \dot{\Pi}_i = (\Pi_i + \mathbf{L}_{r_i}) \times \mathbf{I}_i^{-1} \Pi_i + \mathbf{M}_{ij}, \tag{9}$$

$$\dot{\mathbf{A}}_i = \mathbf{A}_i \widehat{\Omega}_i \quad i \neq j = 1, 2 \quad \mathbf{M}_{ij} = \mathbf{A}_i^T \mathbf{m}_{ij}. \tag{10}$$

The Newton equations have the classical first integrals

$$\mathcal{H} = \mathbf{T}_{21} + \mathbf{T}_{22} + \mathbf{U}, \quad \mathcal{P} = \mathbf{P}_1 + \mathbf{P}_2, \quad \mathcal{L} = \mathbf{L}_1 + \mathbf{L}_2,$$

and $\mathcal{D}_i = \mathbf{A}_i \mathbf{A}_i^T \quad i = 1, 2,$

where $\mathcal{H}$ is the Jacobi integral, $\mathcal{P}$ and $\mathcal{L}$ are integrals of total linear and angular momenta respectively, and $\mathcal{D}_i$ are geometrical integrals, in fact $\mathcal{D}_i = E$ the identity matrix.

## 2.3. HAMILTONIAN STRUCTURE OF THE EQUATIONS OF MOTION

From the expression of attitude matrices (1) we write the gravitational potential in the following form

$$U(\mathbf{R}_1, \alpha_1, \beta_1, \gamma_1; \mathbf{R}_2, \alpha_2, \beta_2, \gamma_2) = \int_{S_1} \int_{S_2} \frac{-\mathbf{G} \, dm_1 \, dm_2}{\|\mathbf{R}_1 + \mathbf{A}_1 \mathbf{q}_1 - \mathbf{R}_2 - \mathbf{A}_2 \mathbf{q}_2\|}. \tag{11}$$

Trivially we obtain

$$\mathbf{f}_{12} = -\frac{\partial U}{\partial \mathbf{R}_1}, \; \mathbf{f}_{21} = -\frac{\partial U}{\partial \mathbf{R}_2}, \; \mathbf{M}_{ij} = \alpha_i \times \frac{\partial U}{\partial \alpha_i} + \beta_i \times \frac{\partial U}{\partial \beta_i} + \gamma_i \times \frac{\partial U}{\partial \gamma_i}. \tag{12}$$

Then the Newton-Euler equations (9)-(10) can be expressed as follows

$$\dot{\mathbf{R}}_i = \frac{\mathbf{P}_i}{m_i}, \quad \dot{\mathbf{P}}_i = -\frac{\partial U}{\partial \mathbf{R}_i}, \tag{13}$$

$$\dot{\Pi}_i = (\Pi_i + \mathbf{L}_{r_i}) \times \mathbf{I}_i^{-1} \Pi_i + \alpha_i \times \frac{\partial U}{\partial \alpha_i} + \beta_i \times \frac{\partial U}{\partial \beta_i} + \gamma_i \times \frac{\partial U}{\partial \gamma_i}, \tag{14}$$

$$\frac{d\alpha_i}{dt} = \alpha_i \times \mathbf{I}_i^{-1} \Pi_i, \quad \frac{d\beta_i}{dt} = \beta_i \times \mathbf{I}_i^{-1} \Pi_i, \quad \frac{d\gamma_i}{dt} = \gamma_i \times \mathbf{I}_i^{-1} \Pi_i. \tag{15}$$

Denote $\mathbf{z} = [\mathbf{z}_1^T, \mathbf{z}_2^T]^T$ with $\mathbf{z}_i = [\mathbf{R}_i^T, \mathbf{P}_i^T, \Pi_i^T, \alpha_i^T, \beta_i^T, \gamma_i^T]^T \in \mathbf{R}^{18}$   $i = 1, 2$ and define a twice contravariant skew-symmetric tensor field $\Lambda$ on $\mathbf{R}^{36}$ that in matrix form is

$$\Lambda[z] = \begin{bmatrix} \Lambda_1[\mathbf{z}] & 0 \\ 0 & \Lambda_2[\mathbf{z}] \end{bmatrix}, \Lambda_i[\mathbf{z}] = \begin{bmatrix} \mathbf{J}_r[\mathbf{z}] & 0 \\ 0 & \mathbf{J}_{\mathbf{II}}[\mathbf{z}_i] \end{bmatrix}, \tag{16}$$

$$\mathbf{J}_r(z) = \begin{bmatrix} 0 & \mathbf{E} \\ -\mathbf{E} & 0 \end{bmatrix} \qquad \mathbf{J}_{\mathbf{II}}(\mathbf{z}_i) = \begin{bmatrix} \widehat{\Pi_i + \mathbf{L}_{ri}} & \widehat{\alpha}_i & \widehat{\beta}_i & \widehat{\gamma}_i \\ \widehat{\alpha}_i & 0 & 0 & 0 \\ \widehat{\beta}_i & 0 & 0 & 0 \\ \widehat{\gamma}_i & 0 & 0 & 0 \end{bmatrix}. \tag{17}$$

This tensor field allows us to define Poisson bracket on $\mathbf{R}^{36}$ $\{\cdot, \cdot\}$, that for $\mathbf{f}, \mathbf{g} \in C^\infty(\mathbf{R}^{36})$ $\{\mathbf{f}, \mathbf{g}\}(\mathbf{z}) = \left(\frac{\partial \mathbf{f}}{\partial \mathbf{z}}\right)^T \Lambda[\mathbf{z}] \left(\frac{\partial \mathbf{g}}{\partial \mathbf{z}}\right)$. Then $\left(\mathbf{R}^{36}, \{\cdot, \cdot\}_\mathcal{M}\right)$ is a Poisson manifold.

Let us consider the Hamiltonian function on $\mathbf{R}^{36}$

$$\mathcal{H}[\mathbf{z}] = \sum_{i=1}^{2} \left\{ \frac{\|\mathbf{P}_i\|^2}{2m_i} + \frac{1}{2} < \Pi_i, \mathbf{I}_i^{-1} \Pi_i > \right\} + \mathbf{U}(\mathbf{z}). \tag{18}$$

Then the Newton-Euler equations of motion (13)-(15) can be written in Hamiltonian form as $\frac{d\mathbf{Z}}{dt} = \{\mathbf{z}, \mathcal{H}\}(\mathbf{z})$.

### 2.3.1. Relative coordinates I

In the first step of the reduction process we use the fact that the interaction between gyrostats depends on the relative position of the bodies. We consider the relative variables

$$\mathbf{r} = \mathbf{R}_1 - \mathbf{R}_2, \qquad \mathbf{p} = \frac{m_2}{m_1 + m_2} \mathbf{P}_1 - \frac{m_1}{m_1 + m_2} \mathbf{P}_2.$$

In these variables we can reduce the number of equations by six. The system of 30 equations that we obtain is Hamiltonian with respect to the Poisson bracket in $\mathbf{R}^{30}$ define below.

We now denote $\mathbf{z} = \left[\mathbf{r}^T, \mathbf{p}^T, \mathbf{z}_1^T, \mathbf{z}_2^T\right]^T$, $\mathbf{z}_i = \left[\Pi_i^T, \alpha_i^T, \beta_i^T, \gamma_i^T\right]^T$   $i = 1, 2$ and define the Poisson bracket $\{., .\}_I$ in $\mathbf{R}^{30}$ by

$$\{\mathbf{f}, \mathbf{g}\}_I (\mathbf{z}) = \left(\frac{\partial \mathbf{f}}{\partial \mathbf{z}}\right)^T \Lambda_I[\mathbf{z}] \left(\frac{\partial \mathbf{g}}{\partial \mathbf{z}}\right) \text{ where} \tag{19}$$

$$\Lambda_I[\mathbf{z}] = \begin{bmatrix} \mathbf{J}_\mathbf{r}(z) & 0 & 0 \\ 0 & \mathbf{J}_{\alpha II}(\mathbf{z}_1) & 0 \\ 0 & 0 & \mathbf{J}_{\mathbf{II}}(\mathbf{z}_2) \end{bmatrix}. \tag{20}$$

In this Poisson structure the new equations are Hamiltonian in the form

$$\dot{z} = \{z, \mathcal{H}_I\}_I,$$

with respect to the Hamiltonian

$$\mathcal{H}_I = \frac{\|\mathbf{p}\|^2}{2m} + \frac{1}{2}\sum_{i=1}^{2}\{< \Pi_i, \mathbf{I}_i^{-1}\Pi_i >\} + \mathrm{U}(\mathbf{z}). \tag{21}$$

The Poisson bracket defined by (19)-(20) is just the effect of reduction of the bracket defined by (16)-(17) with respect to the action of the group of traslations on the phase space.

### 2.3.2. Relative coordinates II

In this step of reduction process we will describe the whole motion of the system with respect to the frame fixed in the gyrostat $S_2$. Then we introduce the new variables

$$\mathbf{R} = \mathbf{A}_2^T \mathbf{r}, \quad \mathbf{P} = \mathbf{A}_2^T \mathbf{p}, \quad \Gamma_1 = \mathbf{A}^T \Pi_1, \quad \Gamma_2 = \Pi_2, \tag{22}$$

$$\mathbf{A} = \mathbf{A}_1^T \mathbf{A}_2, \quad \mathbf{A} = \begin{bmatrix} \alpha^T \\ \beta^T \\ \gamma^T \end{bmatrix}. \tag{23}$$

Now we denote $\mathbf{v}\left[\mathbf{R}^T, \mathbf{P}^T, \Gamma_1^T, \Gamma_2^T, \alpha^T, \beta^T, \gamma^T\right]^T \in \mathbf{R}^{21}$ and define the following Poisson bracket $\{.,.\}_{II}$ in $\mathbf{R}^{21}$, $\{\mathbf{f},\mathbf{g}\}_{II}(\mathbf{v}) = \left(\frac{\partial \mathbf{f}}{\partial \mathbf{v}}\right)^T \Delta[\mathbf{v}]\left(\frac{\partial \mathbf{g}}{\partial \mathbf{v}}\right)$ for $\mathbf{f}, \mathbf{g} \in C^\infty\left(\mathbf{R}^{21}\right)$ where $\Delta[\mathbf{v}]$ is the Poisson tensor defined as follows

$$\Delta[\mathbf{v}] = \begin{bmatrix} 0 & \mathbf{E} & 0 & \widehat{\mathbf{R}} & 0 & 0 & 0 \\ -\mathbf{E} & 0 & 0 & \widehat{\mathbf{P}} & 0 & 0 & 0 \\ \widehat{\mathbf{R}} & \widehat{\mathbf{P}} & \widehat{\mathbf{A}^T \mathbf{L}_{r_1}} - \Gamma_1 & \widehat{\Gamma_1} & -\widehat{\alpha} & -\widehat{\beta} & -\widehat{\gamma} \\ 0 & 0 & \widehat{\Gamma_1} & \Gamma_2 + \mathbf{L}_{r_2} & \widehat{\alpha} & \widehat{\beta} & \widehat{\gamma} \\ 0 & 0 & -\widehat{\alpha} & \widehat{\alpha} & 0 & 0 & 0 \\ 0 & 0 & -\widehat{\beta} & \widehat{\beta} & 0 & 0 & 0 \\ 0 & 0 & -\widehat{\gamma} & \widehat{\gamma} & 0 & 0 & 0 \end{bmatrix}$$

In this Poisson structure the equations of motion described in the new variables (22)-(23) are Hamiltonian with respect the Hamiltonian $\mathcal{H}_{II}$ defined by

$$\mathcal{H}_{II} = \frac{\|\mathbf{P}\|^2}{2m} + \mathbf{H}_r + \mathrm{U}(\mathbf{v}), \tag{24}$$

where $U(\mathbf{v}) = -G \int_{S_1} \int_{S_2} \dfrac{dm_1 \, dm_2}{\left\| \mathbf{R} + \mathbf{A}^T \mathbf{q}_1 - \mathbf{q}_2 \right\|}$

and $H_r = \frac{1}{2} < \Gamma_1, \mathbf{A}^T \mathbf{I}_1^{-1} \mathbf{A} \Gamma_1 > + \frac{1}{2} < \Gamma_2, \mathbf{I}_2^{-1} \Gamma_2 >$. Then in $\left( \mathbf{R}^{21}, \{\cdot, \cdot\}_{II} \right)$ the induced Hamiltonian equations become $\frac{d\mathbf{V}}{dt} = \{\mathbf{z}, \mathcal{H}_{II}\}_{II}$ (v).

Finally, we note that the Poisson bracket $\{\cdot, \cdot\}_{II}$ is degenerated, i. e., there are non-constant Casimir functions: six geometrical integrals and the total angular momentum of the system

$$C_1 = \frac{1}{2} \alpha^T \alpha, \quad C_2 = \alpha^T \beta, \quad C_3 = \alpha^T \gamma, \quad C_4 = \frac{1}{2} \beta^T \beta, \quad C_5 = \beta^T \gamma,$$

$$C_6 = \frac{1}{2} \gamma^T \gamma, \mathcal{F} = \frac{1}{2} \mathbf{L} \mathbf{L}^T, \mathbf{L} = \Gamma_1 + \mathbf{A}^T \mathbf{L}_{r_1} + \Gamma_2 + \mathbf{R} \times \mathbf{P}.$$

## 3. Relative Equilibria

The relative equilibria of our system are the equilibria for the Hamiltonian dynamic generated by $\mathcal{H}_{II}$ on $\mathcal{M}_{II}$, i.e. $\mathbf{v}_e \in \mathcal{M}_{II}$ such that $\{\mathbf{v}, \mathcal{H}_{II}\}_{\mathcal{M}_{II}} (\mathbf{v}_e) = 0$. Using vector algebra in the first equations of the Poisson system we show that there exists $\lambda \in \mathbf{R}$ such that we can write the conditions for the equilibria as

$$\|\Omega\|^2 \mathbf{R} - < \mathbf{R}, \Omega > \Omega - \frac{1}{m} \frac{\partial U}{\partial \mathbf{R}} = 0, \tag{25}$$

$$(\Gamma_2 + \mathbf{L}_{r_2}) \times \Omega + \mu_2 = 0, \tag{26}$$

$$\Gamma_1 + \mathbf{A}^T \mathbf{L}_{r_1} + \Gamma_2 + \mathbf{L}_{r_2} - m < \mathbf{R}, \Omega > \Omega - \lambda \Omega = 0, \tag{27}$$

$$\alpha^T \alpha = \beta^T \beta = \gamma^T \gamma = 1, \ \alpha^T \beta = \alpha^T \gamma = \beta^T \gamma = 0, \tag{28}$$

with $\mu_2 = -G \int_{S_1} \int_{S_2} q_2 \times \left( (q_2 - \mathbf{R} - \mathbf{A}^T q_1) / \left\| \mathbf{R} + \mathbf{A}^T q_1 - q_2 \right\|^3 \right) dm(q_1) dm(q_2)$ and $\mathbf{P} = m\Omega \times \mathbf{R}$, $\Omega_1 = \mathbf{A}\Omega_2$ and $\Omega = \Omega_2$.

Using the expression of $\mathbf{P}$ in the equilibria we obtain trivially $\mathbf{L} = \left( \lambda + m \|\mathbf{R}\|^2 \right) \Omega$, so $\mathbf{L}$ is parallel to $\Omega$.

There are two types of equilibria, the Lagrangian equilibria where $\mathbf{R}$ is orthogonal to $\Omega$ and the non-Lagrangian equilibria where $< \mathbf{R}, \Omega > \neq 0$. In the first case we have $\|\mathbf{r}\| = \|\mathbf{R}\| = cte$ and $< \mathbf{r}, \omega > = < \mathbf{R}, \Omega > = 0$. Then the gyrostat $S_1$ describes a circular orbit with respect to the gyrostat $S_2$ in a plane orthogonal to the vector $\omega$. In the non-Lagrangian case $\|\mathbf{r}\| = \|\mathbf{R}\| = cte$ and $< \mathbf{r}, \omega > = < \mathbf{R}, \Omega > \neq 0$, so the projection of $\mathbf{r}$ over the vector $\omega$ is constant. Then the gyrostat $S_1$ describes, with respect to the gyrostat $S_2$, a circular orbit in a cone of axis $\omega$.

Finally, we note that we have adopted the names of Lagrangian and non-Lagrangian equilibria as Maciejewski suggests in (Maciejewski, 1995). The results obtained here for the two gyrostats problem are similar to the ones obtained by Maciejewski for the two rigid bodies.

## 4. Approximations

In the present section we consider two approximations of the gravitational potential **U** based on Taylor series in a neighborhood of $\varepsilon$ =(nominal dimension of gyrostats)/(orbital radius). The Taylor series expression of **U** is

$$\mathbf{U}(\mathbf{v}) = \mathbf{U}_0 + \mathbf{U}_1 + o\left(\|\mathbf{R}\|^5\right), \quad \mathbf{U}_0 = \frac{-Gm_1m_2}{\|\mathbf{R}\|},$$

$$\mathbf{U}_1 = -\frac{G}{2\|\mathbf{R}\|^3}\{m_2 Tr\mathbf{I}_1 + m_1 Tr\mathbf{I}_2\} +$$

$$\frac{3G}{2\|\mathbf{R}\|^5}\{m_2 < \mathbf{AR}, \mathbf{I}_1\mathbf{AR} > +m_1 < \mathbf{R}, \mathbf{I}_2\mathbf{R} >\}.$$

We will consider therefore two approximate Hamiltonians

$$\mathcal{H}_0 = \frac{\|\mathbf{P}\|^2}{2m} + \mathbf{H}_r + \mathbf{U}_0, \quad \mathcal{H}_1 = \mathcal{H}_0 + \mathbf{U}_1. \tag{29}$$

In the following we will consider that the gyrostatic momentum of $S_2$ is null, i. e., $\mathbf{L}_{r_2} = \mathbf{0}$.

### 4.1. ZERO ORDER APPROXIMATION

Let us consider the dynamic generated by $\mathcal{H}_0$. Then the condition for equilibria (25)-(28) can be written

$$\|\mathit{\Omega}\|^2 \mathbf{R} - < \mathbf{R}, \mathit{\Omega} > \mathit{\Omega} = \frac{Gm_1m_2}{\|\mathbf{R}\|}\mathbf{R}, \mathit{\Gamma}_2 \times \mathit{\Omega} = \mathbf{0},$$

$$\mathit{\Gamma}_1 + \mathbf{A}^T \mathbf{L}_{r_1} + \mathit{\Gamma}_2 - m < \mathbf{R}, \mathit{\Omega} > \mathit{\Omega} - \lambda\mathit{\Omega} = 0, \mathbf{A} \in SO(3).$$

Taking cross product in the above first equation by **R** we obtain that **R** is orthogonal to $\mathit{\Omega}$ in the equilibria (because we exclude the trivial case **P=0**), then in the zero order approximation only the Lagrangian equilibria are possible. Now from the second above equation we deduce that $\mathit{\Gamma}_2$ is parallel to $\mathit{\Omega}$, and so $\mathit{\Omega}$ is an eigenvector of $\mathbf{I}_2$. Then **R**, **P** and $\mathit{\Omega}$ form a triad in the equilibria, and without loss of generality we assign

$$\mathit{\Omega}_e = \|\mathit{\Omega}_e\| e_1^2, \quad \mathbf{R}_e = \|\mathbf{R}_e\| e_3^2, \quad \mathbf{P}_e = m\|\mathit{\Omega}_e\|\|\mathbf{R}_e\| e_1^2 \times e_3^2,$$

where $\{e_1^2, e_2^2, e_3^2\}$ is the basis of the frame $R_2$. We describe the equilibria in this basis and trivialy we have

$$\mathit{\Gamma}_{2e} = \|\mathit{\Omega}_e\| I_1^2 e_1^2, \quad \mathit{\Gamma}_{1e} = \|\mathit{\Omega}_e\| \mathbf{A}_e^T \mathbf{I}_1 \mathbf{A}_e e_1^2, \quad \mathbf{I}_2 = \begin{bmatrix} I_1^2 & 0 & 0 \\ 0 & I_2^2 & J \\ 0 & J & I_3^2 \end{bmatrix},$$

such that $(\mathbf{I}_1\mathbf{A}_e\mathbf{e}_1^2 + \mathbf{L}_{r1}) \times \mathbf{A}_e\mathbf{e}_1^2 = \mathbf{0}$, $\quad m\,\|\Omega_e\|^2 = \frac{Gm_1m_2}{\|\mathbf{R}_e\|}$ where $\mathbf{A}_e \in SO(3)$.

Proceeding as Wang (1990), let us consider now the following solutions of the Hamiltonian system $dv/dt = \{\mathbf{v}, \mathcal{H}_0\}_{\mathbf{II}}(\mathbf{v})$

$$\Gamma_1(t) = \Gamma_{1e}, \ \Gamma_2(t) = \Gamma_{2e}, \ A(tA\alpha) = A_e, \tag{30}$$

$$\mathbf{R}(t) = \exp\left(t \cdot \frac{\mathbf{k}}{\left\|\mathbf{I}_2^{-1}\Gamma_{2e}\right\|}\widehat{\mathbf{I}_2^{-1}\Gamma_{2e}}\right)\mathbf{R}_e, \tag{31}$$

$$\mathbf{P}(t) = m\left(1 + \frac{\mathbf{k}}{\left\|\mathbf{I}_2^{-1}\Gamma_{2e}\right\|}\right)\exp\left(t \cdot \frac{\mathbf{k}}{\left\|\mathbf{I}_2^{-1}\Gamma_{2e}\right\|}\widehat{\mathbf{I}_2^{-1}\Gamma_{2e}}\right)\widehat{\mathbf{I}_2^{-1}\Gamma_{2e}}\mathbf{R}_e, \tag{32}$$

verifying the modified formula $\left(\mathbf{k} + \left\|\mathbf{I}_2^{-1}\Gamma_{2e}\right\|\right)^2 = \frac{Gm_1m_2}{\|\mathbf{R}_e\|}$.

The above solutions represent a pertubation of the above zero order relative equilibria and it is easy to prove that the solutions escape from any small neigborhood of the relative equilibria in finite time. Therefore we have instability.

### 4.2. SECOND ORDER APPROXIMATION

We now take the dynamic generated by the Hamiltonian $\mathcal{H}_1$. The equilibria conditions (25)-(28) read now

$$m\,\|\Omega\|^2\mathbf{R} - m < \mathbf{R}, \Omega > \Omega = \frac{Gm_1m_2}{\|\mathbf{R}\|}\mathbf{R} +$$

$$\frac{3G}{2\,\|\mathbf{R}\|^3}\{\{m_2Tr\mathbf{I}_1 + m_1Tr\mathbf{I}_2\}\,\mathbf{R} + 2m_2\Lambda^T\mathbf{I}_1\Lambda\mathbf{R} \mid 2m_1\mathbf{I}_2\mathbf{R} -$$

$$\frac{15G}{2\,\|\mathbf{R}\|^5}\{m_2 <, I\alpha A\mathbf{R}, I_1 > \alpha A\mathbf{R} > + m_1 < \mathbf{R}, I_2\mathbf{R}>\}\,\mathbf{R}, \tag{33}$$

$$\Gamma_2 \times \Omega + \mu_2 = 0 \quad \text{with} \quad \mu_2 = \frac{3Gm_2}{2\,\|\mathbf{R}\|^5}\mathbf{R} \times \mathbf{I}_2\mathbf{R}, \tag{34}$$

$$[\mathbf{A}^T\mathbf{I}_1\mathbf{A}\Omega + \mathbf{A}^T\mathbf{L}_{r_1} + \mathbf{I}_2\Omega - m < \mathbf{R}, \Omega > \Omega]\lambda\Omega = 0. \tag{35}$$

The above system is a cumbersome non-linear system. Nevertheless, we can always find numerically solutions for specific values fixed of the parameters, or introducing a small parameter, we can use perturbative techniques. An example of Non-Lagrangian equilibria was found numerically by Wang (1990) in the case of a rigid body in a central Newtonian force field. Here, we will restrict to the Lagrangian case.

As in the previous subsection we will denote the equilibria solutions with subscripts e. We choose $\{\Omega_e/\|\Omega_e\|, P_e/\|P_e\|, R_e/\|R_e\|\}$ as the reference frame $R_2$. Using the equations (34) it is easy to show that the inertial tensor $I_2$ is diagonal on this basis.

Now from equation (35) we have that $I_1 A_e \Omega_e + L_{r_1}$ is parallel to $A\Omega_e$. Taking cross product in (33) by $R$ we obtain $I_1 A_e R_e$ parallel to $A_e R_e$. Since $A_e \in SO(3)$ and takes eigenvectors of $I_2$ into eigenvectors of $I_1$ we deduce $A_e$ belong to

$$\left\{ \begin{bmatrix} a & b & 0 \\ -b & a & 0 \\ 0 & 0 & 1 \end{bmatrix}, a^2 + b^2 = 1; \; a, b \in \mathbf{R} \right\}. \tag{36}$$

Taking scalar product in (34) by $R_e$, and since $A_e$ takes $e_3$ into $e_j$ $j \in \{1, 2, 3\}$, we obtain

$$\|\Omega_e\|^2 = \frac{G}{\|R_e\|^3} \left\{ m_1 m_2 + m_2 I_3^1 + m_1 I_3^2 \right\} + $$
$$\frac{3G}{2\|R_e\|^5} \left\{ m_2(Tr I_1 - 5I_j^1) + m_1(Tr I_2 - 5I_j^2) \right\}. \tag{37}$$

Then if $\|R_e\|$ is large enough it is clear that there exist solutions of the above equation. Denoting $L_{r_1} = [a_1, b_1, c_1]^T$, and since $I_1 A_e \Omega_e + L_{r_1}$ is parallel to $A\Omega_e$ we obtain $\|\Omega_e\|(I_2^2 - I_1^1)ab - b_1 a - a_1 b = 0$, $c_1 = 0$.

If we choose $\{P_e/\|P_e\|, \Omega_e/\|\Omega_e\|, R_e/\|R_e\|\}$ as the reference frame $R_2$ we obtain

$$A_e = \begin{bmatrix} 0 & 0 & 1 \\ 1 & 0 & 0 \\ 0 & 1 & 0 \end{bmatrix}, L_{r_1} = [0, b_1, 0]^T,$$

verifying the formula (37) as in the previous case.

Finally, we observe that in the case of two rigid bodies (see (7)) the equilibrium configurations are all the possible combinations between the vector $R_e$, $P_e$, and $\Omega_e$ as elements of the reference frame $R_2$; and the matrix $A_e$ can be any of the matrices which represent a permutation between the elements of the base. However, in the case considered in this paper, for each choice of the base only special cases of the matrix $A_e$ can be choosen.

## References

G. N. Duboshin: 1976, *Celestial Mechanics,* , **14** 239

M. Z. Aboelnaga and Y. V. Barkin: 1979, *Astronom. Zh.* , **56** , 881

Y. V. Barkin: 1980, *Pis'ma Astron. Zh.* , **6**, 377

L. Wang, P. S. Krishnaprasad and J. H. Maddocks: 1991, *Celestial Mech.*, **50**, 349

L. Wang and P. Chen: 1995, *IEEE Transactions on Automatic Control*, **10**, 1732

R. Cid and A. Vigueras: 1985, *Celestial Mech.* , **36**, 135

A. Maciejewski: 1995, *Celestial Mech.*, **63**, 1

# PLANETARY SYSTEMS

PAWEL ARTYMOWICZ

*Stockholm Observatory, S-133 36 Saltsjobaden, Sweden*

**Abstract.** The past decade brought direct evidence of the previously surmised exoplanetary systems. A variety of planetary system types exist: those around pulsars, around both young and old main-sequence stars (as evidenced by planetesimal disks of the Beta Pictoris-type), and the mature giant exoplanets found in radial velocity surveys. The surprising diversity of the exoplanetary systems is addressed by several theories of their origin.

## 1. Planetary Systems around Pulsars

Distant planets have been predicted by the ancient doctrine of atomism. Democritus reportedly said: "In some worlds there is no Sun and Moon, in others they are larger than in our world, and in others more numerous. In some parts there are more worlds, in others fewer (...); in some parts they are arising, in others failing. There are some worlds devoid of living creatures or plants or any moisture." If this was a prescient prediction of planets (called worlds back then), accompanying various stars including binaries, it was correct. The first extrasolar planets have indeed been found in a strange system without a Sun (Wolszczan and Frail 1992, Wolszczan 1994). Nevertheless, the 1992 discovery of 2 Earth-class companions of the pulsar PSR B1257+12 was based neither on Democritus' prediction, nor on the work of N. Copernicus (in whose home town, Toruń, Wolszczan studied astronomy). It was fortuitous.

The pulsar planets are relatively uncommon (at most two systems per $\sim 500$ pulsars known so far), yet the famous prototype remains up to now the only confirmed sensu stricto *planetary system* of several planets. Pulsar planets clearly demonstrate that planet formation occurs under a variety of potentially adverse conditions, e.g., in the aftermath of supernova explosion (Phillips & Thorsett 1994).

## 2. Planetary Material around Normal Stars

Since 1980s, there were very strong indications that planetary systems of some sort exist around a large fraction (now known to be $\sim 30\%$) of normal stars. The IRAS satellite discovered strong IR excess emission from what was later proved to be dust disks around many nearby stars, such as Vega, Fomalhaut, $\beta$ Pictoris ($\beta$ Pic), and others (more than a hundred have been identified subsequently, cf. review by Lagrange et al. 1999). The sizes of the disks, usually several hundred AU across, their central depleted regions (probably by planets located within several dozen AU), the amount of directly observed small meteoroids and dust, the mass of the inferred rock+ice planetesimals needed to replenish the directly observed dust, all the above make such systems resemble closely our Solar System. The strongest case can be made for $\beta$ Pic, which resembles our system at the "early bombardment"

*Celestial Mechanics and Dynamical Astronomy* 73: 313–316, 1999.
© 1999 *Kluwer Academic Publishers.*

epoch (age of $\sim 10^8$ yr). Direct spectroscopic evidence reveals up to several comet-sized planetesimals passing near the star's surface per week. Some of the $\beta$ Pic grains, probably released by colliding and/or evaporating planetesimals, produce the 10-micron emission feature of a great similarity to that of Halley's comet. In addition, the edge-on dust disk imaged in scattered light and thermal radiation (12 and 18 micron bands) seems to be warped at distances $<$ 100 AU from the star by an invisible planetary companion on a slightly inclined orbit (cf. references in a review by Artymowicz 1997, and the most recent HST/STIS observations by Heap et al. 1998). Nevertheless, a direct evidence for a planet is still lacking, even in this best-studied IR-excess system. We expect that some systems might originate from low surface density solar nebulae that are unable to form giant or any planets; those may be long-lived, undeveloped (or failed) planetary systems with only Moon- and comet-sized bodies. We do not know yet if most Vega-type star harbor true or failed planetary systems.

Several recent images taken in the sub-mm wavelengths indicate a possible existence of "blobs" of warm, radiating dust at some distance from $\beta$ Pic, Vega, and Fomalhaut (Holland *et al.* 1998). Until confirmed, this observation must be viewed with caution, as the blobs may be instrumental or caused by background objects.

Imaging and modeling of the 10 Myr-old, IR-excess star HR 4796A by Koerner et al. (1998) and Jayawardhana (1998) revealed a medium sized disk of the $\beta$ Pic–type seen at 18° short of edge-on, where most dust lies between 55 AU and some 120 AU from the star. It appears (Lagrange et al. 1999) that HR 4796A must be a transitional object between the massive and optically thick primordial accretion disks and later-stage, optically thin, post-T Tau disks ($\beta$ Pic disks proper). If so, one should scrutinize the dust density distribution including possible disk density waves and edge features, diagnostic of the properties of planet(s) causing the central gap.

### 3. Exoplanets from Radial Velocity Surveys

By late 1998 13 candidate planetary systems have been found due to radial velocity variations (cf. an up-to-date listing by Schneider 1998, and a recent review by Marcy & Butler 1998), each with one confirmed planet. The curious lack of multiplanetary systems around normal stars may partly reflect observational limitations. Still, the working hypothesis or even a paradigm we used, that the our system is a typical planetary system, has been destroyed. Planets at least as massive 5–9 times the Jupiter mass ("superplanets") exist[1]. Planets on unfamiliar, very elongated orbits not only exist but are fairly common. Four superplanets, and two Jupiter-class planets, are on such elliptic orbits. The "hot Jupiter" or "51 Peg" family of systems has planets so close to the stellar surface that their temperatures exceed 1000 K.

---

[1] Inclinations are not known as a rule, which leaves us with the knowledge of only the minimum masses.

All these features, if not the existence of exoplanets themselves, were unexpected. Only one system remotely matches the sun-Jupiter system (47 UMa).

Existing theories are being verified and new ones created to account for the remarkable variety of planetary systems (e.g., Artymowicz 1992, Lin *et al.* 1996, Lin & Ida 1997, Artymowicz et al. 1998, Levison *et al.* 1998). Two common threads uniting the most promising theories are the processes of disk-planet and planet-planet interactions. Each can significantly affect mass, orbital radius, and eccentricity of a forming planet. Support for primordial disk-planet interaction as the dominant process comes mainly from: its necessity for giant gaseous planet formation, protoplanet migration that it causes (providing the explanation for hot Jupiters), and the natural explanation it offers for the mass-eccentricity correlation (planets much more massive than Jupiter will gain eccentricity from the resonant interaction, while smaller ones are circularized by the resonances and by mass in-flow). Planet-planet (and companion star-planet) interaction may be the best explanation for Jupiter-class systems which have significant eccentricities. Alternative views proposed for extrasolar planet formation involve the early fragmentation of disks (Boss 1998) and the belief that they are not planets after all but brown dwarfs seen at a small inclination angle. Better statistics is needed to exclude or confirm these ideas. (Currently the mass histogram of substellar companions stands in favor of the planetary nature of companions in the sense of the standard core-instability mechanism; e.g., Marcy and Butler 1998).

We do not yet have the possibility to conduct spectroscopy of exoplanets (Burrows et al 1998). The question of how to best distinguish failed stars (brown dwarfs) from overgrown planets, or in fact how to best define a planet (based on mass, differentiated internal structure or origin?) remains wide open for future work, both observational and theoretical. Astrometry, interferometry, photometry of occultations, microlensing, and other techniques promise to paint in the near future a much more detailed picture of exoplanetary systems, possibly extending the list of surprises and wonders. One basic thing is already clear, however. More (perhaps much more) than 4% of sun-like stars have planets. We are not alone, at least not as a planetary system.

## References

Artymowicz, P.: 1992, *PASP*, **104**, 769

Artymowicz, P.: 1997, *Ann. Rev. Earth Pl. Sci.*, **25**, 175

Artymowicz, P., Lubow, S.H., & Kley, W.: 1998, in *Planetary systems - the long view*, Eds. Celnikier, L. et al., Editions Frontières, in print

Boss, A.: 1998, *ApJ*, **503**, 923

Burrows, A., Marley, M., Hubbard, W., Lunine, J. et al.: 1999, *ApJ*, **491**, 856

Heap, S., Lindler, D., Lanz, T., Woodgate, B. et al.: 1998, *ApJ*, in print

Holland, W., Greaves, J., Zuckerman, B., Webb, R., et al.: 1998, *Nature*, **392**, 788

Jayawardhana, R., Fischer, S., Hartmann, L., et al.: 1998, *ApJ*, **503**, L79

Koerner, D. W., Ressler, M. E., Werner, M. W., and Backman, D. E.: 1998, *ApJ*, **503**, L83

Lagrange, A.-M., Backman, D., and Artymowicz, P.: 1999, in*Protostars and Planets IV*, Eds. V. Manning, A. Boss, U. Arizona Press, in print

Levison, H.: Lissauer, J., & Duncan, M., 1998, preprint
Lin, D. N. C.: Bodenheimer, P., & Richardson, D., 1996, *Nature*, **380**, 606
Lin, D. N. C.:, & Ida, S., 1997, *ApJ*, **477**, 781
Marcy, G.W.: & Butler, R.P., 1998, *ARAA*, **36**, 57
Phillips, J.: & Thorsett, A., 1994, *ApSS*, **212**, 1
Schneider, J. (Ed.):, 1998, The Extrasolar Planets Encyclopaedia, on-line database at
    http://www.obspm.fr/planets
Wolszczan, A.: 1994, *Science*, **264**, 538
Wolszczan, A., & Frail, D. A.: 1992, *Nature*, **355**, 145

# COMPLEMENTS TO MOONS' LUNAR LIBRATION THEORY

*Comparisons and fits to JPL numerical integrations*

J. CHAPRONT, M. CHAPRONT-TOUZÉ and G. FRANCOU

*DANOF-Observatoire de Paris, Unité de Recherche Associée au CNRS n° 1125*
*61, avenue de l'Observatoire, 75014, Paris, FRANCE*

**Abstract.** Analytical complements have been brought to Moons' lunar libration theory concerning tidal effects, direct perturbations due to the Earth's figure, and indirect non periodic perturbations. Comparisons to JPL numerical integrations DE245 and DE403 have been performed and the residuals treated by frequency analysis, allowing the determination of fitted free libration parameters and numerical complements.

## 1. Introduction

Elementary descriptions of lunar librations are given in (Danjon, 1959) or in (Hilton, 1992). Since 1975, several precise solutions have been developped in connection with the appearance of Lunar Laser Ranging observations. Moons' analytical theory (Moons, 1981, 1982, and 1984) includes series for the forced libration and series, denoted as "free libration series", which contain pure free libration terms and mixed terms depending on free libration and forced libration. Forced libration series involve :

-   Main problem, in which the Earth is reduced to its mass center, the selenocentric position of the Earth is provided by the main problem of the lunar orbital motion, and the position of the Sun with respect to the Earth-Moon barycenter is approximated by a pseudo-keplerian motion in the mean ecliptic of date ;

-   Indirect planetary perturbations, derived from the periodic part of the planetary perturbations of the lunar orbital motion ;

-   Indirect perturbations due to the Earth's figure, derived from the perturbations of the lunar orbital motion by the Earth's figure ;

-   Direct and indirect perturbations due to the ecliptic motion.

The coefficients of main problem series are literal with respect to increments to nominal values of the parameters $\beta = (C - A)/B$ and $\gamma = (B - A)/C$, and literal with respect to the ratio of third and fourth degree harmonic coefficients $C_{ij}$ and $S_{ij}$ ($i = 3, 4$, $0 \leq j \leq i$) to $C/m_L R_L^2$. $A$, $B$, $C$ are the lunar principal moments of intertia, $m_L$ and $R_L$ the lunar mass and equatorial radius. The coefficients of the indirect planetary perturbation series are literal in a similar way as the main problem, while the coefficients of the other perturbation series are numerical. The arguments are combinations of Delaunay arguments $D$, $F$, $l$, $l'$, and, for perturbations, of planetary longitudes and $\zeta$, the lunar mean mean longitude referred to the mean equinox of date.

*Celestial Mechanics and Dynamical Astronomy* **73**: 317–328, 1999.
© 1999 *Kluwer Academic Publishers.*

The coefficients of free libration series are literal with respect to three free libration parameters $\sqrt{2P}$, $\sqrt{2Q}$, $\sqrt{2R}$ and are otherly similar to the main problem series, except that fourth degree harmonics are not taken into account. The arguments are combinations of Delaunay arguments and of three arguments $p$, $q$, $r$ of the free libration.

The series yield $p_1$, $p_2$, and $\tau$. $p_1$ and $p_2$ are the components of the unit vector pointing toward the pole of the mean ecliptic of date, in the inertial sense as defined by Standish (1981), on the two lunar equatorial principal axes of inertia ; $\tau$ is the libration in longitude referred to the inertial mean ecliptic of date.

Moons' theory takes into account a rigid body. The perturbations due to the deformation of the Moon by the Earth, the Sun and lunar rotation (tidal perturbations) are missing, except those derived from constant perturbations of $\beta$, $\gamma$ and $C/m_L R_L^2$ which may be included. The direct perturbations due to the Earth's figure and to the planets are also missing. At last, the indirect perturbations derived from the part of the perturbations of the lunar orbital motion which contains the time as a factor (Poisson terms), or purely secular terms, have not been computed.

In Sect. 2 we give analytical expressions of some of those missing perturbations for the forced libration only:

-    Tidal perturbations in the case of an elastic model and in two examples of an anelastic model ;

-    Direct perturbations due to the Earth's figure ;

-    A rough estimate of two Poisson terms among the indirect planetary perturbations of $\tau$. They come from Poisson terms of the lunar orbital motion due to Venus action and to secular variation of the solar eccentricity.

Sect. 3 gives a comparison of the so-completed Moons' theory to the numerical integrations DE245 and DE403 of the Jet Propulsion Laboratory (JPL). The residuals are analyzed by means of a frequency analysis which puts into evidence the "free libration terms" and allows to derive fitted values of the free libration parameters.

The frequency analysis also allows to complete the analytical solution by a small number of trigonometric terms whose coefficients, frequencies, and phases are purely numerical. Sect. 4 shows the resulting improvements on residuals of Lunar laser ranging observations.

## 2. Analytical Complements to Moons' Theory

### 2.1. MÉTHOD

The method is similar to Eckhardt's (1981). The column matrix $X$, whose elements are $p_1$, $p_2$, $\tau$, is given by the differential equation:

$$TX = (T - R)X + Y + Y' + \Psi. \tag{1}$$

$R$ is a differential operator function of the components $\omega_i$ of $\omega$, referred to the lunar principal axes of inertia ; $\omega$ is the angular rotational velocity vector of the Moon with respect to the usual reference frame of the lunar motion (mean ecliptic of date and departure point). $Y$ results from the lunar potential and is expressed as a function of the selenocentric coordinates of the Earth $y_i$ referred to the lunar principal axes of inertia. $T$ is a linear differential operator such that $(T - R)X + Y$ does not contain any linear term with constant coefficient in $p_1$, $p_2$, $\tau$ and their derivatives at first order of the small parameters involved. The expressions of $TX$, $RX$, and $Y$ used in this paper can be found in (Chapront-Touzé, 1990) except that $\varepsilon$ is denoted as $\gamma$ in the present paper and $\nu^2 a_0^3$ must be replaced by $Gm_T$, $G$ being the constant of gravitation and $m_T$ the terrestrial mass. $Y'$ is obtained from $Y$ by changing the selenographic coordinates of the Earth $y_i$ to those of the Sun $y_i'$ and $m_T$ to the solar mass $m_S$. $\Psi$ is a vectorial disturbing function.

The leading effect of $\Psi$ is to add to the solution $X_M$ of the main problem of the forced libration the correction $\Delta X$ given by:

$$T\Delta X = \left[\frac{\partial}{\partial X}(T - R)X\right]\Delta X + \Psi + \left[\frac{\partial}{\partial X}Y\right]\Delta X + \left[\frac{\partial}{\partial X}Y'\right]\Delta X. \qquad (2)$$

$\Psi$ and the jacobian matrices $[\partial / \partial X \cdots]$ are computed for $X_M$ by disregarding the contribution of the free libration. Similarly, the contribution of $\Psi$ to the free libration has been disregarded. Eq. (2) is solved by two iterations, $\Delta X$ being set to zero in the right hand member at the first iteration.

## 2.2. DIRECT PERTURBATIONS DUE TO TIDAL EFFECTS

The actions of the Earth, the Sun and lunar rotation induce distortions of the lunar surface which, in turn, induce an additional lunar potential (Lambeck, 1980). This additional potential is equivalent to time dependent corrections $\Delta C_{ij}$, $\Delta S_{ij}$ to the constant harmonic coefficients $C_{ij}$, $S_{ij}$ of the potential of the rigid Moon. Restricting ourselves to harmonics of degree 2 and disregarding the Sun effect, we have:

$$\Delta C_{20} = k_2 \frac{m_T}{m_L} \frac{R_L^3}{r^{*5}} \left(y_3^{*2} - \frac{1}{2}y_1^{*2} - \frac{1}{2}y_2^{*2}\right) + \frac{1}{6} \frac{k_2 R_L^3}{Gm_L} \left(\omega_1^{*2} + \omega_2^{*2} - 2\omega_3^{*2}\right)$$

$$\Delta C_{21} = k_2 \frac{m_T}{m_L} \frac{R_L^3}{r^{*5}} y_3^* y_1^* - \frac{1}{3} \frac{k_2 R_L^3}{Gm_L} \omega_1^* \omega_3^*$$

$$\Delta S_{21} = k_2 \frac{m_T}{m_L} \frac{R_L^3}{r^{*5}} y_2^* y_3^* - \frac{1}{3} \frac{k_2 R_L^3}{Gm_L} \omega_2^* \omega_3^* \qquad (3)$$

$$\Delta C_{22} = \frac{k_2}{4} \frac{m_T}{m_L} \frac{R_L^3}{r^{*5}} \left(y_1^{*2} - y_2^{*2}\right) + \frac{1}{12} \frac{k_2 R_L^3}{Gm_L} \left(\omega_2^{*2} - \omega_1^{*2}\right)$$

$$\Delta S_{22} = \frac{k_2}{2} \frac{m_T}{m_L} \frac{R_L^3}{r^{*5}} y_1^* y_2^* - \frac{1}{6} \frac{k_2 R_L^3}{Gm_L} \omega_1^* \omega_2^*.$$

## TABLE I

Tidal perturbations of the lunar libration for three cases: no time delay, a constant time delay of 0.16485 day, a constant lag angle of 2°.1721. The value of the Love number is 0.02992. The coefficients of cosine and sine terms are given in arcsec. Values between parenthesis are derived from (Yoder, 1979) and given for comparison

| Argument | No time delay cos | sin | Constant time delay cos | sin | Constant lag angle cos | sin |
|---|---|---|---|---|---|---|
| **Variable $p_1$** | | | | | | |
| $F$ | 0 | 0 | 0.2733 | 0.0049 | 0.2647 | 0.0045 |
|  |  | (−0.0009) | (0.2710) | (−0.0009) | (0.2710) | (−0.0009) |
| $F - l$ | 0 | 0.0587 | −0.0153 | 0.0587 | −0.0170 | 0.0586 |
|  |  | (0.0625) | (0.0030) | (0.0625) | (0.0015) | (0.0625) |
| $0$ | −0.0240 |  | −0.0240 |  | −0.0240 |  |
| $2D - F - l$ | 0 | −0.0012 | −0.0003 | −0.0012 | −0.0004 | −0.0012 |
| $F - 2l$ | 0 | 0 | −0.0007 | −0.0001 | −0.0003 | −0.0001 |
|  |  |  | (−0.0019) |  | (−0.0017) |  |
| **Variable $p_2$** | | | | | | |
| $F$ | 0.0134 | 0 | 0.0184 | −0.2718 | 0.0179 | −0.2632 |
|  | (0.0129) |  | (0.0129) | (−0.2710) | (0.0129) | (−0.2710) |
| $F - l$ | 0.0112 | 0 | 0.0112 | 0.0196 | 0.0111 | 0.0219 |
|  | (0.0284) |  | (0.0284) | (0.0314) | (0.0284) | (0.0351) |
| $F - 2l$ | 0 | 0 | 0 | −0.0008 | 0 | −0.0003 |
|  |  |  |  | (−0.0019) |  | (−0.0017) |
| $0$ | 0 |  | −0.0006 |  | −0.0005 |  |
|  | (0.0760) |  | (0.0760) |  | (0.0760) |  |
| **Variable $\tau$** | | | | | | |
| $0$ | 0 |  | 0.3971 |  | 0.3846 |  |
|  |  |  | (0.3974) |  | (0.3974) |  |
| $2F - 2l$ | 0 | −0.0103 | −0.0475 | −0.0104 | −0.0263 | −0.0099 |
|  |  | (0.0024) | (0.0045) | (0.0024) | (0.0192) | (0.0024) |
| $l$ | 0 | −0.0059 | −0.0011 | −0.0059 | −0.0011 | −0.0059 |
|  |  | (−0.0057) |  | (−0.0057) |  | (−0.0057) |
| $2D - 2l$ | 0 | 0.0004 | −0.0058 | 0.0005 | −0.0047 | 0.0004 |
|  |  |  | (0.0057) |  | (0.0067) |  |
| $2D - l$ | 0 | −0.0014 | −0.0004 | −0.0014 | −0.0004 | −0.0014 |
|  |  |  | (−0.0015) |  | (−0.0015) |  |
| $l'$ | 0 | 0 | 0.0003 |  | 0.0092 | 0.0001 |
|  |  |  |  |  | (0.0096) |  |

$r$ is the Earth-Moon distance, $k_2$ is a Love number. The exponent $*$ means that, in the computation of $\Delta C_{ij}$ and $\Delta S_{ij}$ at time $t$, the function must be evaluated at time $t - t_0$ ; $t_0$ is a time delay, equal to zero for an elastic model of the Moon, and constant for a viscous model. Similarly to (Yoder, 1979) a third model, with $t_0$ inversely proportional to the absolute value of the frequency of each term in which it is involved, has also been considered. If the sign of the coefficient of each term is determined so that the frequency is always positive, this case corresponds to a constant lag angle. The terms due to solar action in Eq. (3) are obtained by changing $m_T$ to $m_S$, $y_i$ to $y_i'$, and $r$ to $r'$ (Sun-Moon distance).

Corrections $\Delta C_{ij}$ induce a time dependent corrective tensor $\Delta I$ to the constant tensor of inertia of the rigid Moon $I$. Assuming that the trace of $\Delta I$ is zero, elements of $\Delta I$ are:

$$\Delta I_{11} = m_L R_L^2 \left(\tfrac{1}{3}\Delta C_{20} - 2\Delta C_{22}\right) \qquad \Delta I_{12} = -2m_L R_L^2 \Delta S_{22}$$

$$\Delta I_{22} = m_L R_L^2 \left(\tfrac{1}{3}\Delta C_{20} + 2\Delta C_{22}\right) \qquad \Delta I_{23} = -m_L R_L^2 \Delta S_{21}$$

$$\Delta I_{33} = -\tfrac{2}{3} m_L R_L^2 \Delta C_{20} \qquad \Delta I_{13} = -m_L R_L^2 \Delta C_{21}.$$

$\Delta I$ induces in Eq. (1) the disturbing function $\Psi$ whose components are given by:

$$I_{ii}\Psi_i = -\sum_{j=1}^{3}\left[\Delta \dot{I}_{ij}\omega_j + \Delta I_{ij}\dot{\omega}_j\right] + \sum_{j=1}^{3}\sum_{k=1}^{3}\varepsilon_{ijk}\times$$

$$\left[\Delta I_{jk}\left(\omega_k^2 - \frac{3Gm_T}{r^5}y_k^2 - \frac{3Gm_S}{r'^5}y_k'^2\right) + \Delta I_{ij}\left(\omega_k\omega_i - \frac{3Gm_T}{r^5}y_iy_k - \frac{3Gm_S}{r'^5}y_i'y_k'\right)\right.$$

$$\left.+\Delta I_{jj}\left(\omega_j\omega_k - \frac{3Gm_T}{r^5}y_jy_k - \frac{3Gm_S}{r'^5}y_j'y_k'\right)\right]$$

with $\varepsilon_{ijk} = -\varepsilon_{jik} = -\varepsilon_{ikj}$ and $\varepsilon_{123} = 1$. $I_{ii}$ are respectively the principal moments of inertia $A$, $B$, $C$.

Table I gives the perturbations obtained in the three cases mentioned above. These perturbations involve the complete direct tidal effects by the Earth and the Sun, but the constant contribution of the lunar rotation to $\Delta C_{20}$ and $\Delta C_{22}$, and consequently to $\Delta I_{ii}$, has been removed. This contribution is supposed to be included in the parameters of the rigid Moon. Table I gives also, for comparison, the corresponding quantities derived from Yoder's results (1979) by converting his complex variable $p$ to $p_1$ and $p_2$. Coefficients of terms whose amplitudes are smaller than $0''.001$ in both solutions are not given. The greatest difference concerns a constant term of $0''.0760$ in Yoder's results for $p_2$ which is much smaller in ours. In the opposite, we have a constant term in $p_1$ which does not exist in Yoder's results.

## 2.3. DIRECT PERTURBATIONS DUE TO THE EARTH'S FIGURE

In this section we have supposed that the Earth has a rotational symetry around its polar axis. Furthermore, in the conversion of terrestrial body fixed coordinates to lunar ones we have neglected the libration and the nutation since the corresponding quantities should be mutiplied by the Earth's $J_2$.

By expressing the elements of the Earth's tensor of inertia with respect to the lunar principal axes of inertia, and by subtituting the results in the expressions given by Schutz (1981), we obtain for the components of the disturbing function induced by the Earth's figure in Eq. (1):

$$\Psi_i = -\frac{3Gm_T\alpha_i J_2 R_T^2}{r^5}\left[\frac{35}{2}\frac{y_jy_k}{r^4}D^2 - \frac{5}{2}\frac{y_jy_k}{r^2} - \frac{5}{r^2}(\xi_ky_j + \xi_jy_k)D + \xi_j\xi_k\right].$$

TABLE II

Direct perturbations due to the Earth's figure. The coefficients of cosine and sine terms are given in arcsec. Values between parenthesis are reproduced from (Pešek, 1982) and are given for comparison

|   | Argument | cos | sin |
|---|---|---|---|
| $p_1$ | $\zeta$ | 0 | −0.0729 (−0.0725) |
|   | $F$ | 0 | 0.0121 (0.0108) |
| $p_2$ | $\zeta$ | −0.0729 (−0.0726) | 0 |
|   | $F$ | 0.0121 (0.0108) | 0 |
| $\tau$ | $\zeta - F$ | 0 | −0.0099 (−0.0067) |

$i, j, k$ verify $\varepsilon_{ijk} = 1$ and $D$ stands for $\xi_1 y_1 + \xi_2 y_2 + \xi_3 y_3$. $\xi_i$ are the components of the unit vector pointing towards the Earth's pole referred to the lunar principal axes of inertia (respectively here $-\sin\zeta\sin\varepsilon$, $-\cos\zeta\sin\varepsilon$, and $\cos\varepsilon$, $\varepsilon$ being the mean obliquity of date). $\alpha_i$ stand respectively for $\alpha$, $-\beta$, and $\gamma$.

Table II gives the resulting perturbations on the forced libration, and, for comparison, the results obtained by Pešek (1982). Coefficients of terms whose amplitudes are smaller than $0''.001$ in both solutions are not given. Our results are in good agreement with Pešek's ones for $p_1$ and $p_2$. The difference of $0''.003$ in $\tau$ comes from our second iteration in the resolution of Eq. (2).

## 2.4. NON PERIODIC INDIRECT PERTURBATIONS OF $\tau$

By disregarding all terms of upper orders in Eq. (1), we obtain the following separate equation in $\tau$:

$$\ddot{\tau} + 3\nu^2\gamma\tau = 3\nu^2\gamma(L - \overline{\lambda}) \tag{4}$$

where $L$ is the lunar longitude, $\overline{\lambda}$ the mean mean longitude, and $\nu$ the sidereal mean motion.

For the main problem $\overline{\lambda}$ is a linear function of time $t$. Secular variations of the solar eccentricity and tidal perturbations introduce in $L$ secular terms in $t^2$, $t^3$'s Eq. (4) shows that term $At^n$ in $L$ induces the same term $At^n$ in $\tau$. It induces also terms at lower powers of $t$ which may be disregarded because their coefficients are either zero or quantities much smaller than $A$ for successive divisions by $3\nu^2\gamma$ (about 48 000 rad/cy). The existence of a $t^2$ term in $\tau$ has been mentioned yet by Bois et al. (1996). Nevertheless, since $\tau$ always appears through $\overline{\lambda} + \tau$, e.g. in the expression of matrix $M$ transforming ecliptic coordinates $x_i$ to lunar body fixed coordinates $y_i$ (Chapront-Touzé, 1990), it is simpler to consider that secular terms $At^n$ are involved in $\overline{\lambda}$, constituting the mean mean longitude $W_1$ of the orbital motion, and that $\tau$ contains only periodic and Poisson terms.

Eq. (4) shows that Poisson term $At\sin\varphi$ in $L$ induces in $\tau$ the terms $A't\sin\varphi + B'\cos\varphi$ with :

$$A' = \frac{3\nu^2\gamma A}{3\nu^2\gamma - \dot{\varphi}^2}, \qquad B' = \frac{-6\nu^2\gamma A\dot{\varphi}}{(3\nu^2\gamma - \dot{\varphi}^2)^2}$$

Hence, the Poisson terms

$$\Delta L = 0''.254\,25\,t\sin(18V - 16T - l + 114°.565\,50) + 1''.676\,80\,t\sin l'$$

in the lunar longitude (Chapront-Touzé and Chapront, 1983) induce in $\tau$ Poisson terms which, following Eq. (4), are:

$$\Delta\tau = 0''.2543\,t\sin(18V - 16T - l + 114°.5655) - 0''.2334\,t\sin l' \tag{5}$$

$t$ is the time in century reckoned from J2000.0, $V$ and $T$ are the mean mean longitudes of Venus and the Earth respectively, $l'$ is the solar mean anomaly. Eq. (5), which is only an approximation, shows the interest of computing Poisson terms in the forced libration.

## 3. Comparisons with JPL Numerical Integrations

Several kinds of comparisons with JPL numerical integrations DE245 and DE403 have been performed. Two time spans $\Delta t$ have been chosen to cover the periods of comparison which are of 300 and 600 years in the case of DE245 and DE403 respectively. We have used two different JPL integrations as reference models to insure the numerical consistency of our analysis and provide several sets of libration parameters depending on the model. The general scheme of our analysis is the following: Euler angles in JPL integrations are transformed into the libration variables $p_1$, $p_2$ and $\tau$. The analytical solution (A) is computed using a set of parameters consistent with the JPL numerical integration (N). We compute the differences $\delta = (N) - (A)$ and perform a frequency analysis of the "residuals" $\delta$. Once we have determined the significant frequencies $\omega_i$ of the spectrum, a least square fit of the residuals is done in order to obtain an approximate "solution" for $\delta$ on the time interval $\Delta t$:

$$\delta = \sum_i A_i \sin(\omega_i t + \phi_i)$$

$\delta$ stands for any of the three residuals among the variables $p_1$, $p_2$ and $\tau$ ; $t$ is the time reckoned from J2000 ; $A_i$ and $\phi_i$ are the quantities provided by the least square fit. The quality of the frequency analysis strongly depends on the choice of the filtering. We have used a method proposed by (Laskar et al., 1993) which has been already successfully applied in the case of the construction of planetary ephemerides (Chapront, 1995).

Table III gives the values of the lunar physical parameters substituted in Moons' series for the comparisons, except for the values of the harmonic coefficients of degree 4 which are those of (Ferrari et al., 1980). The values of $\beta$, $\gamma$, $C_{ij}$ and $S_{ij}$ are the one used in the numerical integrations. The values of $C/m_L R_L^2$ are derived from the values of $C_{22}$ used in the numerical integrations by means of the relation for a rigid body $C/m_L R_L^2 = 4C_{22}/\gamma$. $\beta$, $\gamma$, $C/m_L R_L^2$, and $C_{22}$ are assumed to involve the constant tidal perturbation due to the lunar rotation.

TABLE III

Physical parameters adopted for the comparison to DE245 and DE403 (from numerical integrations except for $C/m_L R_L^2$). Units of $10^{-4}$

| DE245 | | DE403 | |
|---|---|---|---|
| $\beta = 6.31619133$ | | $\beta = 6.31610707$ | |
| $\gamma = 2.27885980$ | | $\gamma = 2.27864190$ | |
| $C_{30} = -0.086802$ | | $C_{30} = -0.086474$ | |
| $C_{31} = 0.307083$ | $S_{31} = 0.046115$ | $C_{31} = 0.307083$ | $S_{31} = 0.044875$ |
| $C_{32} = 0.048737$ | $S_{32} = 0.016975$ | $C_{32} = 0.048727$ | $S_{32} = 0.016962$ |
| $C_{33} = 0.017161$ | $S_{33} = -0.002844$ | $C_{33} = 0.017655$ | $S_{33} = -0.002744$ |
| $C/m_L R_L^2 = 3948.72400$ | | $C/m_L R_L^2 = 3950.29692$ | |

A first type of analysis has been done to test the improvements due to the analytical complements described in Sect. 2 (solution SOL2) with respect to Moons' original solution (SOL1). The tidal perturbations introduced in SOL2 correspond to a constant time delay. Missing arguments in SOL1 were detected in the frequency analysis, and compared with those of Tables I and II. These comparisons show a good agreement between numerical $A_i$, $\phi_i$ and $\omega_i$ and the analytical ones for all the arguments of Table I, in particular for the constant terms of $p_1$ and $p_2$ which differ from those of Yoder (1979). This agreement verifies the validity of our spectral analysis. The error is estimated to less than $0''.005$ on amplitudes $A_i$ and less than $10^{-6}$ radian per day on frequencies $\omega_i$. The agreement is not so good for the terms with argument $\zeta$ in $p_1$ and $p_2$, and $\zeta - F$ in $\tau$ (Table II). The discrepancy amounts to $0''.08$ for $p_1$ and $p_2$ and $0''.03$ for $\tau$, but it may be due to the indirect perturbations by the Earth's figure.

A second type of analysis has been done to estimate the accuracy of the free libration series in Moons' solution, and also to determine the numerical values of the free libration parameters $\sqrt{2P}$, $\sqrt{2Q}$ and $\sqrt{2R}$. The general procedure is the following: we compute the residuals $\delta$ with SOL2 but without free libration series. Three terms of importance appear in the spectrum whose related frequencies are close to the libration frequencies $\omega_q$, $\omega_p$ and $\omega_{F+r}$ in Moons' solution. The dominant terms of the residuals are $A[p_2] \sin(\omega_q t + \phi_q)$ in $p_2$, and $B[\tau] \sin(\omega_p t + \phi_p)$ in $\tau$. The argument $F + r$ appears in $p_1$ (and $p_2$) through $C[p_1] \sin(\omega_{F+r} t + \phi_{F+r})$ with smaller amplitude. Moons' solution provides an analytical form for the free libration series which allows to compute the free libration parameters, $\sqrt{2P}$, $\sqrt{2Q}$ and $\sqrt{2R}$ from the coefficients of the above arguments, respectively $B[\tau]$, $A[p_2]$ and $C[p_1]$. The phases $\phi_p$, $\phi_q$, $\phi_{F+r}$ give the values $p_0$, $q_0$, $r_0$ of the free libration fundamental arguments $p$, $q$, $r$ in J2000.0. The "observed" frequencies $\omega_p$ and $\omega_q$ replace the computed ones. $\omega_r$ is poorly determined by frequency analysis through the combination $F + r$. The theoretical computed value is retained.

The free libration parameters obtained repectively from the two integrations, as well as the main terms mentioned above, are gathered in Table IV. A complete numerical evaluation of coefficients of Moons' free libration series is given

TABLE IV

Determination of free libration parameters. The three fundamental libration arguments are: $p = \omega_p t + p_0$ , $q = \omega_q t + q_0$, $r = \omega_r t + r_0$. $t$ is reckoned from J2000.0. The frequencies are in radian per day

| | DE245 | | DE403 | |
|---|---|---|---|---|
| | $B[\tau] = 1''.8235$ | $\phi_p = 224°.303$ | $B[\tau] = 1''.8122$ | $\phi_p = 224°.310$ |
| | $A[p_2] = 8''.1557$ | $\phi_q = 251°.651$ | $A[p_2] = 8''.1825$ | $\phi_q = 251°.777$ |
| | $C[p_1] = 0''.0208$ | $\phi_r = 217°.678$ | $C[p_1] = 0''.0218$ | $\phi_r = 202°.965$ |
| | $\sqrt{2P} = 0.2933$ | $p_0 = 224°.303$ | $\sqrt{2P} = 0.2915$ | $p_0 = 224°.310$ |
| | $\sqrt{2Q} = 5.1924$ | $q_0 = 161°.640$ | $\sqrt{2Q} = 5.2095$ | $q_0 = 161°.766$ |
| | $\sqrt{2R} = 0.0208$ | $r_0 = 124°.394$ | $\sqrt{2R} = 0.0218$ | $r_0 = 109°.681$ |
| | Computed | "Observed" | Computed | "Observed" |
| $\omega_p$ | 0.0060467320 | 0.0059492451 | 0.0060466648 | 0.0059492451 |
| $\omega_q$ | 0.0002281306 | 0.0002304932 | 0.0002281236 | 0.0002304970 |
| $\omega_{F+r}$ | 0.2301836354 | 0.2301811833 | 0.2301836363 | 0.2301820813 |

in (Chapront and Chapront-Touzé, 1997) with free libration parameters fitted to DE245. We mention only here that, after substitution of the values of $\sqrt{2P}$, $\sqrt{2Q}$, $\sqrt{2R}$, $\phi_p$, $\phi_q$, $\phi_r$, $\omega_p$, and $\omega_q$ quoted in Table IV in Moons' free libration series, all the terms are in good agreement with the terms of same frequencies (within the estimated error) in the development of $\delta$ provided by the frequency analysis.

In the following comparisons, the free libration parameters of Table IV and lunar physical constants of Table III in agreement with (N) have been introduced in SOL1 and SOL2, to render the residuals $\delta$ independent of the model as much as possible.

To illustrate the improvements due to our analytical complements, we show on Fig. 1-a a comparison of SOL1 to DE245 for the variable $\tau$ ($\Delta t = 300$ years): $\delta = $ (DE245) − (SOL1). On the contemporary period the main difference is a constant (0''.4) due to tidal effects and oscillations whose maximum amplitudes are about 0''.2. In the past the Poisson terms due to planetary perturbations dominate, and the total differences reach 1.''6. On Fig 1-b, we show the differences evaluated with the improved analytical solution SOL2: $\delta = $ (DE245) − (SOL2). It remains now oscillations whose total amplitudes are less than 0''.15 on the whole time span. Fig. 2-a and 2-b illustrate the same comparisons for variable $p_1$. The gain of precision between SOL1 and SOL2 is not so good in the past because Poisson terms have not yet been introduced in SOL2 for $p_1$ and $p_2$, and the beating effect in Fig. 2-b shows clearly this lack.

## 4. Numerical Complements to the Analytical Libration Series

The solution SOL2 contains now free libration series and again we analyze the residuals $\delta$. The spectrum of $\delta$ is very clean. This means that, at the level of accuracy of 0.''005, few terms are lacking in the analytical series and that the residuals can be represented by complementary series with a small number of sensible components. In Table V we have listed the complementary series as

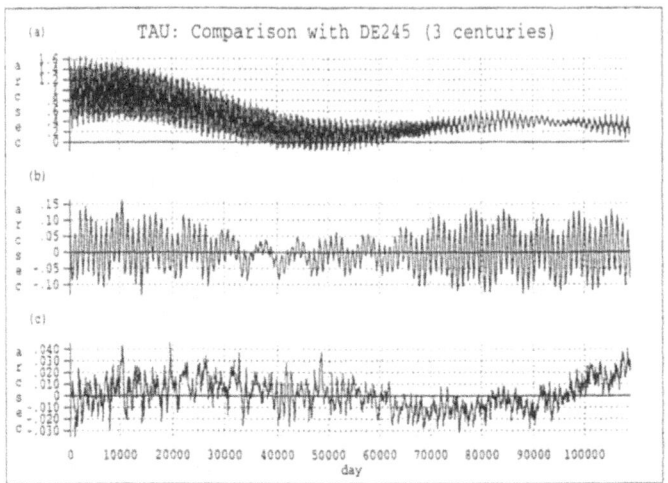

Fig. 1. Residuals on the variable $\tau$ from 1750 to 2050. 1-a : (DE245) – (SOL1). 1-b : (DE245) – (SOL2). 1-c : (DE245) – (SOL3)

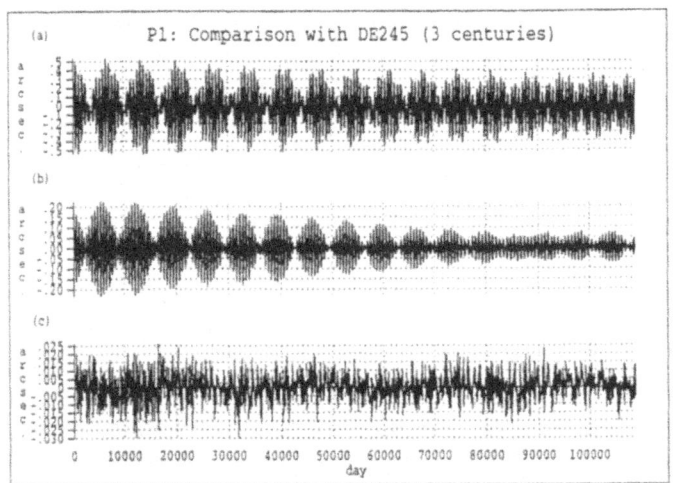

Fig. 2. Residuals on the variable $p_1$ from 1750 to 2050. 2-a : (DE245) – (SOL1). 2-b : (DE245) – (SOL2). 2-c : (DE245) – (SOL3)

they come from the frequency analysis on the residuals $\delta = $ (DE245) – (SOL2). Fig. 1-c and 2-c illustrate the final results of solution SOL3 (SOL2 + numerical complements of Table V) compared with the source DE245. Residuals are below $0''.03$ on the whole time span. The results of the comparison is the same when SOL3 is compared to DE403 over the same time span of three centuries, using constants of DE403 and the free libration series with values listed in Table IV for DE403. Note that numerical complements of Table V are valid only over the time span 1750 – 2050. Outside this time span, residuals slowly diverge because of

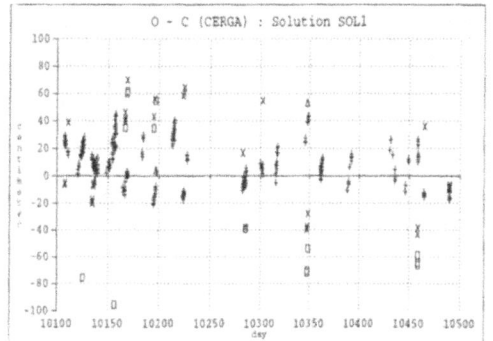

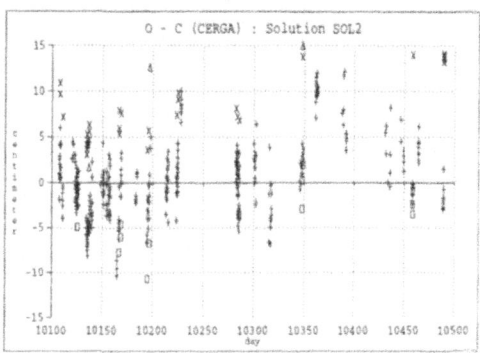

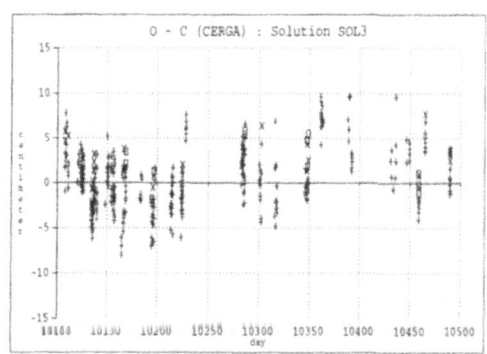

Fig. 3. O-C for the distance Observer-Reflector (Feb. 1997 – March 1998). Reflectors are:

Apollo 11 (×), Apollo 14 (⊡), Apollo 15 (+), and Lunakhod 2 (△)

missing secular and Poisson terms.

As a final test, the solutions SOL1, SOL2 and SOL3 have been compared directly to lunar-laser observations themselves. In the three cases a large set of parameters, including reflector coordinates, has been fitted to the observations as described in (Chapront et al., 1998). We see on Fig. 3, over a time span of 400 days, the residuals O-C (Observation minus Computation) on the one-way range observer-reflector in centimeter, for CERGA observations. The most frequently observed reflector is Apollo 15, that is represented with a sign (+) on the graph; this reflector contributes mainly to the fit of parameters. We observe that the introduction of the analytical

TABLE V

Numerical complements. Series of the differences $\delta = $ (DE245) $-$ (SOL2). Units: $10^{-4}$ arcsec for $A$, degree for $\phi$, and rad/day for $\omega$.

| Variable $p_1$ | | | Variable $p_2$ | | | Variable $\tau$ | | |
|---|---|---|---|---|---|---|---|---|
| $A$ | $\phi$ | $\omega$ | $A$ | $\phi$ | $\omega$ | $A$ | $\phi$ | $\omega$ |
| 73 | 106.98 | 0.22987004 | 52 | 274.41 | 0.00567265 | 137 | 155.61 | 0.00008575 |
| 295 | 7.48 | 0.22990957 | 54 | 331.96 | 0.00580103 | 46 | 223.41 | 0.00089398 |
| 935 | 252.19 | 0.22994149 | 46 | 144.16 | 0.22969520 | 349 | 70.79 | 0.00092222 |
| 821 | 75.97 | 0.22996514 | 150 | 313.56 | 0.22986707 | 61 | 336.73 | 0.00094330 |
| 183 | 226.15 | 0.23001653 | 385 | 218.42 | 0.22991443 | 66 | 262.58 | 0.00432017 |
| 688 | 268.26 | 0.23089745 | 760 | 66.54 | 0.22994551 | 48 | 88.57 | 0.00436703 |
| 164 | 64.93 | 0.23092684 | 508 | 290.12 | 0.23001560 | 134 | 73.84 | 0.00572887 |
| 52 | 252.04 | 0.24813066 | 252 | 172.79 | 0.23004927 | 89 | 176.59 | 0.00585279 |
| | | | 691 | 5.12 | 0.23089738 | 218 | 271.48 | 0.00589170 |
| | | | 123 | 198.97 | 0.23094800 | 637 | 287.48 | 0.00596264 |
| | | | 48 | 32.81 | 0.24807360 | 146 | 241.37 | 0.00601362 |
| | | | 49 | 226.21 | 0.24809876 | 51 | 195.68 | 0.01720365 |

complements of this paper in Moons' solution (SOL2 instead of SOL1) produces a significant decrease of the O-C. The introduction of SOL3 instead of SOL2 makes the dispersion of the O-C with the reflectors smaller.

## 5. Conclusion

This study shows the good quality of Moons' libration theory. Nevertheless it needs to be completed by few missing perturbations. Two kinds of complements are given in this paper. Some futher complements should be achieved in the case of direct planetary perturbations and Poisson term contributions.

## References

Bois, E., Boudin, F., Journet, A.: 1996, *Astron. Astrophys.*, **314**, 989.
Chapront, J.: 1995, *Astron. Astrophys. Suppl. Ser.*, **109**, 181.
Chapront, J., Chapront-Touzé, M.: 1997, *Celest. Mech.*, **66**, 31.
Chapront, J., Chapront-Touzé, M., Francou, G.: 1998, submitted to *Astron. Astrophys.*
Chapront-Touzé, M.: 1990, *Astron. Astrophys.*, **235**, 447.
Chapront-Touzé, M., Chapront, J.: 1983, *Astron. Astrophys.*, **124**, 50.
Danjon, A.: 1959, *Astronomie générale*, Sennac, Paris, p. 343.
Eckhardt, D.H.: 1981, *The Moon and the Planets*, **25**, 3.
Ferrari, A.J., Sinclair, W.S., Sjogren, W.L., Williams, J.G., Yoder, C.F.: 1980, *J.G.R.*, **85**, 3939.
Hilton, J.L.: 1992, in *Explanatory Supplement to the Astronomical Almanac*, P.K. Seidelmann (ed.), University Science Books, California, p. 398.
Laskar, J., Froeschlé, C, Celleti, A.: 1993, *Phys. D*, **67**, 257.
Lambeck, K.: 1980, *The Earth's Variable Rotation*, Cambridge University Press.
Moons, M.: 1981, *Libration Physique de la Lune*, Thesis, Facultés Universitaires de Namur.
Moons, M.: 1982, *The Moon and the Planets*, **27**, 257.
Moons, M.: 1984, *Celest. Mech.*, **34**, 263.
Pešek, I.: 1982, *Bull. Astron. Inst. Czechosl.*, **33**, 176.
Schutz, B.E.: 1981, *Celest. Mech.*, **24**, 173.
Standish, E.M.: 1981, *Astron. Astrophys.*, **101**, L17.
Yoder, C.F.: 1979, in *Natural and Artificial Satellite Motion*, P. Nacozy and S. Ferraz-Mello (eds), Texas University Press, p. 211.

# IMPACT OF THE QUADRUPOLE MOMENT OF THE SUN

# ON THE DYNAMICS OF THE EARTH-MOON SYSTEM

E. BOIS and J.F. GIRARD

*Observatoire de Bordeaux, UMR/CNRS/INSU 5804*
*B.P. 89, F-33270 Floirac, France, E-mail:bois@observ.u-bordeaux.fr*

**Abstract.** Range of values of the Sun's mass quadrupole moment of coefficient $J_2$ arising both from experimental and theoretical determinations enlarge across literature on two orders of magnitude, from around $10^{-7}$ until to $10^{-5}$. The accurate knowledge of the Moon's physical librations, for which the Lunar Laser Ranging data reach an outstanding precision level, prove to be appropriate to reduce the interval of $J_2$ values by giving an upper bound of $J_2$. A solar quadrupole moment as high as $1.1 \ 10^{-5}$ given either from the upper bounds of the error bars of the observations, or from the Roche's theory, is not compatible with the knowledge of the lunar librations accurately modeled and observed with the LLR experiment. The suitable values of $J_2$ have to be smaller than $3.0 \ 10^{-6}$.

As a consequence, this upper bound of $3.0 \ 10^{-6}$ is accepted to study the impact of the Sun's quadrupole moment of mass on the dynamics of the Earth-Moon system. Such an effect (with $J_2 = 5.5 \pm 1.3 \times 10^{-6}$) has been already tested in 1983 by Campbell & Moffat using analytical approximate equations, and thus for the orbits of Mercury, Venus, the Earth and Icarus. The approximate equations are no longer sufficient compared with present observational data and exact equations are required. As if to compute the effect on the lunar librations, we have used our BJV relativistic model of solar system integration including the spin-orbit coupled motion of the Moon. The model is solved by numerical integration. The BJV model stems from general relativity by using the DSX formalism for purposes of celestial mechanics when it is about to deal with a system of $n$ extended, weakly self-gravitating, rotating and deformable bodies in mutual interactions.

The resulting effects on the orbital elements of the Earth have been computed and plotted over 160 and 1600 years. The impact of the quadrupole moment of the Sun on the Earth's orbital motion is mainly characterized by variations of $\Omega$, $\dot{\omega}$, and $\dot{E}$. As a consequence, the Sun's quadrupole moment of mass could play a sensible role over long time periods of integration of solar system models.

**Key words:** Sun, Quadrupole, Moon, Libration, Earth, LLR data
**Abbreviations:** LLR: Lunar Laser Ranging; GR: general relativity; PN: post-Newtonian; mas: milliarcsecond.

## 1. Introduction

Despite recent advances in space (SOHO) or ground-based observations (helioseismology networks) of solar oscillations, we are still unaware of the real value of the Sun's mass quadrupole moment of coefficient $J_2$. Range of values of $J_{2\odot}$ arising both from experimental and theoretical determinations enlarge across literature on two orders of magnitude, from around $10^{-7}$ until to $10^{-5}$. An assessment of various theoretical values compared to available observations has been recently carried out by Rozelot and Bois (1998). The theoretical value strongly depends on the solar model used whereas accurate measurements are very difficult to obtain. If $J_{2\odot}$ exceeds a specific value, then GR alone cannot make up the remainder of the precession rate of the Mercury's perihelion. As an error of 0.1 % is generally admitted on the observed rate, it turns out that this test is not enough revealing.

In the other hand, most of the current observational experiments access to a solar oblateness resulting from combined effects in interactions such as the global

*Celestial Mechanics and Dynamical Astronomy* 73: 329–338, 1999.
© 1999 *Kluwer Academic Publishers.*

gravitational field, the internal magnetic field, the general internal rotation rate, or the mass and angular velocity distributions and the surface rotation of the sun. The $J_{2\odot}$ values derived from an oblateness $\Delta r$ or inferred by inverting some theoretical equations have not to be mistake for a dynamical oblateness directly measured by experiments of gravitational dynamics. Let us stress that a $J_2$ quadrupole moment of mass is a dynamical coefficient strictly related to the gravitational figure. Now, the understanding of the gravitational figure of the Sun is also a way to infer its internal structure and to approach the core rotation problem.

The accurate knowledge of the Moon's physical librations, for which the LLR data reach an outstanding precision level (i.e. 1 cm for the distance Earth-Moon, 1 mas for the lunar librations) prove to be appropriate to reduce the interval of $J_{2\odot}$ values by giving an upper bound of $J_{2\odot}$. The method is explained in the present paper. We find that the suitable values of $J_{2\odot}$ have to be smaller than $3.0\ 10^{-6}$.

As a consequence, this upper bound (coinciding with the largest value that GR can accommodate by fitting to planetary data) is accepted to study the impact of the Sun's quadrupole moment of mass on the dynamics of the Earth-Moon system. A first attempt in this way has been carried out by Campbell & Moffat (1983) who computed the effects obtained by taking into account a $J_{2\odot}$ in the analytical equations used to determine the orbital elements $\varpi$, $\Omega$ and $i$ of Mercury, Venus, the Earth and Icarus (the experiment was performed with $J_2$ equal to $5.5 \pm 1.3 \times 10^{-6}$). Calculations didn't show any significant discrepancy. This was mainly due to a lack of sufficiently accurate observational data in the determination of the planetary orbits, in particular the Icarus' one. However, as predicted by the authors, if the accuracy of the observational data were to improve enough, then using the approximate equations would produce incorrect results. Since the analytical equations are linear in $J_2$, any increase in $J_2$ would magnify this shift. Consequently, we have used a solar system model of complete equations solved by numerical integration and built in accordance with the requirements of current observational accuracy given by the LLR experiment.

## 2. The Model *BJV* of Relativistic Integration of the Solar System

In order to study the impact of $J_{2\odot}$ on the Earth-Moon system dynamics, we have used a gravitational model of the solar system including the Moon's spin-orbit motion. This model, called BJV, was previously constructed by Bois, Journet and Vokrouhlický in accordance with the requirements of LLR observational accuracy (see previous papers: Bois et al. 1992; Bois & Journet 1993; Bois & Vokrouhlický 1995; Bois et al. 1996).

The approach consists in integrating the $n$ - body problem on the basis of the gravitation description given by the Einstein's general relativity theory. The BJV model stems from GR by using the DSX formalism presented in a series of papers by Damour, Soffel and Xu (Damour et al. 1991, 1992, 1993, 1994). It is the most suitable formulation of the post-Newtonian theory of motion of a system of $n$ weakly self-gravitating extended bodies for purposes of celestial mechanics. The DSX formalism is derived from the first post-Newtonian approximation level. Gravitational fields of the extended bodies are parameterized in multipole moment expansions: $(M_L^A, S_L^A)$ define the mass and spin Blanchet-Damour multipoles characterizing the PN gravitational field of the extended bodies while $(G_L^A, H_L^A)$ are tidal gravitoelectric and gravitomagnetic PN fields. Because we do not dispose of dynamical equations for the quadrupole moments $M_{ab}^A$, and although the notion of rigidity faces conceptual problems in the theory of relativity, we have adopted the 'rigid-multipole' model of the extended bodies as known from the Newtonian approach. Practically this is acceptable since the relativistic quadrupole contributions are very small. Consequently and because it is conventional in geodynamical research to use spherical harmonics analysis of the gravitational fields with the corresponding notion of harmonic coefficients $(C_{lm}^A, S_{lm}^A)$, the quadrupole moments $M_{ab}^A$ have been expressed in those terms, according to reasons and assumptions given in Bois & Vokrouhlický (1995). Moreover, internal structures of solid deformable bodies, homogeneous or with core-mantle interfaces, are represented by several terms and parameters arising from tidal deformations of the bodies (both elastic and anelastic). More details and references on these topics are given in the above quoted papers.

The simultaneous numerical integration of the Moon and planets uses a global reference system given by the solar-system barycenter. The lunar rotational motion is evaluated relative to a local dynamically non-rotating reference system whose a slow (de Sitter) rotation is calculated with respect to the global reference frame (kinematically non-rotating). It should be noticed that the Moon's reference frame undergoes a similar de Sitter precession to the Earth one. An alternative way of representing the two effects is to introduce the de Sitter precession of the common Earth-Moon barycentric reference frame, as it can be easily verified that the principal effects originate in the solar action. However, due to the Earth-Moon mutual action, the de Sitter precession of the two reference frames differs slightly. A detailed inspection had shown that the lunar reference frame undergoes an additional precession of the order of 30 mas/cy (Bois & Vokrouhlický 1995).

The model is solved by modular numerical integration and controlled in function of the different physical contributions and parameters taken into account. The $n$-body problem (for the motions of the planets, the Sun and the Moon), the lunar spin motion and the figure-figure and tidal interactions are simultaneously integrated with the choice of the contributions and truncations at our disposal. For instance, the upper limits of the extended figure expansions and mutual interactions may be chosen as follows: up to $l = 5$ in the Moon case, 4 for the Earth, 2 for the Sun while only the Earth-Moon quadrupole-octupole interaction is taken into account.

The model has been especially built to favor a systematic analysis of all the effects and contributions. In particular, it permits the separation of various families of lunar librations. One of the aims in building the model was to include all phenomena up to the precision level resulting from the LLR data (Dickey et al. 1994), and if possible better for reasons of consistency (i.e. at least 1 cm for the distance, 1 mas for the librations). In particular, several phenomena capable of producing effects of at least 0.1 mas in the lunar physical librations have been modeled and analyzed (the resulting libration may be at the observational accuracy level). Other libration effects smaller than this threshold of accuracy have been nevertheless included and studied because of their qualitative interest. Results can be found in previous papers (Bois et al. 1992; Bois & Journet 1993; Bois & Vokrouhlický 1995; Bois et al. 1996). Let us precise that the Earth-Moon figure-figure interaction, i.e. the Earth's quadrupole moment of mass acting as a gravitational torque on the rotational motion of the Moon, produces lunar physical librations with amplitudes of 45 mas over periods of 18.6 and 80.1 years. In the same way, Venus as monopole produces the major impact resulting from direct planetary actions; the resulting lunar libration may reach 3 mas over 18.6 years. The relativistic contributions due to explicit PN terms in the quasi-Newtonian torque related to the Earth as monopole induce lunar librations whose amplitudes reach the observational level (one mas).

## 3. Impact on the Moon

The two modes of lunar motion, spin and orbital, being simultaneously integrated, the spin-orbit couplings of the Moon are naturally taken into account. They play a part in the present calculus. Some particular spin-orbit couplings of the Moon are described in Bois et al. (1996). The solar quadrupole moment modifies the solar gravitational field with respect to a spherical Sun usually reduced in dynamics to a point mass. As a consequence, the orbital motions of the planets are no longer the same (in particular, even without mutual interactions, the semi-major axes of the planets have to take other values according to the classic problem of $J_2$). The motions are then simultaneously integrated in the new barycenter of the solar system. The orbital motion of the Moon is also disturbed but its relative motion with respect to the Earth remains globally of same geometric structure. The gravitational action of the solar quadrupole moment may be therefore considered as a perturbation on the lunar motion. Let us notice that the impact is not taken into account as a gravitational torque directly acting on the rotational motion of the Moon. It should be quite negligible. However, all the figure-figure interactions between the Sun (up to $l = 2$), the Earth (up to $l = 4$), the Moon (up to $l = 5$) and the other planets as monopoles are integrated. Finally, the gravitational action of the solar quadrupole moment of mass is evaluated with the spin motion of the Moon by the way of its lunar spin-orbit coupling. As it is generally the case, the indirect effects on the lunar rotational motion are relatively significant with respect to direct effects (cf. Moons 1984).

The Moon's rotational motion is represented by the classical Eulerian angles (3-1-3 angular sequence). The local reference system is given by the terrestrial equatorial frame (J2000). The indirect signature of the solar quadrupole moment on the Moon's rotational motion has been computed with different values of $J_{2\odot}$,

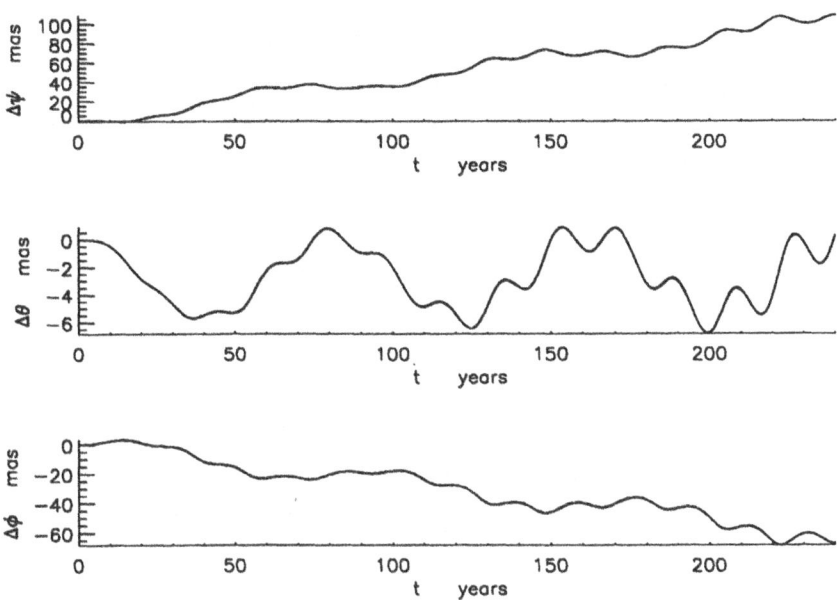

Fig. 1. Indirect effect of the solar quadrupole moment on the lunar physical librations. The integration is performed with $J_2 = 1.1 \ 10^{-5}$. Milliarcseconds are on the vertical axis and years on the horizontal axis (initial date is July/01/1969).

namely $1.1 \ 10^{-5}$ in Figure 1 and $3.0 \ 10^{-6}$ in Figure 2. Computations have been performed in quadruple precision (32 digits). The differences $(\Delta\psi, \Delta\theta, \Delta\phi)$ in Figures 1 and 2 are obtained with respect to the solution free of the gravitational modification related to $J_{2\odot}$.

The non-linearity features of the differential equations, the degree of correlation of the studied effect with respect to its neighbours (in the Fourier space) and the spin-orbit resonance, in the lunar case, make it hardly possible to speak about 'pure' effects with their proper behaviour (even after fitting of the initial conditions). The effects are not absolutely de-correlated but relatively isolated. However, the used technique (modular and controlled numerical integration, differentiation method and frequency analysis) gives the right qualitative behaviour of an effect and a good quantification of this effect relative to its neighbours. When a rotational effect is simply periodic, a fit of the initial conditions for a set of given parameters only refines without changing completely the effect's behaviour (the amplitude variation is lower than five per cent). The amplitudes of librations plotted on Figures 1 and 2 are then slightly upper bounds.

Let us observe $\Delta\theta$ in Figure 1. It represents the nutation variation of the Moon's polar axis relative to the Earth's one due to the indirect effect of the solar quadrupole moment. It shows a periodic dominant term of 80.1 years with an amplitude around

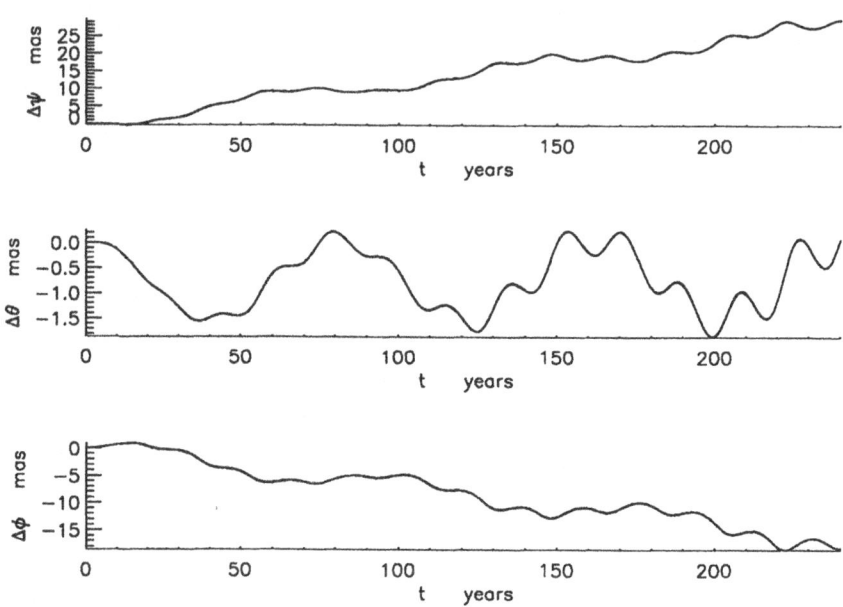

Fig. 2. Indirect effect of the solar quadrupole moment on the lunar physical librations. The integration is performed with $J_2 = 3.0 \ 10^{-6}$. Milliarcseconds are on the vertical axis and years on the horizontal axis (initial date is July/01/1969).

6 mas within the first period. The usual modulation of 18.6 years, related to the nodal precession, takes an amplitude around one mas. The ordinary resonant frequency of 2.9 years for physical librations in longitude is also detectable, by frequency analysis, in the $\theta$ angle with an amplitude around few tenths of mas. Let us note that in the present simulation, our model uses the dynamical parameters of the JPL DE303 ephemeris and the initial conditions come also from DE303 (the initial date is July/01/1969).

Except for the JPL latest ephemeris, DE405, where $J_{2\odot}$ is affected of a low non zero value ($2. \ 10^{-7}$), the recent JPL ephemerides including lunar librations have been adjusted to LLR observations up to the milliarcsecond level of accuracy without taking into account a solar quadrupole moment (Dickey et al. 1994) (in the other hand, Williams, Newhall and Dickey (1995) do not comment on the solar $J_2$ in their determination of relativistic parameters from LLR data analysis). Anyway, the JPL DE303 ephemeris is adjusted to the LLR observations with a zero value of $J_{2\odot}$ and its accuracy is around one mas ($\sigma = 1$ mas). As a consequence, the lunar physical librations being observed up to 1 mas in accuracy, an amplitude of 6 mas, even over a period of 80 years, is too large so that its cause be neglected in the ephemeris. Consequently, Figure 1 shows that the upper envelope determined by the upper limits of the previous mentioned error bars ($1.1 \ 10^{-5}$) is not compatible with the knowledge of the Moon's physical librations.

We accept now that an upper bound of $J_{2\odot}$ has to be suitable with $\frac{3}{2}\sigma$ of the LLR residuals, taking into account a possible reasonable shift in the relevance of residuals derived from a least-squares process; let be a signature lesser than 1.5 mas. In Figure 2, obtained with $J_2 = 3.0 \ 10^{-6}$, the periodic dominant term of 80.1 years takes an amplitude of 1.5 mas within the first period. The other amplitudes are respectively just lesser than 0.5 and 0.1 mas within the periods of 18.6 and 2.9 years. More precisely, the three amplitudes have decreased according to a factor 4 with respect to those of Figure 1. Consequently, taking into account a faint surcharge estimated to few tenth of mas in our simulations performed without fitting in the initial conditions, the present experiment leads in return to give a limit value of $J_2$ equal to $3.0 \ 10^{-6}$. This result coincides with the largest $J_2$ value that general relativity can accommodate by fitting to planetary data (Campbell & Moffat, 1983). In the other hand, an average of all the available data arising from experimental determinations of the solar oblateness yields to $3.6 \pm 2.8 \times 10^{-6}$ (Rozelot and Bois, 1998).

Rather than to isolate the differential signature of $J_{2\odot}$ on the lunar librations as we have done it, an alternative way would consist in performing a big least-squares fit to LLR data in order to try to get $J_{2\odot}$ as an absolutely solve-for parameter. Müller et al. (1996) have carried out this investigation but end up with an upper bound of the value: a realistic error on $J_{2\odot}$ equal to $5. \ 10^{-6}$.

## 4. Impact on the Earth

$J_2 = 3.0 \ 10^{-6}$ being an upper bound, it is an opportunity to evaluate the resulting maximal impact of the Sun's quadrupole moment of mass on the orbital motion of the Earth. Such an effect (with $J_2 = 5.5 \pm 1.3 \times 10^{-6}$) has been already tested in 1983 by Campbell & Moffat using analytical approximate equations, and thus for the orbits of Mercury, Venus, the Earth and Icarus. Their conclusion was that if the planetary data became known with enough accuracy, the exact equations would become necessary. Starting from this, the present calculations have been performed with an accurate model of the solar system, the BJV model as described in the second section.

The resulting effects on the orbital elements of the Earth have been computed and plotted over 160 and 1600 years. The solutions arising from the numerical integration are turned into heliocentric system and expressed in geometrical elements $a, e, i, \Omega, \omega, E$ (respectively the semi-major axis, the eccentricity, the inclination, the longitude of the ascending node, the perihelion relative to the line of nodes, and the eccentric anomaly; the elements refer to the equator and mean equinox J2000). Figures 3 and 4 show the results plotted on 160 years. The differences $\Delta a$, $\Delta e$, $\Delta i$, $\Delta\Omega$, $\Delta\omega$, $\Delta E$ are obtained with respect to the solution free of the gravitational variation related to $J_{2\odot}$. The maximal variations of the Earth's orbital elements related to the solar quadrupole moment give then the following major changes:

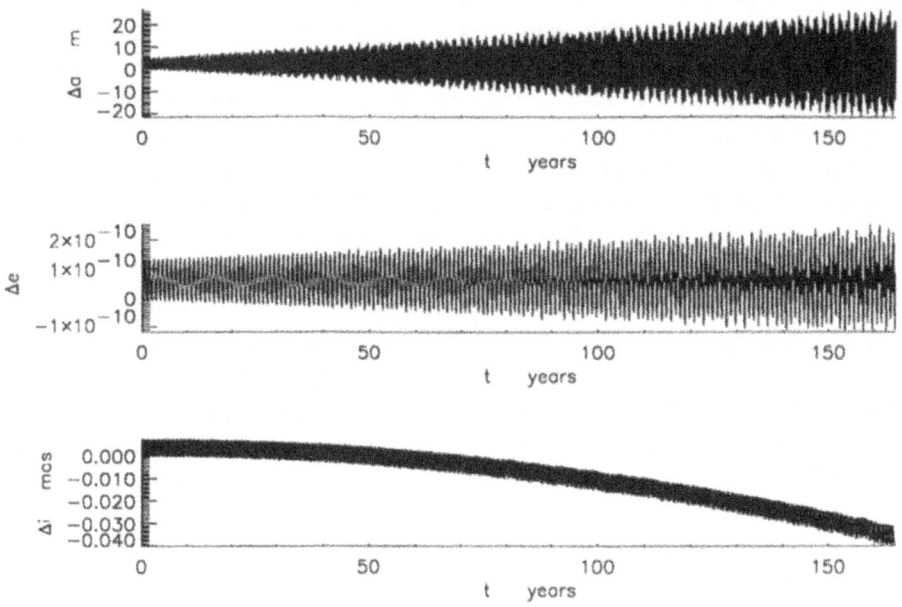

Fig. 3. Maximal variations of the orbital elements $a, e, i$ of the Earth due to a solar quadrupole moment $J_2$ equal to $3.0\ 10^{-6}$. $\Delta a$ is in meters and $\Delta i$ in mas.

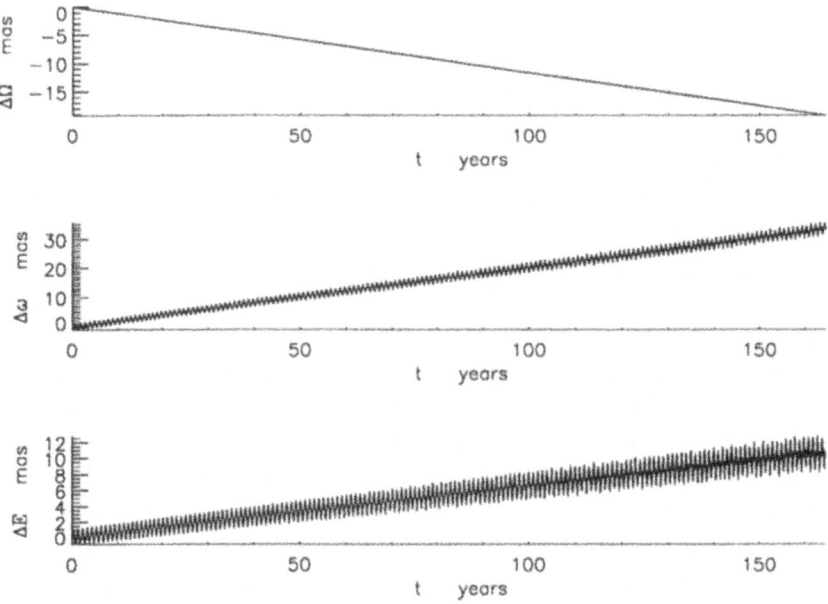

Fig. 4. Maximal variations of the orbital elements $\Omega, \omega, E$ of the Earth due to a solar quadrupole moment $J_2$ equal to $3.0\ 10^{-6}$. The three variables are expressed in mas.

$\Delta a \le 5$ m, i.e. the new mean semi-major axis of the Earth is in fact increased of 2.5 m; $\Delta e = 6. \ 10^{-11}$; the secular variation of the inclination of the Earth's orbital plane is very faint $[\Delta(di/dt) = -0.02$ mas/cy] but it presents a quadratic term $[\Delta(di/dt) = -0.2$ mas/cy when computed over 1600 years] (at this point, it would be worthwhile to take into account the inclination of the solar polar axis); $\Delta(d\Omega/dt) = -12$ mas/cy (which is on the opposite sign of $d\Omega/dt$); $\Delta(d\omega/dt) = 20$ mas/cy; $\Delta(dE/dt) = 6.5$ mas/cy.

Let us note that the nominal values such as $di/dt = -47.5''/$cy, $d\Omega/dt = 11.5''/$cy, and $d\omega/dt = 1250''/$cy deriving from our calculations represent essentially the mutual effects of planets on the orbital motion of the Earth. $\dot{\Omega}$ is then the most disturbed element by the impact of $J_{2\odot}$, let be three orders of magnitude lesser than the planetary effects. It is nevertheless of the same order of magnitude that the impact of the Moon on $\dot{\Omega}$.

## 5. Conclusion

The initial point to emphasize is that values of $J_{2\odot}$ enlarge across literature on two orders of magnitude. It has been shown that a solar quadrupole moment as high as $1.1 \ 10^{-5}$ (such as given either from the upper bounds of the error bars of the observations, or from the Roche's theory) is not compatible with the knowledge of the lunar physical librations accurately modeled and observed with the LLR experiment. The suitable values of $J_{2\odot}$ have to be smaller than $3.0 \ 10^{-6}$. Let us note that this value would be rather slightly an overvalued upper bound. However, the interval of available values of solar $J_2$ is certainly reduced. In the other hand, the recent value, namely $2.18 \pm 0.06 \times 10^{-7}$, inferred from helioseismology by Pijpers (1998), is very probably a minimal realistic value of $J_2$. As a consequence, the interval of possible values of $J_{2\odot}$ should be from now on $[2. \ 10^{-7}, 3. \ 10^{-6}]$. This interval is in fact maybe more reliable than a simple fine value owing to a possible time dependence of $J_{2\odot}$.

Using our BJV relativistic model of solar system integration (including notably the mutual body-body interactions and the spin-orbit coupled motion of the Moon), we have precisely calculated the impact of the quadrupole moment of the Sun on the Earth's orbital motion. The major changes are as follows: $\Delta a = 2.5$ m, $\Delta e = 6. \ 10^{-11}$, $\Delta(d\Omega/dt) = -12$ mas/cy (which is on the opposite sign of $d\Omega/dt$), $\Delta(d\omega/dt) = 20$ mas/cy, and $\Delta(dE/dt) = 6.5$ mas/cy. The variation of $di/dt$ is very faint but it presents a quadratic term. The most disturbed element in relation to its nominal value is $\dot{\Omega}$, let be three orders of magnitude lesser than the effects due to planetary interactions. The impact of $J_{2\odot}$ is globally slight but nevertheless because of the variations of $\dot{\Omega}$, $\dot{\omega}$, and $\dot{E}$, it has to be taken into account in models of long period evolution of the terrestrial orbit and more generally of the solar system. As a matter of fact, the Sun's quadrupole moment of mass (and some more related to its rotational motion) could play a sensible role in the determinations of prediction limits of the solar system stability.

# References

Bois, E., Boudin, F., Journet, A.: 1996, 'Secular Variation of the Moon's Rotation Rate', *A&A* **314**, pp. 989-994.

Bois, E., Journet, A.: 1993, 'Lunar and Terrestrial Tidal Effects on the Moon's Rotational Motion', *Celest. Mech.* **57**, pp. 295-305.

Bois, E., Vokrouhlický, D.: 1995, 'Relativistic Spin Effects in the Earth-Moon System', *A&A* **300**, pp. 559-567.

Bois, E., Wytrzyszczak, I., Journet, A.: 1992, 'Planetary and Figure-figure Effects on the Moon's Rotational Motion', *Celest. Mech.* **53**, pp. 185-201.

Campbell, L., and Moffat, J.W.: 1983, 'Quadrupole Moment of the Sun and the Planetary Orbits', *The Astrophys. Journal* **275**, L77-L79

Damour, T., Soffel, M., Xu, Ch.: 1991, 1992, 1993, 1994, 'General Relativistic Celestial Mechanics', *Phys. Rev.* **D43**, 3273, **D45**, 1017, **D47**, 3124, **D49**, 618.

Dickey, J.O., Bender, P., Faller, J., et al. : 1994, 'Lunar Laser Ranging: A Continuing Legacy of the Apollo Program', *Science* **265**, pp. 482-490.

Moons, M.: 1984, 'Planetary Perturbations on the Libration of the Moon', *Celest. Mech.* **34**, pp. 263-273.

Müller, J., Schneider, M., Soffel, M., Ruder, H.: 1996, 'Determination of Relativistic Quantities by analyzing Lunar Laser Ranging Data', in: Proceedings of the 7th Marcel Grossmann Meeting, eds. R. Jantzen and G. Mac Keiser, 1517.

Pijpers, F.P.: 1998, 'Helioseismic determination of the solar gravitational quadrupole moment', *Mon. Not. R. Astron. Soc.* **297**, pp. L76-L80

Rozelot, J.-P., and Bois, E.: 1998, 'New Results Concerning the Solar Oblateness', Synoptic Solar Physics, *ASP Conf. Series* **140**, pp. 75-82

Williams, J.G., Newhall, X.X., and Dickey, J.O.: 1995, 'Relativity Parameters Determined from Lunar Laser Ranging', *Phys. Rev.* **D53**, pp. 6730-6739.

# THE EFFECT OF $C_{22}$ ON ORBIT ENERGY
# AND ANGULAR MOMENTUM

### D.J. SCHEERES

*Department of Aerospace Engineering & Engineering Mechanics,*
*Iowa State University, Ames, IA 50011-3231, U.S.A., E-mail: scheeres@iastate.edu*

**Abstract.** The effect of the $C_{22}$ gravity field term on a particle is evaluated analytically over one orbit to find the change in orbit energy and angular momentum as an explicit function of the orbital inclination, argument of pericenter, longitude of the ascending node, orbit parameter and eccentricity. Changes in orbit energy and angular momentum are shown to be proportional to a family of integrals which can be parameterized in terms of eccentricity and non-dimensional pericenter radius.

## 1. Introduction

A hallmark effect of particle dynamics close to distended bodies in uniform rotation are the large changes in orbit energy and angular momentum which can occur over one pericenter passage. These changes can be large enough to eject the particle from the body onto a hyperbolic orbit, capture a passing hyperbolic orbit, or cause the particle to impact the surface. Previous studies have established that the $C_{22}$ gravity term of the body, commonly termed the ellipticity, is the main contributor to these effects (Scheeres, 1995; Scheeres *et al.*, 1996; 1998b). This paper investigates the effect of this gravity term, taken in isolation, on an otherwise unperturbed orbit. In particular it investigates the change in orbit energy and angular momentum over one orbit about the central body. The problem of the $C_{22}$ gravity term alone is highly idealistic, but an understanding of its effect is important and can be used to analyze the general and qualitative properties of motion about distended bodies in uniform rotation. This paper does not concern itself with a complete characterization of this problem – such as the computation of periodic orbits, equilibrium points, and zero-velocity curves – as such characterizations have already been performed in detail for a variety of specific bodies (Scheeres, 1994; 1995; Scheeres *et al.*, 1996; 1998a).

## 2. Perturbation Model

The perturbing function for a central body with $C_{22}$ gravity coefficient acting on a particle is:

$$U_{22} = \frac{3\mu}{r^3} R_o^2 C_{22} \cos^2 \delta \cos(2\lambda) \tag{1}$$

where $\mu$ is the gravitational parameter, $r$ is the particle radius, $\delta$ is the body-fixed latitude of the particle, $\lambda$ is the body-fixed longitude of the particle, and $R_o$ is the normalizing radius for the body. We assume that the body is in uniform rotation about its largest moment of inertia with rotation rate $\omega_T$.

*Celestial Mechanics and Dynamical Astronomy* **73**: 339–348, 1999.
© 1999 *Kluwer Academic Publishers.*

The latitude and longitude of the particle in the body-fixed frame can be computed from the osculating orbit elements as:

$$\sin \delta = \sin i \sin u \tag{2}$$

$$\tan \lambda = \frac{\sin(\Omega - \omega_T t) \cos u + \cos(\Omega - \omega_T t) \sin u \cos i}{\cos(\Omega - \omega_T t) \cos u - \sin(\Omega - \omega_T t) \sin u \cos i} \tag{3}$$

where $t$ is the time, $u = \omega + f$ and the orbit elements $a$, $e$, $i$, $\omega$, $\Omega$, and $f$ all have their usual definitions.

To make the problem dimensionless we introduce a new independent parameter, $\tau$, and a length scale, $r_s$:

$$\tau = \omega_T t \tag{4}$$

$$r_s = \left(\frac{\mu}{\omega_T^2}\right)^{1/3} \tag{5}$$

The parameter $\tau$ corresponds to the rotational phase of the body and the length scale $r_s$ corresponds to the radius of a circular 1:1 synchronous orbit with no $C_{22}$ coefficient present. Introducing these scale factors defines the non-dimensional perturbing function:

$$U_{22} = \frac{3}{r^3} \tilde{C}_{22} \cos^2 \delta \cos(2\lambda) \tag{6}$$

where all free parameters have been compressed into the one non-dimensional term:

$$\tilde{C}_{22} = \left(\frac{\mu}{\omega_T^2}\right)^{-2/3} R_o^2 C_{22} \tag{7}$$

Values of the scaling parameter, rotation period and $\tilde{C}_{22}$ for a few select asteroids are shown in Table I.

TABLE I
Scaling radius, rotation period, and normalized $C_{22}$ for some select asteroids.

| Body | $(\mu/\omega^2)^{1/3}$ (km) | $T$ (hours) | $\tilde{C}_{22}$ (-) |
|---|---|---|---|
| Ida | 27.0 | 4.633 | 0.044 |
| Eros | 18.4 | 5.27 | 0.052 |
| Castalia | 0.8 | 4.07 | 0.047 |

## 3. Jacobi Integral

We note that this dynamical problem has a Jacobi integral defined in the body-fixed coordinate system. The derivation of this integral is analogous to the derivation of the integral in the restricted 3-body problem (see, for example, Brouwer and Clemence, 1961, pg 252) and is not developed here. The Jacobi integral for this dimensionless system has the form:

$$J = \frac{1}{2}v^2 - \frac{1}{2}r_{eq}^2 - \frac{1}{r} - U_{22} \tag{8}$$

where $v$ is the particle speed in the body-fixed frame and $r_{eq}$ is the particle radius projected into the equatorial plane.

This integral can be re-expressed in terms of the Keplerian energy and angular momentum, as is commonly done when deriving the Tisserand criterion in the restricted 3-body problem (Brouwer and Clemence, 1961, pg 256). Doing so yields the simplified integral:

$$J = C - H - U_{22} \tag{9}$$

where $C$ is the energy of the particle as measured with respect to the central body (treated as a point mass) and $H$ is the angular momentum of the particle, projected onto the rotation axis of the body. Equation 9 will play an important role later in our analysis.

## 4. Choice of Variables and Problem Restriction

This paper concentrates on changes in orbit energy and angular momentum (i.e., orbit elements $a$, $e$ and $i$) and does not consider secular changes in mean anomaly, argument of pericenter and longitude of the ascending node. This is justified in comparison to the $C_{20}$ problem where the secular rates of these angles are quite large (relative to their change due to $C_{22}$). Given this it is only necessary to focus on a subset of the classical canonical orbit elements and their attendant differential equations. The equations of motion of interest are then a modified set of canonical elements (Brouwer and Clemence, 1961, pg 290):

$$C = -1/(2a) \; ; G = \sqrt{a(1 - e^2)} \; ; H = G\cos i \tag{10}$$

$$C' = \frac{\partial R}{\partial \tau} \; ; G' = \frac{\partial R}{\partial \omega} \; ; H' = \frac{\partial R}{\partial \Omega} \tag{11}$$

where $R$ is the perturbing function and the elements $C$, $G$, and $H$ represent the Keplerian energy, angular momentum magnitude, and angular momentum projected onto the central body rotation axis, respectively.

## 5. Changes in $C, G,$ and $H$ over One Orbit

### 5.1. METHOD OF EVALUATION

To evaluate the change in these elements over one orbit we use the first iteration of Picard's method of successive approximations (for a rigorous discussion of this method see Moulton, 1958). Specifically, given a dynamical system of the form $X' = F(X, \tau)$, the first iteration of Picard's method yields:

$$X_1(\tau_2) = X_o + \int_{\tau_1}^{\tau_2} F(X_o, \tau) d\tau \tag{12}$$

where $X_o$ is assumed constant over the interval (representing the unperturbed orbit elements), and the variation of $F$ with $\tau$ includes both the central body rotation and the true anomaly of the particle orbit. Noting that $X_1(\tau_1) = X_o$ we immediately derive the result:

$$\Delta X_1 = \int_{\tau_1}^{\tau_2} F(X_o, \tau) d\tau \tag{13}$$

The limits of integration, $\tau_1$ and $\tau_2$, are $t_o - T/2$ and $t_o + T/2$, respectively, for elliptic orbits ($t_o$ being the time of pericenter passage and $T$ being the orbital period) and $t_o - \infty$ and $t_o + \infty$, respectively, for parabolic or hyperbolic orbits. We take $t_o = 0$ in general.

### 5.2. APPLICATION OF METHOD

The perturbing function (Equation 6) can be re-expressed as:

$$U_{22} = \frac{3\tilde{C}_{22}}{r^3} \left[ \frac{1}{2} \sin^2 i \left\{ \cos 2\Omega \cos 2\tau + \sin 2\Omega \sin 2\tau \right\} \right.$$
$$+ \cos^4(i/2) \left\{ \cos 2(\omega + \Omega) \cos 2(f - \tau) - \sin 2(\omega + \Omega) \sin 2(f - \tau) \right\}$$
$$\left. + \sin^4(i/2) \left\{ \cos 2(\omega - \Omega) \cos 2(f + \tau) - \sin 2(\omega - \Omega) \sin 2(f + \tau) \right\} \right] \tag{14}$$

which isolates the true anomaly $(f)$ and time $(\tau)$ terms together. Note that the pericenter passage occurs at $f = \tau = 0$, and thus the angles $\omega$ and $\Omega$ represent the argument of pericenter and the ascending node in the body-fixed coordinate frame at pericenter passage.

Directly applying Equation 13 to $C, G,$ and $H$ yields:

$$\Delta C = -\frac{6\tilde{C}_{22}}{p^3} \left[ \frac{3e}{8} \sin^2 i \sin 2\Omega \left( I_1^2 - I_{-1}^2 \right) \right.$$
$$+ \cos^4(i/2) \sin 2(\omega + \Omega) \left( I_2^3 + \frac{3e}{4} \left\{ I_3^2 - I_1^2 \right\} \right)$$
$$\left. + \sin^4(i/2) \sin 2(\omega - \Omega) \left( I_{-2}^3 + \frac{3e}{4} \left\{ I_{-3}^2 - I_{-1}^2 \right\} \right) \right] \tag{15}$$

$$\Delta G = -\frac{6\tilde{C}_{22}}{p^{3/2}}$$
$$\left[\cos^4(i/2)\sin 2(\omega+\Omega)I_2^1 + \sin^4(i/2)\sin 2(\omega-\Omega)I_{-2}^1\right] \tag{16}$$

$$\Delta H = -\frac{6\tilde{C}_{22}}{p^{3/2}}\left[\frac{1}{2}\sin^2 i\sin 2\Omega I_0^1\right.$$
$$\left. + \cos^4(i/2)\sin 2(\omega+\Omega)I_2^1 - \sin^4(i/2)\sin 2(\omega-\Omega)I_{-2}^1\right] \tag{17}$$

These formula allow for explicit computation of the expected change in orbit energy, angular momentum and inclination given the basic parameters of the orbit.

The integrals $I_m^n$ represent the interaction of the rotating body with the particle as it passes through one orbit. They are defined as:

$$I_m^n(e,q) = \int_{-\theta_\infty}^{\theta_\infty} (1+e\cos f)^n \cos(mf-2\tau)\,df \tag{18}$$

$$\tau = \begin{cases} a^{3/2}(E-e\sin E) & e < 1 \\ \tan(E/2) = \sqrt{\frac{1-e}{1+e}}\tan(f/2) & \\ \sqrt{2}q^{3/2}\left[\tan(f/2) + \frac{1}{3}\tan^3(f/2)\right] & e = 1 \\ |a|^{3/2}(e\sinh F - F) & e > 1 \\ \tanh(F/2) - \sqrt{\frac{e-1}{e+1}}\tan(f/2) & \end{cases} \tag{19}$$

where $\theta_\infty = \pi$ if $e \leq 1$, and $\theta_\infty = \arccos(-1/e)$ if $e > 1$. Note that the independent variable of integration is the true anomaly.

## 5.3. ALTERNATE DERIVATION OF $\Delta C$

An alternate derivation of $\Delta C$ exists which provides a simpler form for the expression and reduces the number of $I_m^n$ integrals that must be computed. This derivation uses the Jacobi integral as stated in Equation 9. Evaluating the integral at two subsequent apocenters (or infinities), allowing $C$ and $H$ to change but keeping the items in $U_{22}$ fixed (in accord with the assumptions in deriving Equations 15 – 17), and noting that the Jacobi integral remains constant, yields:

$$\Delta C = \Delta H + \Delta U_{22} \tag{20}$$
$$\Delta U_{22} = U_{22}(\theta_\infty) - U_{22}(-\theta_\infty) \tag{21}$$

Some simple algebraic manipulation will show that:

$$\Delta U_{22} = \frac{6\tilde{C}_{22}}{p^{3/2}} I \left[ \frac{1}{2} \sin^2 i \sin 2\Omega \right.$$

$$\left. + \cos^4(i/2) \sin 2(\omega + \Omega) - \sin^4(i/2) \sin 2(\omega - \Omega) \right] \tag{22}$$

$$I = \begin{cases} \left( \frac{1-e}{1+e} \right)^{3/2} \frac{\sin(2\pi a^{3/2})}{a^{3/2}} & e < 1 \\ 0 & e \geq 1 \end{cases} \tag{23}$$

which leads to a simplified form of $\Delta C$:

$$\Delta C = -\frac{6\tilde{C}_{22}}{p^{3/2}} \left[ \frac{1}{2} \sin^2 i \sin 2\Omega \left( I_0^1 - I \right) \right.$$

$$+ \cos^4(i/2) \sin 2(\omega + \Omega) \left( I_2^1 - I \right)$$

$$\left. - \sin^4(i/2) \sin 2(\omega - \Omega) \left( I_{-2}^1 - I \right) \right] \tag{24}$$

## 6. Elementary Properties of $I_m^n$

The integrals $I_m^n$ have a few simple properties that should be discussed. First, the integrals are completely independent of the central body properties, and thus need only be computed once as a function of non-dimensional $q$ and $e$ to cover all cases of the central body mass, rotation rate, and $C_{22}$.

Second, the integrals are finite and bounded. This is clear by inspection of Equation 18:

$$|I_m^n(e, q)| \leq 2\theta_\infty (1 + e)^n \tag{25}$$

Third, in Equations 16, 17 and 24 we see that the integrals are defined for both positive and negative values of the integer $m$. There is a marked difference in the values of the integrals for these two cases, and in most situations of interest we find that:

$$|I_{-m}^n| \ll |I_m^n| \tag{26}$$

Exceptions to this occur at some small values of $q$ and $e$. This is a significant result as it clearly explains why particles in retrograde orbit about a uniformly rotating body experience relatively small fluctuations in energy and angular momentum, as compared to particles in direct orbits (see Scheeres, 1994; 1995, and Scheeres *et al.*, 1996, for discussions of this effect).

Inequality 26 is most easily understood by noting that at pericenter passage the argument of the $\cos(mf - 2\tau)$ term will remain small over a longer time interval if $m$ is positive, allowing the integrand to contribute more to the integral while in the neighborhood of its maximum value. This effect becomes largest when the angular rate at pericenter passage is equal to the body rotation rate, which occurs at:

$$q^3 = (1 + e)\frac{m^2}{4} \tag{27}$$

Along the lines defined by this condition one sees that the integrals $I_m^n$ take on a larger value. For negative $m$ this condition has no special significance.

Finally, the $I_m^n$ integrals are computed by first recasting them as a differential equation:

$$\frac{dI_m^n}{df} = (1 + e\cos f)^n \cos(mf - 2\tau) \tag{28}$$

with initial condition evaluated at $f = 0$. The equation is then numerically integrated, with error control, from $f = 0$ to $f = \theta_\infty$, the full integral being obtained by doubling the integrated result (since the equation is even about pericenter). In all cases of conic motion this procedure is seen to work well. In the case of parabolic and hyperbolic orbits accurate results are obtained despite the infinite variations in the integrand as $f$ approaches $\theta_\infty$. To understand this we note two items. First, that the contributions of the integrand decrease to zero with increasing true anomaly. Second, for true anomaly close to $\theta_\infty$ the contribution of the integrand to the total integral, taken over any finite interval of true anomaly, rapidly approaches a zero mean due to the swift oscillation of the time argument. These factors combine to allow the differential equation approach to computing the integrals to truncate the tails of the integration as appropriate. The accuracy of this integration method can be checked by comparing the computed value of $\Delta C$ using both Equations 15 and 24, since we know independently that these combinations of integrals should be equal. Performing this comparison we find agreement to within the specified numerical error of the integration.

## 7. Example Computation

To illustrate the utility of this theory we present contour plots showing the normalized change in energy $(C)$ and angular momentum projected onto the rotation axis $(H)$ as a function of dimensionless pericenter radius and eccentricity. The specific results plotted in Figures 1 – 3 are $\frac{6}{p^{3/2}}(I_2^1 - I)$ and $\frac{6}{p^{3/2}}I_2^1$, which correspond to the terms that contribute the most to the change in $C$ and $H$, respectively. To scale these results to a specific body, inclination, and argument of pericenter passage, multiply the contour values by $-\tilde{C}_{22}\cos^4(i/2)\sin 2(\omega + \Omega)$.

In Figures 1 and 2 the contour values for the elliptic case $(e \leq 1)$ are plotted for $\Delta C$ and $\Delta H$, respectively. In Figure 3 the changes for the hyperbolic case $(e \geq 1)$ are plotted, where we recall from Equation 20 that $\Delta C = \Delta H$ when $e \geq 1$.

The results found from this approach have been compared with numerical integrations. We find that agreement is good for high eccentricity elliptic orbits,

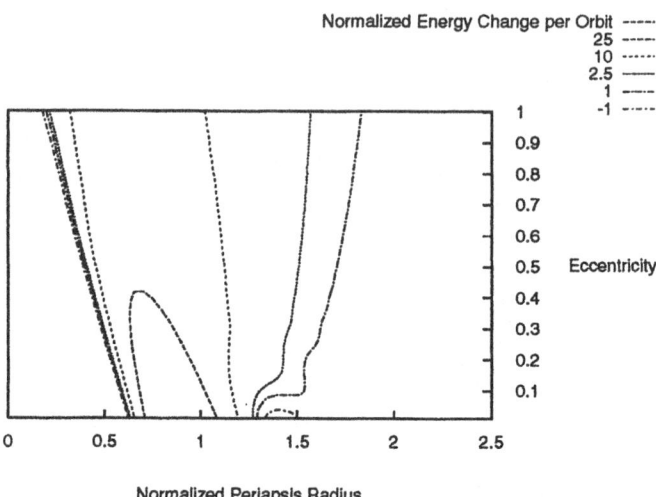

Fig. 1. Normalized $\Delta C$ per orbit for elliptic orbits (dominant terms only). Multiply contour values by $-\tilde{C}_{22}\cos^4(i/2)\sin\left[2(\omega + \Omega)\right]$ to scale to an arbitrary flyby.

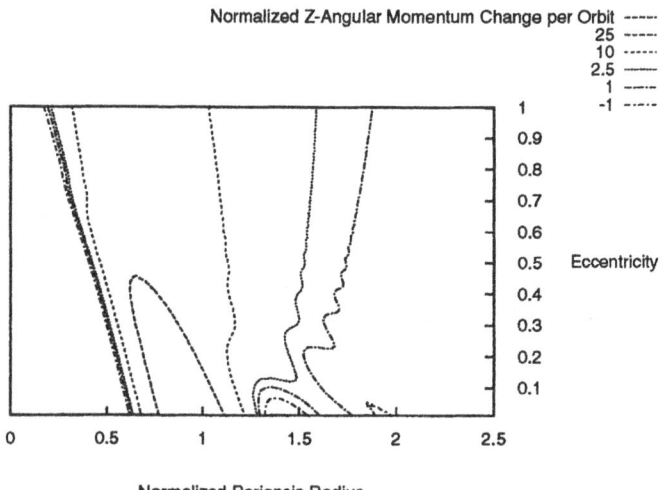

Fig. 2. Normalized $\Delta H$ per orbit for elliptic orbits (dominant terms only). Multiply contour values by $-\tilde{C}_{22}\cos^4(i/2)\sin\left[2(\omega + \Omega)\right]$ to scale to an arbitrary flyby.

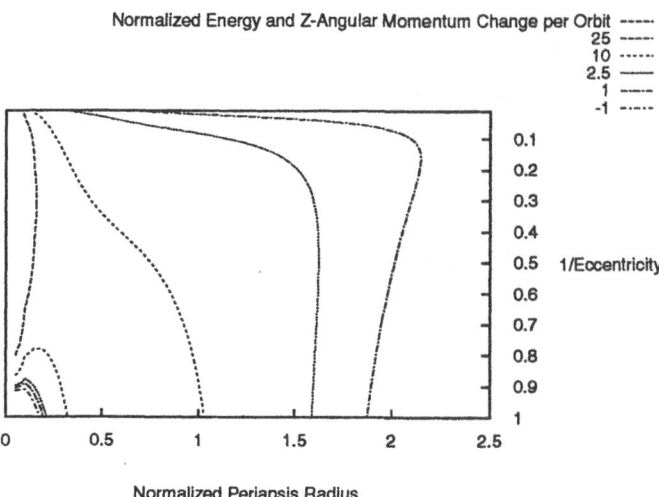

Fig. 3. Normalized $\Delta C$ and $\Delta H$ per orbit for hyperbolic orbits (dominant terms only). Note that $1/e$ is plotted. Multiply contour values by $-\tilde{C}_{22}\cos^4(i/2)\sin\left[2(\omega+\Omega)\right]$ to scale to an arbitrary flyby.

parabolic orbits and hyperbolic orbits, but that accuracy begins to degrade sharply once the eccentricity of an elliptic orbit falls below a few tenths. Possible remedies to this will be investigated in the future.

## 8. Conclusions

The theory presented in this paper applies to all cases of uniformly rotating bodies with a $C_{22}$ gravity term. There are many applications for the theory as derived here. These include the computation of capture and ejection radius about a body, mission design and trajectory planning considerations for a spacecraft mission about an asteroid or comet, and long-term investigations of particle and ejecta dynamics about asteroids and comets. These applications, and others, will be detailed in future papers and reports.

## Acknowledgements

The author thanks an anonymous reviewer for making this article much clearer and more succinct. Portions of this research were performed under contract with the Jet Propulsion Laboratory.

# References

Brouwer, D. and Clemence, G.M.: 1961, *Methods of Celestial Mechanics*, Academic Press.

Moulton, F.R.: 1958, *Differential Equations*, Dover.

Scheeres, D.J.: 1994, Dynamics about uniformly rotating tri-axial ellipsoids, *Icarus*, **110**, 225-238.

Scheeres, D.J.: 1995, Analysis of orbital motion around 433 Eros, *J. Astronautical Sciences*, **43**, 427-452.

Scheeres, D.J., Ostro, S.J., Hudson, R.S. and Werner, R.A.: 1996, Orbits close to asteroid 4769 Castalia, *Icarus*, **121**, 67-87.

Scheeres, D.J., Ostro, S.J., Hudson, R.S., DeJong, E.M. and S. Suzuki: 1998a, Dynamics of Orbits Close to Asteroid 4179 Toutatis, *Icarus*, **132**, 53-79.

Scheeres, D.J., Marzari, F., Tomasella, L. and Vanzani, V.: 1998b, ROSETTA mission: satellite orbits around a cometary nucleus, *Planetary and Space Science*, **46**, 649-671.

# THE USE OF LIE SERIES IN THE CONSTRUCTION OF A PERTURBATION THEORY AND SOME RECENT RESULTS IN THE THEORY OF THE MOTION OF HYPERION

P.J. MESSAGE

*University of Liverpool, L69 3BX, U.K., E-mail: sx20@liverpool.ac.uk*

**Abstract.** This paper begins with a brief review of a form of the Lie series transformation, and then reports some new results in the study, using Lie series methods, of the orbit of Saturn's satellite Hyperion. In particular, improved expressions are given for the long-period perturbations of the orbital elements which describe the motion in the orbit plane, and also first results for expressions for the short-period perturbations in the apse longitude, derived from the Lie series generating function.

## 1. Lie Series Transformations

At this Colloquium, whose topic is the impact of modern dynamics on astronomy, it is appropriate to mark the advances in the study of the motions of celestial bodies which have been made possible by the use of Lie series transformations, with a brief review of which we will begin below. Many applications of Lie series transformations have been in the context of near-commensurability of orbital motions, including the series of investigations into the motion of Saturn's satellite Hyperion, some new results in which are reported in this paper. Let us begin, then, by setting out the main features of the Lie series transformation, in the form in which it is used in this work on the orbit of Hyperion. We suppose that we have a dynamical system, of $n$ degrees of freedom, with co-ordinates $q = (q_1, q_2, \ldots, q_n)$, whose conjugate momenta are $p = (p_1, p_2, \ldots, p_n)$, respectively, and with Hamiltonian function $\mathcal{H}(q, p)$. So the equations of motion are Hamilton's equations:

$$\dot{q}_i = \frac{\partial \mathcal{H}}{\partial p_i}, \quad \text{and}, \quad \dot{p}_i = -\frac{\partial \mathcal{H}}{\partial q_i}, , \quad \text{for}, \quad i = 1, 2, \ldots, n. \tag{1}$$

Suppose we have chosen a function $\mathcal{W}(q, p)$, the *generating function*, then let us define the operator $\mathcal{L}_\mathcal{W}$ so that, for any function $f(q, p)$,

$$\begin{aligned}
\mathcal{L}_\mathcal{W}(f) &= \{f, \mathcal{W}\} \\
&= \sum_{i=1}^{n} \Big\{ \frac{\partial f}{\partial q_i} \frac{\partial \mathcal{W}}{\partial p_i} - \frac{\partial \mathcal{W}}{\partial q_i} \frac{\partial f}{\partial p_i} \Big\},
\end{aligned} \tag{2}$$

which is the Poisson bracket of $f$ and $\mathcal{W}$. In turn define $\mathcal{L}_\mathcal{W}^{(2)}(f) = \mathcal{L}_\mathcal{W}\{\mathcal{L}_\mathcal{W}(f)\}$, and $\mathcal{L}_\mathcal{W}^{(3)}(f) = \mathcal{L}_\mathcal{W}\{\mathcal{L}_\mathcal{W}^{(2)}(f)\}$, and, in general, $\mathcal{L}_\mathcal{W}^{(n)}(f) = \mathcal{L}_\mathcal{W}\{\mathcal{L}_\mathcal{W}^{(n-1)}(f)\}$, for $n = 2, 3, 4, \ldots$, (understanding $\mathcal{L}^{(1)}$ as $\mathcal{L}$). Then the Lie series transformation $(Q, P) \longmapsto (q, p)$ is defined by

$$q_i = Q_i + \sum_{k=1}^{\infty} \frac{1}{k!} \mathcal{L}_\mathcal{W}^{(k)}(Q_i), \text{ and } p_i = P_i + \sum_{k=1}^{\infty} \frac{1}{k!} \mathcal{L}_\mathcal{W}^{(k)}(P_i), \tag{3}$$

*Celestial Mechanics and Dynamical Astronomy* **73**: 349–358, 1999.
© 1999 *Kluwer Academic Publishers.*

for $i = 1, 2, ...., n$. Now in fact $\mathcal{L}_W(Q_i) = \frac{\partial W}{\partial P_i}$, and $\mathcal{L}_W(P_i) = -\frac{\partial W}{\partial Q_i}$, and so

$$q_i = Q_i + \frac{\partial W}{\partial P_i} + \sum_{k=1}^{\infty} \frac{1}{(k+1)!}\mathcal{L}_W^{(k)}\left(\frac{\partial W}{\partial P_i}\right), \tag{4}$$

and $$p_i = P_i - \frac{\partial W}{\partial P_i} - \sum_{k=1}^{\infty} \frac{1}{(k+1)!}\mathcal{L}_W^{(k)}\left(\frac{\partial W}{\partial P_i}\right), \tag{5}$$

for $i = 1, 2, ...., n$. In fact, for any function $f(q, p)$,

$$f(q, p) = f(Q, P) + \sum_{k=1}^{\infty} \frac{1}{k!}\mathcal{L}_W^{(k)}(f(Q, P)). \tag{6}$$

Note that this transformation represents the progression over unit time of a notional dynamical system whose Hamiltonian function is $W$. Therefore it is a contact transformation, and so preserves the Hamiltonian form of the equations of motion. If $\tilde{\mathcal{H}}(Q, P)$ is the Hamiltonian function, in the actual motion, which gives the equations of motion for the $(Q, P)$, then, since the transformation is autonomous, *i.e.* does not involve the time explicitly, we have

$$\tilde{\mathcal{H}}(Q, P) = \mathcal{H}(q, p) \tag{7}$$

$$= \mathcal{H}(Q, P) + \sum_{k=1}^{\infty} \frac{1}{k!}\mathcal{L}_W^{(k)}\{\mathcal{H}(Q, P)\}, \tag{8}$$

where equation (6) has been used for $\mathcal{H}$. Now in a large class of problems encountered in celestial mechanics, we find that the Hamiltonian function may be expressed in the form $\mathcal{H} = \mathcal{H}_0 - \varepsilon\mathcal{R}$, where $\mathcal{H}_0$ is a function only of the momenta $p_i$, and $\varepsilon$ is a small parameter. The case $\varepsilon = 0$ is spoken of as the "unperturbed" motion, and in this case the $p_i$ are the action variables, and the $q_i$ are their conjugate angle variables. Then $\mathcal{R}$ is the *disturbing function*, and it may usually be expanded as a multiple Fourier series:

$$\mathcal{R} = \sum_{\nu} K_{\nu} \cos N_{\nu}, \tag{9}$$

where $\nu = (\nu_1, \nu_2, \nu_3, ....., \nu_n)$, and $N_{\nu} = \sum_{j=1}^{n} \nu_j q_j$, and the summation over $\nu$ is over all sets of integers $\nu_j$ with $\nu_1$ non-negative and with $\sum_{j=1}^{n} \nu_j = 0$. Very often the transformation is to be chosen so that the motion of the transformed variables, $Q_i$ and $P_i$, contains none of the short-period features of the motion, so that these features will be encompassed within the generating function $(W)$, leaving the long-period features to be dealt with in isolation in the transformed system. Usually in celestial mechanics the co-ordinates fall into two sets, the *fast-moving* co-ordinates (e.g. the mean longitudes) and the *slow-moving* (e.g. the apse and node longitudes). In the class of problems where the unperturbed motion is in fact Keplerian elliptic motion, the co-ordinates of the latter set are constant in the unperturbed motion. In the case of motion near to a small-integer commensurability of the mean motions, so that there is a linear combination of the mean longitudes which changes slowly in the unperturbed motion, choose co-ordinates so that some of them (the *critical arguments*) contain the mean longitudes in this particular combination, and so will be classed as *slow-moving* for our purposes. In any case we suppose that we are able to identify which terms of $\mathcal{R}$ are of short-period, and which of long-period. The choice of the transformation is made by choosing the function $\mathcal{W}$. It will generally be necessary to choose it, stage-by-stage, as an expansion in powers of the perturbation parameter $\varepsilon$, of the form: $\mathcal{W} = \sum_{i=1}^{\infty} \varepsilon^i \mathcal{W}_i$, and likewise for the Hamiltonian function for the transformed problem: $\tilde{\mathcal{H}} = \sum_{i=0}^{\infty} \varepsilon^i \tilde{\mathcal{H}}_i$. Then, equating terms independent of $\varepsilon$ in equation (8) gives

$$\tilde{\mathcal{H}}_0(Q, P) = \mathcal{H}_0(Q, P), \tag{10}$$

so that the unperturbed motion in the transformed system is the same as that in the untransformed system. Equating terms of the first order in $\varepsilon$ gives

$$\tilde{\mathcal{H}}_1(Q, P) = -\mathcal{R}(Q, P) + \{\mathcal{H}_0(Q, P), \mathcal{W}_1\}, \tag{11}$$

and equating terms of the second order in $\varepsilon$ gives

$$\tilde{\mathcal{H}}_2(Q, P) = \{\mathcal{H}_0(Q, P), \mathcal{W}_2\} - \{(\mathcal{R}(Q, P) - \frac{1}{2}\{\mathcal{H}_0, \mathcal{W}_1\}), \mathcal{W}_1\}. \tag{12}$$

Now, as noted above, we suppose that $\mathcal{H}_0$ is a function of the momenta $p_i$ only, and we also suppose that we are able to separate the expansion (9) of $\mathcal{R}$ into its long-period part $[\mathcal{R}]_{lp}$, say, and its short-period part $[\mathcal{R}]_{sp}$, say, so that $\mathcal{R} = [\mathcal{R}]_{lp} + [\mathcal{R}]_{sp}$. Then we choose $\mathcal{H}_1(Q, P)$ to be equal to $-[\mathcal{R}]_{lp}(Q, P)$, so that it contains only long-period terms, and so equation (11) becomes

$$\{\mathcal{H}_0, \mathcal{W}_1\} = [\mathcal{R}]_{sp}. \tag{13}$$

If we write $\frac{\partial \mathcal{H}_0}{\partial P_j} = n_j$, for $j = 1, 2, ..., n$, then this becomes

$$\sum_{j=1}^{n} n_j \frac{\partial \mathcal{W}_1}{\partial Q_j} = \sum_{\nu, sp} K_\nu \cos N_\nu, \tag{14}$$

the summation being over the short-period terms, and the solution of it which we take is

$$\mathcal{W}_1 = \sum_{\nu, sp} \frac{K_\nu}{\rho_\nu} \sin N_\nu, \tag{15}$$

where $\rho_\nu = \sum_{j=1}^{n} \nu_j n_j$. Then equation (12), using equation (13), is

$$\tilde{\mathcal{H}}_2(Q, P) = \{\mathcal{H}_0(Q, P), \mathcal{W}_2\} - \{(\mathcal{R} - \frac{1}{2}[\mathcal{R}]_{sp}), \mathcal{W}_1\}, \tag{16}$$

and so we choose $\tilde{\mathcal{H}}_2(Q,P)$ to be the long-period part of $-\frac{1}{2}\{[\mathcal{R}]_{sp}(Q,P),\mathcal{W}_1\}$, which leaves $\{\mathcal{H}_0(Q,P),\mathcal{W}_2\}$ equal to the short-period part of $\{([\mathcal{R}]_{lp}(Q,P)+\frac{1}{2}[\mathcal{R}]_{sp}(Q,P)),\mathcal{W}_1\}$, which we may write as $\sum_{\nu,sp} L_\nu \cos N_\nu$, say, leading to the solution $\mathcal{W}_2 = \sum_{\nu,sp} \frac{L_\nu}{\rho_\nu}\sin N_\nu$. In a similar manner may be found successively $\tilde{\mathcal{H}}_3(Q,P)$, $\mathcal{W}_3$, $\tilde{\mathcal{H}}_4(Q,P)$, $\mathcal{W}_4$, &ce.

## 2. The application of Lie Series Transformations to the Theory of the Motion of Hyperion. Some Recent Results.

Two earlier papers (*Message 1989*) and (*Message 1993*) set out the method being used to study the long-period features of the motion of Saturn's satellite Hyperion, using a Lie series transformation to separate the long-period aspects of the motion. We recall that, very shortly after the discovery of this satellite at Harvard (and, independantly, in Liverpool) in 1848, unusual aspects of its motion drawing attention to it included the retrograde motion of the apse, and the large eccentricity of the orbit. It was first shown by Newcomb (*1891*) that these arose because of the very close near-commensurability of the orbital period of Hyperion with that of Titan, which is by far the most massive satellite of the system. The critical argument $\theta = 4\lambda_H - 3\lambda_T - \varpi_H$ in fact librates about the mean value 180 degrees, the main term having amplitude of about 36 degrees, and period about 21 months. (Here $\lambda_H$ and $\lambda_T$ are, respectively, the mean longitudes of Hyperion and Titan, and $\varpi_H$ is the apse longitude of Hyperion. Another important argument is $\sigma = \varpi_H - \varpi_T$, the difference between the apse longitudes of the two satellites.

Methods were described in the earlier papers to express derivatives of $[\mathcal{R}]_{lp}$, the long-period part of the disturbing function for the action of Titan, and other quantities required for the equations of motion for the long-period motion, as double Fourier series in $\sigma$, and $\omega$, the latter being related to $\theta$ by $\theta = \pi(1+0.3\sin\omega)$. The short-period argument is $\phi = \lambda_H - \lambda_T$, and we find that the equation (13), which we need to calculate the first-order part of the generating function, $\mathcal{W}_1$, takes the form

$$(n_H - n_T)\frac{\partial \mathcal{W}_1}{\partial \phi} + (4n_H - 3n_T)\frac{\partial \mathcal{W}_1}{\partial \theta} = [\mathcal{R}]_{sp}, \qquad (17)$$

where $n_H$ and $n_T$ are the mean motions of Hyperion and Titan, respectively. To use this, in the course of carrying out integrations over $\phi$, we need to use an expansion in powers of the small quantity $\epsilon$, which is defined by $(4n_H - 3n_T) = \epsilon(n_H - n_T)$, and which has a value near to $\frac{1}{120}$. In terms of this, we put

$$\mathcal{W}_1 = \mathcal{W}_{10} + \epsilon\mathcal{W}_{11} + \epsilon^2\mathcal{W}_{12} + \ldots\ldots \qquad (18)$$

Then, equating terms of each power of $\epsilon$ in equation (17), gives in succession

$$(n_H - n_T)\frac{\partial \mathcal{W}_{10}}{\partial \phi} = [\mathcal{R}]_{sp}, \qquad (19)$$

$$\frac{\partial \mathcal{W}_{11}}{\partial \phi} = \frac{\partial \mathcal{W}_{10}}{\partial \theta}, \tag{20}$$

$$\frac{\partial \mathcal{W}_{12}}{\partial \phi} = \frac{\partial \mathcal{W}_{11}}{\partial \theta}, \tag{21}$$

$$\frac{\partial \mathcal{W}_{13}}{\partial \phi} = \frac{\partial \mathcal{W}_{12}}{\partial \theta}, \&\text{ce.} \tag{22}$$

Then the solution of equation (19) is

$$\mathcal{W}_{10}(\phi, \theta) = \mathcal{W}_{10}(\phi_0, \theta) + \frac{1}{(n_H - n_T)} \int_{\phi_0}^{\phi} [\mathcal{R}]_{sp}(\phi', \theta) d\phi', \tag{23}$$

the initial value having been found by use of

$$\mathcal{W}_{10}(\phi_0, \theta) = \frac{1}{(n_H - n_T)} \int_{\phi_0}^{\phi_0 + 2\pi} \phi'.[\mathcal{R}]_{sp}(\phi', \theta) d\phi'. \tag{24}$$

## 2.1. Improved solutions for the parameters of the motion in the orbit plane.

The expressions for the long-period terms in the orbital elements describing the motion of Hyperion in its orbit plane, corresponding to the very-close commensurability type of libration are (*Message 1993*)

$$\theta = 180° + \sum \theta_{i,j}\sin(i\tau + j\zeta), \tag{25}$$

$$\sigma = \zeta + \sum \varpi_{i,j}\sin(i\tau + j\zeta), \tag{26}$$

$$a = a_{0,0} + \sum a_{i,j}\cos(i\tau + j\zeta), \tag{27}$$

$$e = e_{0,0} + \sum e_{i,j}\cos(i\tau + j\zeta), \tag{28}$$

where $a$ is the major-semi axis, $e$ is the eccentricity, of Hyperion's orbit, $\tau = \nu t + \tau_0$ is the argument of the free libration, of period about 21 months, and $\zeta = \chi t + \zeta_0$ is the linear part of the difference between the apse longitudes ($\sigma = \varpi_H - \varpi_T$), which has period about $18\frac{3}{4}$ years. (The summations are over all integer pairs $(i, j)$ with $i$ non-negative.) The earlier papers (*Message, 1989, 1993*) described how a least-squares fit was carried out, beginning with a set of estimates of some of the co-efficients $\theta_{i,j}, \varpi_{i,j}, a_{i,j}, e_{i,j}$, and of the rate of change, $\nu$, of the libration argument, and of the mean motion, $\hat{n}$, and finding a set consistent with the dynamical equations. The work reported in the previous paper (*Message, 1993*), has since been extended, solving for an enlarged set of co-efficients. This was carried out twice, using different sets of data. First this was done using data arising from the main sequences of observations made between 1875 and 1922, which were reduced by Woltjer (*1928*), who gave opposition mean values of the orbital elements, analysis of which gave the following estimates of the co-efficients of the long-period terms, and of $\nu$ and of $\hat{n}$:

$$e_{0,0} = +0.10419 \pm 0.00027, \quad e_{0,1} = +0.02414 \pm 0.00044,$$

$$e_{0,2} = -0.00183 \pm 0.00040, \quad e_{1,0} = -0.00401 \pm 0.00034,$$

$$\varpi_{0,1} = -13.905° \pm 0.273°, \quad \varpi_{0,2} = +0.754° \pm 0.249°,$$

$$\varpi_{1,0} = -0.314° \pm 0.262°, \quad \varpi_{2,0} = -0.795° \pm 0.304°,$$

$$\lambda_{0,1} = -0.054° \pm 0.019°, \quad \lambda_{0,2} = +0.007° \pm 0.018°,$$

$$\lambda_{1,0} = +9.112° \pm 0.018°, \quad \lambda_{2,0} = +0.039° \pm 0.018°,$$

$$\hat{n} = 16.9199890° \pm 0.0000027° \text{ per day},$$

and

$$\nu = 0.562025° \pm 0.000025° \text{ per day.}$$

Using these to provide equations of condition, and solving, as described in (*Message 1993*), but with an extended set of co-efficients in the scheme of solution, gives the following set of estimates of independent parameters:

$$\hat{n} = 16.9199888° \pm 0.0000066° \text{ per day}, \tag{29}$$
$$\theta_{1,0} = 36.877° \pm 0.180°, \tag{30}$$
$$\text{and } m' = 0.0002364220 \pm 0.0000000075, \tag{31}$$

where $m'$ is the mass of Titan in terms of that of Saturn. From these are derived (as described in *Message 1993*) the following dynamically consistent set of expressions:

$$\nu = 0.562024124° \text{ per day},$$

$$e = 0.1046696 + 0.024515\cos\zeta - 0.001427\cos2\zeta + 0.000175\cos3\zeta$$

$$-0.000025\cos4\zeta + 0.000003\cos5\zeta - 0.003888\cos\tau - 0.000049\cos2\tau$$

$$+0.000019\cos3\tau + 0.000182\cos(\tau - \zeta) - 0.000140\cos(\tau + \zeta)$$

$$-0.000041\cos(\tau - 2\zeta) - 0.000024\cos(\tau + 2\zeta) + 0.000007\cos(2\tau - \zeta)$$

$$-0.000007\cos(2\tau + \zeta),$$

$$a = a_{0,0}\{1 - 0.003227\cos\tau - 0.000004\cos2\tau + 0.000082\cos(\tau - \zeta)$$

$-0.000054\cos(\tau + \zeta)\}$,

$\varpi = \varpi_T + \zeta - 13.5880°\sin\zeta + 1.6233°\sin2\zeta - 0.2605°\sin3\zeta$

$+0.0468°\sin4\zeta - 0.0092°\sin5\zeta - 0.4302°\sin\tau - 0.0177°\sin2\tau$

$-0.0004°\sin3\tau + 0.0003°\sin4\tau + 0.3606°\sin(\tau - \zeta)$

$-0.2715°\sin(\tau + \zeta) - 0.0904°\sin(\tau - 2\zeta) + 0.0569°\sin(\tau + 2\zeta)$

$+0.0216°\sin(\tau - 3\zeta) - 0.0133°\sin(\tau + 3\zeta) + 0.0133°\sin(2\tau - \zeta)$

$-0.0066°\sin(2\tau + \zeta)$,

and

$\lambda = \hat{n}t + \lambda_0 + 9.11258°\sin\tau + 0.00300°\sin2\tau - 0.01698°\sin3\tau$

$+0.00006°\sin4\tau - 0.07132°\sin\zeta - 0.00069°\sin2\zeta + 0.00273°\sin3\zeta$

$-0.00067°\sin4\zeta - 0.00036°\sin5\zeta - 0.21473°\sin(\tau - \zeta) + 0.18253°\sin(\tau + \zeta)$

$+0.00125°\sin(\tau - 2\zeta) - 0.00107°\sin(\tau + 2\zeta)$

$+0.00022°\sin(2\tau - \zeta) - 0.00052°\sin(2\tau + \zeta)$.

The contributions of the influence of the Sun, the figure of Saturn, and of the other satellites to the secular motions of the apse and of the mean longitude have been included. These influences will of course also give rise to very small periodic terms, of periods different from those of the terms given here.

The second solution made use of data from the main sequences of observations made between 1967 and 1983, which were fitted to a numerical integration of the equations of motion by Taylor (*1992*), who derived from it estimates of the co-efficients of the long-period terms, and of $\nu$ and $\hat{n}$. (These values were given in *Message 1993*, section 5, though notice that $e_{0,0} = +0.104550 \pm 0.000028$.) A solution for estimates of the independent parameters, carried out in the same way as for the previous (1875 to 1922) set of observational data was carried out also with this set of values (except that the value used for $\hat{n}$ was that from the previous series of observations, which covers a longer time-span). The set of co-efficients in the scheme of solution was extended from that in the solution reported in (*Message 1993*). This gave the following set of estimates of the independent parameters:

$$\hat{n} = 16.9199464° \pm 0.0000396° \text{ per day,} \tag{32}$$
$$\theta_{1,0} = 36.955° \pm 0.096°, \tag{33}$$
$$\text{and } m' = 0.000236398 \pm 0.00000007, \tag{34}$$

The dynamically consistent set of expressions derived from these, proceeding again as described in (*Message 1993*), is:

$\nu = 0.56220934°$ per day,

$e = 0.104506 + 0.024330\cos\zeta - 0.001417\cos2\zeta + 0.000172\cos3\zeta$

$\quad -0.000025\cos4\zeta + 0.000003\cos5\zeta - 0.003904\cos\tau - 0.000049\cos2\tau$

$\quad +0.000019\cos3\tau + 0.000182\cos(\tau - \zeta) - 0.000144\cos(\tau + \zeta)$

$\quad -0.000041\cos(\tau - 2\zeta) - 0.000026\cos(\tau + 2\zeta) + 0.000007\cos(2\tau - \zeta)$

$\quad -0.000008\cos(2\tau + \zeta),$

$a = a_{0,0}\{1 - 0.003234\cos\tau - 0.000004\cos2\tau - 0.000061\cos\zeta$

$\quad +0.000083\cos(\tau - \zeta) - 0.000058\cos(\tau + \zeta)\},$

$\varpi = \varpi_T + \zeta - 13.5006°\sin\zeta + 1.6109°\sin2\zeta - 0.2558°\sin3\zeta$

$\quad +0.0461°\sin4\zeta - 0.0088°\sin5\zeta - 0.4333°\sin\tau - 0.0178°\sin2\tau$

$\quad -0.0004°\sin3\tau + 0.0003°\sin4\tau + 0.3603°\sin(\tau - \zeta)$

$\quad -0.2716°\sin(\tau + \zeta) - 0.0900°\sin(\tau - 2\zeta) + 0.0558°\sin(\tau + 2\zeta)$

$\quad +0.0213°\sin(\tau - 3\zeta) - 0.0136°\sin(\tau + 3\zeta) + 0.0133°\sin(2\tau - \zeta)$

$\quad -0.0067°\sin(2\tau + \zeta), -0.0045°\sin(2\tau - 2\zeta) + 0.0017°\sin(2\tau + 2\zeta),$

$\quad -0.0011°\sin(3\tau - \zeta) + 0.0015°\sin(3\tau + \zeta),$

and

$\lambda = \hat{n}t + \lambda_0 + 9.1311°\sin\tau + 0.0028°\sin2\tau - 0.0169°\sin3\tau$

$\quad +0.0001°\sin4\tau + 0.0001°\sin5\tau - 0.0700°\sin\zeta - 0.0024°\sin2\zeta$

$\quad +0.0002°\sin3\zeta - 0.0005°\sin4\zeta - 0.0004°\sin5\zeta - 0.2167°\sin(\tau - \zeta)$

$\quad +0.1909°\sin(\tau + \zeta) - 0.0008°\sin(\tau - 2\zeta) + 0.0020°\sin(\tau + 2\zeta)$

$\quad +0.0012°\sin(3\tau - \zeta) - 0.0012°\sin(3\tau + \zeta).$

The degree of agreement between these two sets of expressions, which are derived using two quite separate sets of observational data, is an indicator of the reliability of the results obtained so far in this work. (The methods, with analytical basis, being used here, are of course quite different from those of Duriez and Vienne (*1997*), based on numerical integration. Comparison of the results of the two approaches must take account of the use of different parameters.)

## 2.2. Preliminary results for short-period terms in the apse longitude.

As was indicated earlier, the generating function, $\mathcal{W}$, of the Lie series transformation must contain all the information needed to construct expressions for the short-period perturbations. Thus the first equation of (3), with $\varpi$ for $q_i$, gives the short-period perturbations of the apse longitude, and the first-order part is

$$-\frac{\sqrt{1-e^2}}{na^2e}\frac{\partial \mathcal{W}}{\partial e}$$

Preliminary calculations give, for the largest few terms of this:

$+1.0228°\sin\phi + 0.4248°\sin2\phi + 0.2942°\sin3\phi + 0.2813°\sin4\phi$

$+0.0445°\sin5\phi + 0.0307°\sin6\phi + 0.0299°\sin7\phi + 0.0217°\sin8\phi$

$-0.1540°\sin(\phi + \zeta) - 0.0979°\sin(\phi - \zeta) - 0.0874°\sin(\phi + \tau)$

$+0.0941°\sin(\phi - \tau) + 0.0364°\sin(\phi + 2\zeta) + 0.0231°\sin(\phi - 2\zeta)$

$-0.0297°\sin(\phi + \tau - \zeta) + 0.0269°\sin(\phi + \tau + \zeta) - 0.0463°\sin(\phi - \tau + \zeta)$

$-0.0508°\sin(2\phi + \zeta) - 0.0369°\sin(2\phi - \zeta) + 0.0208°\sin(2\phi + \tau)$

$-0.0381°\sin(3\phi + \zeta) - 0.0299°\sin(3\phi - \zeta) + 0.0216°\sin(3\phi - \tau)$

$-0.0448°\sin(4\phi + \zeta) - 0.0288°\sin(4\phi + \tau) + 0.0459°\sin(4\phi - \tau)$

$+0.0201°\sin(5\phi + \tau).$

For comparison, those terms derived by Taylor (*1992*) from Fourier analysis of the results of his numerical integration, are:

$+1.0391°\sin\phi + 0.4209°\sin2\phi + 0.3115°\sin3\phi + 0.2795°\sin4\phi$

$-0.1674°\sin(\phi + \zeta) - 0.0972°\sin(\phi - \zeta) + 0.0833°\sin(\phi + \tau)$

$+0.0946°\sin(2\phi + \tau) - 0.0568°\sin(2\phi + \zeta) + 0.0731°\sin(3\phi + \tau)$

$-0.0482°\sin(3\phi + \zeta) + 0.0636°\sin(4\phi + \tau) - 0.0511°\sin(4\phi + \zeta).$

For comparison with the results of Vienne and Duriez (*1991*), note that 0.1° in $\varpi$ corresponds to about 266 km. along the orbit of Hyperion.

# References

Duriez, L., and Vienne, A.: 1997, Theory of motion and ephemerides of Hyperion, *Astron. & Astrophys.*, **324**, 366-380.

Message, P.J.: 1989, The use of computer algorithms in the construction of a theory of the long-period perturbations of Saturn's satellite Hyperion, *Celestial Mechanics*, **45**, 45-53.

Message, P.J.: 1993, On the second-order long-period motion of Hyperion, *Celestial Mechanics*, **56**, 277-284.

Newcomb, S.: 1891, Hyperion: a new case in celestial mechanics, *Astron. Papers of the American Ephemeris*, **III**.

Taylor, D.B.: 1992, A synthetic theory for the perturbations of Titan on Hyperion, *Astron. & Astrophys.*, **265**, 825-832.

Vienne, A., and Duriez, L.: 1991, A general theory of motion for the eight major satellites of Saturn, *Astron. & Astrophys.*, **246**, 619-633.

Woltjer, J.: 1928, The Motion of Hyperion, *Annalen van de Sterrewacht te Leiden*, **XVI**, Part 3.

# HIGHER ORDER AND ITERATIVE THEORIES TO COMPUTE
# ASTEROID MEAN ELEMENTS

Z. KNEŽEVIĆ

*Astronomical Observatory, Volgina 7, 11160 Belgrade 74, Yugoslavia,*
*E-mail: zoran@aob.bg.ac.yu*

and

A. MILANI

*Department of Mathematics, Via Buonarroti 2, 56127 Pisa, Italy,*
*E-mail: milani@dm.unipi.it*

Mean orbital elements are obtained from their instantaneous, osculating counterparts by removal of the short periodic perturbations. They can be computed by means of different theories, analytical or numerical, depending on the problem and accuracy required. The most advanced contemporary analytical theory (Knežević 1988) accounts only for the perturbing effects due to Jupiter and Saturn, to the first order in their masses and to degree four in eccentricity and inclination. Nevertheless, the mean elements obtained by means of this theory are of satisfactory accuracy for majority of the asteroids in the main belt (Knežević et al. 1988), for the purpose of producing large catalogues of mean and proper elements, to identify asteroid families, to assess their age, to study the dynamical structure of the asteroid belt and chaotic phenomena of diffusion over very long time spans. In the vicinity of the main mean motion resonances, however, especially 2:1 mean motion resonance with Jupiter, these mean elements are of somewhat degraded accuracy.

We analysed a number of algorithms capable of producing mean elements of significantly better quality and much closer to even the strongest mean motion resonances, than it was the case so far (Milani and Knežević 1998). The methods we have tested belong to the two distinct classes: Breiter type (Breiter 1997a, 1997b) methods that take into account part of the effects of the second order in perturbing mass due to the combinations of the first order effects; and iterative methods which compute the convergent fixed frequency theory (Milani 1988). The former methods are based on application of some higher order numerical integration scheme, such as Runge-Kutta order 2, to the system of differential equations with determining function as Hamiltonian, whose solution is the transformation removing the fast angular variables; from this point of view, our current theory is equivalent to the solution of this auxiliary differential equation by means of the classical, first order numerical integration scheme of Euler type. The latter procedures involve fixed point iterative schemes, with current first order theory, or some second order Breiter type theory, as initial iteration step; at convergence they compute the inverse map, i.e. the transformation from mean to osculating elements; a similar method has already been employed in the computation of proper elements from the mean ones (Milani and Knežević 1990).

*J. Henrard and S. Ferraz-Mello (eds.), Impact of Modern Dynamics in Astronomy, 359–360.*
© 1999 *Kluwer Academic Publishers.*

Ten different algorithms were tested on a sample of asteroid orbits taken from the Themis family, covering a range in semimajor axis up to the very edge of the 2:1 resonance. The dispersion and the maximum excursion of the instantaneous mean values with respect to an average value of the mean semimajor axis, computed over 100 000 yr, was used as an accuracy control. The results of these tests led us to a number of interesting conclusions, the most important of which can be summarized as follows.

(i)   The iterative methods turned out to be superior to the Breiter-type methods, in accuracy and in reliability; this can be understood if there are cancellations among second order terms, and some of these terms become unbalanced when the second order effects are only partially accounted for. In other words, unless we take into account a full second order transformation (very difficult in the asteroidal case, due to the complexity of the required second order determining function), the improvement of the results might be compromised to the point that some first order solutions have better performance.

(ii)  As all the iterative methods at convergence supply practically identical results, the simplest Euler type starter (that is the current theory) was the most efficient; using a more sophisticated starter reduces somewhat the number of iterations, but the computational cost involved in deriving a more accurate initial step in practice often exceeds the gain due to the faster convergence.

(iii) A remarkable improvement of the accuracy of the resulting mean elements has been achieved mainly in the vicinity of the strongest mean motion resonances, in particular close to the 2:1, where the accuracy reached that common for the main belt objects even for a semimajor axis up to 3.20 AU. The conjecture proposed by Milani and Knežević (1994), that the further improvements of the existing analytical theory to compute mean and proper asteroid elements can only be of a local character, has thus been confirmed.

## References

Breiter, S., 1997a, *Celest. Mech. Dyn. Astron.*, **65**, 345–354.
Breiter, S., 1997b, *Celest. Mech. Dyn. Astron.*, **67**, 237–249.
Knežević, Z., 1988, *Bull. Astron. Obs. Belgrade*, **139**, 1–6.
Knežević, Z., Carpino, M., Farinella, P., Froeschlé, Ch., Froeschlé, Cl., Gonczi, R., Jovanović, B., Paolicchi, P., and Zappalà, V., 1988, *Astron. Astrophys.*, **192**, 360–369.
Milani, A., 1988, In: *Long-term Dynamical Behaviour of Natural and Artificial N-Body Systems* (A.E. Roy, Ed.), Kluwer Acad. Publ., pp. 73–108.
Milani, A., and Knežević, Z., 1990, *Celest. Mech. Dyn. Astron.*, **49**, 247–411.
Milani, A., and Knežević, Z., 1994, *Icarus*, **107**, 219–254.
Milani, A., and Knežević, Z., 1998, *Celest. Mech. Dyn. Astron.*, in press.

# THE ASTEROID MASS DETERMINATION PROJECT
# AT THE U.S. NAVAL OBSERVATORY

JAMES L. HILTON

*U.S. Naval Observatory, 3450 Massachusetts Ave. NW, Washington, DC 20392, USA*

and

RONALD C. STONE

*U.S. Naval Observatory Flagstaff Station, P.O. Box 1149, Flagstaff, AZ 86002, USA*

Asteroid masses are the largest source of unmodeled forces in current planetary ephemerides research. Williams (1984) showed that the asteroids produce km size perturbations in the position of Mars. However, the masses of only three asteroids are known to better than 10%, and only six other asteroid masses have been determined at all.

Detecting the mass of an asteroid is difficult because the observed quantity is the change in the mean motion of a second, perturbed asteroid. Asteroid masses are small, so the change in the mean motion is typically on the order of $0."015 \text{ yr}^{-1}$. Thus, excellent orbit determinations are needed both before and after the perturbing encounter. This requires high precision observations over as many oppositions as possible.

Hilton (1997) determined the mass of 15 Eunomia to within 25% by detecting perturbations of 1313 Berna. The greatest source of uncertainty in determining the mass of Eunomia was the very poor coverage and accuracy of pre-encounter observations. Hilton (1998) has determined the masses of 1 Ceres, 2 Pallas and 4 Vesta, all based on mutual interactions. The uncertainties in the masses are 1% for Ceres, 3% for Pallas, and 7% for Vesta. The masses of Ceres and Pallas are the best so far, and the mass for Vesta corroborates previous determinations of its mass.

Thirty-six encounters have been chosen from (Hilton *et al.*, 1996) as being the best candidates for asteroid mass determinations. These 40 encounters can determine masses of 13 asteroids. The most important asteroids for observation are: (1) 3946 Shor can give the mass of 10 Hygiea, the fourth largest asteroid. The current uncertainty in the mass of Hygiea is 50%. (2) 263 Dresda can give the mass of 16 Psyche, the largest of the M-type asteroids. (3) 827 Wolfiana is perturbed by 19 Fortuna. (4) 2296 Kugultinov can give the mass of 24 Themis. The oppositions in 1998 and 1999 are particularly important to improve the pre-encounter orbit. (5) 2873 Binzel can significantly improve on the mass of Vesta. Other potential targets for mass determination are: 45 Eugenia, 52 Europa, 65 Cybele, 324 Bamberga, 511 Davida, and 704 Interamnia. Improving the masses of 1 Ceres and 15 Eunomia is also likely.

As previously discussed, accurate positions are needed before and after the mutual encounter of two asteroids. Most of the existing positions for asteroids have uncertainties of $0."5$ or larger and contain systematic errors on the order of

*J. Henrard and S. Ferraz-Mello (eds.), Impact of Modern Dynamics in Astronomy, 361–362.*

a few tenths of an arcsecond inherited from the reference frame or star catalog used in making the reductions. Starting in 1984 observations taken with either the Carlsberg Automatic Meridian Circle (Helmer and Morrison, 1985) or the Flagstaff Astrometric Scanning Transit Telescope (FASTT) (Stone, 1997) have reduced the errors in these positions to ±0."1 to ±0."2 in each coordinate.

Even better observations are now possible using a large format CCD detector and on-chip differential reductions using ACT Reference catalog reference stars (Urban et al., 1998). An observing program using this technique has been started with the FASTT telescope. Positional accuracies of ±0."06 (s.e.) are routinely obtained. Systematic errors in these observations are believed to be less than ±0."02. The FASTT observing program includes high-accuracy observations of all the asteroids involved in the 40 encounters. These data will be necessary to provide new and improved asteroid masses during the next few years. Other observatories will soon be starting similar astrometric programs.

Thus, determinations of accurate masses for several asteroids are planned over the next several years. More information, including the list of target asteroids, is available at http://aa.usno.navy.mil/hilton/.

## References

Helmer, L., and Morrison, L.V.: 1985, 'Carlsberg Automatic Meridian Circle', *Vista Astr.*, **28**, 505

Hilton, J. L.: 1997, 'The Mass of 15 Eunomia from Observations of 1313 Berna and 1284 Latvia', *Astron. J.*, **114**, 402

Hilton, J. L.: 1998, 'U.S. Naval Observatory Ephemerides of the Largest Asteroids', submitted to *Astron. J.*

Hilton, J. L., Seidelmann, P. K., and Middour, J.; 1996, 'Prospects for Determining Asteroid Masses', *Astron. J.*, **11**, 2319

Stone, R.C.: 1997, 'CCD Astrometry of Asteroids as Determined in the Extragalactic Reference Frame', *Astron. J.*, **113**, 2317

Urban, S.E., Corbin, T., and Wycoff, G.: 1998, 'The ACT Reference Catalog', *Bull. Am. Astr. Soc.*, **29**, 1306

Williams, J. G.: 1984, 'Determining Asteroid Masses from Perturbations on Mars', *Icarus*, **57**, 1

# THE ASTEROID IDENTIFICATION PROBLEM

M.E. SANSATURIO

*E.T.S.I.I., Univ. Valladolid, Spain; E-mail: marsan@wmatem.eis.uva.es*

and

ANDREA MILANI and A. LA SPINA

*Dip. Matematica, Università di Pisa, Italy; E-mail: milani@dm.unipi.it*

A large fraction of the asteroids observed so far are to be considered lost, that is they cannot be recovered by pointing the telescope at the predicted position only. The catalogues of asteroid orbits are therefore polluted by large numbers of low accuracy orbits, which cannot be easily improved by observation. Two of these inaccurate orbits can belong to the same physical object, and indeed identifications are regularly found when the orbital elements are close; an unknown number of real identifications have not been found in the existing catalogues because the orbital elements as computed are far apart. It is impossible to systematically check for possible identification of all the couples of orbital element sets; an algorithm is required to single out a comparatively small number of proposed identifications, and to find a suitable first guess within the region of convergence of the differential corrections procedure to be used to confirm the identification.

In the linear approximation the target function to be minimized is a quadratic form; thus a fully analytic algorithm can be given to compute both the first guess for an identification and the cost, that is the increase in the target function resulting from joining the two sets of observations. This allows to define a number of distances in the elements space, which are functions, not only of the two orbital element sets, but also of their uncertainties, as measured by the Gaussian covariance matrices. Tests on the values of these distances are used as preliminary filters, to allow more extensive computations to be performed on a fraction of the total number of possible couples.

The difficulties in the application of this algorithm arise from the fact that poorly observed orbits have badly conditioned covariance matrices: all the computations performed with these matrices are affected by large numerical errors. Moreover, it is clear that the larger the eigenvalues of the covariance matrix, the larger the nonlinear effects will be. For asteroids observed over a very short arc, and lost since a long time, the linear approximation fails. Nonlinear optimisation algorithms can be used but are computationally expensive (Sansaturio *et al.*, 1996). Identifications based upon some of the orbital elements, e.g., excluding the mean longitude, can be effective in reducing the relevance of the nonlinear effects; the corresponding distances are defined by the marginal covariance matrices, using a formalism developed in (Milani, 1997).

We propose an orbit identification procedure involving a cascade of filters, using first a distance computed using only the inclination and the node, then a distance using all elements but the mean longitude, and finally a full 6-elements distance.

*J. Henrard and S. Ferraz-Mello (eds.), Impact of Modern Dynamics in Astronomy, 363–364.*

In each stage the distance depends also upon the relevant covariance matrices. We have tested this algorithm on a set of 100 already identified orbits, and found that all of them could have been obtained with a fully automated procedure. We suggest that either our procedure or an equivalent one should be applied as a matter of routine to all newly discovered asteroids. The search for orbit identifications in the existing catalogues can be done in a similar way, but requires to recompute—with the utmost care in guaranteeing numerical accuracy—all the orbits and their normal/covariance matrices, and this in turn requires unrestricted access to the data set of all the astrometric observations.

## References

Milani, A.: 1997, *The Identification Problem I: recovery of lost asteroids, Icarus*, in press.
Sansaturio, M. E., Milani, A. and Cattaneo, L.: 1996, Nonlinear optimisation and the asteroid identification problem, in *Dynamics, Ephemerides and Astrometry of the Solar System*, (S. Ferraz Mello et al., Eds.), Kluwer, pp. 193–198.

# INTERACTION OF RESONANCES AND YARKOVSKY
# NON-GRAVITATIONAL EFFECTS IN THE ASTEROID BELT

DAVID VOKROUHLICKÝ

*Institute of Astronomy, Charles University, V Holešovičkách 2,*
*CZ – 18000 Prague 8, Czech Republic, E-mail:vokrouhl@mbox.cesnet.cz*

and

PAOLO FARINELLA

*Dipartimento di Matematica, Università di Pisa, Via Buonarroti 2,*
*I – 56127 Pisa, Italy, E-mail: paolof@dm.unipi.it*

**keywords:** non-gravitational forces, meteorites, NEAs

The interaction between resonances and dissipative effects has long been investigated in relationship with the dynamics of dust particles or the evolution of planetesimals in the primordial solar nebula. More recently, it has been shown that another important application of this kind of dynamics is the delivery of asteroid fragments (precursors of meteorites and NEAs) to Earth–crossing orbits. The relevant dissipative mechanism, the so–called Yarkovsky effect (Öpik 1951; Peterson 1976; Burns *et al.* 1979) has been known for a long time, but its "seasonal" and "diurnal" variants, as well as its interaction with the collisional process taking place in the asteroid belt, have been studied quantitatively only in the last few years (Rubincam 1995, 1998; Hartmann *et al.* 1997, 1998; Farinella *et al.* 1998; Vokrouhlický and Farinella 1998a,b; Vokrouhlický 1998a,b; Farinella and Vokrouhlický 1998a,b).

In the Farinella *et al.* (1998) paper we have provided a unified discussion of the Yarkovsky effect in both its variants. After computing the rate of the corresponding semimajor axis drift as a function of size and spin rate, and comparing the relevant timescales with those for collisional disruption and spin axis reorientation, we have rediscussed some issues in meteorite science which are put in a new light by the relevance of the Yarkovsky effect. In particular, this mechanism provides a good explanation for the fact that meteorite cosmic ray exposure ages (in particular for irons) are much longer than the dynamical lifetimes of objects delivered to the Earth–crossing region through resonances. Thanks to the Yarkovsky effect, small asteroid fragments in the belt undergo a slow drift in semimajor axis (with a random–walk component related to their rotational state) and therefore have enough mobility to reach the resonances after comparatively long times spent in nonresonant main–belt orbits. Metal–rich fragments have slower Yarkovsky drift rates than stones, but their much longer collisional lifetimes can explain why iron meteorites appear to sample a larger number of asteroid parent bodies compared to ordinary chondrites.

Then, we have analyzed in more detail the dynamical evolution of asteroid fragments released in the Flora region, near the inner edge of the main asteroid belt,

*J. Henrard and S. Ferraz-Mello (eds.), Impact of Modern Dynamics in Astronomy, 365–368.*

and drifting into the $\nu_6$ secular resonance due to Yarkovsky effects (Vokrouhlický and Farinella 1998a). We have found that fragments 5 to 20 m in size evolve under the "seasonal" Yarkovsky effect, which causes a secular semimajor axis decay; they reach $\nu_6$ after a time shorter than their collisional lifetime when they start within about 0.05 to 0.2 AU out of the resonance. Metal–rich fragments drift slower but have have much longer lifetimes than stony ones, so they drift farther from their formation site. Fragments around 100 m in size are mainly influenced by the "diurnal" Yarkovsky effect if their surface is covered by a (thin) insulating regolith layer, and as a result their semimajor axis undergoes a random walk controlled by impacts which reorient the spin axis. Within their lifetime of $\approx$ 100 Myr these fragments can move throughout the inner part of the belt, episodically crossing $\nu_6$, until their orbit starts to be strongly perturbed by close encounters with the terrestrial planets. Meter–sized stony fragments, which probably deliver most meteorite falls, may also drift into the resonance under the "diurnal" effect, provided their surfaces have low thermal conductivities and/or their rotation is unusually slow.

Whereas in Vokrouhlický and Farinella (1998a) we have focused on the dynamical aspects of the Yarkovsky effects (by investigating individual fragment orbits), in Farinella and Vokrouhlický (1998b) we have modeled in a statistical way the evolution of large "swarms" of fragments released by catastrophic break-up events in the main belt. Their dynamics has been represented in proper element space by a simple, two–component model, including (i) the secular semimajor axis drift due to Yarkovsky effects, and (ii) the effects of random impact events resulting in the cascade–like generation of new populations of fragments. We have found that the combination of these mechanisms can feed efficiently the main resonances with small asteroid fragments from nearly all the locations in the main belt, and that only very close to the resonance edges direct injections are important. For instance, according to our model some 50–80 % (in mass) of the material released from a Flora–like asteroid is transported into the $\nu_6$ resonance, either as first–generation fragments or through their collisional offspring (this should be compared to only a few percent chance of direct injection in this case). Another important result from this model is that the distribution of accumulated cosmic–ray exposure (CRE) ages in the population of fragments reaching the Earth is in fair agreement with the observations. Relatively old events are likely to generate the background CRE age profiles, peaked at 20–50 Myr for stones and 200–500 Myr for irons, while comparatively recent events may create discrete peaks in the CRE age distributions (such as the 7–8 Myr prominent peak for the H–chondrites). In the latter case, the bulk of the original fragment population may still reside in the main belt, and will supply a significant flux of meteorites in the near future. We plan to perform a detailed quantitative comparison with the observed CRE age data in order to constrain the surface thermal conductivity of fragments from different source asteroids, as this is currently the main unknown parameter of the model. Hopefully, the forthcoming interplanetary missions (such as NEAR) will allow to directly measure/estimate the surface thermal conductivity of NEAs, providing us with an independent source of information.

In Farinella and Vokrouhlický (1998a) we have noted that Yarkovsky effects are capable of providing some semimajor axis mobility even to km–sized small asteroids in the main belt. Typically, bodies in the 1 to 20 km diameter range may move in semimajor axis by $\approx 0.01$ AU within their collisional lifetimes of 0.1 to 1 Byr. Interestingly, this result is almost independent of the thermal properties of the surface material. This mobility may be a key mechanism for feeding the high–order and/or mixed resonances in the inner asteroid belt, which have been recently identified as the most likely dynamical routes for multi–km sized Mars–crossers and NEAs (Migliorini et al. 1998). Other likely consequences are the eventual fall into the main resonances of fragments generated "on the brink" (Knežević et al. 1997) and the gradual spreading in semimajor axis of the small members of the most compact asteroid families.

Finally, Yarkovsky effects may provide a natural mechanism for explaining the observed overabundance of 10–100 m bodies among NEAs (see e.g. Rabinowitz 1993, 1994). The Yarkovsky mobility of a population of bare–rock (or iron–rich) fragments, dominated by the seasonal Yarkovsky effect, is maximum for bodies of size comparable to the penetration depth of the seasonal thermal wave (about 10 m for stones and 20 m for irons). Thus, we may expect that these bodies are preferentially removed from the main belt, and eventually show up in a relative overabundance within the population of Earth-crossing objects (Vokrouhlický and Farinella 1998a, Hartmann et al. 1998). Moreover, their removal from the main–belt population would imply a longer collisional lifetime for the bodies $\approx 100$ m in size (which can be fragmented by impacts with the 10–m bodies), allowing them to drift over a wider portion of the belt and eventually to feed the resonances. Therefore the overabundance in the near–Earth population may well extend to sizes larger than those corresponding to the maximum efficiency of the seasonal Yarkovsky effect, possibly to about 100 m. We plan to develop detailed quantitative models for these complex feedback effects between Yarkovsky mobility and collisional evolution in the near future.

# References

Burns, J.A., Lamy, P.L. and Soter, S.: 1979, *Icarus*, **40**, 1.

Farinella, P. and Vokrouhlický, D.: 1998a, *Science*, submitted.

Farinella, P. and Vokrouhlický, D.: 1998b, in preparation.

Farinella, P., Vokrouhlický, D. and Hartmann, W.K.: 1998, *Icarus*, **132**, 378.

Hartmann, W.K. *et al.*: 1997, in Proc. Lunar and Planet. Sci. Conf. XXVIII (Houston: Lunar Planet. Inst.), p. 517.

Hartmann, W.K. *et al.*: 1998, *Meteoritics Planet. Sci.*, in press.

Knežević, Z. *et al.*: 1997, *Planet. Space Sci.*, **45**, 1581.

Migliorini, F. *et al.*: 1998, *Science*, **281**, 2022

Öpik, E.J.: 1951, *Proc. Roy. Irish Acad.*, **54**, 165.

Peterson, C.: 1976, *Icarus*, **29**, 91.

Rabinowitz, D.L.: 1993, *Astrophys. J.*, **407**, 412.

Rabinowitz, D.L.: 1994, *Icarus*, **111**, 364.

Rubincam, D.P.: 1995, *J. Geophys. Res.*, **100**, 1585.
Rubincam, D.P.: 1998, *J. Geophys. Res.*, **103**, 1725.
Vokrouhlický, D.: 1998a, *Astron. Astrophys.*, **335**, 1093.
Vokrouhlický, D.: 1998b, *Astron. Astrophys.*, **338**, 353.
Vokrouhlický, D. and Farinella, P.: 1998a, *Astron. Astrophys.*, **335**, 351.
Vokrouhlický, D. and Farinella, P.: 1998b, *Astron. J.*, **116**, 2032.

# TARGET PLANE CONFIDENCE BOUNDARIES:

## MATHEMATICS OF THE 1997 XF11 SCARE

ANDREA MILANI

*Dip. Matematica, Università di Pisa;* `milani@dm.unipi.it`

and

GIOVANNI B. VALSECCHI

*IAS-Planetologia, Roma;* `giovanni@ias.rm.cnr.it`

The uncertainty of the close approach distance of a Potentially Hazardous Object (PHO), either an asteroid or a comet, can be represented on the Modified Target Plane (MTP), a modification of the one used by Öpik. The MTP is orthogonal to the geocentric velocity at the closest approach along the nominal orbit, solution of the least square fit to the observations. The confidence regions of this solution in the 6-D space of orbital elements (for an epoch close to the observations) are well approximated by a family of concentric ellipsoids, if the observed arc is not too short. In the linear approximation these ellipsoids are mapped on the MTP into concentric ellipses, which can be computed by solving for the state transition matrix.

For a PHO observed at only one opposition, with a close approach expected after many revolutions, the ellipses on the MTP become extremely elongated and the linear approximation may fail. In this case the confidence boundaries on the MTP, i.e. the nonlinear images of the confidence ellipsoids, may not be well approximated by the ellipses. The Monte Carlo method (Muinonen and Bowell, 1993) can be used to find nonlinear confidence regions, but the computational load is very heavy: to estimate a low probability event the number of test cases must be larger than the inverse of the probability. We propose a new method to compute semilinear confidence boundaries on the MTP (Milani and Valsecchi, 1998), based on the theory developed to compute confidence boundaries for predicted observations (Milani, 1999). This method is a good compromise between reliability and computational load, and can be used for real time risk assessment.

We apply this technique to the case of asteroid *1997 XF*$_{11}$, which appeared, with the observations available as of March 1998, to be on an orbit with a near miss to the Earth in 2028. Although the least squares solution had a close approach at 1/8 of the lunar distance, the linear confidence regions corresponding to acceptable observational errors are very elongated ellipses not including collision. The semilinear confidence boundaries differ in a significant way from the linear ellipses but this happens only far from the Earth: an impact by asteroid *1997 XF*$_{11}$ was not compatible with the observations from the discovery to March 4, 1998, unless it could be admitted that the RMS observation error was about 4 arcsec. Anyway the use of the 1990 pre-discovery observations has confirmed the impossibility of a 2028 impact; with the additional data, the semilinear confidence regions are

*J. Henrard and S. Ferraz-Mello (eds.), Impact of Modern Dynamics in Astronomy,* 369–370.

reduced to a much smaller subset of those estimated previously; in the linear approximation, the MTP prediction with more data is not a subset of the one with less data. Even worse, in a simulated example of Earth impacting asteroid (Bowell and Muinonen, 1992), the semilinear confidence boundary can have a completely different shape from the linear ellipse, and indeed for orbits determined with only few weeks of data the semilinear confidence boundary includes possible collisions, while the linear one does not.

We conclude that the procedure to exclude the possibility of an impact involves the following steps: (1) find the least squares fit to the existing observations, discard the outliers and select weights; (2) propagate the orbit for the time span to be monitored, and detect the dangerous close approaches; (3) given a close approach and its MTP, compute the linear confidence ellipse and analyse its size and orientation; (4) if the linear ellipse gets close to the Earth, especially when this occurs far from the nominal orbit, use the semilinear approximation; (5) if the semilinear approximation indicates the possibility of an impact resort to a fully nonlinear exploration by the Monte Carlo method (Muinonen, 1993). The algorithms described in this paper are implemented in the free software package OrbFit; thus, the next time the problem of a worrisome close approach occurs, we are ready, as well as anybody else: see `ftp://copernico.dm.unipi.it/pub/orbfit/`.

## References

Bowell, E. and Muinonen, K.: 1992, *BAAS*, **24**, 965.
Milani, A.: 1999, The asteroid identification problem I: recovery of lost asteroids, *Icarus*, **137**
Milani, A. and Valsecchi, G.B.; 1998, The asteroid identification problem III: target plane confidence boundaries, submitted to *Icarus*.
Muinonen, K. and Bowell: E. 1993, *Icarus*, **104**, 255–279.

# A DYNAMICAL SURVEY OF INNER SOLAR SYSTEM ASTEROIDS

MARC A. MURISON

*Astronomical Applications Dept., U.S. Naval Observatory,*
*3450 Massachusetts Ave NW, Washington DC USA*

**Abstract.** Results from a numerical integration survey of all 179 currently-known inner solar system asteroids with $a \leq a_{Mars}$, $q \geq a_{Mercury}$ are presented. A surprising number of asteroids are currently in, or very near, mean-motion resonances with Mercury, Venus, Earth, or Mars. Some of the resonance associations are of high order. Most of the resonance associations are relatively short-lived, with the asteroids wandering in and out of resonance on timescales of hundreds to several thousand years.

Integrations and analyses were performed with Newton, a solar system numerical integration computer program that is able to automatically identify mean-motion resonances of any order (http://aa.usno.navy.mil/Newton/). Initial conditions for the planets were taken from DE405. The integration method used in this study is a variable step size Bulirsch-Stoer. All planets – Mercury through Neptune – were included in the purely gravitational force calculations. Earth and Moon were treated as one body.

Asteroid initial conditions were obtained from the Lowell Obs. list and recent MPC Circulars. In total, there were 179 asteroids that satisfied the selection criteria $a \leq a_{Mars}$, $q \geq a_{Mercury}$. The latter condition was imposed in order to exclude the asteroids that would potentially require general relativistic corrections.

Four stages of resonant asteroid filtering were employed. The initial integration was for 500 Earth years. Asteroids that appeared to be in, or could not be considered excluded from, one or more mean-motion resonances with any of the planets Mercury through Saturn were then integrated for 1000 years. Similarly, survivors went on to 3000 years, then to 10,000 years. An asteroid is here considered *not excludable* from a $(p+q) : p$ mean-motion resonance if the critical angle does not circulate within the integration time span.

The algorithm employed by Newton does not actually detect libration *per se* of the critical angle, but rather it detects circulation. Possible libration is inferred from a demonstrated lack of circulation over some interval. Hence, the algorithm

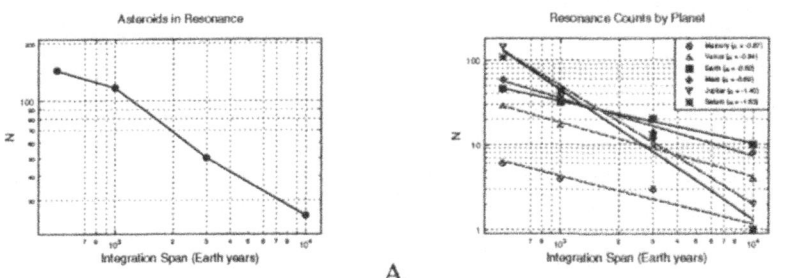

Fig. 1. A. Number of asteroids not classifiable as not being in resonance as a function of integration time. B. Resonance counts per planet as a function of integration time.

*J. Henrard and S. Ferraz-Mello (eds.), Impact of Modern Dynamics in Astronomy, 371–372.*

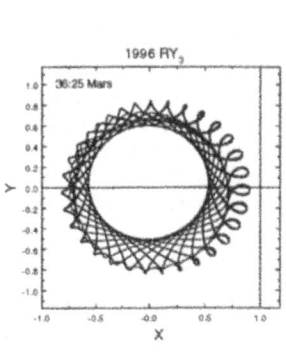
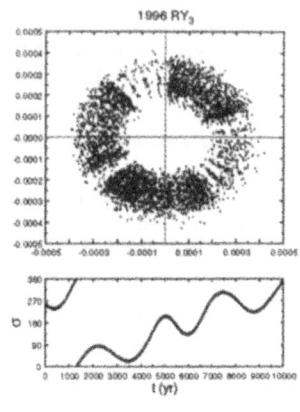

A.                                              B.

Fig. 2. A. 500-year orbit trace of 1996 $RY_3$ projected onto the rotating xy plane. Mars is fixed at (1,0). B. The resonance critical angle for 1996 $RY_3$ for 10,000 years.

cannot say for certain whether or not an asteroid is *in* a mean-motion resonance, but it can conclusively tell if an asteroid is *not*.

Figures 1 show the main results. The number of asteroids that are not excludable from being in resonance, as a function of integration time, is shown in Fig. 1A. The decline is fit well by a power law $N = N_0 t^\alpha$ with slope $\alpha = -0.58$ and $N_0 = 5250$. In Fig. 1B (note the ordinate scale change) we have, for each planet separately, the number of mean-motion resonances associated with that planet as a function of integration time. Again, we see a power law dependence, but with two groupings. Resonance counts associated with the inner planets Mercury through Mars exhibit one behavior, with slopes in the range $-0.7 < \alpha < -0.5$, while those of the two outer planets Jupiter and Saturn behave differently, their slopes being much larger, $\alpha \approx -1.5$. Due to multiple resonances for some asteroids, resulting from "slow drifters" contamination (see below), there appear to be more resonance associations (Fig. 1B) than there are asteroids in resonance (Fig. 1A). The outer planet grouping turns out to consist almost exclusively of slow drifters. As the integration time increases, more and more of these are shown to be circulating in the critical angle. By $t = 10,000$ yr, all but a few have disappeared. Contamination for the inner planets is much less severe, since the planetary periods are shorter, leading to quicker identification of circulators.

Figure 2A shows a 500-year orbit trace of a resonating asteroid projected onto the orbital plane of the reference planet. Figure 2B shows the corresponding critical angle for an integration covering a span of 10,000 years. The radius in the upper panel is $\mathbf{r} = \left[ (a - a_{\min}) + \frac{1}{2} (a_{\max} - a_{\min}) \right] \cdot (\cos \sigma, \sin \sigma)$

1996 $RY_3$ is an example of a "slow drifter". It is very near the 36:25 resonance with Mars, and it is strongly influenced by the close proximity of that resonance, but the critical angle clearly circulates.

# DETAILLED STUDY OF THE DYNAMICS
# OF FRAGMENTS OF COMET C/1996 B2 HYAKUTAKE

E. DESVOIVRES AND J. KLINGER

*Laboratoire de Glaciologie et de Géophysique de l'Environnement,*
*UJF-CNRS, 54, rue Molière, BP 96, Saint Martin d'Heres, France*

and

A.C. LEVASSEUR-REGOURD

*Université Paris 6, Service d'Aéronomie, CNRS, BP 3, 91371 Verrières, France*

The fragmentation of cometary nuclei is a frequent phenomenon, but the dynamics of the fragments is not yet well understood. During the close approach of comet C/1996 B2 Hyakutake to the Earth (0.1 AU) on late March 1996, images were taken with the 1 meter telescope of Pic du Midi observatory. Bright condensations were observed near the nucleus on images taken between March, 22, 1996 and March, 31, 1996. It was suggested that these features were *mini-comæ* surrounding fragments receding from the nucleus (Lecacheux *et al.*, 1996). A model was developed for the motion of cometary fragments in the orbital plane of the comet, and the simulations were compared with the observations (Desvoivres et al, 1998).

In the model, we consider that the nucleus of the comet and a fragment are under the influence of the gravity of the Sun, of their mutual gravity, and of non-gravitational forces (NGF) due the loss of mass induced by solar heating. From an estimation of those NGF, we compute numerically the trajectories of the fragment and of the nucleus with respect to their common center of mass (CoM). Then, the motion of the center of mass is studied in an heliocentric reference frame using the theory of perturbed keplerian motion.

The energy balance at the cometary surface is:

$$S(1 - al) = \epsilon \sigma T^4 + L_I \dot{Z}_I(T) \tag{1}$$

where $S$ is the solar flux, $al$ the albedo, $\epsilon$ the emissivity, $L_I$ the latent heat of sublimation of the water ice, and $\dot{Z}_I(T)$ the production rate of water vapour. The surface temperature is deduced from equation 1 by an iterative method. The value of the NGF is derived using the conservation of momentum.

The vectorial sum of the gravity and of the NGF is integrated numerically in order to obtain the momentum equation of both the nucleus and a fragment in the reference frame of the CoM. The modification of the motion of the CoM around the Sun is described using the *Gauss equations* (Danjon, 1959).

On several images, up to 4 fragments are visible at the same time. The evolution of the distance of the fragments to the nucleus as a function of time is known for each fragment. The values of the physical parameters are discussed in Desvoivres et al, 1998. It has been shown that the density and the size of the fragment are the only parameters which influence significantly the motion. Here, we have assumed

*J. Henrard and S. Ferraz-Mello (eds.), Impact of Modern Dynamics in Astronomy, 373–374.*

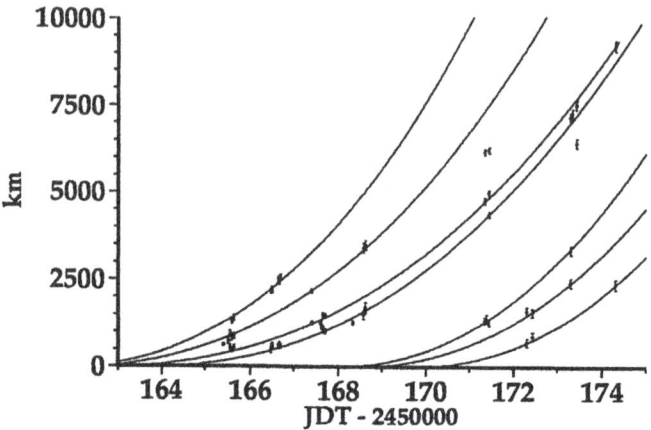

Fig. 1. Evolution of the distance to the nucleus of the simulated fragments (continuous lines), and comparison to the observations.

TABLE I
Characteristics of the different simulated fragments.

| Fragment | 1 | 2 | 3 | 4 | 5 | 6 | 7 |
|---|---|---|---|---|---|---|---|
| Fragment size (meters) | 13 | 18 | 22 | 20 | 16 | 19 | 18 |
| Emission date (March, 1996, TU) | 19.7 | 20.0 | 20.8 | 21.7 | 25.7 | 26.2 | 27.5 |

the density to be constant and equal to $300 \, kg.m^{-3}$. Different numerical simulations were performed, and a $\chi^2$ method is used to select the best fit for each fragment. Seven different fragments are needed to reproduce the observations. The results are presented in figure 1 and the characteristics of the fragments are presented in table I. As can be noticed on figure 1, among 56 detected events, only 3 are not reproduced by the simulations. They might be due to fragmentation of fragments.

## Conclusion

A model for the dynamics of cometary fragments was developped. It confirms the hypothesis that the bright condensations that were observed near comet C/1996 B2 Hyakutake are nucleus fragments. In agreement with the observations, the model shows that the fragments receed in the anti-solar direction.

## References

Danjon A., 1959, *Astronomie Générale*, Albert Blanchart, Paris

E. Desvoivres, J. Klinger, A.C. Levasseur-Regourd, J. Lecacheux, L. Jorda, A. Enzian, F. Colas, E. Frappa, P. Laques, 1998, *M.N.R.A.S.*, accepted for publication.

Lecacheux J., Jorda L., Enzian A., Klinger J., Colas F., Frappa E., Laques P., 1996, IAU Circ. No 6354

# PROPER ELEMENTS AND STABILITY OF THE TROJAN ASTEROIDS

CH. BURGER, E. PILAT-LOHINGER, R. DVORAK
*Institute for Astronomy, University of Vienna,
Türkenschanzstrasse 17, A-1180 Vienna, AUSTRIA*

and

A. CHRISTAKI
*Department of Theoretical Mechanics, University of Thessaloniki
GR-540 06 Thessaloniki, GREECE*

Up to now $\sim$ 400 asteroids are known which move close to the Lagrangian equilibrium points $L_4$ (246) and $L_5$ (167) of Jupiter. In this investigation the orbits of all known Trojans were integrated numerically for 10 million years using the Lie Series integrator with adaptive stepsize (Hanslmeier and Dvorak, 1984) in the dynamical model of the outer planetary system. The goal of the study was to extend the computation of the proper elements for all known Trojans for a longer time interval; the respective results are compared to already existing ones (e.g. Bien & Schubart, 1987; Milani, 1993).

The determination of the characteristic quantities proper eccentricity $e_p$, proper inclination $i_p$ and libration $D$ was done by numerical filtering techniques using the method of Labrouste (Burger, 1998) and by a very precise frequency analysis (Chapront, 1997). Our results differ from the ones calculated for a shorter time interval by Milani (1993) between 5 % and 10%. In Fig.1 we plotted the nomogram for the three proper elements mentioned above for $L_4$ and the $e_p$ for $L_5$. As a new interesting result we found Trojans with $0.15 < e_p < 0.22$ for $L_4$ (upper left) and $L_5$ (upper right) which answers an open question of an existing gap in this range of the $e_p$. For $L_4$ Trojans there is a well defined maximum in the nomogram for $e_p \sim 0.06$; for the $L_5$ Trojans the maximum is flat and shifted versus $e_p \sim 0.07$. The proper elements $\sin(i_p)$ for $L_4$ (bottom left graph) show a maximum for Trojans moving in the plane of Jupiter's orbit and a decrease versus larger inclinations, although a plateau for $14° < i < 27°$ is visible; this confirms that the Trojan orbits are more inclined that the orbits of the main belt asteroids.

One more point to report is, that about 10 % of the orbits escaped from their orbits close to the Lagrangian points within the 10 million years integration, but this happened primarily for asteroids with orbital elements determined after one single opposition.

*J. Henrard and S. Ferraz-Mello (eds.), Impact of Modern Dynamics in Astronomy, 375–376.*

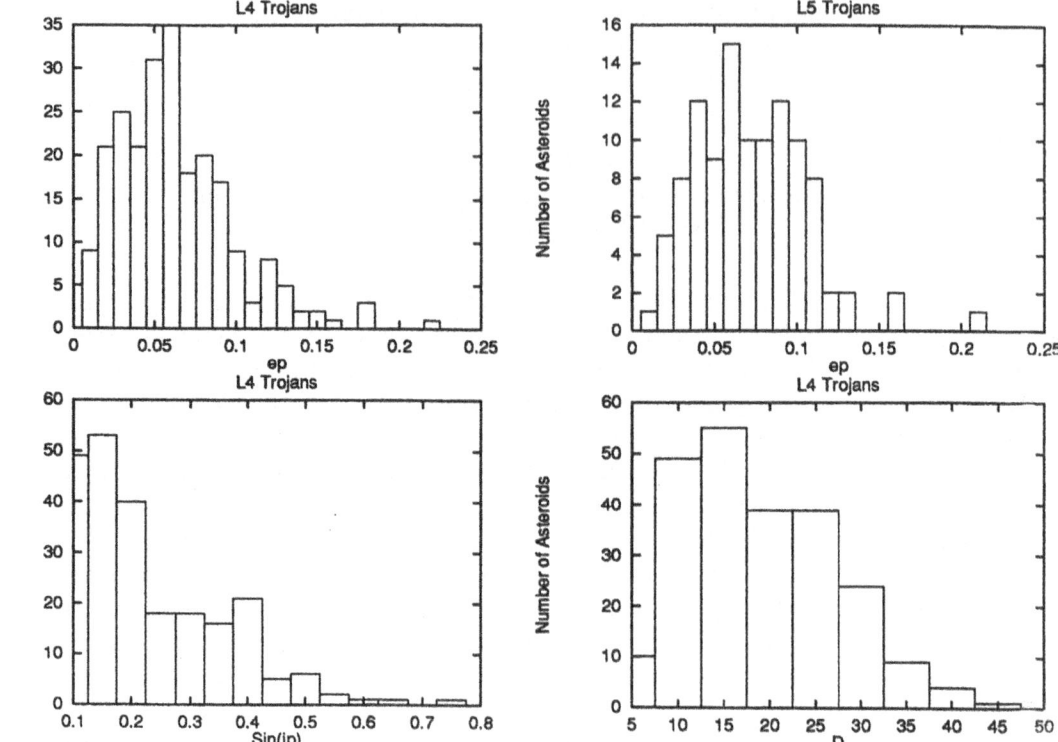

Fig. 1. Nomogram of the proper elements for $L_4$ Trojans: $e_p$ (upper left), $sin(i_p)$ (lower left), libration $D$ (lower right) and for $e_p$ of the $L_5$ Trojans (upper right).

## Acknowledgements

The work presented here was carried out within the framework of the *Jubiläums-fondsprojekt* Nr. 6446 of the *Österreichische Nationalbank*. Two of the authors (CB and EP-L) have to thank Prof. J.Henrard for the financial support; EP-L received a travel grant from the *Österreichische Forschungsgemeinschaft*, whereas AC accomplished the work as Erasmus scholar in Vienna.

## References

Burger, Ch.: 1998, Diploma Thesis, University of Vienna
Milani, A.: 1993, *Cel.Mech.Dyn.Astr.*, **57**, 59-94
Bien, R., Schubart, J.: 1987, *Astron. Astrophys.*, **175**, 292-298
J. Chapront: 1995, *Astron.Astrophys.*, **109**, 181
Hanslmeier, A., Dvorak, R.: 1984, *Astron. Astrophys.*, **132**, 203-210

# DYNAMICAL BEHAVIOUR OF ASTEROIDS IN THE 4:1 RESONANCE

FRANCISCO LÓPEZ-GARCÍA

*Observatorio Astronómico, Av. Benavidez 8175-oeste, 5413 Chimbas,San Juan, Argentina*

and

ADRIAN BRUNINI

*Fac. de Ciencias Astron.y Geofís., Paseo del Bosque s/n, 1900 La Plata, Argentina*

We study the dynamics of mean motion resonance with Jupiter in the 4:1 gap using only gravitational methods. This mechanism is capable of explaining this Kirkwood gap in an uniform way (see Ferraz-Mello, 1994; Ferraz-Mello et al., 1994; Moons, 1997; Yoshikawa, 1989). We considered the asteroidal motion in two and three dimensions and we carried out our investigations integrating numerically the full equations of motion and taking into account Mars, Jupiter and Saturn as disturbing planets. The orbital evolution of asteroids was obtained considering the elements variation. The numerical investigations were carried out using symplectic integrators. These integrations were stopped when the asteroid had close encouters with Mars or Jupiter, this occurs when the distance between the planet and the asteroid is of the order of 0.01 AU or less, or when the eccentricity increases up to 0.9. We studied real and fictitious asteroids on a time scale of $5 \times 10^7$ yr. The initial osculating elements of perturbing planets and their inverse masses were taken from the Ephemerides of Minor Planets (EMP) at the epoch of JD 2450000.5. The initial data corresponding to the real asteroids were also taken from the EMP. The starting elements of fictitious asteroids were, in all analyzed cases, $a = a_{crit} = 2.064 AU$, $i = 2°.5$ and $e = 0.01$ (in the majority of cases). The other initial elements are shown in Table II. We have also studied fictitious asteroids with $i = 0°$, $a = a_{crit}$ and $e = 0.01$ (Table I). The present analysis leads to the following results: (1) The motions are unstable. The eccentricity, in the majority of cases, has very large increase. It may grow up to 0.9 in $10^6$ yr. The semi major axis has large variations then, owing to both effects some fictitious asteroids reach the 3:1 resonance while others reach 7:2 resonance in a few million years, they are very chaotic regions. (2) The eccentricities of fictitious asteroids become large by the effect of the secular resonance $\nu_6$, i.e. when $(\varpi - \varpi_{Sat}) \cong 0$, the rate

## TABLE I

All bodies, planets and fictitious asteroids, with i = 0°, $a_{crit}$ = 2.064 AU

| Fict. Ast. | e | $\Omega°$ | $\omega°$ | $M°$ | T(f) yr. | Comments |
|---|---|---|---|---|---|---|
| BF01 | 0.01 | 0.0 | 195.0 | 0.0 | < 2.5E+6 | $(\varpi - \varpi_{jup}) = \pi$, Mars crosser, reaches the 3:1 resonance. |
| BF02 | 0.01 | 0.0 | 190.0 | 0.0 | < 2.2E+6 | Ejection, reaches the 3:1 resonance. |
| BF03 | 0.01 | 0 .0 | 185.0 | 0.0 | 2.0E+6 | Ejection, temporarilly captured by Mars. |
| BF04 | 0.01 | 0.0 | 180.0 | 0.0 | 2.1E+6 | Ejection, reaches the 3:1 resonance. |
| BF05 | 0.01 | 0.0 | 175.0 | 0.0 | 3.0E+6 | Ejection, reaches the 3:1 resonance. |
| BF06 | 0.01 | 0.0 | 185.0 | 180.0 | 7.2E+6 | Irregular motion, a jumps to 2.2567 AU, agrees with 7:2 resonance. Ejection. |
| BF07 | 0.01 | 90.0 | 90.01 | 80.0 | 2.2E+6 | Ejection. |
| BF08 | 0.30 | 0.0 | 185.0 | 180.0 | 2.0E+6 | Ejection. |

377

*J. Henrard and S. Ferraz-Mello (eds.), Impact of Modern Dynamics in Astronomy, 377–378.*
© 1999 *Kluwer Academic Publishers.*

TABLE II
Fictitious asteroids, $a_{crit}$ = 2.064 AU, all with i = 2°.5

| Fict. Ast. | e | $\Omega^o$ | $\omega^o$ | $M^o$ | T(f) yr. | Comments |
|---|---|---|---|---|---|---|
| AF01 | 0.01 | 90.0 | 105.71 | 250.0 | 1.E+7 | $(\varpi - \varpi_{jup}) = \pi$, full stable motion |
| AF02 | 0.01 | 90.0 | 105.71 | 70.0 | 5.E+7 | Regular motion up to 1.E+7 yr, later the motion finishes in 7:2 resonance,$a_{crit}$ =2.25 UA., ejection. |
| AF03 | 0.01 | 90.0 | 105.71 | 0.0 | < 8.6 E+6 | $(\varpi - \varpi_{jup}) = \pi$, chaotic motion, reaches the 3:1 resonance. |
| AF04 | 0.01 | 90.0 | 105.71 | 180.0 | 6.9 E+6 | $(\varpi - \varpi_{jup}) = \pi$, ejection |
| AF05 | 0.30 | 90.0 | 90.0 | 180.0 | 2.E+7 | Temporarily captured by Mars, e approachs to 1 for some time. |
| AF06 | 0.25 | 90.0 | 90.0 | 180.0 | < 2.3 E+6 | Ejection |
| AF07 | 0.01 | 8.214 | 7.5 | 0.0 | 2. E+7 | $(\varpi - \varpi_{jup})= 0$, stable motion, a jumps to 2.36 AU, e < 0.35 |
| AF08 | 0.01 | 8.214 | 7.5 | 180.0 | 5.E+7 | $(\varpi - \varpi_{jup}) = 0$, no regular motion, temporarily captured by Mars. i < 0.4 rad. |
| AF09 | 0.30 | 8.214 | 7.5 | 0.0 | < 1. E+7 | $(\varpi - \varpi_{jup}) = 0$, chaotic motion, reaches the resonance |
| AF10 | 0.30 | 8.214 | 7.5 | 180.0 | 3.E+7 | $(\varpi - \varpi_{jup}) = 0$, temporarily captured by Mars, e < 0.5, i < 0.25 rad. |
| AF11 | 0.01 | 100.0 | 275.0 | 250.0 | 5.E+7 | $\Omega$, $\omega$, M equal to Jupiter's, regular motion, e < 0.06, i < 0.05 rad. |
| AF12 | 0.01 | 100.0 | 275.0 | 70.0 | 5.E+7 | $\Omega$, $\omega$ equal to Jupiter's, the motion is fully regular, a nearly constant, e < 0.06 |
| AF13 | 0.01 | 100.0 | 275.0 | 180.0 | < 7.2E+6 | $\Omega$, $\omega$ equal to Jupiter's, e → 1, close approach to Jupiter's. $\Delta$ = 4.974 AU. |
| AF14 | 0.01 | 100.0 | 275.0 | 215.0 | < 6.8E+6 | Ejection, close approach to Jupiter. |
| AF15 | 0.01 | 100.0 | 275.0 | 270.0 | 5.E+7 | The motion is nearly regular |
| AF16 | 0.01 | 100.0 | 275.0 | 260.0 | < 2.E+6 | Ejection |
| AF17 | 0.01 | 100.0 | 275.0 | 280.0 | < 9.2.E+6 | Ejection |
| AF18 | 0.01 | 100.0 | 275.0 | 290.0 | 5.E+7 | Regular motion, after 2.E+6 yr. a = 2.235 UA, e < 0.25, i < 0.07 rad. |
| AF19 | 0.01 | 7.2 | 8.5 | 180.0 | < 5.2 E+6 | Ejection |
| AF20 | 0.15 | 90.0 | 90.0 | 180.0 | < 2.5 E+6 | Chaotic motion, reaches the 3:1 resonance |
| AF21 | 0.01 | 0.0 | 185.0 | 0.0 | < 2.6E+7 | (i = 0). Up to 1.E+7 yr. the motion is regular, then a decreases to 1.5 UA and finally ejection |
| AF22 | 0.01 | 90.0 | 90.0 | 180.0 | 1.5E+7 | Chaotic motion, reaches the 3:1 resonance. |
| AF23 | 0.01 | 25.0 | 25.0 | 36.05 | 1.E+7 | Full stable motion. $\Delta$ a $\simeq$ 0.01 AU, $\Delta$ e $\simeq$ 0.03 |

of this resonance is 26.217 "/year with period $\sim 4.9 \times 10^4$ years (Bretagnon, 1974). (3) The fictitious asteroids studied with $a = a_{crit}$, $e < 0.05$ and $i < 3°$ are removed of this gap mainly by the effects of the secular resonances $\nu_6$ and $\nu_{16}$ (see Moons and Morbidelli, 1995; Williams, 1969). (4) There are close encounters with Mars or eventually with the Earth (not considered here) in a time scale of $10^6$ - $10^7$ yr. (5) For certain initial conditions some fictitious asteroids are temporally captured by Mars and in some cases for a long time. (6) If $a = a_{crit}$, $e = 0.3$ and the inclination is less than 3°, Mars and asteroid's perihelion are very close ( $\sim 0.06AU$ ). This situation helps the capture. (7) The (a,e)-plane was used to determine the dynamical behaviour of all asteroids and we found that the 4:1 resonance is very strong. The Lyapunov times are very short.

**Acknowledgements:** FLG thanks Mrs. Luisa Navarro for typing the manuscript.

## References

Bretagnon, P.: 1974, *Astron. Astrophys.*, **30**, 141-154.
Ferraz-Mello, S.: 1994, "Kirkwood gaps and resonant groups", in Asteroids, Comets, Meteors 1993 (A.Milani, M. Di Martino ans A. Cellino, eds.), Kluwer, Dordrecht.
Ferraz-Mello, S., Dvrack, R. and Michtchenko, T.: 1994, "Depletion of the asteroidal belt at resonances", in From Newton to Chaos (A.E. Roy and B.A. Steves, eds), Plenun Press.
Moons, M.: 1997, *Celest. Mech*, **65**
Moons, M. and Morbidelli, A: 1995, *Icarus*, **114**
Williams, J.G.: 1969, Ph. D. Dissertation, Univ. of California, Los Angeles.
Yoshikawa, M.: 1989, *Astron. Astrophys.*, **213**

# PERTURBATIONS OF SMALL MOONS ORBITS DUE TO THEIR ROTATION: THE MODEL PROBLEM

K.GOŹDZIEWSKI and A. J.MACIEJEWSKI

*Toruń Centre for Astronomy, N. Copernicus University, Poland*

We consider here the spin–orbit coupling influence on the relative orbital motion of two bodies interacting gravitationally. We assume that one of the bodies is spherically symmetric and the other possesses a plane of dynamical symmetry. In the full non-linear settings, this problem permits coplanar motion when the mass center of the spherically symmetric body moves in the plane. We used this simple model for a qualitative estimation of the changes of the relative orbit in two cases: A) the Sun-asteroid case (the fast rotating rigid body), B) a small satellite of a big planet in resonant rotation.

The motion is described in the rigid body fixed frame. An appropriate change of physical units (Goździewski,1998a) leads to nondimensional dynamical variables and parameters. After that the Hamiltonian of the problem, written in polar variables, is the following

$$H = \frac{1}{2}\left(p_r^2 + \frac{p_\phi^2}{r^2}\right) - \frac{1}{r} + \frac{1}{2I_3}(G_3 - p_\phi)^2$$
$$- \frac{\epsilon}{2r^3}\left[I_1 - 2I_2 + I_3 - 3(I_1 - I_2)\cos^2\phi\right],$$

where $(I_1, I_2, I_3)$ are the principal moments of inertia, $(r, \phi)$ are the relative polar coordinates of the point mass in the body frame, $(p_r, p_\phi)$ are the canonical momenta, $G_3$ represents the constant of total angular momentum, $\epsilon = (r_0/r)^2$, and $r_0$ is the mean radius of the body.

In the case of model A the disturbing part of the potential may be averaged with respect to the fast variable $\psi$. This gives rise to the perturbed Kepler problem with perturbing potential of the form $\tilde{V}(r) = -\alpha/r^3$, $\alpha > 0$. Thus, as a result we obtain the well known effect of the pericenter precession, however for a typical asteroid it is negligible. The semi-major axis and the eccentricity do not change secularly.

In the resonant case the orbit does not exhibit evolutional changes either. The theory describing this situation was developed by Barkin (1975). Assuming the spherical harmonics expansion of the gravitational potential Barkin wrote out the equations of motion averaged through the Delaunay scheme. He showed that the osculating semi-major axis and the eccentricity change periodically in time.

Our aim was to look closely at the short time behaviour of the relative orbit in the resonant case. Here we present an example. For the numerical study we selected Amalthea. As $r_0 \simeq 100$ km, $r \simeq 1.8 \times 10^5$km, the small parameter is of the order $\simeq 10^{-7}$. In the case of the 'real' eccentricity $e \simeq 0.003$, the semi-major axis varies periodically with time. This is shown in the upper part of Fig. 1. Its right

*J. Henrard and S. Ferraz-Mello (eds.), Impact of Modern Dynamics in Astronomy, 379–380.*
© 1999 *Kluwer Academic Publishers.*

K. GOŹDZIEWSKI and A.J. MACIEJEWSKI

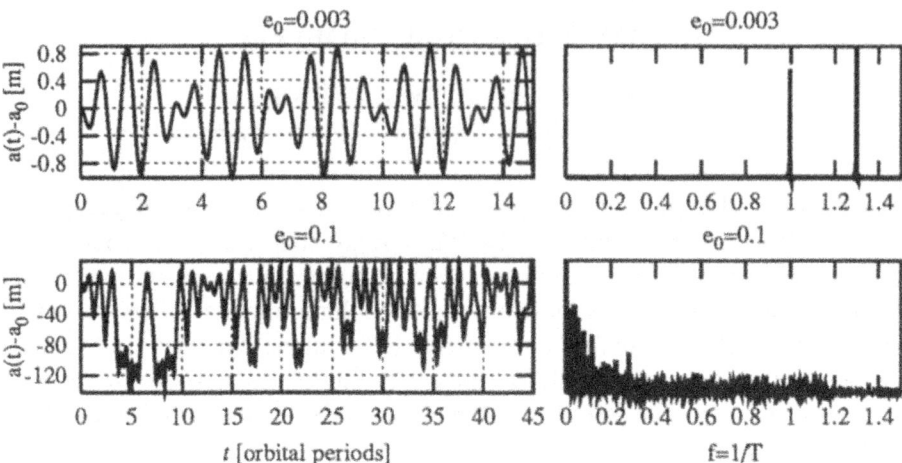

Fig. 1. Short time variation of the osculating semi–major axis of Amalthea in the case B (left panels) and Lomb periodograms of the changes (right panels).

panel shows the Lomb periodogram of the variation. Two dominant frequencies are clearly visible. One of them, equal to 1, is easily identified with the orbital frequency, the other one is almost the same as the librational, longitudal frequency of the moon (Goździewski, 1998b). The increase of the initial eccentricity up to 0.1 leads to an unexpected, interesting result: the osculating semi-major axis varies chaotically. The corresponding Lomb periodogram shows that the spectrum is almost continuous with maximum of power near the zero frequency. Let us note very large amplitude of variations, of the order of 200 m. The scale of this effect justifies that it is worth detailed study.

This work was supported by N. Copernicus Grant No. 315-A.

## References

Barkin, Y.W. (1952) Intermediate plane motion of a rigid body in the gravitational field of a sphere. *Astron. Zh.*, **52**(5), pp. 1076–1083.

Goździewski K., and Maciejewski, A.J. (1998a) Equations of motion and Lagrangian equilibria in the special version of the three body problem , *in preparation.*

Goździewski K., and Maciejewski, A.J. (1998b) Semi–analytical model of libration of a rigid moon orbiting an oblate planet. *Astron. Astroph.*, (in print).

# MEAN MOTION RESONANCES IN THE ASTEROID BELT

D. NESVORNÝ AND A. MORBIDELLI

*Observatoire de la Côte d'Azur, B.P. 4229 – Nice Cedex 4, France*

The Kirkwood gaps in the main asteroidal belt (2 – 3.5 AU) coincide with the mean motion resonances with Jupiter (4/1, 3/1, 5/2, 7/3, 2/1). Similarly, several narrower gaps are observed in the outer asteroid belt (3.5 – 4 AU) at places of 11/6, 9/5, 7/4 and 5/3 Jovian resonances (Holman and Murray 1996). As it is now generally accepted, the formation and preservation of these gaps is due to the chaos of the resonant space and efficient ejection of the primordial and collisionaly injected bodies towards high eccentricities and planet-crossing orbits.

The Jovian mean motion resonances are not the most important in what concerns the chaos of the observed (i.e. remaining) asteroid population. It was estimated by Šidlichovský and Nesvorný (1998) that about 40% of known objects have the Lyapunov time less than $10^5$ years. It was later found (Nesvorný and Morbidelli 1998, 1999; Morbidelli and Nesvorný 1999) that the resonances responsible for this chaos are, in decreasing order of importance: 1) three-body resonances with Jupiter and Saturn, 2) exterior resonances with Mars, 3) moderate order Jovian resonances, and 4) three-body resonances with Mars and Jupiter. All these mean motion resonances are narrow ($10^{-3}$ – $10^{-4}$ AU) and dense in the semimajor axis so that they fill a significant part of the phase space. The resonant width typically grows with eccentricity and above the critical eccentricity, resonances overlap and make the belt globally chaotic. In the inner part of the belt (2.1 – 2.5 AU) this happens little under the curve defined by $q = a(1 - e) = a_M$ (i.e. the perihelion equal to semimajor axis of Mars), and in outer part of the belt (3 – 3.2 AU) close to $e = 0.25$. The asteroids with higher eccentricity than the critical one have the Lyapunov time less then $10^5$ years (and frequently less than $10^4$ years). The most regular is the central part of the main belt (2.5 – 3 AU).

The strength of Martian resonances in the inner belt might be surprising as the mass of planet Mars is very small (less than 1/3000 of Jupiter's mass), but as an analytic evaluation of the resonant terms shows (Morbidelli and Nesvorný 1999) the low mass is compensated by higher eccentricity of Mars (when compared to that of Jupiter) and, in the inner belt, by a small distance between the asteroid and the planet.

The 16/27 exterior resonance with Mars ($a_{res} = 2.1596$ AU) is at about the same place as the 15/4 resonance with Jupiter ($a_{res} = 2.1547$ AU). Both these resonances are of the 11th order. Assuming $e = e_M = 0.1$ and $e_J = 0.05$, the estimated width of the resonant area of the Martian resonance is $\Delta a = 1.5 \times 10^{-4}$ AU, while the Jovian resonance is much narrower having a width of only $\Delta a = 6.5 \times 10^{-6}$ AU. This example shows the importance of the Martian resonances in the inner asteroid belt. At a given order, they are not only more dense in semimajor axis but also wider than the Jovian resonances.

As it was described by Nesvorný and Morbidelli (1999), the chaos related to

*J. Henrard and S. Ferraz-Mello (eds.), Impact of Modern Dynamics in Astronomy, 381–382.*

the weak mean motion resonances is due to the overlap of 'sub-resonances'. In fact, every mean motion resonance is composed from several sub-resonances (also called multiplet resonances) of about the same strength, each having a different combination of perihelion (and nodes) in the resonant argument. As these sub-resonances are not sufficiently separated and usually (at least for resonances $10^{-3}$ – $10^{-4}$ AU wide) entirely or partially overlap, the trajectory is temporaly guided by one sub-resonance and then by another. This alternation between the sub-resonances (accompanied by small jumps in semimajor axis) gives rise to the chaos (Morbidelli 1993).

The fast exponential divergence of nearby trajectories in the weak mean motion resonances does not imply a fast macroscopic change of the resonant orbit. Indeed, in the numerical simulations over several 100 Myr, the semimajor axis remains in the range of the resonance (or at most wanders into a nearby resonance if this is sufficiently close), and the eccentricity and inclination change only moderately. This is the reason, why there are so many objects present in weak mean motion resonances: their dynamical lifetime is sufficiently large. Even if there are some uncertain indications of density distribution decrease by at most a factor of 4 in some of the Martian and Jovian resonances of $10^{-3}$ AU in size, the collisional evolution of the asteroid belt is apparently efficient in replenishing the escaped objects.

There are several astronomical consequences of the above theoretical work: 1) The asteroids in inner belt are continuously leaking through high eccentricities and sustaining the populations of large Mars and Earth-crossers. A quantitative argument in favor of this idea was provided by numerical simulations by Migliorini et al. (1998). 2) The erosion of the inner belt by this process on the age of the solar system is estimated to have removed a number of objects roughly equivalent to that being in the inner belt at present (Morbidelli and Nesvorný 1999). It is not straightforward to see how this conclusion will be reconciled with the cratering record which suggests a constant cratering rate of the inner planets and the Moon during last 3 Gyr. 3) Many asteroid families all over the belt are crossed by narrow tracks where orbital eccentricity can significantly change. Collision fragments injected into these places may have evolved with respect to their initial orbit. What signatures of this effect will be actually found in asteroid families is a matter of future project recently initiated for the Flora family.

## References

Migliorini, F., Michel, P., Morbidelli, A., Nesvorný, D. and Zappalà, V.: 1998, Origin of Multikilo-meter Earth and Mars-Crossing Asteroids: A Quantitative Simulation, *Science*, **281**, 2022-2024

Morbidelli, A.: 1993, On the succesive elimination of perturbation harmonics, *Celest. Mech.*,**55**, 101–130

Morbidelli, A. and Nesvorný, D.: 1999, Numerous weak resonances drive asteroids towards terrestrial planets orbits, *Icarus*, in press

Nesvorný, D. and Morbidelli, A.: 1998, Three-body mean motion resonances and the chaotic structure of the asteroid belt, *Astron. J.*, **116**, 3029-3037

Nesvorný, D. and Morbidelli, A.: 1999, An analytic model of three-body mean motion resonances, *Celest. Mech.*, in press

# ON THE CRITICAL PHENOMENA IN THE DYNAMICS OF ASTEROIDS

IVAN I. SHEVCHENKO

*Institute of Theoretical Astronomy, Russian Academy of Sciences*

**Abstract.** Two statistical effects in the long-term chaotic asteroidal dynamics are considered, namely the power-law character of the dependence of recurrence times on local Lyapunov times and the power-law decay in the tails of the recurrence distributions. The dependences in both cases are shaped by effects of anomalous transport, due to the presence of the chaos border in phase space, and by statistical selection effects.

In the chaotic asteroidal dynamics, two long-term effects in the statistics of sudden orbital changes are known. The first one consists in the power-law character of the dependence of times of sudden orbital changes on Lyapunov times (Soper *et al.*, 1990; Lecar *et al.*, 1992; Levison and Duncan, 1993; Murison *et al.*, 1994; Ferraz-Mello, 1997), while the second one in the power-law decay in the tails of distributions of such times (Shevchenko and Scholl, 1996; 1997). Both effects are considered here as critical phenomena. The critical motion, i.e. the chaotic motion in the vicinity of the chaos border, does not represent normal diffusive process. It is said that transport is anomalous (Chirikov, 1996). In the following note, particular effects of anomalous transport in the behaviour of the standard map are demonstrated and discussed. They correspond to the considered effects in the asteroidal dynamics and provide better understanding of the latter.

Let us associate the time of a sudden orbital change with a suitably defined recurrence time $T_r$. By "recurrences" we imply sequential returns of a trajectory to some arbitrary domain or surface in phase space. If phase space is divided, i.e. if chaotic and regular components are both present, longest recurrences of a chaotic trajectory are due to stickings to the chaos border. Sporadic stickings result in intermittent behaviour (Shevchenko, 1998a).

Consider the standard map

$$
\begin{aligned}
y_{n+1} &= y_n + \frac{K}{2\pi}\sin(2\pi x_n), \\
x_{n+1} &= x_n + y_{n+1}.
\end{aligned}
\tag{1}
$$

Let us choose $K = 2$. In fact, the studied effects can be recovered for any non-zero $K$ not too large, i.e. when the regular component is adequately present.

Integral distribution of recurrence times for a single chaotic orbit is shown in Fig. 1. The quantity $F(T_r)$ is the fraction of recurrences longer than $T_r$. The recurrences are counted at the line $y = 0 \bmod 1$. Stickings to the island of stability around the integer resonance situated at this line lead to the initial steep short-scale drop in the distribution. Then, on some limited interval, namely at $0.7 < \log T_r < 1.2$, the distribution follows the power law with index equal to $-0.56$. This is close to the inverse square root law, which is inherent to free diffusion

*J. Henrard and S. Ferraz-Mello (eds.), Impact of Modern Dynamics in Astronomy, 383–386.*

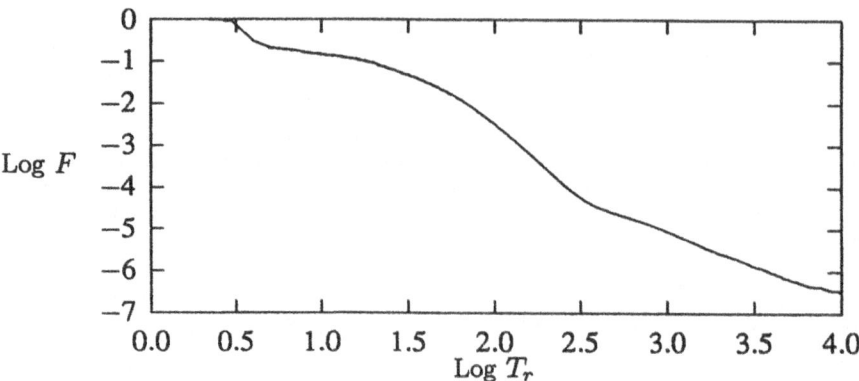

Fig. 1. Integral distribution of recurrence times. $K = 2$; number of iterations $N_{it} = 10^9$. Initial short-scale drop, major exponential decay and subsequent power-law decay are prominent. Logarithms are decimal in Figs. 1—2.

in the central part of a stochastic layer until finite width of the layer becomes important (Chirikov and Shepelyansky, 1981, p. 9). Note that in our case the borders $y = 0 \bmod 1$ of the "layer" are conventional. At $\log T_r = 1.2-1.3$ the dependence becomes exponential, because finite width of the layer starts to be important. Indeed, according to (Chirikov and Shepelyansky, 1981, p. 10), when the time of diffusion across a stochastic layer is finite, the distribution of recurrences decays exponentially due to fluctuations of diffusion. The tail of the distribution in Fig. 1 follows the power law with index $\alpha = -1.48$. Generally, in its over-all shape, the distribution is strikingly close to those presented in Fig. 4 of (Shevchenko and Scholl, 1997) for intervals between eccentricity bursts of model intermittent trajectories in the 3/1 Jovian resonance. According to (Shevchenko and Scholl, 1997), the values of power-law indices in the tails of the latter distributions are close to the value $-1.5$ theoretically considered and explained by Chirikov (1990, 1996) for critical motion. As our experiment with the standard map indicates, all prominent features of the distributions in the asteroidal case (initial steep short-scale drop, major exponential decay, and subsequent power-law decay) are present in case of the standard map, and have straightforward universal explanations.

In Fig. 2, the dependence "$\log T_L - \log T_r$" is shown. The Lyapunov time $T_L$ is the inverse of the LLCE (largest Lyapunov characteristic exponent). A formal mathematical approach requires the LLCE to be measured on an infinite time scale. In real computations this cannot be achieved. Henceforth the LLCE is measured for a recurrence (for detailed definitions and discussion, see Shevchenko, 1998b). As adopted already, the recurrences are counted at the line $y = 0 \bmod 1$. For convenience of handling large arrays of data, the field of Fig. 2 is partitioned in pixels. The figure represents a kind of a density plot. Each pixel contains no more than one symbol. Recurrences with $T_L, T_r$ in the area covered with dots are relatively frequent. They take place already on the adopted minimum time span

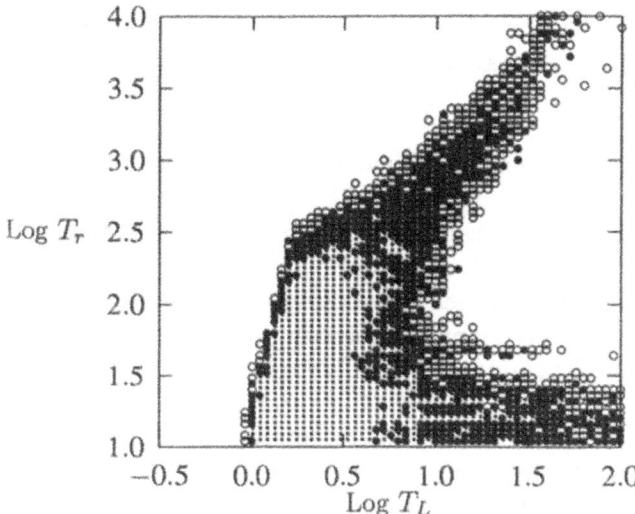

Fig. 2. Statistical dependence "$\log T_L$ — $\log T_r$". $K = 2$. Dots: $N_{it} = 10^6$; dots plus full circles: $N_{it} = 10^8$; dots plus full and open circles: $N_{it} = 10^9$.

of computation $N_{it} = 10^6$. Increasing $N_{it}$ allows one to recover recurrences with less frequent values of the pair $T_L$, $T_r$. The area increments corresponding to the increase of $N_{it}$ up to $10^8$ and then up to $10^9$ are covered respectively with full and open circles. Note that short recurrences (with $T_r < 10$) are not considered on the reason of large statistical fluctuations in evaluations of the LLCE on the short time scale. Such fluctuations are still clearly seen for recurrences with somewhat greater $T_r$ in the form of jumps of $T_L$ to high values.

One can see that recurrences with $\log T_r > 2.5$ are rare if $N_{it} < 10^6$, and the diffusion is normal: the mean $\log T_L$ does not depend on the duration $T_r$ of a recurrence. At $\log T_r > 2.5$ the dependence is a power law, with index $\beta = 1.5–2$. One can graphically demonstrate, e.g. by means of construction of a spectrum of winding numbers (Shevchenko, 1996), that the recurrences with $\log T_r > 2.5$ are due to stickings to the chaos border. The winding number is formally defined for a recurrence. Indeed, at $\log T_r > 2.5$ the broad spectrum degenerates into a sharp peak corresponding to the half-integer resonance.

Theoretical dependence "$T_L — T_r$" for critical motion was derived in (Shevchenko, 1998b). In particular, on the condition that the LLCE were computed on finite time intervals (corresponding to the recurrences), the dependence was shown to be close to quadratic: $T_r \propto T_L^2$. The quadratic relationship is set forth by the fact that the transport near the chaos border is anomalous and by the selection effect following from the limitation of the time of computation of the LLCE from above, since the LLCE correspond to recurrences.

Another important selection effect distorts statistical evaluations of the exponent $\beta$ of the "$T_L — T_r$" relationship. The distortion is due to a strong statistical predominance of short recurrence times. Therefore, when calculating mean observed values of $\beta$, longer recurrences should be taken with greater weight; see discussion in (Shevchenko, 1998b).

One more selection effect is often present and, on the opposite, enhances appearance of the generic relationship. This is the effect of sparsity of the data set. In order to construct a relationship of the kind "$T_L - T_r$" for a set of trajectories, one should choose a corresponding array of starting data. When the grid of the data is fine, the presence of narrow chaotic layers disconnected from the main chaotic domain would lead to distortions in the observed relationship, due to the apparition of the chaotic orbits which never exhibit "sharp" orbital changes. In this way, cases of "stable chaos" in the asteroidal dynamics are naturally explained.

Conclusions are as follows.

(1) Two known long-term effects in the statistics of sudden changes of asteroidal orbits, namely the power-law character of the dependence of times of sudden orbital changes on Lyapunov times (Soper *et al.*, 1990; Lecar *et al.*, 1992; Levison and Duncan, 1993; Murison *et al.*, 1994; Ferraz-Mello, 1997) and the power-law decay in the tails of distributions of such times (Shevchenko and Scholl, 1996; 1997), are both plausibly explained as critical phenomena, i.e. effects of anomalous transport near the chaos border. Our experiments with the standard map unambiguously recover similar dependences. (2) The "$T_L - T_r$" relationship can indeed be used to statistically predict sudden changes in the orbital behaviour of asteroids, if the initial part of the power-law dependence is recovered numerically. (3) When interpreting the observed dependences, it is necessary to take into account selection effects. The main selection effects, in case of the "$T_L - T_r$" relationship, are: limitations on the time of computation of the LLCE, concentration of data points to the lower time edge, sparsity of statistical data.

It is a pleasure to thank Sylvio Ferraz-Mello for useful remarks on the manuscript.

## References

Chirikov, B.V.: 1979, *Phys. Reports*, **52**, 263.
Chirikov, B.V.: 1990, *Patterns in Chaos*, Preprint 90–109, Inst. of Nucl. Phys. Novosibirsk: Novosibirsk; see also: 1991, *Chaos, Solitons and Fractals*, **1**, 79.
Chirikov, B.V.: 1996, *JETP*, **83**, 646 [*Zh. Eksp. Teor. Fiz.*, **110**, 1174].
Chirikov, B.V. and Shepelyansky, D.L.: 1981, *Statistics of Poincaré recurrences and the structure of the stochastic layer of the non-linear resonance*, Preprint 81–69, Inst. of Nucl. Phys. Novosibirsk.
Ferraz-Mello, S.: 1997, *Celest. Mech. & Dyn. Astron.*, **65**, 421.
Lecar, M., Franklin, F. and Murison, M.: 1992, *Astron. J.*, **104**, 1230.
Levison, H. F. and Duncan, M. J.: 1993, *Astrophys. J. Lett.*, **406**, L35.
Murison, M., Lecar, M. and Franklin, F.: 1994, *Astron. J.*, **108**, 2323.
Shevchenko, I. I.: 1996, in J. C. Muzzio et al. (eds.), *Chaos in Gravitational N-Body Systems*, Kluwer: Dordrecht, p. 311.
Shevchenko, I. I.: 1998a, *Physica Scripta*, **57**, 185.
Shevchenko, I. I.: 1998b, *Phys. Letters A*, **241**, 53.
Shevchenko, I. I. and Scholl, H.: 1996, in S. Ferraz-Mello et al. (eds.), *Dynamics, Ephemerides and Astrometry of the Solar System*. Kluwer: Dordrecht, p. 183.
Shevchenko, I. I. and Scholl, H.: 1997, *Celest. Mech. & Dyn. Astron.*, **68**, 163.
Soper, M., Franclin, F. and Lecar, M.: 1990, *Icarus*, **87**, 265.

# DYNAMICS OF REAL ASTEROID AT THE HECUBA GAP

F. ROIG and S. FERRAZ-MELLO

*Instituto Astronômico e Geofísico, Universidade de São Paulo*
*Av. Miguel Estéfano 4200, São Paulo, 04301, SP, Brazil*

**Abstract.** We study the dynamics of real asteroids at the 2/1 resonance with Jupiter, and find that Griqua-like objects have a short life-time inside the resonance, while Zhongguo group can survive for many $10^8$yr.

The depletion of the Hecuba gap has been recently explained by the existence of a global stochasticity in the region with significant diffusion rates (Nesvorný & Ferraz-Mello, 1997). This stochasticity is generated by several mechanisms, like the presence of low-order secular resonances (Morbidelli & Moons, 1993), the overlap of secondary resonances (Henrard *et al.*, 1995), and the overlap of high-order secular resonances with the forced oscillations of the Great-Inequality (Ferraz-Mello *et al.*, 1998). However, the gap is not completely depleted and a few asteroids are found today to be at 2/1 resonance with Jupiter.

In this work, we study the dynamical behaviour of these real asteroids by numerical integration of the equations of motion. The integration scheme was based on the mixed variable symplectic subroutines SWIFT_MVS and SWIFT_RMVS (Levison & Duncan, 1994). We analyze 15 real asteroids with well-known orbits having a librating critical argument (Table I). Initial conditions for the date 02-21-97 were taken from Bowell's asteroids database and from JPL ephemerides DE403. The simulated Solar System models included perturbations from Venus to Neptune.

The studied objects can be divided in three major groups. Group I contains those asteroids that we call "Griqua-like" objects. They are located at high-eccentricities ($e > 0.25$), have high amplitude of libration ($> 90°$), and normally also high inclination ($I > 15°$). These objects were integrated backwards and forwards for $10^7$yr. Their evolution was characterized by a chaotic variation of the semi-major axis, eccentricity and inclination. The asteroids had several close encounters, within one Hill's sphere, with the inner planets and also with Jupiter, but they did not abandon the resonance ($a$ stayed librating around the resonant value) until a strong planetary encounter, within 1/10 of Hill's sphere, occurred. After such an encounter (generally with Jupiter), the asteroid was either driven to hit the Sun or to escape from the Solar System. The life-time of these objects inside the resonance is a few $10^6$yr or less. The only exception is 1362 Griqua, that can survive inside the resonance for more than $5 \times 10^7$yr.

Group II contains the "Zhongguo group". These objects are located at small eccentricities ($e < 0.25$), have low inclinations ($I < 5°$) and moderate amplitudes of libration. Our simulations showed that their behaviour is rather regular and they are able to survive inside the resonance for at least $4 \times 10^8$yr (or even $10^9$yr as found by Morbidelli,1996). Finally, Group III contains objects that have a low

*J. Henrard and S. Ferraz-Mello (eds.), Impact of Modern Dynamics in Astronomy*, 387–388.
© 1999 *Kluwer Academic Publishers.*

TABLE I
Real asteroids at the Hecuba gap

| I | | II | | III |
|---|---|---|---|---|
| 1362 | Griqua | 3789 | Zhongguo | 4177 | 1987 SS1 |
| 1921 | Pala | | 1975 SX | | 1981 EX11 |
| 1922 | Zulu | | 1990 TH7 | | |
| 3688 | Navajo | | 1993 SK3 | | |
| 5370 | Taranis | | 1994 UD1 | | |
| | 1977 OX | | | | |
| | 1992 AB | | | | |
| | 1995 QN3 | | | | |

eccentricity but high inclination. Their life-time is greater than Group I but smaller than Group II, and they are not able to survive inside the resonance for more than $10^8$yr.

Due to their short life-times it is clear that Griqua-like asteroids were injected in the resonance at some time. This process should be very recent and probably not related to the formation of asteroidal families, which are older. It is more probable that Griqua-like asteroids came from the neighbourhood of the resonance, in which case it is likely that a continuous process of low-rate injection of objects into the resonance may exist. This hypothesis is now under study.

## References

Ferraz-Mello S., Michtchenko T.A. and Roig F.: 1998, *Astron. J.*, **116**, 1491-1500
Henrard J., Watanabe N. and Moons M.: 1995, *Icarus*, **115**, 336-346
Levison H. and Duncan M.: 1994, *Icarus*, **108**, 18-36
Morbidelli A.: 1996, *Astron. J.*, **111**, 2453-2461
Morbidelli A. and Moons M.: 1993, *Icarus*, **102**, 316-332
Nesvorný D. and Ferraz-Mello S.: 1997, *A& A*, **320**, 672-680

# MODELLING VERY-HIGH-ECCENTRICITY ASTEROIDAL
# LIBRATIONS WITH THE ANDOYER HAMILTONIAN

A. SIMULA AND S. FERRAZ-MELLO

*Instituto Astronômico e Geofísico. Universidade de São Paulo*
*Av. Miguel Stéfano 4200, São Paulo, SP, Brazil*

and

C. GIORDANO

*Universid Nacional de La Plata - La Plata, Argentina*

High-eccentricity asteroidal librations are modelled using the high-eccentricity non-planar asymmetric expansion (Roig *et al* 1997). This second-degree expansion gives us the potential of the perturbing forces acting on a resonant asteroid in a first order resonance in explicit form, as a quadratic polynomial in the canonical non-singular variables. Secular and short periodic perturbations are introduced in the model, giving a more realistic description of the dynamics.

The reducing Sessin's transformation (Sessin, 1981; Sessin & Ferraz-Mello, 1984) is used to include the main effect of Jupiter's ecc entricity in the main part of the Hamiltonian. It leads to an integrable first-order approximation known as the second fundamental model for resonance (Henrard & Lemaitre 1983) or Andoyer Hamiltonian (Andoyer 1903).

The solution of Andoyer Hamiltonian may be computed in closed form either with elliptic integrals or with truncated, but explicit, Fourier series with coefficients analytically known. The canonical transformation which gives the action integral is also explicit. The semi-analy tical averaging method of Henrard (Henrard 1990, Henrard & Lemaitre 1986) is then applied to the model in the region of high asteroidal eccentricity. Our approach takes into account the fact that the perturbation is known explicitly as a function of the angles and actions. The planar and spatial parts of the problem are well separated, at the order of approximation used. The planar part of the averaged Hamiltonian depends only on one angle and is, thus, reduced to one degree of freedom.

The construction of this semi-analytical model is useful to allow the identification of all kinds of secular and non-secular resonances acting in the central part of the Hecuba gap; as we can switch on and off the different perturbations, the different rôle of each one in the depletion of the gap may be assessed.

## References

Andoyer H.: 1903, Bulletin Astronomique, **20**,321-356
Henrard J., Lemaitre A.: 1983, 'A second fundamental model for resonance', *Cel. Mech. Dyn. Astr.*, **30**, 197-218.
Henrard J.: 1990, 'A semi-numerical perturbation method for separable Hamiltonian systems', *Cel. Mech. Dyn. Astr.*, **49**, 43-67.

*J. Henrard and S. Ferraz-Mello (eds.), Impact of Modern Dynamics in Astronomy, 389–390.*

Henrard J., Lemaitre A.: 1986, 'A semi-numerical perturbation method for problems with two critical arguments', *Cel. Mech. Dyn. Astr.* , **39**, 213-238.

Roig F., Simula A., Ferraz-Mello, S. and Tsuchida M.: 1997, 'The high-eccentricity asymmetric expansion of the disturbing function for non-planar resonant problems', *Astron. Astrophys.*, (in press).

Sessin W.: 1981, *'Estudo de um Sistema de dois Planetas com Periodos commensurav eis'*, Ph.D. Thesis, IAG-USP.

Sessin W., Ferraz-Mello S.: 1984, 'Motion of two planets with periods commensurable in the ratio 2:1. Solutions of the Hori auxiliar system', *Cel. Mech. Dyn. Astr.* , **32**, 307-332.

# RESONANT RELAXATION

SCOTT TREMAINE

*Princeton University Observatory, Princeton, NJ 08544;*
*E-mail: tremaine@astro.princeton.edu*

The two main arenas of astrophysical dynamics are celestial mechanics and stellar dynamics. The former deals with the motion of few bodies in a near-Kepler potential; the latter with the motion of many bodies in a non-Kepler potential. I would like to discuss the hybrid problem of many bodies in a near-Kepler potential, which is relevant to a number of astrophysical systems, including protoplanetary disks and the centers of galaxies containing massive black holes.

Consider a spherical system of radius $r$, containing $N$ bodies of mass $m$ orbiting in the gravitational field of a body of mass $M \gg Nm$. Assume that the orbits have moderate eccentricities and random orientations and imagine taking a time exposure of the system over several orbits. Each body is then smeared into an approximate Kepler ellipse, which precesses slowly on a timescale $t_{\text{prec}}$. Each ellipse exerts a force on other bodies at comparable radius, $f \sim Gm/r^2$. The mean force from all the ellipses is $F_m \sim Nf$. The mean force determines the precession time through the relation $L \sim rF_mt_{\text{prec}}$, where $L \sim (GMr)^{1/2}$ is the specific angular momentum of an orbit. Thus $t_{\text{prec}} \sim (M/Nm)t_{\text{cr}}$, where $t_{\text{cr}} \sim (r^3/GM)^{1/2}$ is the crossing time.

There is also a stochastic force $F_s \sim N^{1/2}f$, which depends on the orientation and shape of individual orbits and hence changes on a timescale $t_{\text{prec}}$. Thus the angular momentum of an individual star random walks, with steps of duration $t_{\text{prec}}$ during which the angular momentum changes by $\Delta L \sim rF_st_{\text{prec}}$. This random walk leads to relaxation of the angular momentum on a timescale

$$t_{\text{rel}} \sim t_{\text{prec}} \left(\frac{L}{\Delta L}\right)^2 \sim t_{\text{cr}}\frac{M}{m}, \tag{1}$$

which, remarkably, is independent of $N$ so long as $N$ is large enough that the forces are stochastic. The *resonant relaxation rate* $1/t_{\text{rel}}$ is faster than the usual Chandrasekhar or *non-resonant* relaxation rate by a factor $M/(Nm \ln \Lambda)$ where $\ln \Lambda$ is the usual Coulomb logarithm. In contrast, the resonant relaxation rate for the energy is negligible, since the stochastic forces we have described vary only slowly with time.

A massive body travelling through a system of $N$ lighter bodies experiences dynamical friction, a drag force due to its gravitational wake. Similarly, massive bodies orbiting in a near-Kepler $N$-body system experience resonant friction. Resonant friction can lead to either decay or growth of the orbital eccentricity of the massive body, depending on the mean orbital eccentricity of the light bodies.

Resonant relaxation plays an important role in determining the steady-state star distribution around a massive central black hole in a galaxy or star cluster.

*J. Henrard and S. Ferraz-Mello (eds.), Impact of Modern Dynamics in Astronomy, 391–392.*
© 1999 *Kluwer Academic Publishers.*

Resonant relaxation and friction are also present in near-Kepler disks. The most interesting case is a disk whose precession rate is dominated by its own self-gravity (e.g. protoplanetary disks, stellar disks close to a massive black hole, but not planetary rings). Resonant relaxation and friction in thin disks can be understood by the marriage of two classic calculations of mathematical physics: Lagrange's secular perturbation theory, which is taken in the continuum limit, and Landau's integration contour, which is used to handle the singular integrands that arise in this continuum limit. Just as the secular interactions of the planets in the solar system are tightly coupled even though the individual planet masses are small, the bodies in a near-Kepler disk are coupled to the low-order, low-frequency $m = 1$ normal modes of the disk even when the disk mass is small. As a result, resonant relaxation rates generally cannot be calculated accurately without solving for the disk normal modes and their frequencies; however, we expect that equation (1) remains correct to order of magnitude. Resonant relaxation affects eccentricities and inclinations (but not semi-major axes) and leads to a Rayleigh distribution of these quantities. Resonant relaxation is likely to dominate over non-resonant relaxation in the late stages of planet formation.

Further discussion is given by Rauch and Tremaine (1996), Rauch and Ingalls (1997), and Tremaine (1998).

## References

Rauch, K. P., and Ingalls, B.: 1997, 'Resonant tidal disruption in galactic nuclei', *Mon. Not. Royal Astron. Soc.*, to be published (astro-ph/9710288).
Rauch, K. P., and Tremaine, S.: 1996, 'Resonant relaxation in stellar systems', *New Astronomy*, **1**, 149–170
Tremaine, S.: 1996, 'Resonant relaxation in protoplanetary disks', *Icarus*, to be published

# PERIODIC ORBITS IN THREE-ARMED GALAXIES

PREBEN J. GROSBØL

*European Southern Observatory, Karl-Schwarzschild-Str. 2, D-85748 Garching, Germany*

and

PANOS A. PATSIS

*Astronomisches Rechen Institut, Mönchhofstraße 12-14, D-69120 Heidelberg, Germany*

## 1. Introduction and Models

Although interarm features in the Population I disk of spiral galaxies frequently give an impression of a three-fold symmetry (Patsis *et al.*, 1997), true three-armed spiral structures in the old stellar disk are seldomly seen. Such systems are of special interest since they display unique conditions which favor the growth of m=3 modes. The face-on spiral NGC 7137 shows a clear three-armed pattern on K-band images and was used as a prototype for the potential of these systems.

A K-band map of NGC 7137 was decomposed in axisymmetric components and a synthetic rotation curve was generated [see (Grosbøl and Patsis, 1998)]. The maximum rotational velocity was taken to be $\approx 150$ km sec$^{-1}$ corresponding to the mean value given by (Rubin *et al.*, 1982) for this type of galaxy.

The potential used by (Contopoulos and Grosbøl, 1986) was adopted for the dynamic models substituting the m=2 perturbation with a three-armed spiral. The rotation curve associated to the axisymmetric potential was fitted to the synthetic curve. The spiral pattern had a pitch angle of $-33.2°$ while the pattern speed $\Omega_p$ was set to 31.7 km sec$^{-1}$ kpc$^{-1}$ corresponding to the 6:1 resonance being at the end of the strong symmetric spiral. Two sets of models were calculated namely: a weak spiral with a relative radial force perturbation $f_r = 1.5\%$, and a strong spiral ($f_r = 9.0\%$) to show non-linear effects.

## 2. Periodic orbits

The central family of periodic orbits were found for each model. The stability index $\alpha$ is shown on Fig. 1a where also the location of resonances are indicated. The orbits are all stable (i.e. $|\alpha| < 1$) and display the typical variation as function of energy expressed as the radius $r_c$ in an axisymmetric model. The central family of periodic orbits splits at the resonances 3:1, 6:1, 9:1 etc. similar to the behavior for two-armed spiral models at the resonances 2:1, 4:1 and 6:1.

The actual periodic orbits for the strong spiral case are plotted in Fig. 1b where also locations of the spiral potential minima are shown. It is clear from this figure that the orbits develop cusps at the main resonances (i.e. 3:1, 6:1, 9:1 etc.) and that their orientations shift (e.g. by 60° at 3:1 and 30° at 6:1). Further, the response is only in phase with the perturbation in the region between 3:1 and 6:1 [see

393

*J. Henrard and S. Ferraz-Mello (eds.), Impact of Modern Dynamics in Astronomy, 393–394.*

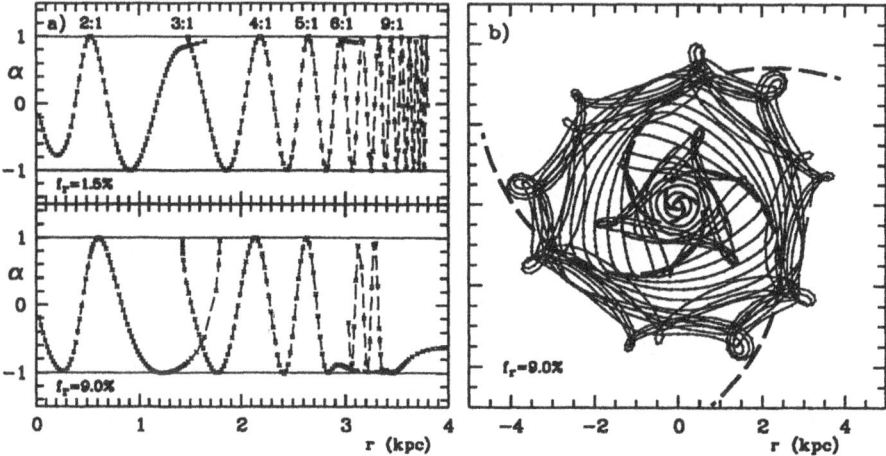

Fig. 1. a) Stability index $\alpha$ for the weak and strong spiral models as function of energy expressed as $r_c$. b) Central family of periodic orbits for the strong spiral model. The dashed lines show the locations of the spiral potential minima.

(Contopoulos and Grosbøl, 1988) for the two-armed case]. This suggests that self-consistent models of strong three-armed spirals can be constructed only between the 3:1 and 6:1 resonances while weak spiral perturbation can exist outside this region. It confirms the general conclusions of (Contopoulos and Grosbøl, 1988) in the case of three-armed spiral galaxies where the 3:1 and 6:1 resonances replace the 2:1 and 4:1 resonances as the most important.

## References

Contopoulos, G., Grosbøl, P.: 1986, *A&A*, **155**, 11
Contopoulos, G., Grosbøl, P.: 1988, *A&A*, **197**, 83
Grosbøl, P. Patsis, P.A.: 1998, *A&A*, **336**, 840
Patsis, P.A., Grosbøl, P., Hiotelis, N.: 1997, *A&A*, **323**, 762
Rubin, V.C., Ford, W.K.Jr., Thonnard, N., Burstein, D.: 1982, *ApJ*, **261**, 439

# ORBITS IN A STÄCKEL APPROXIMATION

V. DE BRUYNE, F. LEEUWIN AND H. DEJONGHE

*Sterrenkundig Observatorium, Universiteit Gent, Belgium*

Because of their analytical simplicity and regularity, Stäckel potentials are attractive tools for modelling galaxies. The third integral $I_3$ is explicitly known in a Stäckel potential, and can be used as an approximation to the effective third integral, in order to construct three-integral models (cf. Dejonghe, *et al.*, 1996, A&A 306, 363).

Moreover, Stäckel potentials turn out to yield good global descriptions for either axisymmetric or triaxial systems without central mass concentration (de Zeeuw 1985, MNRAS 216, 273, de Zeeuw & Lynden-Bell 1985, MNRAS 215, 713), and even for some systems with a black hole included (Sridhar & Touma 1997, MNRAS 292, 657).

One long-standing concern though, is that Stäckel potentials form only a very small subspace in the family of all potentials. The main orbit families found by numerical integration in general triaxial potentials are present in a Stäckel potential (Schwarzschild 1981, ApJ 232, 236, de Zeeuw 1985, MNRAS 216, 273), but there is obviously no place in an integrable potential for smaller orbital families or stochastic orbits. However, since regular orbits are the rule rather than the exception, a potential which yields a good representation of those orbits is certainly a good basis for building models.

We want to improve the generality of models based on Stäckel potentials, by using a set of Stäckel potentials, each of which fits the true galactic potential in a spatially limited region. These potentials then provide an explicit expression for the integrals that will allow us to construct semi-analytical distribution functions expressed as $f(E, L_z, I_3)$. In a Stäckel potential, there is a function of 1 variable that can be freely chosen. This flexibility is advantageous, and will, of course, be exploited at the fullest when performing the fit.

This work is a preliminary study on the feasibility and effectiveness of such an approach. This we do by comparing orbit integrations with their counterparts in the Stäckel-set representation.

As a test case, we consider an axisymmetric Miyamoto-Nagai (MN) model (1975, PASJ 27, 533) with intermediate flattening ($\epsilon \sim 4.5$). In this model, the diskiness is largely exaggerated, so we are considering a specially demanding case. The fitting of Stäckel potentials to the MN potential in spatially limited regions is done on a grid, using quadratic programming. In brief, a number of basic potentials $\psi_i$ are chosen out of a library and combined to yield a $\psi_S = \sum_i c_i \psi_i$, in order to minimize the quantity $\chi^2 = \sum_l (\Psi - \sum_i c_i \psi_i)^2$, with $l$ an index covering the points in the grid. A complete description of this fit method can be found in Mathieu & Dejonghe, 1996, A&A 314, 25. The integration of orbits in both potentials, is performed using a fourth-order Runge-Kutta with variable time-step

*J. Henrard and S. Ferraz-Mello (eds.), Impact of Modern Dynamics in Astronomy, 395–396.*

and with energy conservation better than $10^{-6}$ (relative error) over 100 azimuthal periods.

Checking the representation of orbits in both potentials, we find that typical orbits have very similar surfaces of section. Minor resonances trapping some of the orbits are well reproduced by the Stäckel potentials. Other resonances present in the MN potential, but absent from the Stäckel fit, are found for small values of $I_3$ (these are orbits remaining close to the equatorial plane, where diskiness is important). They represent $\sim 4\%$ of our orbit library. Those orbits could be represented by an alternative method (such as a frequency decomposition).

The volume of phase-space occupied by the orbit is important when assembling orbits to reproduce a given density. Therefore, we also computed the orbital densities $\rho(R, z; E, L_z, I_3)$ which are functions of $(R, z)$ for each given orbit. The mass fraction correctly located in the Stäckel potential compared to the original MN potential is $M = 1 - \delta M$, where we computed $\delta M = \sum_{k,\ell} |(\rho_{MN}(R_k, z_\ell) - \rho_S(R_k, z_\ell))|/2$. The average $M$ is 92%, the lowest values for $M$ are found for orbits with resonances that were not well fitted.

An important question is how well the $I_3$ is conserved along the orbits, so that it can be used as a label in modelling procedures. We calculated the variation of $I_3$ along the orbits and find that it is usually of order the error in the potential fit (at worst a few percents). For orbits that remain close to the equatorial plane, the variations reach a few tenths. This shows that, for a real galaxy, a strong diskiness would certainly require an additional local Stäckel potential for these regions.

We can conclude that, using a small set of Stäckel potentials as an approximation to a trial MN potential, we are able to reproduce most of its orbits with satisfactory accuracy, except for very few resonances. We find that for the vast majority of orbits, the Stäckel $I_3$ does provide a valid approximation for the effective third integral of our trial potential. Therefore, this value of $I_3$ can be used for labelling those orbits when constructing dynamical models.

# MEASURING MASS LOSS RATES FROM GALACTIC SATELLITES

KATHRYN V. JOHNSTON
*Institute for Advanced Study,*
*Olden Lane, Princeton, NJ 08540, USA*

STEINN SIGURDSSON
*Institute of Astronomy, Madingley Road, Cambridge, CB3 0HA*

and

LARS HERNQUIST
*Board of Studies in Astronomy and Astrophysics,*
*University of California, Santa Cruz, CA 95064*

Number count profiles of many Galactic and some extra-galactic satellite systems show evidence for associated stars beyond the cut-off in density that is identified as the point of tidal limitation (e.g. Irwin & Hatzidimitriou 1995, Grillmair et al. 1995). These "extra-tidal" stars are assumed to be debris lost from the satellite due to heating or stripping by the Galactic tidal field or (in the case of globular clusters) evaporation of stars over the tidal boundary. In this contribution we present a method for using these features to measure the mass loss rate from the satellite, and test it on the results of numerical simulations of satellite disruption. A more detailed discussion of all aspects of this work can be found in Johnston, Sigurdsson & Hernquist (1998).

In the numerical simulations, the satellite's evolution along an orbit in a three component rigid model of the Galaxy is followed using a self-consistent field (SCF) code (developed by Hernquist & Ostriker 1992) to calculate the mutual interactions of stars in the satellite. Figure 1 shows the annularly averaged number surface density from one example simulation "observed" from the viewpoint of the center of the Galaxy after several Gigayears. The closed (open) symbols show the profile recovered if only bound (all) stars are considered. Clearly there is a break in the open symbols at the radius $r_{\text{break}}$ where the analysis becomes dominated by unbound stars.

Tremaine (1993) pointed out that the change in the orbital frequency of a star torn from the satellite at $r_{\text{break}}$ should approximately be given by $\Delta\Omega \sim r_{\text{break}} d\Omega/dR$, where $\Omega(R)$ is the frequency of a circular orbit in the parent galaxy at radius $R$. Hence, debris will spread over an angular distance comparable to the size of the cluster ($r_{\text{break}}/R$) in a time ($r_{\text{break}}/R$)/$2\Delta\Omega \sim T_{orb}/\pi$, where $T_{orb}$ is the azimuthal time period of the orbit. This suggests that we can estimate the average surface density of stars in an annulus between $r_{\text{break}}$ and $r$ from the centre of the cluster to be

$$\langle \Sigma_{\text{xt}}(r) \rangle = \left[ \frac{dm}{dt} \frac{(r - r_{\text{break}})}{r_{\text{break}}} \frac{T_{\text{orb}}}{\pi} \right] \Big/ \left[ \pi(r^2 - r_{\text{break}}^2) \right] , \qquad (1)$$

where $dm/dt$ is the mass loss rate from the cluster. We can differentiate equation (1) to find an approximate expression for the absolute surface density. This estimate

397

*J. Henrard and S. Ferraz-Mello (eds.), Impact of Modern Dynamics in Astronomy, 397–398.*

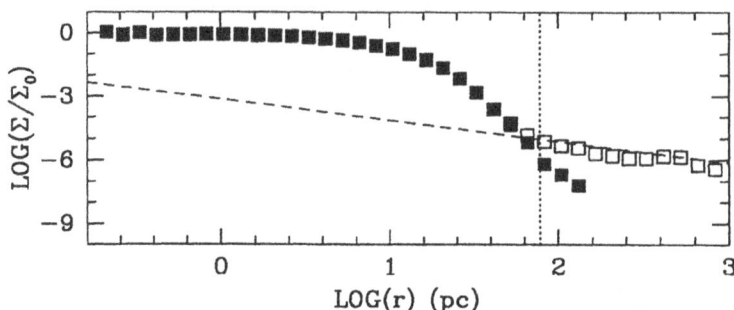

Fig. 1. Number count profiles from a simulation of satellite disruption, as viewed from the center of the Galaxy. The closed (open) symbols are for the bound (total) population of stars. The dotted line shows the position of $r_{break}$ used to model the extra-tidal population. The dashed line shows the model.

is overlaid in dashed lines on the profile shown in Figure 1 using $dm/dt$ averaged over the simulation, and $r_{break}$ indicated by the vertical dotted line. This method reproduces the extra-tidal features in all simulations with similar success.

Equation (1) can be used to measure the mass loss rate from a Galactic satellite from the density of extra-tidal stars and assuming a value for $T_{orb}$. When tested on the simulations, this method recovered the mass loss rate to within a factor of two. When applied to observations it can be used to provide constraints on dynamical models of individual satellites and to directly measure the current destruction rate of the Galactic satellite system. Tests on simulations of mass-segregated globular clusters also indicate that this mass-loss estimate can be used to place limits on the initial mass function of stars in the cluster from local and global measurements of the present day mass function (see Johnston, Sigurdsson & Hernquist, 1998).

## References

Grillmair, C.J., Freeman, K.C., Irwin, M. and Quinn, P.J.: 1995, *Astron. J.* **109**, 2553
Hernquist, L. & Ostriker, J.P.: 1992, *Astrophys. J.* **386**, 375
Irwin, M. J. and Hatzidimitriou, D.: 1995, *MNRAS* **277**, 1354
Johnston, K.V., Sigurdsson, S. and Hernquist, L.: 1998, *MNRAS*, in press

# EFFECT OF CHAOTIC ORBITS ON DYNAMICAL FRICTION

SOFÍA A. CORA*, M. MARCELA VERGNE and JUAN C. MUZZIO

*Facultad de Ciencias Astronómicas y Geofísicas, Universidad Nacional de La Plata,
Paseo del Bosque, 1900 La Plata, Argentina*

When a body moves through a medium of smaller particles, it suffers a deceleration due to dynamical friction (Chandrasekhar 1943). Dynamical friction is inversely proportional to the relaxation time, which can be defined as the time needed for the orbits to experiment an energy exchange of the order of their initial energies, as a result of the perturbations produced by stellar encounters. Chaotic orbits, present in non-integrable systems, have exponential sensitivity to perturbations, a feature that makes them to relax in a time much shorter than regular ones, which suggests that dynamical friction would increase in the presence of chaotic orbits (Pfenniger 1986). We present preliminary results of numerical experiments used to check this idea, investigating the orbital decay, caused by dynamical friction, of a rigid satellite which moves within a larger stellar system (a galaxy) whose potential is non-integrable. Triaxial models with similar density distributions but different percentages of chaotic orbits are considered. This last quantity depends on the central concentration of the models. If the potential corresponds to *triaxial mass models with smooth cores*, the regular orbits have shapes that can be identified with one of the four families of regular orbits in Stäckel potentials (box and three types of tubes). Chaotic orbits behave very much like regular orbits for hundreds of oscillations at least. In this case, the galaxy is represented by the triaxial generalization of the $\gamma$-models with $\gamma = 0$ (Merritt & Fridman 1996). However, the situation is very different in *triaxial models with divergent central densities (cusps) or black holes*, a feature that is in agreement with the observations. While the tube orbits are not strongly affected by central divergencies, the boxlike orbits are often rendered chaotic (Gerhard & Binney 1985). The timescale in which the chaos manifests itself in the orbital motion is short compared to a Hubble time. In this models, the compact object is taken as a Plummer sphere.

The percentage of regular and chaotic orbits in a given model is obtained computing the six Liapunov exponents for each orbit of the particles that make up the galaxy.

The initial conditions of the particles that make up the galaxy ($N = 10,000$) are obtained generating them in the corresponding spherical model with isotropic velocity distribution, and integrating them for a few dynamical times in the potential corresponding to the triaxial model with core (axis ratios $c/a = 0.5$, $b/a = 0.79$) and black hole (mass $m_h = 0.02M$ and softening parameter $\varepsilon_h = 0.08$); the adopted units are such that the gravitational constant $G$, the total mass $M$ and the x-axis scale length $a$ are unity.

* e-mail: sacora@fcaglp.fcaglp.unlp.edu.ar

*J. Henrard and S. Ferraz-Mello (eds.), Impact of Modern Dynamics in Astronomy, 399–400.*

The satellite is modelled by a softened point mass, with mass $m_s = 0.01 M$ and softening parameter $\varepsilon_s = 0.1$. Its initial position corresponds to a distance slightly larger than the radius containing the half mass of the galaxy. Its velocity is such that it would move on a circular orbit in the spherical model.

To integrate the equations of motion, a non-self-consistent code is used: the particles move in the fixed potential that represents the galaxy and do not interact with each other; the satellite interacts with the particles in order to simulate the dynamical friction process. This method allows one to know whether an orbit is regular or chaotic in a given potential.

The satellite decay rates in models with a black hole are greater in some cases, and smaller in others, than in models without it. It is possible, however, that to affect the orbital decay, chaotic orbits should have a Liapunov time ($T_{lia}$) shorter than the time required by the satellite to reach the center of the galaxy ($T_{dec} \simeq 18$ crossing times). The lack of a clear behaviour of the satellite may be attributed to the fact that the perturbation produced by the black hole is not strong enough to produce a great difference in the percentages of chaotic orbits with $T_{lia} < T_{dec}$, in the models with and without black hole (25% and 16% the total number of particles of the system, respectively). This analysis will be repeated with other galaxy models that present larger differences in their percentages of chaotic orbits, in order to elucidate how the presence of those orbits affects the orbital decay of the satellite.

We are grateful to D. Pfenniger for kindly making his code available to us. This work was supported through grants from CONICET.

## References

Chandrasekhar, S.: 1943, *ApJ*, **97**, 255
Gerhard, O. E., & Binney, J. J.: 1985, *MNRAS*, **216**, 467
Merritt, D., & Fridman, T.: 1996, *ApJ*, **460**, 136
Pfenniger, D.: 1986, *A&A*, **165**, 74

# NONPARAMETRIC STATISTICAL MODELS OF GLOBULAR CLUSTER DYNAMICS

WILLIAM D. HEACOX

*University of Hawaii at Hilo, Hilo, HI 96720-4091 USA*

Observations only partially constrain dynamical models of globular clusters and similar systems, in large part because distances and velocities are seen only in projection, and only at a single (unknown) orbital phase. As a result, some dynamical assumptions are necessary if the cluster dynamics and mass distribution are to be inferred from kinematic data – typically that mass follows light, or that the stars observed are the sole source of the gravitational field in which they move, or that orbital energies are completely thermalized, or that orbital angular momenta are exponentially distributed – most often some combination of these is invoked. Such assumptions, reasonable as they may seem, are almost never justified by the data *per se* for the cluster in question, and thus diminish the credibility of resulting estimates of cluster dynamics and, especially, of the presence and extent of dark matter.

Apparently one can do better, at least in principle. The use of sophisticated statistical models should allow the determination of dynamical properties of clusters as statistical distributions of orbital energies and angular momenta, even where the individual values are not observable; and of the radial distribution of gravitational potential. This is so because the cluster dynamics and potential are mapped onto the observed kinematics in a completely deterministic manner, given only the assumption of spherical symmetry (an apparently reasonable choice in many cases). The requisite statistical models take the generic form

$$f_{\mathbf{k}}(\mathbf{k}; \Phi) = \int f_{\mathbf{k}|E,L;\Phi}(\mathbf{k}|E, L; \Phi) f_{E,L}(E, L) \, dE \, dL \qquad (1)$$

where $\mathbf{k}$ is a multivariate kinetic variable, $E$ and $L$ are the orbital specific energy and angular momentum, $\Phi(r)$ is the radial run of gravitational potential within the cluster, and $f$ is the probability density function (pdf) of the indicated variables ($f_{x|y}$ is the pdf of $x$ conditional upon $y$). The left-hand side of this equation is the observed joint statistical distribution of selected kinematic variables (projected velocities and distances) among cluster members; the function $f_{E,L}$ is the underlying true distribution of orbital energies and angular momenta. The integral kernels $f_{\mathbf{k}|E,L;\Phi}$ may be *a priori* computed from principles of mathematical statistics and the assumption of spherical symmetry. As an example: for the bivariate combination $\mathbf{k} = (r_p, v_t)$ of projected distance from the center of the cluster $r_p$ and the component of proper motion velocity transverse to the projected radial vector, $v_t$, the model kernel is

J. Henrard and S. Ferraz-Mello (eds.), Impact of Modern Dynamics in Astronomy, 401–402.

$$f_{k|E,L;\Phi}\left(r_p, v_t | E, L; \Phi\right) \propto$$

$$r_p \int \frac{dr}{r\sqrt{\left[2\left(E - \Phi\left(r\right)\right)^2 - \left(L/r\right)^2\right]\left[r^2 - r_p^2\right]\left[\left(L/r\right)^2 - v_t^2\right]}} \tag{2}$$

When substituted as the integral kernel into Equ. 1 the result is the connection between the observed joint distribution of $r_p$ and $v_t$, and the underlying gravitational potential and statistical mix of orbital energies and angular momenta. Connections of this sort, for any chosen mixture of observed kinematics **k** (projected position, radial velocity, plane-of-sky velocities from individual proper motions), must be satisfied by the same gravitational potential $\Phi\left(r\right)$ and (if the observed stars are the same in all cases) the same mixture of orbital energies and angular momenta, $f_{E,L}$. The requisite models (of the form of Equ. 2 for all likely combinations of observable variables) may be found in (Heacox, 1998).

In principle such relations may be numerically inverted to determine the potential and dynamics required to produce the observed joint distributions of all projected kinematics. In practice the calculations appear to require a fast parallel processor and sophisticated optimization algorithms, and there is yet no guarantee that the solution is unique. The computational problem is currently being pursued. As for uniqueness, Heacox (1997) demonstrates that proper motion velocities (available for some clusters) are absolutely required for anything like a unique solution; their absence admits of a mass ambiguity in the solution of at least a factor of 2. It seems likely that obvious physical constraints – such as non-negative mass densities – will effectively limit the possible solutions to permit, *e.g.*, a resolution of the dark matter problem in spherical globular clusters.

## References

Heacox, W. D.: 1997, Statistical Dynamics of Nonrotating Globular Clusters, *Astrophys. J.*, **490**, 263-266

Heacox, W. D.: 1998, Nonparametric Statistical Models of Spherically Symmetric Kinematics, *Astrophys. J. Supp. Ser.*, **114**, 121-132

# ORBITS OF HIPPARCOS METAL-POOR STARS WITH WELL-DEFINED SPECTROSCOPIC ABUNDANCES

F. BANCKEN and E. JEHIN

*Institut d'Astrophysique de l'Université de Liège, 5 avenue de Cointe, Liège, Belgium*

## 1. Introduction

Orbital characteristics of 80 HIPPARCOS mildly metal-poor stars are derived assuming a Galactic model slightly modified from that of Ostriker and Caldwell (1979). A comparison with other works is performed and a relationship between the dynamical properties of the stars and their well-determined spectroscopic abundances is investigated.

## 2. Stars Sample

The sample was built from 3 different sources and has a mean star metallicity of [Fe/H] $\sim -1$, which corresponds more or less to the transition between the halo and the disk.

- Jehin *et al.* (1998) have determined high precision abundance ratios of 21 nearby unevolved metal-poor stars.
- Edvardsson *et al.* (1993) derived accurate abundances of a dozen elements for 189 F and G disk stars among which we selected the 40 most metal-poor. The authors originally computed the orbital parameters with a three-component Myamoto potential.
- 25 disk and halo metal-poor stars were also taken from Nissen and Schuster (1997) who originally computed orbits with the model of Allen and Santillàn (1991).

We derived orbital parameters for the stars of Jehin *et al.* (1998) and reanalysed the two last samples with our Galactic model using new parallaxes and proper motions from HIPPARCOS (ESA, 1997).

## 3. Mass Model

We adopted a Galactic model made up of a central mass, a flattened disk, a spherical bulge-halo and a spherical dark corona. The density laws were taken from Ostriker and Caldwell's Galaxy mass model (1979) except for the disk whose isodensity surfaces are oblate similar concentric spheroids. It was preferred to Ostriker and Caldwell's surface disk because it makes orbit integration possible.

Following a classical scheme we adjusted the contribution of each component of the model in order to reproduce the observed Galactic rotation curve.

*J. Henrard and S. Ferraz-Mello (eds.), Impact of Modern Dynamics in Astronomy, 403–404.*

For each star our calculations were carried on during 1000 crossings of the Galactic plane. We used time steps of $10^4$ years and at the end of most orbital computations the total energy was conserved to $|\Delta E/E| < 10^{-4}$. The z-component of the angular momentum was kept to its initial value.

Our star sample gave rise to both regular (box, tube, ...) and chaotic orbits. Orbital characteristics such as $R_{min}$, $R_{max}$, eccentricity and $z_{max}$ (the maximum height reached by the star) were computed along with their mean errors.

## 4. Results

With these new orbital parameters we confirm the correlation found by Nissen and Schuster (1997) between [Ni/Fe] abundance ratio and $z_{max}$. Other possible relationships are currently under investigation and may help to better understand the transition between disk and halo stars.

## References

Allen, C., Santillàn, A.: 1991, *Rev. Mex. Astron. Astrofis.*, **22**, 255
Edvardsson, B., *et al.*: 1993, *A&A*, **275**, 101.
ESA, 1997: 'The HIPPARCOS Catalogue', *ESA SP-1200, ESTEC*, Nordwijk
Jehin, E., *et al.*: 1998, *A&A*, **330**, L33.
Nissen, P.E., Schuster, W.J.: 1997, *A&A*, **326**, 751.
Ostriker, J.P., Caldwell, J.A.R.: 1979, in IAU Symposium 84, 441.

# HST-FGS ASTROMETRY OF THE LOW MASS BINARY L722-22

L. G. TAFF

*Department of Physics and Astronomy, The Johns Hopkins University*
*3400 North Charles Street, Baltimore, MD 21218 U.S.A.*

and

JOHN L. HERSHEY

*Astronomy Programs, Computer Sciences Corporation,*
*Space Telescope Science Institute, 3700 San Martin Drive, Baltimore, MD 21218 U.S.A.*

The M dwarf L722-22 (= LHS 1047) was discovered to be a binary system by Ianna 20 years ago. The analysis of the ground- based data indicated a mass $0.06 M_\odot$ for the secondary. This is below the nominal stellar mass limit of $0.08 M_\odot$. The importance of potential "brown-dwarf" candidates, and the fact that the masses of both components place them near the end of the main sequence, made this system a prime object for further, intensive, study.

This close (separation 0."3), faint (V = 11.m5, 14.m4) binary was near the limit for ground-based work. The residuals of an individual night's photographic data were typically at the 50% level. Also, the photographic images are completely blended. The few one-dimensional speckle data points yielded a merged, asymmetric image profile. Finally, this system is too faint for HIPPARCOS. Our proposal for Hubble Space Telescope Fine Guidance Sensor (FGS) observing was approved in 1992.

By using the two FGS observing modes in tandem—"TRANSfer" mode to determine the separation of the pair projected onto the plane of the sky and "POSition" mode measurements to ascertain the trigonometric parallax and the motion of the primary about the center-of-mass of the system—the relative orbit and the fractional masses can be determined. Our combined use of the full facilities of the FGSs has become the standard binary star observing methodology. Anticipating this, and coupled with the importance of low mass objects near the end of the main sequence, our original HST observing proposal included a list of the six most promising systems possibly containing a sub-stellar member. All the others were subsequently awarded to other observers.

The relative orbital motion of the binary has been observed utilizing the astrometer FGS in its interferometric TRANS mode with the F583W filter. Such data can be rectified to within a milli-arc second (mas) or less and provides the angular separation of the components in two orthogonal dimensions. The absolute parameters of the orbit were determined by analyzing FGS POS mode measurements. The repeatability of this kind of observation is significantly worse, typically at the 3 mas level. Finally, the sky surrounding L722-22 is relatively empty. This makes the transformation from relative to absolute parallax uncertain.

Further details on the electro-optical characteristics of the FGSs and the reduction of TRANS and POS mode data are already published, by us, elsewhere. The

*J. Henrard and S. Ferraz-Mello (eds.), Impact of Modern Dynamics in Astronomy, 405–407.*
© 1999 *Kluwer Academic Publishers.*

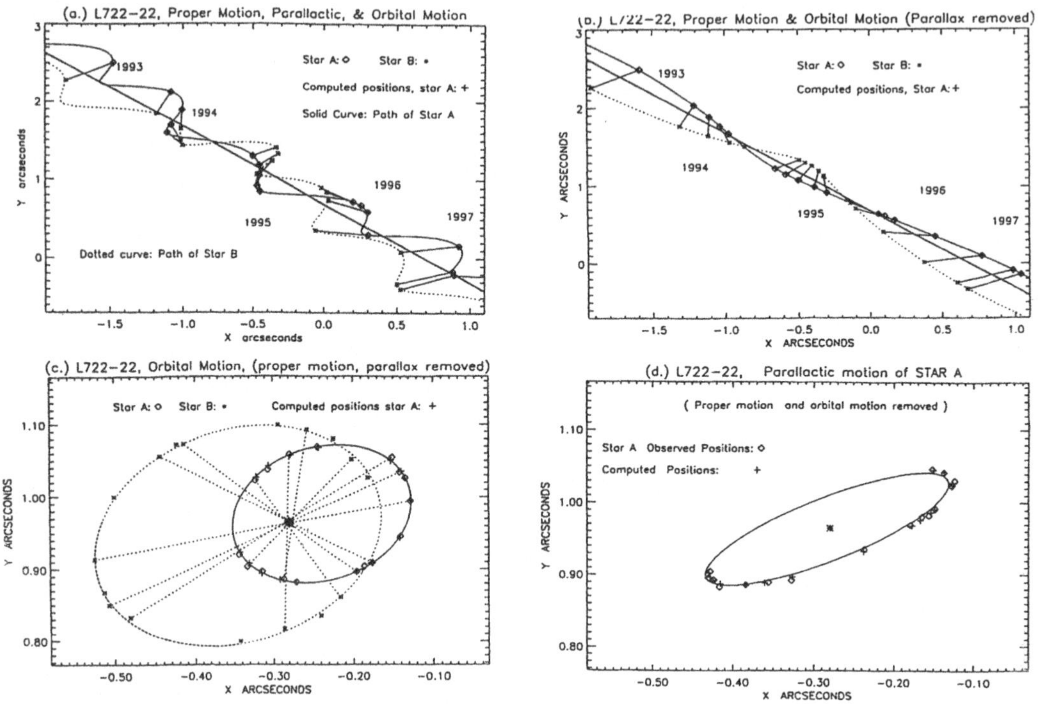

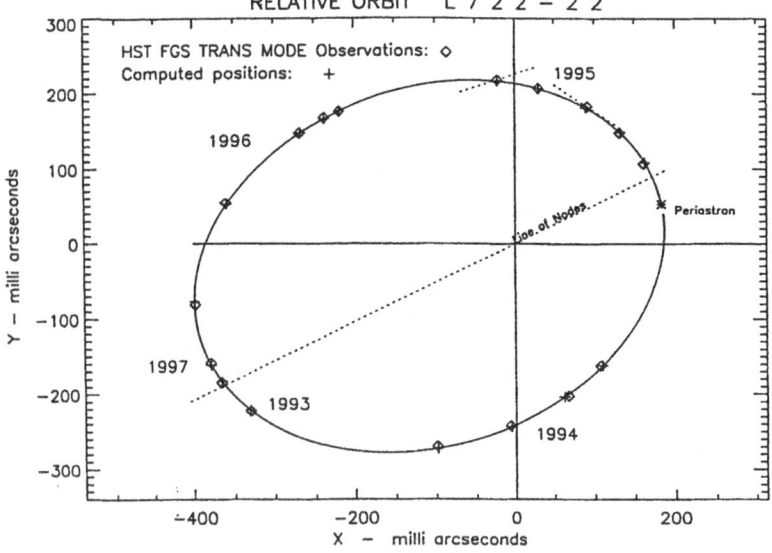

Fig. 1. The four figures show various view of the motions of the binary on the plane of the sky. In addition to the orbital motion, the binary has noticeable proper motion compounded by its path along its parallactic ellipse.

Fig. 2. Relative orbit of the binary.

relatively red color of the system, B - V = 1.m74 required the use of an appropriately red calibration star. (See our forthcoming Astron. J. paper for the complete details and references.)

One can analyze TRANS mode measurements in one of two ways: by fitting the "sum" of two single star Transfer Functions to match the observed result or by Fourier transform de-convolution. We employed both procedures and they consistently provided the same results for the separations of the pair and the magnitude difference. We also utilized two methods to determine the relative orbit—both the classical orbital element set $(P, T, e, a, i, \omega, \Omega)$ and the Thiele-Innes constants were obtained. (The astrometric reduction of the POS mode data proceeded using standard STScI processing methods.) The results from our analysis, for this relatively eccentric orbit, are contained below along with the kinematic parameters.

**Classical Orbital Elements:**

$$P = 4.566 \pm 0.009 \text{yr} \qquad T = 1995.366 \pm 0.003 \text{yr}$$
$$a = 303.7 mas \pm 0.7 mas \qquad e = 0.364 \pm 0.001$$
$$i = 146.0 \pm 0.3 (J2000.0) \qquad \omega = 62.6 \pm 0.6 (J2000.0)$$
$$\Omega = -13.6 \pm 0.6 (J2000.0)$$

**Thiele-Innes Constants:**

$$A = 0."0831 \pm 0."0010 \quad B = 0."2894 \pm 0."0007$$
$$F = 0."2522 \pm 0."0007 \quad G = -0."0492 \pm 0."0015$$

**Kinematic Parameters:**

$$\Pi_{rel} = 0."1656 + 0."0008 \qquad \Pi_{abs} = 0."1666$$
$$\mu_x = 0."6009/\text{yr} \pm 0."0006/\text{yr} \quad \mu_y = -0."5993/\text{yr} \pm 0."0005/\text{yr}$$

The analysis of the 17 sets of FGS data, over nearly a complete orbital period (8/93 to 12/97), definitely eliminates the possibility that the secondary component has a sub-stellar mass. The values we found are $0.179 M_\odot$ and $0.112 M_\odot$ with a (formal) random error of only 1.5%.

This work has been based on observations with the NASA/ESA HST obtained at the STScI which is operated by the AURA, Inc. under NASA contract NAS5-26555. Support for this work was provided, in part, by NASA through grant numbers 4283, 5510, 6063, and 6641 from the STScI.

# CAN COMBINATION OF 'KOZAI EFFECT' AND TIDAL FRICTION PRODUCE CLOSE STELLAR AND PLANETARY ORBITS?

L.G. KISELEVA AND P.P. EGGLETON

*Institute of Astronomy, Madingley Road, Cambridge CB3 0HA, UK*

In binary stars, tidal friction dissipates a fraction of the orbital energy at constant angular momentum and will circularise binary orbits on a rather short timescale compared with the nuclear timescale, provided that at least one star of the binary has a radius comparable to the separation between binary components. This dissipation effectively ceases once the orbit is circularised. In a hierarchical triple system such dissipation cannot cease entirely, as neither inner nor outer orbit can become exactly circular because of the perturbation of the third distant body. Thus in such systems tidal friction can lead to a steady secular decrease of the inner semimajor axis, accompanied by transfer of angular momentum from the inner to the outer pair, persisting over the whole nuclear lifetime of the system. The situation can be even more dramatic if two orbits have high relative inclination $i > 40°$ It can be shown analytically and numerically (see e.g., Kozai 1962, Marchal 1990, Kiseleva 1996 and references therein) that for triple systems with high relative inclination there is a quasi-periodic change of the inner eccentricity (on a timescale $\sim P_{out}^2/P_{in}$) during which it reaches a maximum value $e_{in}^{max}$. This value only depends on the inclination $i$ between the two orbital planes; other parameters affect only the timescale. For example, if we approximate a triple stellar system like $\beta$ Per (Algol) ($m_1 = 0.8M_\odot, m_2 = 3.7M_\odot, P_{in} = 2.87$ days; $m_3 = 1.7M_\odot, P_{out} = 1.86$ yr, $e_{out} = 0.23; i = 100°$) as three point masses, then the inner eccentricity $e_{in}$ cycles rather smoothly between 0 and 0.985, while $i$ fluctuates between $100°$ and $140°$. We call these fluctuations 'Kozai cycles'. Such 'Kozai cycles' do not actually occur in this semi-detached system: they can be damped to a small value by tidal friction, but in fact they are also strongly reduced by the non-dissipative effect of the quadrupole moments of the two stars in the inner pair. This effect produces apsidal motion which is much more rapid than the apsidal motion due to the third star, and so prevents the Kozai cycles from operating.

However, the situation may have been different at an early stage in Algol's evolution if we assume that (i) the stars in the inner semi-detached system were smaller at zero age, and (ii) the inner period was longer. In this case, the Kozai cycles are not damped by the quadrupole distortion, as in actual Algol. Instead, after one or less commonly a few Kozai cycle, which brings the inner pair of stars close together at periastron at the peak of the cycle, the semimajor axis is rapidly decreased. In all cases the orbit only shrinks to the size which allows tidal friction to kill the next Kozai cycle and to circularise the orbit. After that the orbit remain nearly of the same size with $e_{in}$ close to zero.

Fig. 1 shows the evolution of the orbit of a Jupiter-mass ($M \sim 10^{-3}M_\odot$) planet around a solar type companion in a rather wide binary stellar system. The

*J. Henrard and S. Ferraz-Mello (eds.), Impact of Modern Dynamics in Astronomy, 409–410.*

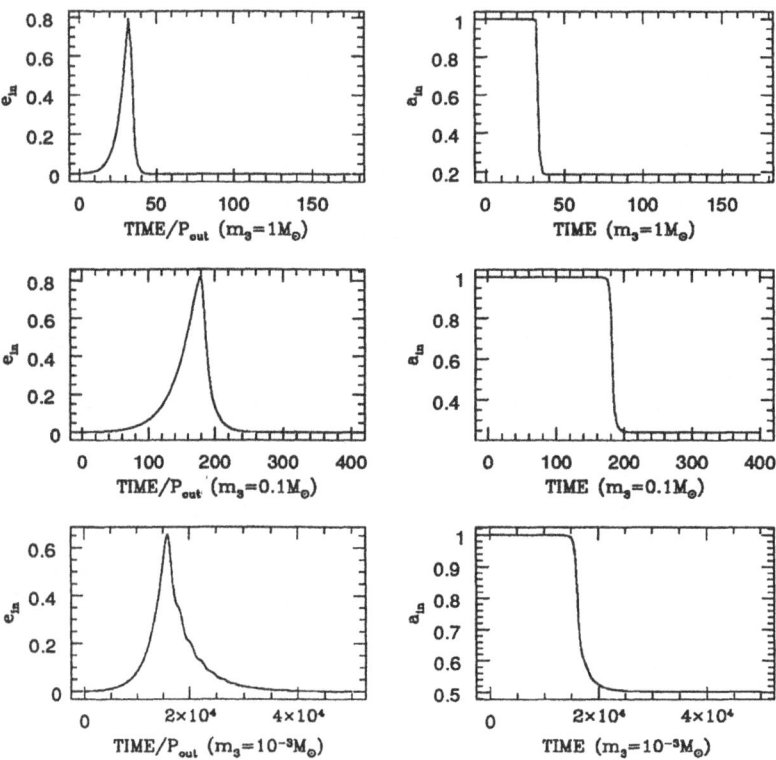

Fig. 1. A Kozai cycle, and its destruction by TF, in a 'Sun-Jupiter' system with a third body of $1 M_\odot$ (top panels), $0.1 M_\odot$ (middle panels) and $10^{-3} M_\odot$. The orbits are inclined at $100°$. The cycle developed to the much the same peak of eccentricity in all three cases (left panels), but more slowly with the lower-mass third body. Tidal friction shrank the orbit drastically at the peak of the cycle (right panels).

inclination of the planetary orbital plane with respect to the binary orbit is taken to be the same as in $\beta$ Per - $100°$. Such a high inclination does not seem very improbable (e.g. Holman et al. 1997). For each mass of third body considered, we see that a single cycle takes place which brings the 'Sun-Jupiter' pair close together and allows tidal friction to recircularise the orbit at a considerably shorter period. Thus, although the combination of third body (which might perhaps be of very low, even planetary, mass) and tidal friction may only be important within a limited range of parameters, it may produce some fraction of *close* binary-star or star-planet systems, such as 51 Peg or $\tau$ Boo.

### References

Holman, M., Touma, J. and Tremaine, S.: 1997, *Nature*, **386**, 254

Kiseleva, L. G.: 1996, in *Dynamical Evolution of Star Clusters*, eds. P. Hut & J. Makino, p233

Kozai, Y.; 1962, *Astron. J.*, **67**, 591.

Marchal, C:. 1990, *The Three-Body Problem*, Elsevier, Amsterdam

# SIMPLE PERIODIC ORBITS IN ELLIPTICAL GALAXIES MODELLED BY HAMILTONIANS IN 1-1-1 RESONANCE

J. PALACIÁN and P. YANGUAS

*Dpto. de Matemática e Informática, Universidad Pública de Navarra, 31006 Pamplona, Spain*

and

S. FERRER

*Dpto. de Matemática Aplicada, Universidad de Murcia, 30071 Murcia, Spain*

**Abstract.** We consider elliptical galactic models, whose dynamical system consists of a three–dimensional isotropic harmonic oscillator plus a potential given by a homogeneous polynomial of degree four with an additional discrete symmetry. We identify families of simple periodic orbits by studying the reduced phase space.

## 1. Triaxial Galaxies

Most galaxies do not show a violent activity; on the contrary, they are supposed to exhibit a stationary behaviour. Only a few years ago, it was thought that the elliptical galaxies were rather simple axisymmetric systems. However, it is not completely true. Thus, the study of the dynamics of elliptical galaxies has become a very interesting subject of research. As a consequence of the observations made during the last two decades, astronomers have learned that many galactic components are not spherical nor do they possess an axial symmetry, but they are, indeed, triaxial objects. There is also an evidence about the fact that many galaxy bulges are triaxial structures. Even barred galaxies evolve towards non–symmetric objects.

Three–dimensional oscillators are used to model the dynamics of the elliptical galaxies. Considering an idealized non–rotating elliptical galaxy and choosing an appropriate reference frame, we can take the gravitational potential $V$ as a smooth function which can be expanded in power series around the origin. In addition to that, we restrict ourselves to the isotropic case. Observations strongly suggest that most triaxial potentials can be described as having equal frequencies (de Zeeuw, 1985). Thus, the unperturbed Hamiltonian function in Cartesian variables reads as $\mathcal{H}_2 = \frac{1}{2}(X^2 + Y^2 + Z^2) + \frac{1}{2}\omega^2(x^2 + y^2 + z^2)$ where $\omega$ has the physical dimension of a frequency. The model we take is a prototype of a galaxy. The perturbation contains only quartic terms but such that $V$ is symmetric with respect to the three principal planes. That is, $\mathcal{H}_4 = b_0\, x^4 + b_1\, y^4 + b_2\, z^4 + b_4\, x^2\, y^2 + b_7\, x^2\, z^2 + b_{10}\, y^2\, z^2$, where $b_i$ are small–size and real parameters with dimensions $[L\, T]^{-2}$.

## 2. The Reduced Phase Space

The oscillator symmetry permits to convert the original system $\mathcal{H}$ into a normalized Hamiltonian. The reduction is regular since $\mathcal{H}$ is isotropic (for details see Yanguas,

*J. Henrard and S. Ferraz-Mello (eds.), Impact of Modern Dynamics in Astronomy, 411–412.*

1998). Moreover, the reduced space is the fourth–dimensional space $\mathbf{C\,P}^2$ (Moser, 1970). It is generated by the nine quadratic generators:

$$
\begin{aligned}
\pi_1 &= \omega^2\,x^2 + X^2, & \pi_2 &= \omega^2\,y^2 + Y^2, & \pi_3 &= \omega^2\,z^2 + Z^2, \\
\pi_4 &= \omega^2\,x\,y + X\,Y, & \pi_5 &= \omega^2\,x\,z + X\,Z, & \pi_6 &= \omega^2\,y\,z + Y\,Z, \\
\pi_7 &= x\,Y - y\,X, & \pi_8 &= x\,Z - z\,X, & \pi_9 &= y\,Z - z\,Y\ .
\end{aligned}
$$

The first six describe the solution of the equations for harmonic oscillators, that is, the ellipse in three dimensional space. The other three invariants $\pi_7$, $\pi_8$ and $\pi_9$ give the position of the plane in space, as they are the components of the angular momentum vector (except for the sign of $\pi_8$). Apart from the constraint of the energy $\pi_1 + \pi_2 + \pi_3 = 2\,h$, the other independent relations are:

$$
\begin{aligned}
\pi_1\,\pi_2 &= \pi_4^2 + \omega^2\,\pi_7^2, & \pi_1\,\pi_3 &= \pi_5^2 + \omega^2\,\pi_8^2, \\
\pi_2\,\pi_3 &= \pi_6^2 + \omega^2\,\pi_9^2, & \pi_4\,\pi_6 &= \pi_2\,\pi_5 + \omega^2\,\pi_7\,\pi_9, \\
\pi_1\,\pi_6 &= \pi_4\,\pi_5 + \omega^2\,\pi_7\,\pi_8, & \pi_3\,\pi_4 &= \pi_5\,\pi_6 + \omega^2\,\pi_8\,\pi_9\ .
\end{aligned}
$$

The reduction procedure is accomplished by the Lie–Deprit method using symplectic variables (Yanguas, 1998). Up to third order, the reduced Hamiltonian reads as $\mathcal{K}_\pi = \frac{1}{2}\,(\pi_1 + \pi_2 + \pi_3) + \varepsilon^2\,\bar{\mathcal{K}}_2/(48\,\omega^6) + \mathcal{O}(\varepsilon^4)$ where $\bar{\mathcal{K}}_2$ is

$$
3\,b_0\,\pi_1^2 + b_4\,\pi_1\,\pi_2 + 3\,b_1\,\pi_2^2 + b_7\,\pi_1\,\pi_3 + b_{10}\,\pi_2\,\pi_3 + 3\,b_2\,\pi_3^2 + 2\,b_4\,\pi_4^2 + 2\,b_7\,\pi_5^2 + 2\,b_{10}\,\pi_6^2.
$$

Now we set up the differential system and apply Liouville's Theorem: $\dot{\pi}_i = \{\pi_i\,,\bar{\mathcal{K}}_2\}$, $i = 1, \ldots, 9$. Previously we had computed (Yanguas, 1998) the Poisson brackets $\{\pi_i\,,\pi_j\}$ to have an explicit expression of the $\dot{\pi}_i$. The critical points of this system are the simple periodic orbits in terms of the $b_i$. We reproduce the results given in (de Zeeuw, 1985) obtaining fourteen families of periodic orbits. The advantage of our procedure is that we can make the analysis using the generators of $\mathbf{C\,P}^2$ and covering therefore, the whole reduced system. Besides, we manipulate quadratic polynomial equations instead of Poisson series. This is very adequate for a commercial symbolic processor.

## Acknowledgements

Research has been partially supported by CICYT PB 95–0795 (Spain). First and second authors also benefitted partially from a Project of Departamento de Educación y Cultura, Gobierno de Navarra (Spain).

## References

de Zeeuw, T.: 1985, Motion in the Core of a Triaxial Potential, *M.N.R.A.S.*, **215**, 731–760

Moser, J.: 1970, Regularization of Kepler's Problem and the Averaging Method on a Manifold, *Comm. Pure and Appl. Math.*, **23**, 609–636

Yanguas, P.: 1998, Integrability, Normalization and Symmetries of Hamiltonian Systems in 1–1–1 Resonance, Ph.D. Thesis, Universidad Pública de Navarra, Pamplona (1998)

# MOON'S PLANETARY PERTURBATIONS

P. BIDART and J. CHAPRONT

*Observatoire de Paris, 61 avenue de l'Observatoire, 75014, Paris, France*

In ELP, the computation of planetary perturbations is about 20 years old. A better knowledge of lunar and planetary parameters, new planetary solutions under construction and progresses in numerical tools, are factors that should contribute to their improvements. The construction of planetary perturbations takes widely its inspiration from Brown's method. In a first step, we only consider the main problem (Earth, Moon, and Sun with a Keplerian motion). The solution of the main problem is actually of a high precision and is used as a reference (Chapront-Touzé, 1980). This solution is expressed in Fourier series of the 4 Delaunay arguments, with numerical coefficients, and partials with respect to integration constants.

## 1. Integration Method

The method based on the variation of arbitrary constants is described in (M.Chapront-Touzé, J.Chapront, 1980). Equations of Moon's motion are written in a rotating frame where the reference plane is the mean ecliptic. In this frame, the absolute acceleration is expressed by means of disturbing forces acting on the Moon, by the Sun, the Earth and a planet. It is the gradient of $F$ which can be divided into several components : $F_c$ related to the main problem, $F_D$ and $F_I$ giving rise to direct and indirect planetary perturbations. Equations of motion are initialy expressed in rectangular coordinates $(x_i, u_i)$, $u_i = \dfrac{dx_i}{dt}$ :

$$\frac{dx_i}{dt} = \frac{\partial \mathcal{H}_c}{\partial u_i}, \qquad \frac{du_i}{dt} = -\frac{\partial \mathcal{H}_c}{\partial w_i}$$

where $\mathcal{H}_c$ is the main problem's Hamiltonian. Then we perform a change of variables more appropriate to the lunar problem. Equations are thus expressed with the set of variables $\{z_1^0 = \frac{n'}{\nu}, z_2^0 = E, z_3^0 = \Gamma, w_1, w_2, w_3\}$, $n'$ and $\nu$ being respectively Sun's and Moon's mean motions, $\Gamma$ and $E$ halves of the coefficients of the $\sin \overline{F}$ term in latitude and $\sin \overline{l}$ term in longitude. $w_i = b_i t + w_i^0$ are respectively the mean longitude, the longitude of the perigee and the longitude of the node of the Moon. The system becomes :

$$\left(\frac{dz_j^0}{dt}\right) = C^{-1}\left(\frac{\partial \mathcal{R}}{\partial w_i}\right)$$

$$\left(\frac{dw_j}{dt}\right) = (b_j) - [C^{-1}]^T\left(\frac{\partial \mathcal{R}}{\partial z_i^0}\right)$$

*J. Henrard and S. Ferraz-Mello (eds.), Impact of Modern Dynamics in Astronomy, 413–414.*
© 1999 *Kluwer Academic Publishers.*

$\mathcal{R}$ represents in either case, the direct or indirect disturbing force function. The $C$ matrix is composed of $c_{i,j} = \dfrac{\partial c_i}{\partial z_j^0}$ where $c_i = \left\langle \sum_k u_k \dfrac{\partial x_k}{\partial w_j} \right\rangle$, the brackets meaning the constant term of the series. After integrating the system, the integration constants $\delta z_j^0$ are chosen to force the solution to be fitted on the main problem constants $(\nu, \Gamma, E)$.

## 2. Direct and Indirect Perturbations

For direct perturbations, function $F_D$ is developed in Legendre polynomials of the variable $\Theta_P = \dfrac{1}{rD}\overrightarrow{EM} \cdot \overrightarrow{GP}$. Next a separation of the variables depending on the Moon from those depending on the planet is performed. This makes the integration easier in the case of small divisors. In $F_D$, the distance $D$ between a planet and the Earth-Moon barycenter is developped in Fourier series of the planetary longitudes. For indirect perturbations, $F_I$ is now concerned : Sun's elements are not any longer Keplerian but contain planetary perturbations. We use a planetary solution for the set : $\{\sigma_k, k = 1,\ldots,6\} = \{a', \lambda', k' = e'\cos\varpi', h' = e'\sin\varpi', q' = \sin\frac{i'}{2}\cos\Omega', p' = \sin\frac{i'}{2}\sin\Omega'\}$. Note that : $\sigma_k = \sigma_k^{(0)} + \Delta\sigma_k$. As in the direct case, $\mathcal{R}_c = F_c - k\dfrac{m_E + m_M}{r}$ is developed in Legendre polynomials of the variable $\Theta' = \dfrac{1}{rr'}\overrightarrow{EM} \cdot \overrightarrow{GS}$.

## 3. Computation and Precision

Integrating the product of Moon and planet series, small divisors can appear and blow up the coefficients. Hence we sort coefficients and their associated frequencies in order to keep long period terms with small coefficients. However, terms with periods exceeding five thousand years are linearized. Once the series for the six variables $z_j^0$ and $w_j$ are determined, we construct the coordinates substituing these results into longitude, latitude and radius vector expressions. Our computations let us hope a final precision for planetary perturbations of about $10^{-5}$".

## References

Chapront-Touzé M.: 1980, *Astron. Astrophys.*, **83**, 86
Chapront-Touzé M. and Chapront J.: 1980, *Astron. Astrophys.*, **91**, 233

# ON CONSTRUCTION OF A LONG-TERM MOON'S THEORY

T.V. IVANOVA

*Institute of Applied Astronomy of the Russian Academy of Sciences,*
*St.Petersburg, E-mail: itv@ita.spb.su*

An analytical long-term theory of the motion of the Moon is constructed within the framework of the general planetary theory (Brumberg, 1995). A method, different from the one of (Ivanova, 1997) designated below as (*), for the determination of the perturbations depending on the eccentricities and inclinations of lunar and planetary orbits is used which allows to obtain the solution of the problem in the purely trigonometric form up to any order with respect to the small parameters.

The aim of this paper is to construct the long-term Lunar theory in the form consistent with the general planetary theory (Brumberg, 1995). For this purpose the Moon is considered as an additional planet in the field of eight major planets (Pluto being excluded). In the result the coordinates of the Moon may be represented by means of the power series in the evolutionary eccentric and oblique variables with trigonometric coefficients in mean longitudes of the Moon and the planets. The long-period perturbations are determined by solving a secular system in Laplace-type variables describing the secular motions of the lunar perigee and node and taking into account the secular planetary inequalities. This paper is a development of results obtained in (*). Unlike (*) the equations of the Moon's motion are described in the different form

$$\ddot{p} + 2\sqrt{-1}\,n\dot{p} - \frac{3}{2}n^2(p+q) + n^2\sum_{i=1}^{9}(K_i\delta p_i + L_i\delta q_i) = n^2 P',$$

$$\ddot{w} + n^2 w + n^2\sum_{i=1}^{9} M_i w_i = n^2 W'. \tag{1}$$

Here the variables $p, q, w$ are introduced by $x + \sqrt{-1}\,y = \mathrm{a}(1-p)\exp\sqrt{-1}\,\lambda$, $q = \bar{p}$, $z = \mathrm{a}\,w$, $w = \bar{w}$ instead of geocentric rectangular coordinates $\mathbf{r} = (x, y, z)$ and represent the small deviations from the plane circular lunar motion. The bar means a conjugate value. $\mathrm{a}, n, \lambda$ are the semi-major axis, mean motion and longitude of the Moon, respectively. The heliocentric coordinates $\mathbf{r_i} = (x_i, y_i, z_i)$ $(i = 1, 2, \ldots, 8)$ of the principal planets are subjected to similar transformation. For the Moon the values both without any indexes and with index 9 are used. The right-hand members are expressed in the form

$$P' = P - P^{(0)} + \sum_{i=1}^{9}(K_i\delta p_i + L_i\delta q_i), \qquad W' = W + \sum_{i=1}^{9} M_i w_i,$$

$$P = -1 - \frac{1}{2}p - \frac{3}{2}q + \frac{2}{n^2\,\mathrm{a}^2}\frac{\partial U}{\partial q}, \qquad W = w + \frac{1}{n^2\,\mathrm{a}^2}\frac{\partial U}{\partial w},$$

*J. Henrard and S. Ferraz-Mello (eds.), Impact of Modern Dynamics in Astronomy, 415–416.*
© 1999 *Kluwer Academic Publishers.*

$U$ being the force function. In accordance with the general planetary theory one has $p_i = p_i^{(0)} + \delta p_i$. The intermediary $p = p^{(0)}$, $w = 0$ is a particular planar quasi-periodic solution of equations (1) with the the right-hand members $P = P^{(0)}$, $W = 0$ provided that the major planets move in their intermediate orbits. This solution generalizes Hill's variational curve and includes all solar and planetary inequalities independent of eccentricities and inclinations of all bodies. It is built in the similar way as in (*). The coefficients $K_i$, $L_i$, $M_i$ are functions of the intermediate solution of the bodies under consideration. The perturbations $\delta p$ and $w$ depending on the eccentricities and inclinations of the planetary and lunar orbits are determined by different method than in (*) by means of separating linear terms in the eccentricities and the inclinations of all bodies from the right-hand members. The right-hand members $P$ and $W$ used in (*) contain the terms of first order with respect to $\delta p_i$ and $w_i$ while $P'$ and $W'$ start with the second degree terms of $\delta p_i$ and $w_i$. It guarantees the convergence of iterations. Finally, the solution for $\delta p$ and $w$ is sought in the form of power series

$$\delta p = \sum p_{klmn} \prod_{i=1}^{9} a_i^{k_i} \bar{a}_i^{l_i} b_i^{m_i} \bar{b}_i^{n_i} , \qquad w = \sum w_{klmn} \prod_{i=1}^{9} a_i^{k_i} \bar{a}_i^{l_i} b_i^{m_i} \bar{b}_i^{n_i} \quad (2)$$

with the initial approximation

$$\delta p = -\frac{1}{2} a_9 + \frac{3}{2} \bar{a}_9 + \sum_{i=1}^{9} (c_i a_i + d_i \bar{a}_i), \qquad w = b_9 + \bar{b}_9 + \sum_{i=1}^{9} (f_i b_i + \overline{f_i b_i}),$$

where $a_i$ and $b_i$ are the complex variables of Laplace-type proportional to the eccentricity and inclination of the body with number $i$, respectively. The functions $c_i$, $d_i$, $f_i$ depend on the intermediary of all bodies. The summation is performed over all non-negative values of 9-indices $k$, $l$, $m$, $n$. The coefficients in (2) are quasi-periodic functions of time. The method of $\delta p$ and $w$ construction is based on the separation of the fast and slowly changing variables and on the Birkhoff normalization. All analytical calculations are performed by the specialized Poisson series processor PSP (Ivanova, 1996). This research is supported by the Russian Foundation of Fundamental Researches (grant No. 97-01-01176) and the Ministry of Science of Russia (Astronomy).

## References

Brumberg V. A.: 1995, *Analytical Techniques of Celestial Mechanics*, Springer, Heidelberg.
Ivanova, T. V.: 1996, "PSP: A New Poisson Series Processor", in: *Dynamics, Ephemerides and Astrometry of the Solar System* (eds. S.Ferraz-Mello, B.Morando and J.-E.Arlot), 283, Kluwer.
Ivanova,T. V.: 1997, "A Trigonometric Solution of the Secular Lunar System Intended for Long-Term Theory of the Earth's Rotation", in: *Journees 1997 "Systemes de Reference Spatio-Temporels"* (eds. J. Vondrak and N. Capitaine), 99, Prague.

# A NEW MODEL OF MARS ROTATION FROM PATHFINDER DATA

SEBASTIEN BOUQUILLON & JEAN SOUCHAY

*Observatoire de Paris, France, E-mail:seb@danof.obspm.fr*

**Abstract.** Following efforts to construct an accurate modelisation of Mars rotation starting from canonical equations in an Hamiltonian theoretical frame (Bouquillon and Souchay, 1996), we use recent results from radio tracking data of the Mars Pathfinder mission (Folkner *et al.*, 1997) to modelize in the best way the motion of precession and nutation of the planet. A complete set of coefficients related to these two motions is presented, including the main effect due to the Sun and also those due to the two satellites Phobos and Deimos as well as to the planets. Morever, Oppolzer terms are calculated and included.

## 1. The Materialization of the Motion of Rotation

The canonical variables chosen here in order to solve the equations of motion of Mars by the way of the Hamiltonian equations are the Andoyer variables equivalent to those used in the case of the study of the motion of rotation of the Earth (Kinoshita, 1977). The basic plane $(P_M^t)$ is the mean orbit of Mars for the date $t$, which is slightly moving with respect to an inertial plane, which is the mean orbit of Mars $(P_M^0)$ at the epoch J2000.0. The basic point used in order to measure the motion of the precession and nutation in longitude of Mars is the point called the "non-rotating origin" $D_t$ along $(P_M^t)$ which is described in detail by Guinot(1979) and Capitaine et *al.* (1986).We choose the non-rotating origin $D_0$ at J2000.0 in coincidence with the equinox $\Gamma_0$ of Mars at this epoch. For more details on our materialization of the motion of rotation of Mars, see (Bouquillon and Souchay, 1996).

## 2. Pathfinder Data Used in this Study

With only the two years of Viking data, the observational determination of the precession constant of Mars cannot be done very accurately, as it was shown by Pitjeva(1995), who found a value of $-750"\pm36"/cy$. Nevertheless the recent results from the Mars Pathfinder mission lead to a much more accurate determination of the precession, that is to say -757.6 $\pm3.5"/cy$ (Folkner *et al.*,1997). So, this more accurate determination reduces the uncertainly of the dynamical ellipticity: $H_d = 0.005363 \pm 25$ which has important effects for the accuracy of the determination of nutation coefficients.

## 3. Results

The series of nutations are calculated with the help of VSOP87, (see Bretagnon and Francou, 1988) for the motion of Mars and the other planets, and with ESAPHO and ESADE (see Chapront-touzé, 1990) for the motion of Phobos and Deimos. These series are truncated at 0.01 mas. These series are available from the authors ( Bouquillon and Souchay, 1998).

*J. Henrard and S. Ferraz-Mello (eds.), Impact of Modern Dynamics in Astronomy, 417–418.*

## References

Bouquillon S., Souchay J.:1996, *Proceeding of I.A.U colloquium 165*.

Bouquillon S., Souchay J.:1998, *Astron. Astrophys.*, in preparation

Bretagnon, P., Francou, G.,1988, *Astron. Astrophys.*, **202**,309

Capitaine, N., Guinot, B., and Souchay, J.: 1986, *Celest. Mech.*, **39**, 283.

Chapront-Touzé, M.: 1990, *Astron. Astrophys.*, **240**, 159.

Folkner, W.M., Yoder, C.F., Yuan,D.N., Standish, E.M., Preston, R.A., 1997, *Science*, **278**, 1749

Guinot, B.: 1979, "Basic problems in the kinematics of the rotation of the Earth", in: *Time and the Earth's Rotation 7-8* (D.D. McCarthy, J.D. Pilkington, eds), Reidel, Dordrecht.

Kinoshita, H.: 1977, *Celest. Mech.*, **26**, 296.

Pitjeva E.V.: 1995, *I.A.U. symposium 172*, (S.Ferraz-Mello, B.Morando and J.-E.Arlot, eds), 45.

# DYNAMICS OF THE MIMAS-TETHYS SYSTEM

SYLVAIN CHAMPENOIS

*Laboratoire d'Astronomie, 1 Impasse de l'Observatoire, 59000 LILLE,*
*E-mail: Sylvain.Champenois@bdl.fr*

**Abstract.** The role of the 200 yr-long period found recently in the mean longitude of Mimas (Vienne and Duriez 1992) is investigated through numerical integrations. It is shown that it has a deciding effect on the descriptions of the resonance motion of the Mimas-Tethys system, as considered up to now. As a result, Mimas's inclination before capture may have been higher (up to 0.7°) or lower (down to 0.03°) than the value previously considered (0.42°). Also, Tethys's eccentricity on capture may have been quite higher ($\approx 0.008$ versus 0). Moreover, the probability of capture is found to be very sensitive to Tethys's eccentricity, and possibly much higher (up to 1) than the value considered before (0.04).

## 1. Introduction

Mimas and Tethys are the first and third satellite of saturn. We use the following notations: $a, n, e, i, \gamma, \varpi, \Omega, \lambda$ and $m$ are Mimas's orbital semimajor axis, mean motion, eccentricity, inclination, sine of semi-inclination, longitude of periapse, longitude of ascending node, mean longitude and mass (in units of Saturn's mass), respectively; the corresponding primed quantities refer to Tethys. These two satellites are connected by an inclination-type resonance: the argument $\varphi = 2\lambda - 4\lambda' + \Omega + \Omega'$ slowly oscillates around zero with a great amplitude (95° with a period of 70 years). Since the argument $\varphi$ is present in terms factored by $ii'$ in the expansion of the disturbing potential, we shall call this resonance the $ii'$ resonance.

Up to now, the dynamics of this resonance had been studied assuming that, close to the resonance, the period of the resonant argument is much longer than the periods of all the other arguments present in the right-hand side of Lagrange's equations. This allows to average out all the non-resonnant arguments in those equations. To this averaged equations were added tidal effects on the satellites' mean motions, due to dissipation in Saturn.

According to this analysis, $i$ and $i'$ are constant before capture, while the ratio $\alpha = a/a'$ increases under tidal effects, causing the system to come closer and closer to the resonance. Once the system is captured, $\alpha$ starts oscillating around the value corresponding to strict resonance, while $i$ and $i'$ start increasing. Tethys being 17 times more massive than Mimas, its inclination has remained almost unchanged since capture up to the present time, ($\approx 1.0°$). On the other hand, Mimas's one has increased fourfold in the same time, from 0.4° to 1.6° (Allan 1969; Champenois and Vienne 1999a).

The recent discovery, by Vienne and Duriez (1992), of long-period terms in the mean longitude of Mimas, lead us to reconsider this classical analysis. As a matter of fact, we realized that we could not retain the only resonnant argument in the right-hand side of the equations of the Mimas-Tethys system, since the periods of

*J. Henrard and S. Ferraz-Mello (eds.), Impact of Modern Dynamics in Astronomy,* 419–424.
© 1999 *Kluwer Academic Publishers.*

some of the non-resonant arguments were not, close to the resonance, much lower than the period of the resonant argument.

## 2. A crucial Low Frequency

Among the long periods discovered, one of them, with period 200 years, is particularly interesting because of its relative closeness to the period of the resonant argument (which is currently 70 years, and was higher in the past). This period has its origin in the argument $\sigma = \frac{1}{2}\Omega - \frac{3}{2}\Omega' + \varpi'$, which appears in the right-hand side of Lagrange's equation through the following three main arguments :

$$\frac{3\varphi}{2} + \sigma = 3\lambda - 6\lambda' + 2\Omega + \varpi'$$

$$\frac{\varphi}{2} + \sigma = \lambda - 2\lambda' + \Omega - \Omega' + \varpi'$$

$$-\frac{\varphi}{2} + \sigma = -\lambda + 2\lambda' - 2\Omega' + \varpi'$$

The terms of these arguments are factored by the respective quantities $i^2 e'$, $ii' e'$ and $i'^2 e'$ in the expansion of the disturbing potential. The value of $e'$ is very badly known. The TASS1.6 theory (Vienne and Duriez 1995) gives $e' = 0.000235$, but with such uncertainties that it lies actually in the interval $[0, 0.001]$. Hence, we shall consider in what follows several possible current values. A certain thing, however, is that $e'$ is very small at the present time. But this was not the same in the past, if we take into account its damping by the tides raised on Tethys by Saturn. Dermott et al. (1988) give for the time scale $\tau$ of the damping of Tethys' eccentricity the value $\tau = 10^8$ yr. This is smaller than the age of the Mimas-Tethys resonance, which can be deduced from Allan's study to be $\approx 2.4\tau$ (Champenois and Vienne 1999a). Therefore $e'$ may have been sensibly higher on time of capture into the present resonance than at the present time (up to 0.01).

Hence, we can expect the presence of $j/k$ secondary resonances between the libration or circulation frequency $\omega$ of the primary $ii'$ resonance, and the frequency $\dot{\sigma}$ (defined by the relation $j\omega + k\dot{\sigma} = 0$), as well as the presence of chaotic zones due to the overlapping of these secondary resonances near the border of the $ii'$ resonance.

## 3. Numerical Integrations

Averaged Lagrange's equations are considered: only the following four arguments, with periods of the order of $10^2$ years, are retained in the expansion of the disturbing potential: $\varphi$, $\frac{3}{2}\varphi + \sigma$, $\frac{1}{2}\varphi + \sigma$ and $-\frac{1}{2}\varphi + \sigma$. In particular, secular terms depending on longitudes of nodes or pericenters (like $ee' \cos(\varpi - \varpi')$) are removed, because their periods are of the order of the year. Moreover, only the largest terms have been retained in the right-hand side of the averaged equations. Then the main terms coming from the oblateness of Saturn are added (up to $J_2^3$), as well as tidal terms on $n$, $n'$ and $e'$. Regular variables are considered. We integrated these equations backwards in time, for various initial values $e_0'$ of $e'$ (subscript 0 is for the present time). The initial conditions are taken from TASS1.6 for J2000.

$\dot{\sigma}$ was found to be only slowly varying since capture time (from $\approx 2\pi/185$ yr to $\approx 2\pi/200$ yr at the present time). Therefore, in order to locate secondary resonances in an $(\alpha, y = \sqrt{\gamma\gamma'})$ plane, we may consider it constant throughout that period. Using a pendulum model for the $ii'$ resonance, we thus obtain the curves labelled $j/k$, shown on Fig 1. The V-shaped zone on these figures is obtained from the same model. This is the libration zone of the $ii'$ resonance: inside it, $\alpha$ oscillates around the value $\alpha_r = 0.629308514$ corresponding to strict resonance, between two extreme values. During one period of these oscillations, $y$ increases in an imperceptible way. Only the borders of these librations are plotted on these figures.

Figure 1 shows, on the left, the results obtained for $e_0' = 0$ (1), 0.00235 (2) and 0.001 (3), the other initial conditions being taken from TASS1.6. We notice that $y$ on capture increases with $e_0'$. But $i'$ is found to be insensitive to $e_0'$. Thus, $i$ on capture increases with $e_0'$. We have: $0 < e_0' < 0.001 \Rightarrow 0.4° < i < 0.7°$ and $8 \, 10^{-5} < e' < 0.008$ at capture time. On the right is shown a run also obtained for $e_0' = 0.001$, but with a slight change in Mimas's initial mean motion compared to the previous run (by $5 \, 10^{-4}$ rad yr$^{-1}$: this remains within the range of uncertainties). This run reveals a capture in the 1/2 secondary resonance: the system enters the primary resonance with $i = 0.18°$ and $e' = 0.008$ (1). After capture, there are large oscillations in the libration amplitude (2), followed by a jump to about 30°. Then the trajectory follows the adiabatic invariant theory (3; see Champenois and Vienne 1999a), until it gets trapped in the 1/2 secondary resonance (4). After escaping from this secondary resonance, the libration amplitude decreases again smoothly up to the present time (5). Other runs also reveal the possibility for a capture in the 1/1 secondary resonance.

## 4. Surfaces of Section

Surfaces of section are computed thanks to the following equation (Champenois and Vienne 1999a):

$$\ddot{\varphi} = R_0\Big(\sin(\varphi) + R_1 \sin(\frac{3}{2}\varphi + \dot{\sigma}t + \sigma_0)$$

$$+ R_2 \sin(\frac{1}{2}\varphi + \dot{\sigma}t + \sigma_0) + R_3 \sin(-\frac{1}{2}\varphi + \dot{\sigma}t + \sigma_0)\Big) \tag{1}$$

where the coefficients $R_i$, ($0 \leq i \leq 3$) essentially depend on $e'$ and $i$. If $i > 0.1°$, then they may also be viewed as parameters, and Eq. (1) becomes a periodically time-dependent one-degree-of-freedom equation, in which $\dot{\sigma}$ is fixed at the time at which we want to compute our surfaces of section.

Figure 2 shows the surfaces of section $(\varphi, \dot{\varphi})$ of the Mimas-Tethys system for trajectories 2 (left) and 3 (right), shown on Fig. 1, both at present time (up) and

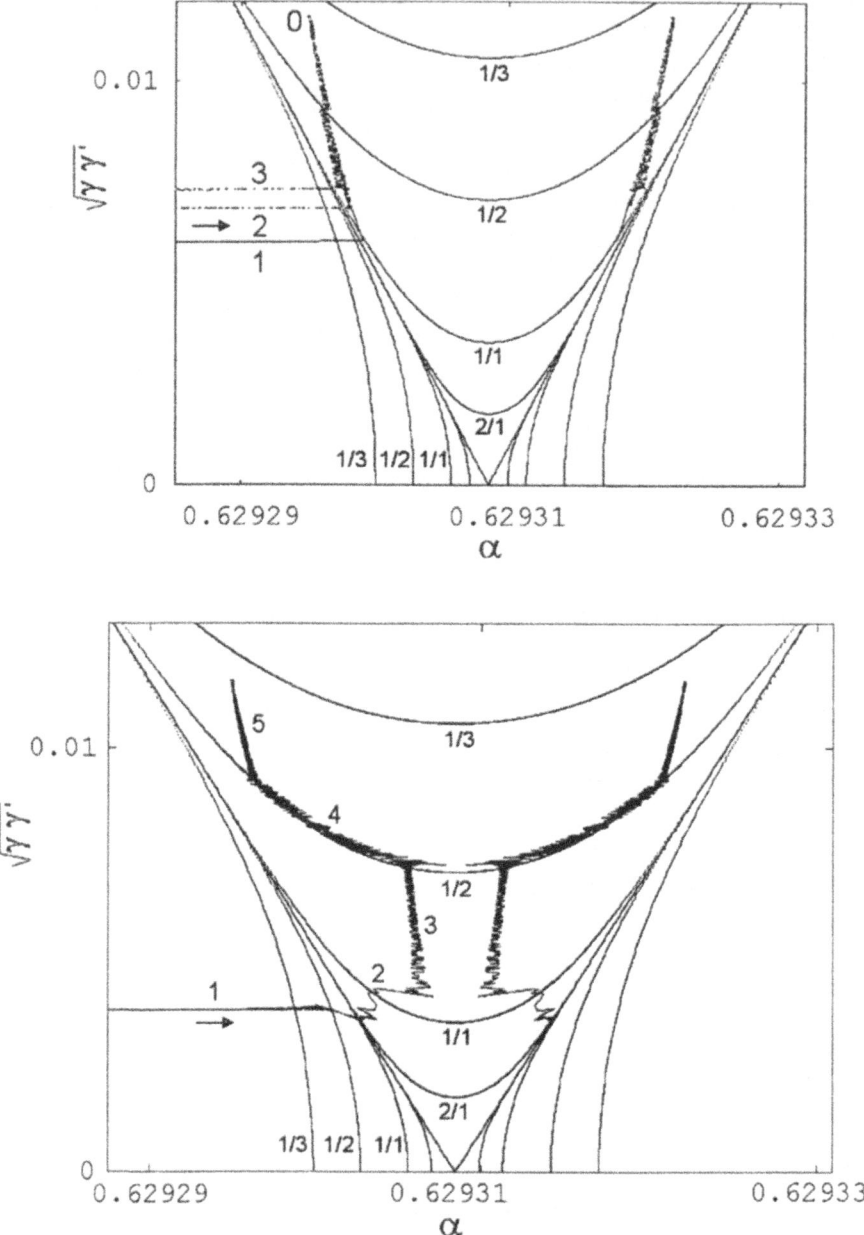

Fig. 1. Different possible trajectories for the past evolution of the Mimas-Tethys system. in the bottom panel is shown a possible capture in the 1/2 secondary resonance. The arrow indicates the direction of evolution, and 0 is for the present time. The $j/k$ curves correspond to secondary resonances, the V-shaped zone to the primary libration zone. See text.

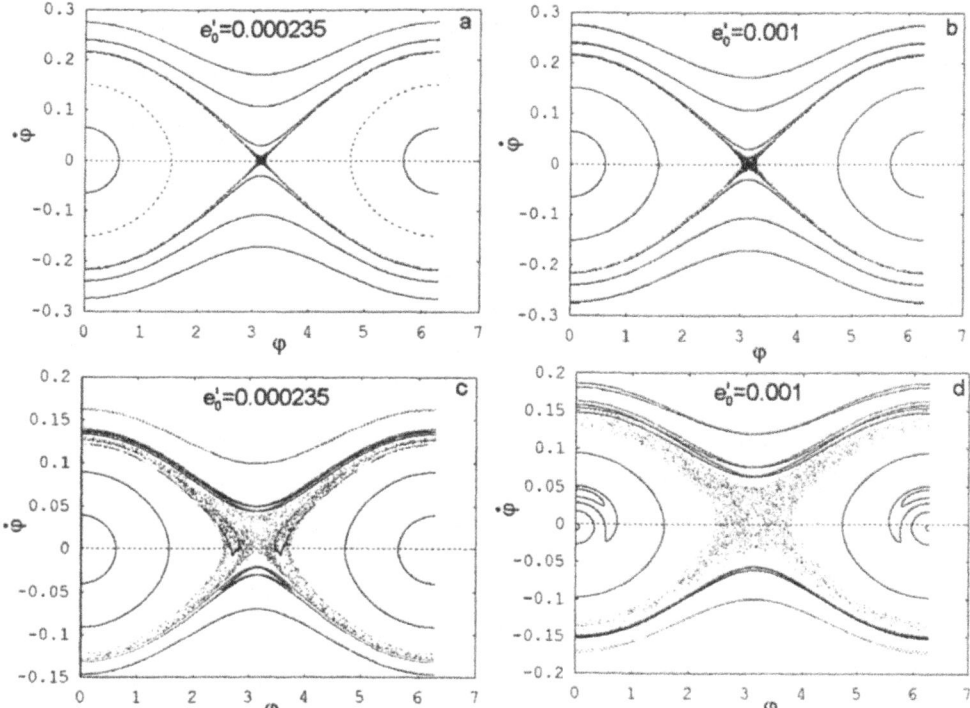

Fig. 2. Surfaces of section of the Mimas-Tethys system at present time (up) and capture time (down), for $e'_0 = 0.000235$ (left) and $e'_0 = 0.001$ (right), corresponding to trajectories 2 and 3 shown on Fig. 1. See text for comments.

capture time (down). This figure reveals narrow chaotic borders of the $ii'$ resonance at present time, but large ones at capture time (when the eccentricity of Tethys is bigger), as well as islands associated to the 1/1 and 1/2 secondary resonances.

## 5. Capture Probability

To evaluate the probability of capture, Sinclair (1972) uses for the Mimas-Tethys resonance a pendulum model, whose "restoring force" increases with time. He then uses for the probability of capture the following formula: $P = (u/v)^2$, where $u$ is the value of $\varphi$ (above the unstable equilibrium point, near $\varphi = -\pi$) separating capture from escape, and $v$ is the value separating escape from direct circulation. He thus obtains $P = 0.04$.

Taking now $\sigma$ into account, there is now one more choice in addition to capture and inverse circulation: the system may also librate and circulate in turn, in a chaotic way. However, tidal effects place bounds on it, urging the system to choose for good between capture and inverse circulation. If $m(\mathcal{C})$ is the measure of the set of values for which the system is captured for good, and $m(\mathcal{E})$ the measure of the set of values for which it escapes from resonance, we can then generalize Sinclair's formula in the following way (Champenois and Vienne 1998b):

$$P = \left( \frac{m(\mathcal{C})}{m(\mathcal{C}) + m(\mathcal{E})} \right)^2 \tag{2}$$

Applying formula (2), $P$ appears to vary strongly with $e'$, increasing from $P = 0.04$ for $e'_0 = 0$ to $P = 1$ for $e'_0 = 0.0016$. The present uncertainty on $e'_0$ therefore results in a big uncertainty on the capture probability into the present resonance. However, it seems likely to be higher than that determined by Sinclair.

## 6. Conclusion

We can see from this study that the consideration of the low frequency $\dot{\sigma}$ upsets the vision of the dynamics of the Mimas-Tethys system that we had so far. As a matter of fact, the system may have been trapped in a secondary resonance or may have behaved in a chaotic way on capture in the present $ii'$ resonance. As a result, the inclination of Mimas may have been higher (up to 0.7°), or lower (down to 0.03°) than derived by Allan (1969). Moreover, the eccentricity of Tethys have been higher in the past (up to 0.008), and the probability of capture into the present resonance may be much higher (up to 1) then that found by Sinclair (1972).

## References

Allan, R.R.:1969, *Astron.J.*, **74**, 497-506
Champenois, S. and Vienne, A.: 1999a, *Icarus*, accepted
Champenois, S. and Vienne, A.: 1999b, *Cel. Mech. & Dyn. Astro.*, submitted
Dermott, S.F., Malhotra, R. and Murray, C.D.:1988, *Icarus*, **76**, 295-334
Sinclair, A.T.: 1972, *Mon. Not. R. Astron. Soc.*, **160**, 169-187
Vienne, A. and Duriez, L.: 1992, *Astron. Astrophys.*, **257**, 331-352
Vienne, A. and Duriez, L.: 1995, *Astron. Astrophys.*, **297**, 588-605

# A NECESSARY CONDITION TO THE SIMULTANEOUS CORRECTION
## OF CELESTIAL REFERENCE FRAMES AND MINOR PLANETS
## ORBITAL ELEMENTS

M.J. MARTÍNEZ USÓ

*Departamento de Matemática Aplicada. ETSII, Universidad Politécnica de Valencia,
SPAIN.*

and

F. MARCO CASTILLO and J.A. LÓPEZ ORTÍ

*Departamento de Matemáticas. Universitat Jaume I, Castellón. SPAIN.*

**Abstract.** In order to obtain corrections to orbital elements of minor planets and to determine parameters of rotation of a reference frame, a geometrical model with restrictions in the declination, longitude of the ascending node and argument of the perihelion to minimize a residual function involving all named parameters was presented (Marco et al., 1996, 1997). On the other hand, such rotational model of correction should not modify the semiaxis, eccentricity nor the mean anomaly, so new restrictions on these parameters seem to be necessary. These conditions should reflect the invariability in the size of the orbit and they should be included in the model to complete a consistent set of restrictions.

## 1. Introduction

Let $\overrightarrow{X}(\alpha, \delta)$ be an unitary vector position in equatorial coordinates for a body in elliptic motion around the Sun. We suppose the existence of errors on its position which are modeled by means of a rotational model given by

$$\Delta\overrightarrow{X} = R\overrightarrow{X} \text{ with } R = \begin{bmatrix} 1 & -\Delta\xi & -\Delta\eta \\ \Delta\xi & 1 & -\Delta\varepsilon \\ \Delta\eta & \Delta\varepsilon & 1 \end{bmatrix} \tag{1}$$

From (1) the incremental errors induced by the rotational model and relating the spherical coordinates at whatever time $t$ are given by

$$\Delta\alpha = \Delta\xi + \Delta\eta \sin\alpha \tan\delta - \Delta\varepsilon \cos\alpha \tan\delta \tag{2}$$
$$\Delta\delta = \Delta\eta \cos\alpha + \Delta\varepsilon \sin\alpha$$

## 2. Rotation Model Applied to the Orbital Plane at $t_0$

We denote the different unitary vectors ( at $t_0$) defining the orbit orientation by $\overrightarrow{p_1^0}$ (in the perihelion direction), $\overrightarrow{p_3^0}$ (normal to the orbit) and $\overrightarrow{p_2^0}$ the necessary one to built a direct triad. The application of the rotational model to the triad is carried out by means of the expressions

$$R_1(\varepsilon)RR_1(-\varepsilon)\overrightarrow{p_i^0} = \overrightarrow{p_i^0} + \Delta\overrightarrow{p_i^0} \qquad \text{for i=1,2,3} \tag{3}$$

*J. Henrard and S. Ferraz-Mello (eds.), Impact of Modern Dynamics in Astronomy, 425–426.*

where $R$ is the rotation matrix from (1) and $R_1(\pm\varepsilon)$ the rotation matrix relating equatorial and ecliptic coordinates. The relation (4) follows from (3)

$$\overrightarrow{\Delta p_i^0} = \overrightarrow{p_i^0} \times R_1(\varepsilon)\overrightarrow{Y} \text{ being } \overrightarrow{Y} = [\Delta\varepsilon, -\Delta\eta, \Delta\xi]^t \tag{4}$$

If we express these $\overrightarrow{p_i^0}$ vectors in ecliptic heliocentric coordinates, due to the fact that they are functions of the incremental values $\Delta\Omega_0$ (argument of ascending node), $\Delta\omega_0$ (argument of perihelion) and $\Delta i_0$ (inclination of the orbit), we can obtain from (3)

$$\begin{bmatrix} \Delta\varepsilon \\ -\Delta\eta \\ \Delta\xi \end{bmatrix} = \begin{bmatrix} 0 & \sin i_0 \sin\Omega_0 & \cos\Omega_0 \\ -\sin\varepsilon & \cos i_0 \sin\varepsilon + \sin i_0 \cos\Omega_0 \cos\varepsilon & -\sin\Omega_0 \cos\varepsilon \\ \cos\varepsilon & \cos i_0 \cos\varepsilon - \sin i_0 \cos\Omega_0 \sin\varepsilon & \sin\Omega_0 \sin\varepsilon \end{bmatrix} \begin{bmatrix} \Delta\Omega_0 \\ \Delta\omega_0 \\ \Delta i_0 \end{bmatrix} \tag{5}$$

### 3. Necessary Compatibility Conditions for the Rotational Model

Let $\overrightarrow{W} = P\overrightarrow{v}$ be the ecliptic coordinates of the unitary vector position $\overrightarrow{v} = [\cos u, \sin u, 0]^t$ in the orbital plane at $t$, where $u$ represents the true anomaly and let $\overrightarrow{X} = R_1(-\varepsilon)\overrightarrow{W}$ be its equatorial representation. Then, we have the incremental vector $\overrightarrow{\Delta W} = R_1(\varepsilon)\overrightarrow{\Delta X}$,

$$\overrightarrow{\Delta W} = \left[\overrightarrow{\Delta p_i}v_i + \overrightarrow{p_i}\Delta v_i\right] \tag{6}$$

Also, from (4) this infinitesimal vector is expressed as follows

$$\overrightarrow{\Delta W} = R_1(\varepsilon)\left[\overrightarrow{X} \times \overrightarrow{Y}\right] = \overrightarrow{W} \times R_1(\varepsilon)\overrightarrow{Y} = P\overrightarrow{v} \times R_1(\varepsilon)\overrightarrow{Y} \tag{7}$$

Through the expression of $P\overrightarrow{v}$ in vectorial form and due to the fact that $R_1(\varepsilon)\overrightarrow{Y}$ is an infinitesimal vector we obtain

$$\overrightarrow{\Delta W} = \sum_{i=1}^{2} \overrightarrow{p_i}v_i \times R_1(\varepsilon)\overrightarrow{Y} \simeq \sum_{i=1}^{2} \overrightarrow{p_i^0}v_i \times R_1(\varepsilon)\overrightarrow{Y} \tag{8}$$

applying (5) and (6) , we obtain

$$\overrightarrow{\Delta W} = \sum_{i=1}^{2} \left\{ \left[\frac{\overrightarrow{\partial p_i}}{\partial\Omega}\Delta\Omega + \frac{\overrightarrow{\partial p_i}}{\partial\omega}\Delta\omega + \frac{\overrightarrow{\partial p_i}}{\partial i}\Delta i\right] v_i + \overrightarrow{p_i}\Delta v_i \right\} \tag{9}$$

which give us together with (8) the complete set of restrictions.

### References

F.J. Marco, J.A. López, M-J. Martínez: 1996, A time-dependent extension to Brower's method of orbital elements correction, *Dynamics, Ephemerides and Astrometry of the Solar System*, pp.199-202

F.J. Marco, J.A. López, M.J. Martínez: 1997, Temporal variation of perturbed elliptic elements: a semi-analytical approach, *Celest. Mech.*, **68**, 193-198

# SECULAR PERTURBATIONS NEAR MEAN MOTION RESONANCES

A. A. CHRISTOU

*Astronomy Unit, Queen Mary & Westfield College, London E1 4NS, United Kingdom,
E-mail:A.Christou@qmw.ac.uk*

In this work we attempt to make progress into assessing the importance of secular interactions between planetary satellites. In recent years, discrepancies have been observed in the expected positions of small planetary satellites (Bosh & Rivkin, 1996; Roddier *et al.*, 1998). The existing ephemerides-producing algorithms for these objects assume fixed, elliptical and inclined orbits whose rate of precession is determined by oblateness alone. Even though the masses of these satellites are quite small relative to the planet ( $\sim 10^{-9} - 10^{-10}$ ) their small mutual separations and the existence of much larger satellites further out leaves open the possibility that in some cases at least the fixed-orbit assumption is only a crude approximation to reality. Two important dynamical mechanisms through which these orbits may evolve are resonant or secular interactions. In order to explore the possibility of the latter we have set up a simple planar system where an satellite in a circular orbit around a spherical planet is perturbing a massless particle which moves in proximity to various mean motion resonances. We aim to examine the effect of the resonance on the particle's reference orbit by measuring the secular frequency. The effects of oblateness have not been taken into account as they are adequately modeled by orbit-fitting theories and can thus be readily subtracted.

TABLE I

Initial conditions for the numerical integrations near the 2:1 resonance (upper row) and the 1:1 resonance (lower row). The symbol $\Delta \alpha$ denotes the resolution of the sampling in the initial value of $\alpha$ and $e$ the initial eccentricity.

| Range in $\alpha$ | $\Delta \alpha$ | $e$ | Samples |
|---|---|---|---|
| 0.58–0.68 | 0.005 | 0.05 | $2 \times 21$ |
| 0.80–0.90 | 0.002 | 0.001 | 51 |

We have integrated the motion of a particle for a range of values of the semimajor axis $\alpha$ (Table I) using the SWIFT_RMVS2 symplectic code (Levison & Duncan, 1994). The output was reduced to a list of frequencies, amplitudes and phases using a program based on the Frequency Modified Fourier Transform algorithm (Šidlichovský & Nesvorný , 1997). We then compare the resulting frequencies with analytical estimates derived with a second order secular theory (Christou, 1998).

In this manner we have investigated (a) the vicinity of the 2:1 resonance (b) orbital evolution in proximity to the planet (Christou & Murray, 1998). Our results can be summarized in Table II where we show numerical and analytical estimates

427

*J. Henrard and S. Ferraz-Mello (eds.), Impact of Modern Dynamics in Astronomy, 427–428.*
© 1999 *Kluwer Academic Publishers.*

of the secular frequency for values of the semi-major axis which were chosen to be close to specific first order $(p+1:p)$ mean motion resonance locations.

TABLE II

Estimates of the secular frequency for values of the semimajor axis close to specific first order mean motion resonance locations. The suffixes (A), (P) give the initial position of the particle relative to the satellite (pericentric or apocentric conjunction). For all cases apart from the first two rows the particle was tested with a pericentric conjuction start only. Units used are rad per perturber revolution.

| $\alpha$ | Resonance | Order 1 Est. | Order 2 Est. | Numer. Est. |
|---|---|---|---|---|
| 0.610 | 2:1 (P) | 0.00540443 | 0.00931030 | 0.00933683 |
| 0.610 | 2:1 (A) | 0.00537956 | 0.00899243 | 0.00914993 |
| 0.824 | 4:3 | 0.000309033 | 0.00104631 | 0.000996402 |
| 0.860 | 5:4 | 0.000493989 | 0.00134269 | 0.00128978 |
| 0.883 | 6:5 | 0.000721684 | 0.00210074 | 0.00201287 |
| 0.900 | 1:1 | 0.000969363 | 0.00196977 | 0.00198843 |

We clearly see that the Laplace–Lagrange approximation (third column) is insufficient for describing the secular evolution of the particle evolving in proximity to a resonance. The difference between that and the numerical estimate is, at best, of the order of the estimate itself. In contrast, the second order analytical estimate manages to stay within 10% of the numerical value demonstrating the need for such high order theories. It is interesting to note that the observed deviations from the first order approximation are sufficient to produce a substantial lag (at least several tens of degrees) in the orbital longitude of a satellite on a 1/2 day orbit around a giant planet over a period of a few years. In order to discover whether such mechanisms can account for observable variations in the properties of *actual* satellite systems we intend to extend our investigation to existing small satellite configurations in the near future.

## References

Bosh, A. S. and Rivkin, A. S.: 1996, 'Observations of Saturn's inner satellites during the May 1995 ring-plane crossing', *Science*, **272**, 518–521.

Christou, A. A.: 1998, 'An investigation of secular perturbations in planetary and satellite systems', *Ph. D. Thesis*, University of London.

Christou, A. A. and Murray, C. D.: 1998, 'The effect of near-resonances on the secular precession of close orbits', *Mon. Not. Royal Astron. Soc.*, submitted.

Levison, H. F. and Duncan, M. J.: 1994, 'The long-term dynamical behavior of short-period comets', *Icarus*, **108**, 18–36.

Roddier C., Roddier F., Graves J. E., Guyon, O. and Northcott M. J.: 1998, 'Satellites of Neptune', *IAU Circ.*, **6987**.

Šidlichovský, M. and Nesvorný, D.: 1997, 'Frequency Modified Fourier Transform and its application to asteroids', *Cel. Mech. Dyn. Astron.*, **65**, pp. 137–149.

# DYNAMICS OF FICTITIOUS EARTH'S SATELLITES
# WITH POSSIBLE PAST VALUES OF THE ECLIPTIC

N. CALLEGARI JR, T. YOKOYAMA AND E. P. MARINHO

*Universidade Estadual Paulista-IGCE-DEMAC*
*C.Postal 178 CEP 13.500-970 Rio Claro Brasil*

**Abstract.** According to many authors, there is no reason to believe that inner planets never had satellites in the past. Indeed, Alfvén and Arrhenius (1972), Burns (1973), Ward and Reid (1973), etc, explain their disappearance using essentially tidal effects which cause secular variations in the semimajor axes. Here, taking into account only gravitational forces, we show that the extinction or the non existence of Earth's satellites could also be related to some strong instabilities which increase the eccentricities $e$ to prohibitive values.

## Variation Of The Eccentricity

According to Laskar et all (1993), depending on the spin, the past obliquities $\epsilon$ of the inner planets could have varied in a large interval. In the case of Earth, this was a possibility in the absence of the Moon. First, let's see the dynamics of fictitious Earth's satellites, in this situation (without Moon). The main perturbations on a single massless satellite are the Sun and the oblateness of the planet. At the distance when the magnitude of these two perturbations balance, the satellite's semimajor axis $a$ has a critical value ($a_{crit}$). In order to study the dynamics of this problem we did a lot of numerical integrations of equatorial satellites placed in almost circular orbits, considering several values of $a$ and $\epsilon$. Our analysis is similar to the case described in Yokoyama (1998). Then we show that when $\epsilon$ is high ($\geq 40^0$) and depending on the initial $a$, the eccentricity $e$ suffers strong variation, which cause collision of the satellite with Earth and usually this occurs in a few hundred of years. We also show that, if $\epsilon$ is $\geq 40^0$ and the semimajor axis $a$ of the satellite is near $a_{crit}$, a large chaotic zone appears which again gives prohibitive values for $e$. Finally it is worth mentioning that this chaotic zone can occur at different distances from the planet, since $J_2$ (oblateness coefficient), depends on the Earth's spin. Therefore considering this fact and also possible high values of $\epsilon$ we show the existence of significant regions where $e$ can reach prohibitive values. Now, let's see a new situation when we include the Moon. This time, we study the behaviour of a massless satellite disturbed by the Moon, Sun and the Earth's oblateness. In spite of the mass ratio, note the similarity between this problem and Triton-Neptune and some of its inner satellites (Triton's retrograde orbit is spiraling, Chyba et all 1989). The exact Earth-Moon history is not known (Touma and Wisdom 1998) but here, for the past values of $\epsilon$ and Moon's orbit, we basically consider the results given in Goldreich (1966) and Touma and Wisdom (1994). Taking several different values for Moon's semimajor axis $a_L$ and also for $a$, we integrated a grid of initial $a, a_L$ where satellites are placed on the Earth's equator in circular orbits. As we said before, Moon's inclination and $\epsilon$ at several Earth-Moon distances, are given

*J. Henrard and S. Ferraz-Mello (eds.), Impact of Modern Dynamics in Astronomy, 429–430.*
© 1999 *Kluwer Academic Publishers.*

in Goldreich (1966) or Touma and Wisdom (1994). The numerical integrations of this grid of semimajor axes, give a series of interesting figures showing very large regions where the eccentricity attains values about 0.8, 0.9, in less than 500 years. Some of these instabilities are related to Kozai resonances and others to orbital resonances. As a conclusion, like in the previous case, there are significant regions of the phase space where satellites cannot survive. We also derived the averaged equations for this problem and some additional integrations (also for Triton-Neptune system) is being carried out. The full results should be reported elsewhere.

## Acknowledgements

T. Yokoyama thanks FUNDUNESP (proc.142/98-DFP) and E. P. Marinho thanks FAPESP (proc.97/06157-4).

## References

Alfvén H. and Arrhenius G.:1972 , *The Moon*, **5**, 210-230.
Burns J.: 1973, *Nature Physical Science*, **242**, 23-25.
Canup R.M. *et al.*:1999, *preprint*
Chyba C.F.*et al.*: 1989, *Astron. Astrophys.*, **219**, L23-L26.
Goldreich P.: 1966, *Rev. of Geophysics*, **4**, 411-434.
Kinoshita H. and Nakai H.: 1991, *Cel. Mech.& Dyn. Ast.*, **52**, 293-303.
Laskar J. and Robutel P.: 1993, *Nature*, **361**, 608-612.
Touma J. and Wisdom J.: 1994, *Ast. J.*, **108**, 1943-1961.
Touma J. and Wisdom J.:1998, *Ast. J.*, **115**, 1653-1663.
Ward W. and Reid M.: 1973, *Mon. Not. R. Astr. Soc.*, **164**, 21-32.
Yokoyama T.:1998, *Planet.Sp.Sci*, accepted

# AN INTERMEDIARY PERIODIC ORBIT FOR HYPERION

I. STELLMACHER

*Bureau des longitudes, Université Lille I*

**Abstract.** We have seen (Stellmacher, 1996) that the long-period terms for Hyperion's motion are very well represented by a second kind and second genius periodic orbit, after Poincaré's classification.

We have shown how to construct such an orbit with, as only data, the observed periods, which characterise the resonance of the Titan-Hyperion couple, and the Titan's motion which is an elliptical one. The physical quantities as the masses and the $J_2$ term of Saturn's flattening are given. We will present the results that we obtained, and compare them with those that other authors obtained by fitting the series to the observations.

## 1. Description of the method

Let $T_H, T', T_\omega$ be the observed periods for Hyperion's and Titan's mean anomalies and for the longitude of Hyperion's pericenter and $n_H, n', n_\omega$ the corresponding frequencies. We have $4n_H - 3n' + 3n_\omega = 0$, $-n_H/n_\omega \equiv N_H = 332$ and $-3n/n_\omega \equiv N_T = 1325$. We put : $T_H = k\overline{T}/N_H$, $T' = 3k\overline{T}/N_H$ and $T_\omega = k\overline{T}$ with $k \in N$ and $\overline{n} = 2\pi/\overline{T}$. We have then $4N_H - N_T - 3 = 0$ and $N_H, N_T$ and $k$ characterise the resonance.

If $n'$ is known, all the other frequencies can be calculated and must fit to the observations. With $n' = 22.576973850/j$ (see Garcia, 1972) then $n_\omega = -3n'/N_T = -0.051117550/j$, $n_H = 3n'N_H/N_T = 16.9710686450/j$ and if $k = 11, \overline{n} = 0.56229440/j$. These calculated data are very close to the observed ones.

In order to determine the semi-major axis, we define: $n_H = n_0(1 + \varepsilon\gamma_H)$, $n_0$ the mean motion corresponding to the observed semi major axis of Hyperion $a_{0H}$, $a_H$ the semi major axis of a keplerian orbit corresponding to $n_H$, $a'$ the semi major axis of Titan's orbit, $m', M$ the masses of Titan and Saturn, $\varepsilon = m'/M$, $n'^2a'^3 = n_H^2a_H^3 = n_0^2a_{0H}^3 = \mu$, $n_H^2a_H^3 = \mu(1 + \varepsilon\gamma_H)^2 \neq \mu$ and $\rho_{0H} = a'/a_{0H}, \rho_H = a'/a_H$. Hyperion's perturbed motion does not verify the third Keplerian law and the semi major axis $a_{0H}$ of the real orbit will be determined after having calculated the quantity $\varepsilon\gamma_H$.

Let $R_1 = Gm'/\Delta = (Gm'/a_{0H}) \cdot (a_{0H}/\Delta)$ (with $\Delta$ = mutual distance Hyperion-Titan), be the principal part of the disturbing function. Let $r, v, 1, \omega$ be the radius vector, the true and mean anomalies, and the longitude of the pericenter of Hyperion (the primed quantities are those for Titan), and put $\rho = r/r'; \theta = \omega - \omega'$ and $\psi = 4\ell - 3\ell' + 3\theta$.

As we explained in (Stellmacher, 1996), the disturbing function is developped by taking Saturn's flattening and Titan's action upon Hyperion into account. As far as Titan's action upon Hyperion is concerned, we only consider the $n\psi \pm m\theta$ ($n \le 12$ if $m = 0; n \le 9$ if $m = 1$ or 2).

## 2. Results

Let $\tau = \overline{n}(t - t_0), \xi = n_\omega t + \omega_0 - \omega'_0, \rho_{0H} = a'/a_{0H}$ and $a, e, \omega, \lambda$ the semi major axis, the eccentricity, the longitude of the pericenter, and the longitude.

*J. Henrard and S. Ferraz-Mello (eds.), Impact of Modern Dynamics in Astronomy, 431–432.*

In the first case (I) we have calculated $\rho_{0H}$, $e_0$ with only the $n\psi$ arguments, $\beta_0$ and $\bar{n}$ are calculated by replacing $\theta$ by $\xi$ for the $n\psi \pm \theta$ and $n\psi \pm 2\theta$ terms. In the second case (II), we have calculated $\rho_0$, $e_0\beta_0$ and $\bar{n}$ with $n\psi$, $n\psi \pm \theta$; $n\psi \pm 2\theta$ arguments with $\theta = \xi - \beta_0 \sin\xi$. The quantities in brackets (III) are the minimal and maximal values determined by different authors (see references).

(I) $\rho_{0H} = 0.824846$

$a = a_0(1 - 0.00327\cos\tau)$

$e = 0.10440 - 0.00395\cos\tau + 0.0245\cos\xi - 0.00183\cos 2\xi$

$\omega = n_\omega t + \omega_0 - 13^0 47\sin\xi + 1^0 64\sin 2\xi - 0^0 49\sin\tau - 0^0 32\sin(\xi-\tau) - 0^0 27\sin(\xi+\tau)$

$\lambda = (n_\omega + n_H)t + \lambda 0 + 9^0 05\sin\tau + 0^0 25\sin(\xi-\tau) + 0^0 20\sin(\xi+\tau)$

(II) $\rho_{0H} = 0.824825$

$a = a_0(1 - 0.00332\cos\tau)$

$e = 0.10384 - 0.00403\cos\tau + 0.0260\cos\xi - 0.00160\cos 2\xi$

$\omega = n_\omega t + \omega_0 - 13^0 09\sin\xi + 1^0 50\sin 2\xi - 0^0 50\sin\tau - 0^0 31\sin(\xi-\tau) - 0^0 26\sin(\xi+\tau)$

$\lambda = (n_\omega + n_H)t + \lambda 0 + 9^0 18\sin\tau + 0^0 25\sin(\xi-\tau) + 0^0 21\sin(\xi+\tau)$

(III) $\rho_{0H} = 0.824942$

$a = a_0(1 - [0.00323, 0.00354]\cos\tau)$

$e = [0.10346, 0.10473] - [0.00389, 0.00410]\cos\tau + [0.0230, 0.0245]\cos\xi - [0.00110, 0,0014]\cos 2\xi$

$\omega = n_\omega t + \omega_0 - [12^0 87, 13^0 58]\sin\xi + [1^0 56, 2^0 16]\sin 2\xi - [0^0 43, 0^0 45]\sin\tau - [0^0 354, 0^0 360]\sin(\xi-\tau)$
$\quad - [0^0 263, 0^0 265]\sin(\xi+\tau)$

$\lambda = (n_\omega + n_H)t + \lambda 0 + [9^0 089, 3^0 142]\sin\tau + [0^0 210, 0^0 220]\sin(\xi-\tau) + [0^0 190, 0^0 228]\sin(\xi+\tau)$

## References

Blitzer, L.: 1977, *Cel. Mech. & Dyn. Astro.*, **16**, 87–95.

Blitzer, L., Anderson, J.P.: 1981, *Cel. Mech. & Dyn. Astro.*, **25**, 65–78.

Brouwer, D. and Clemence, G. M. :1961, *Celestial Mechanics*, Academic Press, New York.

Dourneau, G.: 1993, *Astron. Astrophys.*, **267**, 292–299.

Duriez, L.:1988, *Astron. Astrophys.*, **194**, 309–318.

Ferraz Mello, S.: 1987, *Astron. Astrophys.*, **183**, 397–402.

Ferraz Mello, S., Sato, M.,:1989, *Astron. Astrophys.*, **225**, 541,47.

Garcia, H.A.: 1972, *A.J.*, **77**, . 684.

Haag, J.: 1948, *J. Ann. Sci. de l'E.N.S.*, **64** , 299.

Hagihara, Y.: 1971, *Celestial Mechanics*, **2** part 1., M.I.T. Press.

Message, P. J.: 1987-1988, *Cel. Mech. & Dyn. Astro.*, **43**, 119–125.

Message, P. J.: 1988-1989, *Cel. Mech. & Dyn. Astro.*, **45**, 45–53.

Message, P. J.: 1993, *Cel. Mech. & Dyn. Astro.*, **56**, 277–284.

Poincaré, H.: 1987, *Méthodes nouvelles de la mécanique céleste*, réimpression par A. Blanchard de l'édition Gauthiers-Villars, Tome **1**, p. 79, Tome **3**, p. 201.

Roseau, M.: 1966, *Vibrations non linéaires et théorie de la stabilité*, Springer Tracts in Natural philosophy, Vol. 8, Berlin, Heidelberg, New-York.

Stellmacher, I.: 1996, IV International Workshop on Positional Astronomy and Celestial Mechanics; Peniscola, Spain, Oct.7.11. (to appear).

Taylor, D. B.: 1984, *Astron. Astrophys.*, **141**, 151–158.

Taylor, D. B., Sinclair, A.D., Message, P.J.: 1987, *Astron. Astrophys.*, **181**, 383–390.

Taylor, D. B.: 1992, *Astron. Astrophys.*, **265**, 825–832.

Woltjer, J., Jr.: 1928, *Annalen van de Sterrewachte Leiden XV*.

# AVERAGING OF EARTH-CROSSING ORBITS

G.F. GRONCHI and   A. MILANI

*Department of Mathematics, University of Pisa*

The orbits of planet-crossing asteroids (and comets) can undergo close approaches and collisions with some major planet. This introduces a singularity in the N-body Hamiltonian, and the averaging of the equations of motion, traditionally used to compute secular perturbations, is undefined. We have shown (Gronchi and Milani, 1998) that it is possible to define in a rigorous way some generalised averaged equations of motion, in such a way that the generalised solutions are unique and piecewise smooth, with corners on the node crossing lines.

The model is the averaged equations of motion first introduced by Kozai (1962): the perturbing planets are assumed to move in circular, coplanar orbits, and the equations of motion are averaged over the anomalies of the asteroid and of the planets. In the non-crossing case the averaging is integrable; in the planet-crossing case there is a polar singularity of order two in the equations of motion, and

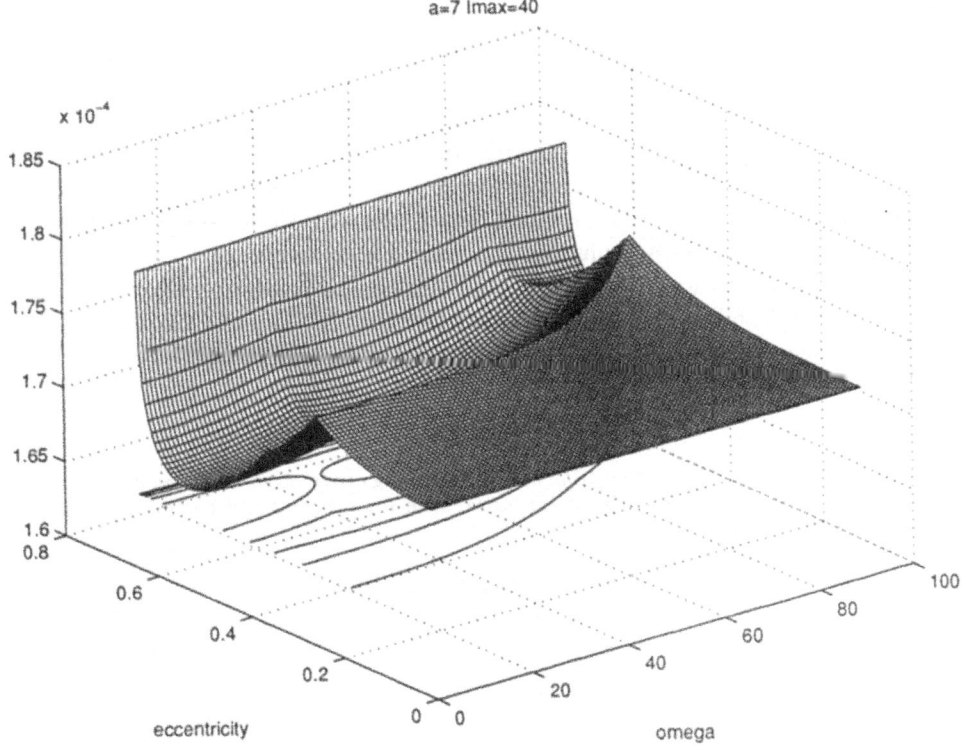

Fig. 1. Graph of the averaged perturbing function for a Centaur type orbit. The node crossing lines with Jupiter appear as crests, the double crossing point as a pyramid-like maximum, and the stable Kozai librations as minima.

*J. Henrard and S. Ferraz-Mello (eds.), Impact of Modern Dynamics in Astronomy, 433–434.*

averaging is not possible. To define a generalized solution, we decrease the order of the polar singularity by the method of extraction of the singularities by Kantorovich. The singularity of the perturbing function is approximated by a modified inverse distance, the one between the straight lines tangent to the two orbits at the nodal points. In this approximation the averaged perturbing function has an analytical expression, allowing explicit computation with elliptic integrals and elementary functions.

There are two main applications of this generalized averaging. First, the Kozai librations of the argument of perihelion $\omega$, protected from node crossings, can be identified by means of the level lines of the averaged perturbing function: the libration centers correspond to minima of the averaged perturbing function, which can not occur on node crossing lines. There are also librations around maxima belonging to node crossing lines, with node crossings taking place in each cycle. Thus we have been able to prove (Gronchi and Milani, 1999) that some stable center of oscillation, either libration or circulation of $\omega$, always exists for every given value of the integrals $a$ and $\sqrt{1 - e^2} \cos I$, and this for an arbitrary number of perturbing planets. For objects which can cross the orbits of several planets, there are usually many such stable states, free from node crossings, with both symmetric and asymmetric librations.

Second, for both librations and circulations of the argument of perihelion $\omega$, it is possible to define proper frequencies for the angles $\Omega$ and $\varpi = \Omega + \omega$, and proper elements such as the maxima and/or minima of $e$ and $I$ over one cycle. To compute proper elements for given initial conditions we need to compute a solution of the generalized averaging equations, not only a level curve; this requires to find a numerical method to approximate the piecewise smooth solution of a differential equation with a singularity; at least one method exists (Milani and Gronchi, 1998), but its efficiency has yet to be investigated.

## References

Gronchi, G.F., Milani, A.: 1998, 'Averaging on Earth-crossing orbits', *Celest. Mech.*, in press.
Gronchi, G.F., Milani, A.: 1999, 'The stable Kozai state for asteroids and comets', *Astron. and Astrophys.*,, 341, 928–935.
Kozai, Y.: 1962, *Astron.J.*, 67, 591-598
Milani, A., Gronchi, G.F.: 1998, 'Proper elements for Earth-crossers', *IAU Colloquium 173*, submitted.

# ROTATIONAL MOTION OF CELESTIAL BODIES
# IN THE RELATIVISTIC FRAMEWORK

SERGEI A. KLIONER and
MICHAEL SOFFEL
*Lohrmann Observatory, Dresden Technical University, 01062 Dresden, Germany*
*e-mail: klioner(soffel)@rcs.urz.tu-dresden.de*

There are several important reasons to consider relativistic effects in rotational motion of celestial bodies. General Relativity is now recommended by the International Astronomical Union and International Union of Geodesy and Geophysics as a theoretical framework for modeling of high-precision observational data. On the other hand, various geodynamical observations provide data which are widely used for testing General Relativity itself.

In Newtonian mechanics it is well known how to describe rotational motion of an extended body. In General Relativity this is a rather subtle issue. The concept of a precessing extended rigid body in general relativity encounters fundamental difficulties and cannot be introduced even in the first post-Newtonian approximation. From a practical point of view, however, the rotational motion of the Earth even at the Newtonian level is defined operationally through the time-dependence of geocentric quasi-inertial coordinates of observing sites. An analogous operational definition can be applied in general relativity. To this end, we need a set of physically adequate reference systems.

Nowadays there are two well-developed formalisms for the construction of relativistic astronomical reference systems: the Brumberg-Kopeikin formalism (see, e.g., Brumberg, 1991) and the DSX formalism (Damour, Soffel, Xu, 1991, 1992, 1993). The two reference systems needed to model the Earth rotation are the barycentric reference system of the solar system and the local geocentric reference system, where the influence of the external masses reduces to tidal effects. Each reference system is defined by the structure of its metric tensor.

In the local geocentric reference system one can derive rotational equations of motion of the extended deformable arbitrarily-shaped Earth, which take the same form as in Newtonian physics

$$\frac{d}{dt}S^i = L^i, \tag{1}$$

where $S^i$ is the post-Newtonian spin, and $L^i$ is the post-Newtonian tidal torque (Damour, Soffel, Xu, 1993; Klioner, 1996). Eq. (1) is sufficient to discuss precession and nutation of the spin. To consider precession and nutation of the angular velocity and figure axis, we need a relativistic definition of the tensor of inertia and angular velocity. A variety of theoretical approaches (from restricted rigid body models (Soffel, 1994) to relativistic Tisserand-like axes of deformable Earth (Klioner, 1996)) lead to the same post-Newtonian definition of the tensor of inertia $C^{ij}$. The

*J. Henrard and S. Ferraz-Mello (eds.), Impact of Modern Dynamics in Astronomy, 435–436.*
© 1999 *Kluwer Academic Publishers.*

spin can be then split into a product of the tensor of inertia and the corresponding angular velocity $\omega^i$

$$S^i = C^{ij}\,\omega^j. \tag{2}$$

The definition of angular velocity of a deformable Earth is not unique already in Newtonian mechanics (Tisserand axes, principal axes of inertia, etc.). This ambiguity is aggravated in general relativity by the ambiguity in the definition of the relativistic spin $S^i$.

There are three main observational consequences of general relativity in Earth rotation: (a) The kinematically nonrotating geocentric reference system used in practice rotates relative to a locally inertial reference system. This results in Coriolis and centrifugal forces in the right-hand side of (1) resulting in geodetic precession ($\sim 1.9''$ per century) and nutation (with a period of one year and an amplitude of 0.15 mas). (b) Explicit relativistic terms in $L^i$ produce additional periodic relativistic effects amounting to $\sim 1$ microarcsecond (Bizouard, et al., 1992). (c) The key relation of Newtonian mechanics

$$\left(M^{ij}\right)_{\text{Newton}} = \left(\frac{1}{3}\,\delta^{ij}\,C^{ss} - C^{ij}\right)_{\text{Newton}} \tag{3}$$

between the quadrupole moment of the gravitational field $M^{ij}$ and the tensor of inertia $C^{ij}$ is violated in general relativity. This means that $J_2$, derived from $M^{ij}$, cannot be easily related to the dynamical ellipticity $H$, which is derived from $C^{ij}$, the effect being of order of $\delta J_2/J_2 \sim 10^{-9}$ (Klioner, 1997).

## References

Bizouard C., Schastok, J., Soffel M.H., Souchay J.: 1992, Étude de la rotation de la Terre dans le cadre de la relativité générale: première approche. In: N. Capitaine (ed.), *Journées 1992*, Observatoire de Paris, 76

Brumberg V.A.: 1991, *Essential Relativistic Celestial Mechanics*, Adam Hilger, Bristol

Damour T., Soffel M., Xu Ch.,: 1991, *Phys.Rev. D*, **43**, 3273

Damour T., Soffel M., Xu Ch.,: 1992, *Phys.Rev. D*, **45**, 1017

Damour T., Soffel M., Xu Ch.,: 1993, *Phys.Rev. D*, **47**, 3124

Klioner, S.A.: 1996, Angular Velocity of Rotation of Extended Bodies in General Relativity. In: S. Ferraz-Mello, B. Morando, J.E. Arlot (eds.), *Dynamics, ephemerides and astrometry in the solar system*, Kluwer, Dordrecht, 309

Klioner, S.A.: 1997, On the problem of post-Newtonian Rotational Motion In: I.M. Wytrzysczak, J.H. Lieske, R.A. Feldman (eds.), *Dynamics and Astrometry of Natural and Artificial Celestial Bodies*, Kluwer, Dordrecht, 383

Soffel, M.: 1994, The problem of rotational motion and rigid bodies in the post-Newtonian framework, *unpublished notes*

# LOMMEL FUNCTIONS IN SOME DRAG-PERTURBED PROBLEMS

SŁAWOMIR BREITER

*Astronomical Observatory of the A. Mickiewicz University, Słoneczna 36, PL 60-286 Poznań, Poland. E-mail: breiter@phys.amu.edu.pl*

and

ALBERT A. JACKSON

*Lunar and Planetary Institute, 3600 Bay Area Blvd, Houston, TX 77058, USA.*

Let us consider the orbital problem in which a particle is subject to the force (per unit mass)

$$\mathbf{F} = - \frac{\mu}{r^3}\mathbf{r} - \frac{\alpha}{r^2}\left(2\,\gamma\,\mathbf{v}_r + \mathbf{v}_t\right). \tag{1}$$

The force consists of the Newtonian two body attraction term and a drag part which is linear in both components of velocity ( radial $\mathbf{v}_r$ and transverse $\mathbf{v}_t$). Depending on a physical interpretation of the parameters $\mu$ and $\alpha$, as well as on the particular choice of the dimensionless constant $\gamma$, the model (1) can match various two body problems with dissipation. They include the classical Poynting-Robertson effect (Robertson, 1936) with $\gamma = 1$, Poynting's (1903) version with $\gamma = \frac{1}{2}$ and the two body drag cases recently studied by Mittleman and Jezewski (1982) and by Mavraganis and Michalakis (1994) under the name of Danby's drag.

Equations of motion in polar variables $r$, $\vartheta$ for the general model (1) can be reduced to the linear form (Breiter and Jackson 1998)

$$x^2 \frac{d^2 y}{d x^2} + x \frac{d y}{d x} + \left(x^2 - \gamma^?\right) y - x^{-\gamma}, \tag{2}$$

$$\frac{d t}{d x} = \frac{-\alpha^3}{\mu^2\, y^2\, x^{2\gamma+1}}. \tag{3}$$

The reduction is achieved by means of the "generalized Robertson transformation" (Breiter and Jackson 1998)

$$x = \frac{h}{\alpha} - \vartheta\,, \quad y = \frac{\alpha^2}{\mu\, r\, x^{\gamma}}, \tag{4}$$

where $h$ is the initial value of the angular momentum $r^2\,\dot{\vartheta}$. The equation of orbit (2) is a Lommel equation (Watson 1958) and its solution can be expressed in terms of the Bessel functions $J_\gamma(x)$, $Y_\gamma(x)$ and of the Lommel function $S_{-\gamma-1,\gamma}(x)$

$$y(x) = A\, J_\gamma(x) + B\, Y_\gamma(x) + S_{-\gamma-1,\gamma}(x), \tag{5}$$

437

J. Henrard and S. Ferraz-Mello (eds.), *Impact of Modern Dynamics in Astronomy*, 437–438.

where $A$, $B$ are arbitrary constants. The time equation (3) is reduced to a quadrature, but it does not seem to be expressible by known special functions in a closed form.

Even without the knowledge of the explicit dependence of $x(t)$ the orbit solution (5) can lead to some interesting conclusions. Klačka and Kaufmannová (1992) observed that the eccentricity of some orbits in Poynting-Robertson problem can grow systematically. Using equations (3) and (5), and taking the limits at $x \to 0$, one can verify, that osculating eccentricity tends asymptotically to $e = 1$ for **all** orbits in the general problem (1). The true anomaly $f$ is a fast variable only at the initial stage of the orbital evolution, when the eccentricity decreases. But once the eccentricity starts increasing, $f$ tends slowly towards the asymptotic value $f = \frac{3}{2}\pi$ (Breiter and Jackson 1998). Another interesting conclusion is that the classical solution for the Poynting-Robertson problem of Wyatt and Whipple (1950) should be used with caution. The quasi-integral $p\,e^{-4/5}$, which is the cornerstone of this solution, reveals a significant secular trend during the final stage of motion.

A definite advantage of the solution (5) is its general validity for all types of motion, including capture or escape problems from open and closed orbits.

## References

Breiter S., Jackson A. A.: 1998, 'Unified analytical solutions to two-body problems with drag', *MNRAS*, **299**, 237-243.

Klačka J., Kaufmannová J.: 1992, 'Poynting-Robertson Effect: "Circular" Orbit', *Earth, Moon and Planets*, **59**, 97-102.

Mavraganis A. G., Michalakis D. G.: 1994, 'The Two-Body Problem with Drag and Radiation Pressure', *Celest. Mech. & Dynam. Astron.*, **58**, 393-403.

Mittleman D., Jezewski D.: 1982, 'An Analytic Solution to the Classical Two-Body Problem with Drag', *Celest. Mech.*, **28**, 401-413.

Poynting J. H.: 1903, 'Radiation in the Solar System: its Effect on Temperature and its Pressure on Small Bodies', *Phil. Trans. Roy. Soc.*, **A 202**, 525-552.

Robertson H. P.: 1937, 'Dynamical Effects of Radiation in the Solar System', *MNRAS*, **97**, 423-438.

Watson G. N.: 1958,*A Treatise on the Theory of Bessel Functions*, Cambridge Univ. Press, Cambridge.

Wyatt S. P., Whipple F. L.: 1950, 'The Poynting-Robertson Effect on Meteor Orbits', *ApJ*, **111**, 558-565.

# TWO CONJECTURES OF G.D. BIRKHOFF

CHRISTOPHER K. MCCORD and KENNETH R. MEYER*
*Department of Mathematical Sciences, University of Cincinnati, Cincinnati, Ohio
45221-0025,
E-mail:CHRIS.MCCORD@UC.EDU and KEN.MEYER@UC.EDU*

The spatial (planar) three-body problem admits the ten (six) integrals of energy, center of mass, linear momentum and angular momentum. Fixing these integrals defines an eight (six) dimensional algebraic set called the integral manifold, $\mathfrak{M}(c, h)$ ($\mathfrak{m}(c, h)$), which depends on the energy level $h$ and the magnitude $c$ of the angular momentum vector. The seven (five) dimensional reduced integral manifold, $\mathfrak{M}_R(c, h)$ ($\mathfrak{m}_R(c, h)$), is the quotient space $\mathfrak{M}(c, h)/SO_2$ ($\mathfrak{m}(c, h)/SO_2$) where the $SO_2$ action is rotation about the angular momentum vector. We want to determine how the geometry or topology of these sets depends on $c$ and $h$. It turns out that there is one bifurcation parameter, $\nu = -c^2 h$, and nine (six) special values of this parameter, $\nu_i, i = 1, \ldots, 9$.

At each of the special values the geometric restrictions imposed by the integrals change, but one of these values, $\nu_5$, does not give rise to a change in the topology of the integral manifolds $\mathfrak{M}(c, h)$ and $\mathfrak{M}_R(c, h)$. The other *eight special values give rise to nine different topologically distinct cases.* We give a complete description of the geometry of these sets along with their homology. These results confirm some conjectures and refutes several others.

Birkhoff in his classic book *Dynamical Systems*, (1927) states as a theorem that these sets change their topological type only at values of $\nu$ corresponding to relative equilibrium solutions. This is the first "Birkhoff Conjecture". In the seventies Easton, Iacob and Smale (Easton 1971; Iacob, 1973; Smale, 1970a, 1970b) prove this theorem for the planar problem, but in (McCord *et al.*, 1998) we prove that there are other bifurcation values due to "critical points at infinity". Thus the first Birkhoff conjecture is true for the planar problem, but false for the spatial problem.

In the same book Birkhoff notes that $\mathfrak{m}_R(c, h)$ is a codimension two invariant set in $\mathfrak{M}_R(c, h)$ and wonders if $\mathfrak{m}_R(c, h)$ is the boundary of a general cross section for the flow on $\mathfrak{M}_R(c, h)$. This is the second "Birkhoff conjecture". If this were to be the case the Euler-Poincaré characteristics of $\mathfrak{M}_R(c, h)$ and $\mathfrak{m}_R(c, h)$ would have to agree (McCord and Meyer, 1998). Our computations of the homologies show this not to be the case and so the second Birkhoff conjecture is false also.

We not only study the integral manifolds, but also the six dimensional Hill's regions, $\mathfrak{H}(c, h)$, and the five dimensional reduced Hill's region

$$\mathfrak{H}_R(c, h) = \mathfrak{H}(c, h)/SO_2.$$

* This research partially supported by grants from the National Science Foundation and the Taft Foundation.

J. Henrard and S. Ferraz-Mello (eds.), Impact of Modern Dynamics in Astronomy, 439–440.

The Hill's region is the projection of the integral manifold onto position space. We determine the homotopy type of the reduced Hill's region, and the homology groups of $\mathfrak{H}(c, h)$ and $\mathfrak{H}_R(c, h)$. This does not detect all the changes in the homeomorphism type of $\mathfrak{H}(c, h)$, so we investigate the topology of the boundary points of these Hill's regions. From this finer analysis, we conclude that these Hill's regions undergo bifurcations at eight of the special values of the bifurcation parameter, but none of the topological invariants that we calculate detect a bifurcation at the parameter value $\nu_5$.

# References

A. Albouy: 1993, Integral manifold of the N-body problem, *Invent. Math.*, **14**, 463-488.

G.D. Birkhoff: 1927, *Dynamical Systems* , Amer. Math. Soc., Providence, RI.

R. Easton: 1971, Some topology of the three-body problem, *J. Dif. Eqs.*, **10**, 371-377.

A. Iacob: 1973, *Metode Topologice î Mecanica Clasică*, Editura Academiei Socialiste România, Bucure sti,.

C. McCord, K. Meyer and Q. Wang: 1998, The integral manifolds of thre three-body problem, *Mem. Amer. Math. Soc*, **628**, 1-91.

C. K. McCord and K. R. Meyer: 1998, Cross sections in the three-body problem, to appear.

S. Smale: 1970a, Topology and mechanics,*Invent. Math.*, **10**, 305-331.

S. Smale: 1970b, Topology and mechanics II,*Invent. Math.*, **11**, 45-64.

# THREE-BODY STABILITY CRITERIA

J.R. DONNISON

*Department of Mathematical and Computing Sciences,*
*Goldsmiths College, University of London, New Cross, London SE14 6NW, UK*

Progress has been made in understanding the stability of hierarchical three-body systems where the third body moves on an approximately Keplerian orbit about the centre of mass of the binary, at a distance large compared to the binary separation. Harrington (1968,1969) showed analytically that provided the third body was sufficiently distant from the binary no secular terms appeared in the semi-major axis and the system was stable. Harrington (1972,1975,1977) established numerically a critical minimum separation distance (or period) for a stable system in terms of the masses, unaffected by the relative inclinations of the orbits, except for angles close to $90^0$. Most subsequent investigations have therefore used planar configurations. Graziani & Black (1981), Black (1982) and Pendleton & Black (1983) again using long-term integration of the orbits obtained a criterion for high and low mass binaries. Donnison & Mikulskis (1992,1994,1995) carried out numerical integrations on prograde, retrogade, planetary and stellar triple systems and found for prograde systems very good quantitative agreement with the $c^2 H$ method. Eggleton & Kieselva (1995) suggested a critical distance ratio approximation determined by the masses in the system. Systems with eccentric orbits are covered using the period ratio determined by Kepler's third law.

An analytic approach for determining the stability of coplanar three-body systems was outlined by Golubev (1967,1968) and independently by Szebehely & Zare (1977) and Marchal & Saari (1975). This extends the toplogy of the zero-velocity surfaces of Hill to general coplanar three-body motion and compares $c^2 H$ ($c$ is the angular momentum, $H$ the energy of the system) for a particular system with the critical values at the collinear Lagrange points. This gives a sufficient condition for stability against exchange of the outer mass and the binary masses. A new analytic criterion was recently developed by Aarseth & Mardling using chaos theory.

To compare the various criteria consider where all the masses are equal and both the inner and outer orbits are circular, then the critical distance ratio $(q_{out}/a_{in})_0$ for the different criteria are shown below ($q_{out}$ is the closest approach of the third mass, $M_3$, to the centre of mass of the binary and $a_{in}$ is the semi-major axis of the binary orbit).

| Harrington | Eggleton & Kieselva | Aarseth & Mardling | Szebehely & Zare | Donnison & Mikulskis | Graziani & Black |
|---|---|---|---|---|---|
| 3.5 | 3.12 | 3.3 | 3.2 | 3.3 | 5.0 |

Even for this critical case the various criteria give diverse results with the majority lying in the range $3.1 \leq (q_{out}/a_{in})_0 \leq 3.5$, with one of 5. For a more detailed comparison consider the criteria when $M_1 > M_2$ and $M_2 = M_3$, the orbits remaining circular. The separation into two predicted ranges is evident. The most

*J. Henrard and S. Ferraz-Mello (eds.), Impact of Modern Dynamics in Astronomy, 441–442.*
© 1999 *Kluwer Academic Publishers.*

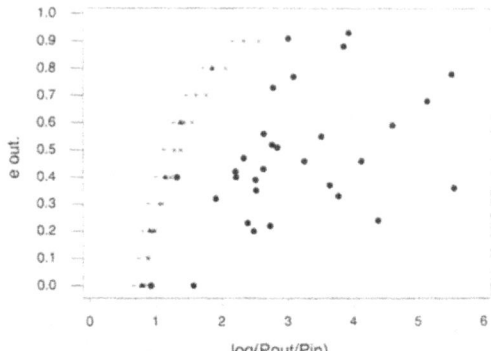

Triple systems: solid circle data; + Eggleton et al; x Aarseth et al ; * Harrington; solid triangle Donnison et al.

Fig. 1. $e_{out}$ against the log of the period ratio for triple systems (data from Fekel, 1981 and Chambliss, 1922). The equal mass criteria curves are shown superimposed.

likely reason for a variety of values is the definition and interpretation of stability in the manner of Laplace, that is the orbits of the masses are bounded so that the orbital elements (the semi-major axes and eccentricities) show no secular or large periodic variations during the time covered by the integrations. At high values of period ratio the criteria tend to converge, and most are also similar at small $e_{out}$ but significantly different from that of Black.

Multiple star systems can be divided into Trapezium systems with three or more stars whose separations are large and roughly equal and are dynamically unstable and hierarchical systems consisting of masses whose successive separations increase by large factors and can be described by a series of two-body motions. About 5-15% of all stellar systems are at least triple and nearly all of the observed multiple systems are hierarchical. Triple systems have recently been identified in open clusters. A plot of $e_{out}$ against the log of the period ratio for triple systems with spectroscopic binaries (spectroscopic-visual triples) the data being taken from Fekel (1981) and Chambliss (1992) are shown in fig.1 with the equal mass criteria curves shown superimposed. The real systems all lie to the right of the values given by the various criteria and should be stable.

## References

Aarseth,S.J.& Mardling,R.A.:1998 (personal communication).
Chambliss C.R.:1992, *PASP*, **104**,663
Donnison J.R. & Mikulskis D.F.:1992, *MNRAS*, **254**, 21.
Donnison J.R. & Mikulskis D.F.:1994, *MNRAS*, **266**, 25.
Donnison J.R. & Mikulskis D.F.:1995, *MNRAS*, **272**, 1.
Eggleton P.P., Kiseleva L.G.:, 1995, *ApJ*, **455**, 640.
Fekel F.C.:1981, *ApJ*, **246**, 879
Golubev V.G.: 1967, *Doklady Akad. Nauk SSSR*, **12**, 529.
Golubev V.G.: 1968, *Doklady Akad. Nauk SSSR*, **13**, 373.
Harrington, R.S.:1968, *Astr.J.*, **73**, 190.
Harrington, R.S.:1969, *Celest.Mech.*, **1**, 200.
Harrington, R.S.:1972, *Celest.Mech.*, **6**, 322.
Harrington, R.S.:1975, *Astr.J.*, **80**, 1081.
Harrington, R.S.:1977, *Astr.J.*, **82**, 753.
Marchal,C. & Saari,D.G.:1975, *Celest.Mech.*,**12**, 115.
Szebehely,V. & Zare,K.:1977, *A& A*, **58**, 135.

# NEKHOROSHEV STABILITY IN

## QUASI–INTEGRABLE

## DEGENERATE HAMILTONIAN SYSTEMS

MASSIMILIANO GUZZO

*Università di Padova, Dipartimento di Matematica Pura e Applicata*
*Via G. Belzoni 7, 35131 Padova, Italy*

Many classical problems of Mechanics can be studied regarding them as perturbations of integrable systems; this is the case of the fast rotations of the rigid body in an arbitrary potential, the restricted three body problem with small values of the mass–ratio, and others. However, the application of the classical results of Hamiltonian Perturbation Theory to these systems encounters difficulties due to the presence of the so-called 'degeneracy'. More precisely, the Hamiltonian of a quasi–integrable degenerate system looks like

$$H(I, \varphi, p, q) = h(I) + \varepsilon f(I, \varphi, p, q),$$ (1)

where $(I, \varphi) \in U \times \mathbf{T}^n$, $U \subseteq \mathbf{R}^n$, are action–angle type coordinates, while the degeneracy of the system manifests itself with the presence of the 'degenerate' variables $(p, q) \in \mathcal{B} \subseteq \mathbf{R}^{2m}$. The KAM theorem has been applied under quite general assumptions to degenerate Hamiltonians (Arnold, 1963), while the Nekhoroshev theorem (Nekhoroshev, 1977) provides, if $h$ is convex, the following bounds: there exist positive $\varepsilon_0, a_0, t_0$ such that if $\varepsilon < \varepsilon_0$ then $|I(t) - I(0)| \leq a_0 \varepsilon^{\frac{1}{2n}}$ if $|t| \leq \min\{T_e, t_0 \exp(\varepsilon_0/\varepsilon)^{\frac{1}{2n}}\}$ where $T_e$ is the escape time of the solution from the domain of (1). An escape is possible because the motion of the degenerate variables can be bounded in principle only by $(\dot{p}, \dot{q}) = \mathcal{O}(\varepsilon)$, and so over the time $\exp(\varepsilon_0/\varepsilon)^{\frac{1}{2n}}$ they can experience large variations. Therefore, there is the problem of individuating which assumptions on the perturbation and on the initial data allow to control the motion of the degenerate variables over long times. The main assumptions we consider are: (I) the so–called secular Hamiltonian of $H$, that is

$$\mathcal{K}(I, p, q) = h(I) + \frac{\varepsilon}{(2\pi)^n} \int_0^{2\pi} \cdots \int_0^{2\pi} f(I, \varphi, p, q) d\varphi_1 \ldots d\varphi_n$$

is integrable, or quasi–integrable. Then, one can choose the $(p, q) \in V \times \mathbf{T}^m$ (with $V \subseteq \mathbf{R}^m$) to be action–angle coordinates and there is a second small parameter $\eta$ such that $\mathcal{K} = h(I) + \varepsilon(K_0(I, p) + \eta K_1(I, p, q))$. The problem poses even when the secular Hamiltonian is integrable. (II) the secular Hamiltonian provides convexity with respect to all actions $I, p$ (though with a convexity constant of order $\varepsilon$). (III) the perturbation is analytic in a complex neighbourhood of the real domain. We can state now the result

*J. Henrard and S. Ferraz-Mello (eds.), Impact of Modern Dynamics in Astronomy, 443–444.*

**Theorem** *Let $H$ be as in (1) satisfying* I, II, III. *There exist positive $\varepsilon_*$, $\eta_*$, $r_*$, $R_0$, $a$, $T_*$ such that if $\varepsilon \leq \varepsilon_*$ and $\eta \leq \eta_*$ then any motion with initial datum such that $|\nu \cdot \omega(I(0))| > \varepsilon^a$ for any $\nu \in \mathbf{Z}^n$ with*

$$|\nu| \leq r_* \ln \frac{1}{\varepsilon} \, ,$$

*satisfies*

$$|I(t) - I(0)| \leq \varepsilon^{\frac{5}{6}} R_0 \zeta^{\frac{1}{2n(n+m)}} \, , \quad |p(t) - p(0)| \leq R_0 \zeta^{\frac{1}{2n(n+m)}} \, ,$$

*where $\zeta = \max \left\{ \varepsilon^{\frac{n(n+m)}{n(n+m)+3n-2}} , \eta \right\}$, for any time $t$ with*

$$|t| \leq T_* \sqrt{\zeta} \min \left\{ \exp \left[ \left( \frac{\varepsilon_*}{\varepsilon} \right)^{\frac{1}{2n(n+m)+6n-4}} \right] , \exp \left[ \left( \frac{\eta_*}{\eta} \right)^{\frac{1}{2n(n+m)}} \right] \right\} .$$

An analogous stability result was proved for the first time in ( Guzzo and Morbidelli, 1997 – see also Morbidelli and Guzzo, 1997) in connection with stability problems in the Asteroid Main Belt. The above theorem is proved with all details in (Guzzo, 1998).

## References

Arnold, V.I.: 1963, "Small denominators and problems of stability of motion in classical and celestial mechanics.", *Russ. Math. Surveys*, **18**, 85–191.

Nekhoroshev, N., N.: 1977, "Exponential estimates of the stability time of near–integrable Hamiltonian systems.", *Russ. Math. Surveys*, **32**, 1-65.

Morbidelli, A., and Guzzo, M.: 1997, "The Nekhoroshev theorem and the asteroid belt dynamical system.", *Celest. Mech.*, **65**, 107-136.

Guzzo, M. and Morbidelli, A.: 1997, "Construction of a Nekhoroshev like result for the asteroid belt dynamical system.", *Celest. Mech.*, **66**, 255–292.

Guzzo, M.: 1998, :"The Nekhoroshev stability in quasi–integrable degenerate Hamiltonian systems". Preprint.

# NEKHOROSHEV–STABILITY OF $L_4$ and $L_5$ IN THE SPATIAL RESTRICTED PROBLEM

GIANCARLO BENETTIN, FRANCESCO FASSÒ and MASSIMILIANO GUZZO

*Università di Padova, Dipartimento di Matematica Pura e Applicata, Via G. Belzoni 7, 35131 Padova, Italy*

The Lagrangian equilateral points $L_4$ and $L_5$ of the restricted circular three–body problem are elliptic for all values of the reduced mass $\mu$ below Routh's critical mass $\mu_R \approx .0385$. In the spatial case, because of the possibility of Arnold diffusion, KAM theory does not provide Lyapunov–stability. Nevertheless, one can consider the so–called 'Nekhoroshev–stability': denoting by $d$ a convenient distance from the equilibrium point, one asks whether

$$d(0) \leq \epsilon \quad \Longrightarrow \quad d(t) \leq \epsilon^a \quad \text{for} \quad |t| \leq \exp \epsilon^{-b}$$

for any small $\epsilon > 0$, with positive $a$ and $b$. Until recently this problem, as more generally the problem of Nekhoroshev–stability of elliptic equilibria of Hamiltonian systems, was studied only under some arithmetic conditions on the frequencies, and thus on $\mu$ (see e.g .Giorgilli, 1989). Our aim was instead considering all values of $\mu$ up to $\mu_R$. As a matter of fact, Nekhoroshev–stability of elliptic equilibria, without any arithmetic assumption on the frequencies, was proved recently under the hypothesis that the fourth order Birkhoff normal form of the Hamiltonian exists and satisfies a 'quasi–convexity' assumption (Fassò *et al*, 1998; Guzzo *et al*, 1998; Niedermann, 1998). However, in the case of $L_4$ and $L_5$ such assumption is not satisfied for any $\mu < \mu_R$. Therefore, our study rests in a crucial way on two extensions of the above result:

(i) The first extension replaces quasi–convexity by a weakened requirement, called *directional quasi–convexity* (DQC), which is specific for the case of an equilibrium point, and has no analogue in the general Nekhoroshev theorem. This property consists in testing quasi–convexity not in the whole plane of fast drift, but only in its intersection with the action space of the elliptic equilibrium, that is, the 'first octant' where all actions are nonnegative. DQC is a natural hypothesis for the case of elliptic equilibria, which appears to play the same role as quasi–convexity in Nekhoroshev theorem, and leads to good stability estimates (e.g. $b = 1/n$).

(ii) The second extension relaxes instead quasi–convexity to a simple 'steepness' condition on the 3–jet of the sixth order Birkhoff normal form, much in the line of Nekhoroshev's original theorem. For a system with three degrees of freedom it is necessary that the Birkhoff normal form can be constructed up to order eight at least, and we obtain $b = 1/20$; this result improves whenever it is possible to construct higher order normal forms.

Precise statements of these results are given in (Bennetin *et al.*, 1998), where one also finds the application to $L_4$ and $L_5$. The result of such an analysis, which

*J. Henrard and S. Ferraz-Mello (eds.), Impact of Modern Dynamics in Astronomy*, 445–446.

was performed by numerically constructing the Birkhoff normal forms, is that the system is DQC for all $\mu$ outside a relatively small interval $[\mu_1, \mu_2]$, with $\mu_1 \approx .0109$ and $\mu_2 \approx .0164$, but at one point, namely $\mu_{(1,2,0)} \approx .0243$ where the Birkhoff normal form does not exist because of the resonance $(1, 2, 0)$. The hypotheses for the applicability of extension (ii) are satisfied everywhere in the interval $[\mu_1, \mu_2]$, but at three points $\mu_{(1,3,0)} \approx .0135$, $\mu_{(0,3,1)} \approx .0149$, and $\mu_{(3,3,-2)} \approx .0116$ where certain resonances prevent the existence of normal forms of order eight, and at a point $\mu_3 \approx .01478$ where the considered steepness property is violated. Summarizing, $L_4$ and $L_5$ are Nekhoroshev–stable for all values of $\mu < \mu_R$ but possibly five isolated values. However, the estimated stability times are not uniform in $\mu$, depending on whether the system is DQC ($b = 1/3$) or not ($b = 1/20$, in the absence of special properties); it may be interesting to observe that the reduced mass of the Earth–Moon system falls inside the interval $[\mu_1, \mu_2]$ so that, on the basis of our analysis, rather poor stability properties might be found.

## References

G. Benettin, F. Fassò and M. Guzzo: 1998, *Nekhoroshev–stability of L4 and L5 in the spatial restricted three–body problem*, to appear in Regular and Chaotic Dynamics. Postscript file available at http://www.math.unipd.it:80/~fasso/#Publications.

F. Fassò, M. Guzzo, G. Benettin: 1998, Nekhoroshev–stability of elliptic equilibria of Hamiltonian systems, *Comm. in Math. Phys.*, **197**, 347-360.

A. Giorgilli, A. Delshams, E. Fontich, L. Galgani, C. Simó: 1989, *J. Diff. Eq.*, **77**, 167-198

M. Guzzo, F. Fassò and G. Benettin: 1998, *Mathemathical Physics Electronic Journal*, **4**, Paper 1

L. Niedermann: 1998, Nonlinear stability around an elliptic equilibrium point in an Hamiltonian system, *Nonlinearity*, **11**, 1465-1479.

# EFFECTIVE LYAPUNOV NUMBERS AND CORRELATION DIMENSIONS IN A 3-D HAMILTONIAN SYSTEM

K. TSIGANIS, A. ANASTASIADIS AND H. VARVOGLIS

*Section of Astrophysics, Astronomy & Mechanics, Department of Physics, Aristotle University of Thessaloniki, 54006 Thessaloniki, GREECE*

Transport in Hamiltonian systems, in the case of strong perturbation, can be modeled as a *diffusion process*, with the diffusion coefficient being constant and related to the maximal Lyapunov number (Konishi 1989). In this respect the relation found by Lecar et al. (1992) between the escape time of asteroids, $T_E$, and the Lyapunov time, $T_L$, can be easilly recovered (Varvoglis & Anastasiadis 1996). However, for moderate perturbations, chaotic trajectories may have a peculiar evolution, owing to *stickiness* effects or migration to adjacent stochastic regions. As a result, the function $\chi(t)$, which measures the exponential divergence of nearby trajectories, changes behaviour within different time intervals. Therefore, trajectories may be divided into segments, $i = 1, ..., n,$ each one being assigned an "Effective" Lyapunov Number (ELN), $\lambda_i = \chi(t_i)$.

We study trajectories in a well known 3-D Hamiltonian system (Contopoulos & Barbanis 1989)

$$H = \frac{1}{2}(p_x{}^2 + p_y{}^2 + p_z{}^2) + \frac{1}{2}Ax^2 + \frac{1}{2}By^2 + \frac{1}{2}Cz^2 - \epsilon x z^2 - \eta y z^2 = h \quad (1)$$

By noting $H_0$ the sum of the first four terms of the Hamiltonian at every timestep $t_i$, one constructs the "quasi-integral" timeseries, $H_0(t_i)$, which characterizes a trajectory, in analogy to a canonical coordinate timeseries.

For every trajectory we calculate $\chi(t)$ along with $H_0(t_i)$. After defining an appropriate number of segments, we calculate the value of the correlation dimension, $D^{(2)}$, for each of them. For this we calculate the correlation integral, $C_2(r)$, through a formula proposed by Isliker (1994), using state vectors generated from $H_0(t_i)$ via the time-delay reconstruction scheme (Takens 1981). Practical details concerning this subject can be found in Isliker (1994).

During our numerical experiments we have found trajectories exhibiting different types of behaviour; two of them are shown in Fig. (1). The ELN and $D^{(2)}$ values for all the segments studied are shown in the bottom-right panel of Fig. (1), from which a positive correlation between the ELN's and the $D^{(2)}$'s can be deduced. This fact indicates that the statistical properties of transport are not the same throughout the phase space (see also Zaslavsky 1994) but depend on the measure of surviving invariant sets within this region. In this respect, $D^{(2)}$ can be used as a comparative measure for the stochasticity level of different regions.

The two trajectories shown here demonstrate a behaviour similar to that of *stable chaos* (Milani & Nobili 1992). The ELN of Trajectory II (bottom of Fig. 1) has a larger value than any of the first three segments ($t \leq 5 \times 10^5$) of Trajectory I

*J. Henrard and S. Ferraz-Mello (eds.), Impact of Modern Dynamics in Astronomy, 447–448.*

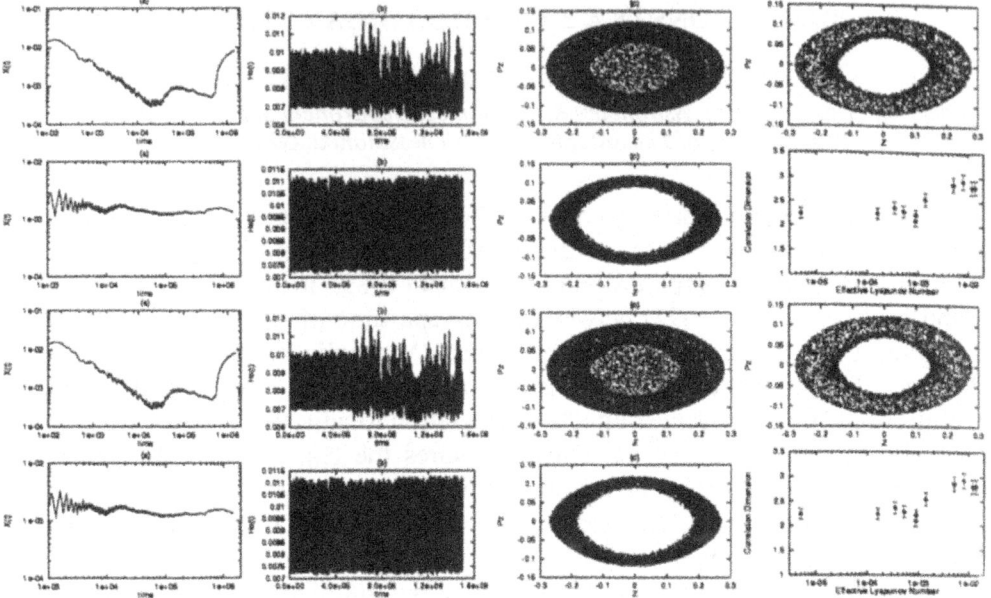

Fig. 1. Top: $\chi(t)$, $H_0(t)$ and $z - p_z$ projections for trajectory $I$. The last figure is the projection up to $t = 5 \times 10^5$. Bottom: The same plots for trajectory $II$. Bottom right: graphic representation of the results found the relation between ELN and $D^{(2)}$ for several trajectory segments.

(top) and, yet, the latter is the one which manages to escape towards a more chaotic region. This indicates that $T_E$ must depend also on some quantity other than the Lyapunov number. If we assume a dynamical system whose action space is a "box" divided into two regions by an imperfect barrier, then $T_E$ should be inversely proportional to the "collision frequency" (or $\lambda$) as well as to the measure of the "holes" (or the Hausdorff codimension) of the barrier (see also Gaspard & Baras 1995). Accordingly, in our model, $D^{(2)}$ changes when motion takes place close to a barrier (low ELN values). We believe that this quantity may prove important for a statistical description of chaotic motion in problems of astronomical interest.

## References

Contopoulos G. and Barbanis B., 1989, A& A 222, 329
Gaspard P. and Baras F., 1995, Phys. Rev. E 51, 6, 5332
Isliker H., 1994, Ph.D. Thesis, ETH: No 10495, Swiss Federal Institute of Technology, Zürich
Konishi T., 1989, Prog. Theor. Phys. Suppl. 98, 19
Milani A. and Nobili A. M., 1992, Nature 357, 569
Takens F., 1981, Lecture Notes in Mathematics, vol. 898 (Springer, New York, 1981)
Varvoglis H. and Anastasiadis A., 1996, AJ 111, 1718
Zaslavsky G., (1994), Phys. D 76, 110

# A SYMPLECTIC INTEGRATION SCHEME THAT ALLOWS CLOSE ENCOUNTERS BETWEEN MASSIVE BODIES

J. E. CHAMBERS

*Armagh Observatory, Armagh, North Ireland, UK*

Mixed-variable symplectic integrators provide a fast, moderately accurate way to study the long-term evolution of a wide variety of N-body systems (Wisdom & Holman 1991). They are especially suited to planetary and satellite systems, in which a central body contains most of the mass. However, in their original form, they become inaccurate whenever two bodies approach one another closely. Here, I will show how to overcome this difficulty using a hybrid integrator that combines symplectic and conventional algorithms.

A symplectic integrator works by splitting the Hamiltonian, $H$, for an N-body system, into two or more parts $H = H_0 + H_1 + \cdots$, where $\epsilon_i = H_i/H_0 \ll 1$ for $i = 1, 2 \ldots$. An integration step consists of several substeps, each of which advances the system due to the effect of one part of the Hamiltonian only. The error incurred over the whole step is $\sim \epsilon \tau^{n+1}$, where $\tau$ is the timestep, $n$ is the order of the integrator, and $\epsilon$ is the largest of $\epsilon_i$.

A symplectic algorithm is efficient provided that the $\epsilon$ factors are small. In the planetary system, this is usually achieved by making $H_0$ the unperturbed Keplerian motion of the planets about the Sun, and $H_1$, $H_2$ etc. the perturbations between planets. For example, using mixed coordinates (heliocentric positions and barycentric velocities), $H$ is split up as

$$H_0 = \sum_{i=1}^{N} \left( \frac{p_i^2}{2m_i} - \frac{Gm_\odot m_i}{r_{i\odot}} \right)$$

$$H_1 = -G \sum_{i=1}^{N} \sum_{j=i+1}^{N} \frac{m_i m_j}{r_{ij}}$$

$$H_2 = \frac{1}{2m_\odot} \left( \sum_{i=1}^{N} \mathbf{p}_i \right)^2 \tag{1}$$

where $N$ is the number of planets, $m$ denotes mass, and $r$, $p$ are position and momentum respectively. Note that each part of the Hamiltonian can be integrated analytically in the absence of the others, and $H_0$ contains all of the large terms, provided that the planets are widely separated.

Now consider a close encounter between bodies $a$ and $b$. During the encounter, the distance $r_{ab}$ is small, and the corresponding term in $H_1$ is large. This means that $\epsilon_1$ is no longer small and the integrator becomes inaccurate. An approximate solution to this difficulty is to transfer the offending term from $H_1$ to $H_0$ for the duration of the encounter. This ensures that $\epsilon_1$ is always small. However, each time

*J. Henrard and S. Ferraz-Mello (eds.), Impact of Modern Dynamics in Astronomy,* 449–450.

$H_0$ and $H_1$ are changed in this way, the system undergoes a shift in energy, and the integrator's symplectic property is lost.

A better solution is to split each of the interaction terms between $H_0$ and $H_1$ as follows

$$H_0 = \sum_{i=1}^{N} \left( \frac{p_i^2}{2m_i} - \frac{Gm_\odot m_i}{r_{i\odot}} \right) - G \sum_{i=1}^{N} \sum_{j=i+1}^{N} \frac{m_i m_j}{r_{ij}} [1 - K(r_{ij})]$$

$$H_1 = -G \sum_{i=1}^{N} \sum_{j=i+1}^{N} \frac{m_i m_j}{r_{ij}} K(r_{ij}) \tag{2}$$

where the function $K$ is chosen so that $K \rightarrow 0$ when $r_{ij}$ is small, and $K \rightarrow 1$ when $r_{ij}$ is large. This ensures that $\epsilon_1$ is always small, without requiring that terms move from one part of the Hamiltonian to another.

When all of the separations $r_{ij}$ are large, $H_0$ can be advanced analytically as before (since $1 - K = 0$). If two bodies undergo a close encounter, the terms in $H_0$ due to these objects must be integrated numerically, but all the remaining terms can still be advanced analytically. By trial and error, I find that a good expression for $K$ is

$$K = \begin{cases} 0 & \text{for } y < 0 \\ y^2/(2y^2 - 2y + 1) & \text{for } 0 < y < 1 \\ 1 & \text{for } y > 1 \end{cases}$$

where

$$y = \left( \frac{r_{ij} - 0.1\, r_{crit}}{0.9\, r_{crit}} \right)$$

and $r_{crit}$ is the larger of 3 Hill radii and $0.5\tau v_{max}$, where $v_{max}$ is the maximum likely orbital velocity of any of the objects.

## References

Wisdom J., Holman M.: 1991, *Astron.J.*, **102**, 1528.

# ANALYTICAL APPROXIMATION
## OF THE DISSIPATIVE STANDARD MAP

ALESSANDRA CELLETTI

*Dipartimento di Matematica, Università di L'Aquila, I-67010 L'Aquila (Italy)*

and

GABRIELLA DELLA PENNA and CLAUDE FROESCHLÉ

*Observatoire de Nice, BP 229, F-06304 Nice Cedex 4 (France)*

## 1. Analytical Approximation

We investigate the dynamics of a dissipative standard mapping defined by the equations

$$y' = y + \varepsilon \sin(x) - \alpha(y - \psi)$$
$$x' = x + y',$$

$$(1)$$

where $y \in \mathbf{R}$, $x \in \mathbf{T}$ and $\varepsilon$ is a real parameter; we refer to $0 < \alpha < 1$ as the "dissipative parameter" and to $\psi$ as the "dissipative coefficient" ($\varepsilon = \alpha = 0$ provides an integrable mapping). Notice that the dynamics is contractive, since the jacobian of the above mapping equals to $1 - \alpha$. In particular, we want to compare (see Celletti *et al.*, 1997) the solutions associated to the conservative map (i.e., $\alpha = 0$) with that related to (1) ($\alpha \neq 0$). For simplicity, we consider the case when $\alpha = \varepsilon^2$ and construct explicit approximate solutions to the conservative and dissipative systems, using a suitable parametrization like in (Celletti and Chierchia, 1988).

In particular, we consider the curves described by the parametric equations

$$x = \theta + u(\theta; \varepsilon) + \alpha D w(\theta; \varepsilon)$$
$$y = \omega + u(\theta; \varepsilon) - u(\theta - \omega; \varepsilon) + \alpha[D w(\theta; \varepsilon) - D w(\theta - \omega; \varepsilon)]$$

$$(2)$$

($\theta \in \mathbf{T}$) where $u(\theta; \varepsilon)$ and $w(\theta; \varepsilon)$ are suitable functions. Here $D$ is the operator acting on a function $v(\theta; \varepsilon)$ as

$$D v(\theta; \varepsilon) = v(\theta + \frac{\omega}{2}; \varepsilon) - v(\theta - \frac{\omega}{2}; \varepsilon).$$

Inserting (2) in (1) one gets

$$D^2 \ u \ (\theta; \varepsilon) + \alpha D^3 w(\theta; \varepsilon) = \varepsilon \sin[\theta + u(\theta; \varepsilon) + \alpha D w(\theta; \varepsilon)] \qquad (3)$$
$$- \alpha(\omega - \psi + [u(\theta; \varepsilon) - u(\theta - \omega; \varepsilon)] + \alpha[D w(\theta; \varepsilon) - D w(\theta - \omega; \varepsilon)]).$$

451

J. Henrard and S. Ferraz-Mello (eds.), *Impact of Modern Dynamics in Astronomy*, 451–452.

Using analyticity properties, we expand $u$, $w$ and $\psi$ in Taylor series around $\varepsilon = 0$ as

$$u(\theta; \varepsilon) = \sum_{j=0}^{\infty} u_j(\theta)\varepsilon^j, \quad w(\theta; \varepsilon) = \sum_{j=0}^{\infty} w_j(\theta)\varepsilon^j, \quad \psi = \psi(\varepsilon) = \sum_{j=0}^{\infty} \psi_j\varepsilon^j; \quad (4)$$

inserting the above expansions in (3) and equating same orders of $\varepsilon$ one obtains explicit formulae for the functions $u_j$, $w_j$ and $\psi_j$. A finite truncation of the series expansions (4) provides an approximate solution of the equations of motion (see Celletti et al., 1997, for more details).

## 2. Results

We analyzed the behaviour of conservative invariant curves with irrational frequency $\omega$ and the corresponding dissipative solutions as $\varepsilon$ is varied. The approximate solutions look similar for low values of $\varepsilon$. In most cases the dynamics showed marked differences as $\varepsilon$ approaches the critical break–down value $\varepsilon_c$, due to analyticity properties as $\varepsilon < \varepsilon_c$ which are lost as $\varepsilon$ approaches the critical break–down threshold. Few exceptions were found in which conservative and dissipative dynamics are different for values much bigger than $\varepsilon_c$. A higher order solution should be computed in order to investigate these peculiar cases. We report in Table 1 the values $\varepsilon_c(\omega)$ representing the critical break–down threshold of the invariant curve with frequency $\omega$ and the values $\varepsilon_{CD}(\omega)$ at which the conservative and dissipative curves become different. In most cases these values may be thought as indicators of the onset of chaos. To provide some results, we report the values for the following rotation numbers: $\beta = \frac{\sqrt{5}-1}{2}$, $\omega_1 = \frac{1}{2} + \frac{1}{100+\beta}$, $\omega_2 = \frac{2}{3} + \frac{1}{20+\beta}$, $\omega_3 = 1 - \frac{1}{10+\beta}$, $\omega_4 = 1 - \frac{1}{100+\beta}$.

TABLE I

| Rotation number $\omega$ | $\varepsilon_c(\omega)$ | $\varepsilon_{CD}(\omega)$ |
|---|---|---|
| $\beta$ | 0.971 | 0.32 |
| $\omega_1$ | 0.3 | 0.28 |
| $\omega_2$ | 0.7 | 0.5 |
| $\omega_3$ | 0.4 | 0.4 |
| $\omega_4$ | 0.05 | 0.5 |

## References

Celletti A., Chierchia L.: 1988, Construction of analytic KAM surfaces and effective stability bounds, Commun. in Math. Physics, **118**, 119

Celletti A., Della Penna G., Froeschlé C.: 1997, Analytic approximation of the solution of the dissipative standard map, International Journal of Bifurcation and Chaos, in press.

# SECOND ORDER PERTURBATIONS OF ELLIPTIC ELEMENTS
# WITH RESPECT TO THE INITIAL ONES

F.J. MARCO CASTILLO

*Depto de Matemáticas, Universitat Jaume I, Castellón, SPAIN.*

M.J. MARTÍ NEZ USÓ

*Depto de Matemática Aplicada, ETSII, Universidad Politécnica de Valencia, SPAIN.*

and

J.A. LÓPEZ ORTÍ

*Depto de Matemáticas, Universitat Jaume I, Castellón, SPAIN.*

**Abstract.**
   The following paper is devoted to the theoretical exposition of the obtention of second order perturbations of elliptic elements and is a follow-up of previous papers (Marco *et al.*, 1996; Marco *et al.*, 1997) where the hypothesis was made that the matrix of the partial derivatives of the orbital elements with respect to the initial ones is the identity matrix at the initial instant only. So, we must compute them through the integration of Lagrange planetary equations and their partial derivatives.
   Such developments have been applied to the individual corrections of orbits together with the correction of the reference system through the minimization of a quadratic form obtained from the linearized residual. In this state two new targets emerged:
   1. To be sure that the most suitable quadratic form was to be considered.
   2. To provide a wider vision of the behavior of the different orbital parameters in time.
   Both aims may be accomplished through the consideration of the second order partial derivatives of the elliptic orbital elements with respect to the initial ones.

## 1. First and second order derivatives

Let

$$\dot{\sigma}_i = f_i(\overrightarrow{\sigma}, t) \tag{1}$$

be the $i^{th}$ Lagrange planetary equation, where $\sigma_i$ $1 \leq i \leq 6$ denotes the orbital elements at a given time, say $t$ and let $\sigma_{ij}(t) = \partial\sigma_i/\partial\sigma_j^0$ $(t)$ be its first partial derivative. As it was shown in( Marco *et al.*, 1997) these variables verify the differential equations

$$\frac{d\sigma_{ij}}{dt} = \sum_{m=1}^{6} \left.\frac{\partial f_i}{\partial\sigma_m}\right|_t \sigma_{mj}(t) \tag{2}$$

with the initial conditions $\sigma_{ij}(t_0) = \delta_{ij}$. Also, we denote the second partial derivatives as $\sigma_{ij}^k(t) = \left.\frac{\partial^2\sigma_k}{\partial\sigma_i^0\partial\sigma_j^0}\right|_t$ and its temporal derivatives are given by

$$\frac{d\sigma_{ij}^k}{dt} = \sum_{m=1}^{6} \sigma_{mi}(t) \left.\frac{\partial^2 f_k}{\partial\sigma_m\partial\sigma_r}\right|_t \sigma_{rj}(t) + \sum_{r=1}^{6} \left.\frac{\partial f_k}{\partial\sigma_r}\right|_t \sigma_{ij}^r(t) \tag{3}$$

453

*J. Henrard and S. Ferraz-Mello (eds.), Impact of Modern Dynamics in Astronomy, 453–454.*

with the initial conditions $\sigma_{ij}^k(t_0) = 0$. Finally, the equations (1), (2) and (3) must be integrated all together in order to obtain the desired values $\sigma_{ij}^k(t), \sigma_{ij}(t)$ and $\sigma_i(t)$.

## 2. Third order derivatives of the perturbation function

In the present section we devote our efforts in presenting a formulation to obtain the derivatives of the perturbation function which are necessary to evaluate the derivatives of the Lagrange equations. To compute the first and second derivatives of the perturbation function with respect to the orbital elements we follow (Simon, 1988)

$$\frac{\partial^2 \Re}{\partial \sigma_i \partial \sigma_m} = \vec{V_i}.(\partial^2 \Re).\vec{V_m} + \vec{V_{i,m}}.\partial \vec{\Re} \tag{4}$$

where $\vec{V}$ is the rectangular ecliptic vector of position, $\vec{V_i}$ the partial derivative with respect to the $\sigma_i$ element and $\partial \Re$ and $\partial^2 \Re$ are, respectively, the gradient and the matrix of the derivatives in second order of the perturbation function in ecliptic rectangular coordinates. Considering the derivatives of (4) we obtain $\Re_{kjl} \equiv \frac{\partial^2 \Re_k}{\partial \sigma_j \partial \sigma_l}$

$$\Re_{kjl} = \vec{V_{kjl}}.\partial \vec{\Re} + \vec{V_{kj}}(\partial^2 \Re)\vec{V_l} + \vec{V_{kl}}(\partial^2 \Re)\vec{V_j} + \vec{V_k}(\partial^3 \Re \otimes \vec{V_l})\vec{V_j} + \vec{V_k}(\partial^2 \Re)\vec{V_{jl}} \tag{5}$$

where the following notation is employed

$$\partial_i^2 \Re = \frac{\partial \Re}{\partial x_i} \text{ i=1,2,3} ; \partial^3 \Re = \left[\partial_1^2 \Re, \partial_2^2 \Re, \partial_3^2 \Re\right] ; \vec{V_{kj}} = \frac{\partial \vec{V_k}}{\partial \sigma_j}; \vec{V_{kjl}} = \frac{\partial \vec{V_{kj}}}{\partial \sigma_l} \tag{6}$$

## References

F.J. Marco, J.A. López, M.J. Martínez: 1996, A time-dependent extension to Brouwer's method of orbital elements correction, *Dynamics, Ephemerides and Astrometry of the Solar System*, pp. 199-202

F.J. Marco, J.A. López, M.J. Martínez: 1997, Temporal variations of perturbed elliptic elements: A semi-analytical approach, *Celest. Mec.*, **68**,193-198

Simon, J.L.: 1988, Calcul des dérivées premières et secondes des équations de Lagrange par analyse harmonique, *Astron. Astrophys.*, **175**, 303-308

# ZONAL HARMONICS OF THE GRAVITY FIELD

# IN DEF–VARIABLES

IGNACIO APARICIO AND LUIS FLORÍA

*Grupo de Mecánica Celeste I, Departamento de Matemática Aplicada a la Ingeniería,*
*ETSII. Universidad de Valladolid. E – 47 011 Valladolid, Spain.*

To take advantage of the *linear and regular* formulation and treatment of Celestial Mechanics problems (Kustaanheimo & Stiefel 1965; Stiefel & Scheifele 1971; Deprit, Elipe & Ferrer 1994), Sharaf & Saad (1997) have given an analytical expansion of the Earth's gravitational *zonal potential* in terms of Kustaanheimo–Stiefel (KS) regular elements (Stiefel & Scheifele 1971, §19), with special emphasis on its application to *elliptic–type two–body orbits* and, consequently, using a generalized (elliptic) eccentric anomaly as the independent variable.

Motivated by these and other considerations based on the definition and use of KS elements, and following a treatment similar to that of Stiefel & Scheifele (1971, §19), we develop *element equations corresponding to a DEF–formulation* of the satellite problem under the effect of the zonal potential.

The weakly canonical DEF–transformation to focal–type variables (Deprit, Elipe & Ferrer 1994, §§4.1) was originally devised to *exactly* linearize the equations of motion governing the conservative, spatial Kepler problem, giving it the form of a *4–dimensional harmonic oscillator*. By analogy with Stiefel & Scheifele (1971, §19), the *DEF–elements of the motion* will be constants of integration occurring in the general solution of the unperturbed harmonic oscillator equations generated by the pure Kepler problem in DEF–variables, whereas for the perturbed problem they satisfy a system of first–order differential equations. Elements vary almost linearly if the motion is subjected to weak perturbations. The equations of motion corresponding to the perturbed problem can be treated by the *method of variation of constants*, which leads to the said system of differential equations for these quantities (with a true–like anomaly as the independent variable). In certain cases, these equations must also be *supplemented* by the equations for the variation of other dynamical quantities (e.g., the law of variation of the angular momentum, or the variation of the energy) and by the differential time transformation. All these relations constitute the *element equations*.

We carry out an application of the *focal method* (Burdet 1969, §2) to derive the analytical expression, in terms of elements attached to the linearizing DEF–variables, of any zonal harmonic of the gravitational field created by a central body, and obtain the corresponding equations of motion *for any value of the eccentricity.* To this end, we will follow a variant of the *focal method canonical approach* based on the (weakly) canonical extension of the projective–decomposition point–transformation proposed by Deprit *et al.*

The focal method enjoys the property that the reciprocal of the distance (the quantity $1/r$) is represented by a simple polynomial expression in trigonometric

*J. Henrard and S. Ferraz-Mello (eds.), Impact of Modern Dynamics in Astronomy,* 455–456.

functions of the true–like anomaly, which facilitates the expansion of negative powers of the distance in the form of *Fourier polynomials in that independent variable*. Accordingly, our expression for any zonal harmonic turns out to be simpler than those (truncated, but in principle infinite) proposed by Sharaf & Saad (1997, §3) in terms of an eccentric–like anomaly. As a consequence, we expand the zonal harmonics as Fourier polynomials in the true–like anomaly with coefficients depending on the oscillator DEF–elements, the orbital eccentricity and semi–latus rectum. Our description is not restricted to the case of elliptic–type orbits, and so our developments and results are *uniformly valid* for any type of two–body conic–section orbit.

Research partially supported by Junta de Castilla y León (Grant VA61/98).

## References

Burdet, C. A.: 1969, *J. reine angew. Math.*, **238**, 71–84.
Deprit, A., Elipe, A. & Ferrer, S.: 1994, *Celest. Mech. and Dyn. Astron.*, **58**, 151–201.
Kustaanheimo, P., & Stiefel, E. L.: 1965, *J. reine angew. Math.*, **218**, 204–219.
Sharaf, M. A. & Saad, A. S.: 1997, *Celest. Mech. and Dyn. Astron.*, **66**, 181–190.
Stiefel, E. L. & Scheifele, G.: 1971, *"Linear and Regular Celestial Mechanics."*, Springer–Verlag, Berlin–Heidelberg–New York.

# ON THE STABILITY OF STATIONARY SOLUTIONS OF THE
# TWICE-AVERAGED HILL'S PROBLEM TAKING INTO ACCOUNT
# THE PLANET OBLATENESS

M.A. VASHKOVYAK

*Keldysh Institute of Applied Mathematics, Russian Academy of Sciences, Miusskaya pl., 4,*
*Moscow, 125047 Russia, E-mail: vashkov@applmat.msk.su*

The problem of satellite orbital evolution with the combined influence of a distant perturbing body and the planet oblateness is well known (Laplace, 1805; Lidov, 1962, 1973; Kozai, 1963; Kudielka, 1994, 1997). The case of near-circular orbits is investigated in more details in (Sekiguchi, 1961; Allan and Cook, 1964; Vashkovyak, 1974).

In the full report (Vashkovyak,1998) the stability of the stationary solutions of the secular equations is investigated for elliptic orbits. The analytical and graphical dependences for the stationary values of the satellite orbital elements and the stability conditions are given. The main results are:

- the solution $\cos i = \cos \Omega = \sin \omega = 0, e = e(a)$ is stable in the linear approximation;

- the solution $\sin \Omega = \sin \omega = 0, e = e(a, \beta), i = i(\Omega, \beta)$ is unstable;

- for the solution $\sin \Omega = \cos \omega = 0, e = e(a, \beta), i = i(\Omega, \beta)$ both a region of unstability and a region of linear stability exist.

Here $(a, e, i)$ are the standard notations of the Keplerian elements of the satellite orbit, $\Omega$ is the longitude of accending node, $\omega$ is the argument of pericenter; $\beta$ is the angle between the orbital plane of the perturbing body (as a fundamental plane) and the equatorial plane of the planet.

The comparision with the results both of numerical integrations of the evolutionary system and of the paper (Kudielka, 1997) is carried out in the full report.

## References

Allan R.R., Cook G.E.: 1964, *Proc. Roy. Soc. A*, **280**, 97.
Kozai Y.: 1963, *Publ. Astron. Soc. Japan*, **15**, 301.
Kudielka V.: 1994, *Celest. Mech.*, **60**, 455.
Kudielka V.: 1997, in *The Dynamical Behavior of our Planetary System*, ( Dvorak and Henrard eds), Kluwer A. P., 243.
Laplace P.S.: 1805, Mécanique Céleste, t. IV.
Lidov M.L.: 1962, in *Dynamics of satellites, Symp. Paris*, ( Roy ed), Springer-Verlag, 169.
Lidov M.L.: 1974, in *The Stability of the Solar System and of Small Stellar Systems*, (Kozai ed), 117.
Sekiguchi N.: 1961,*Publ. Astron. Soc. Japan*, **13**, 207.
Vashkovyak M.A.: 1974, *Kosmicheskie Issledovaniya*, **12**, 834.
Vashkovyak M.A.: 1998, *Astronomy Letters*, **24**, 682.

*J. Henrard and S. Ferraz-Mello (eds.), Impact of Modern Dynamics in Astronomy,* 457.

# STUDY OF THE NORMAL VARIATIONAL EQUATION IN AN HOMOGENEOUS FIELD OF DEGREE FIVE

A. MAKHLOUF AND N. DEBBAH

*Institute of mathematics, BP 12 El-Hadhar, University of Annaba, Algeria*

We study the normal variational equation of degree five:

$$\frac{d^2\xi}{dt^2} + \lambda \cdot C_5^3(t) \cdot \xi = 0 \tag{1.1}$$

where $\lambda$ is a parameter and $C_5(t)$ verifies

$$\begin{cases} \ddot{x} + x^4 = 0 \,, \\ x(0) = 1 \,, \quad \dot{x}(0) = 0 \,, \end{cases} \qquad x = x(t) \,. \tag{1.2}$$

The solution $C_5(t)$ of (1.2) is the inverse function of

$$t = F(x) = \sqrt{\frac{5}{2}} \int_x^1 \frac{du}{\sqrt{1 - u^5}} \,.$$

We prove that the pole $t_5$ of $C_5(t)$ defined by $C_5(t) \xrightarrow[t \to t_5]{} -\infty$ is given by

$$t_5 = K \left[ 1 + \frac{1}{\cos\left(\frac{\pi}{5}\right)} \right] \,,$$

where $K = \sqrt{\frac{5}{2}} I_1$ , $I_1 = \int_0^1 \frac{du}{\sqrt{1-u^5}}$ .

We give the development of $C_5(t)$ in the neighborhood of the pole $t_5$ by using the Siegel theorem (Yoshida, 1986):

$$C_5(t_5 + \tau) = \sqrt[3]{\frac{-10}{9}} \, \tau^{-2/3} [1 - x + a_1\tau^2 + \cdots + a_n x^n + \cdots]$$

where $x = \frac{9^{5/3}}{13 \cdot 10^{5/3}} \cdot \tau^{10/3}$ , $a_1, \cdots, a_n$ are constants to be determined. We raccord the solution $C_5(t)$ in the neighborhood of $t_5$ . We prove by using the lacets method that the solution $C_5(t)$ has two independent real periods and two independent imaginary periods:

$$\begin{cases} P = 8\sqrt{\frac{5}{2}} \cdot I_1 \cdot (1 - \alpha^2) \,, & Q = 4\sqrt{\frac{5}{2}} \cdot I_1 \cdot (1 + \alpha) \,, \\ P' = i\,8\sqrt{\frac{5}{2}} \cdot I_1 \cdot \alpha \cdot \sqrt{1 - \alpha^2} \,, & Q' = i\,4\sqrt{\frac{5}{2}} \cdot I_1 \cdot \sqrt{1 - \alpha^2} \,, \end{cases}$$

where $\alpha = \cos\left(\frac{\pi}{5}\right)$ . By putting $z = [c_5(t)]^5$ , we transform the N.V.E. (1.1) to the Gauss hypergeometric equation:

$$z(1 - z)\frac{d^2y}{dz^2} + \left(\frac{4}{5} - \frac{13z}{10}\right)\frac{dy}{dz} + \frac{\lambda}{10} y = 0$$

where $y = y(z)$ . We obtain the solutions of (1.1) in form of series.

*J. Henrard and S. Ferraz-Mello (eds.), Impact of Modern Dynamics in Astronomy, 459–460.*
© *1999 Kluwer Academic Publishers.*

# References

A. Makhlouf and F. Nahon: 1991, On a family of Hill's equation in the complex field, *Celest. Mech.*, **52**, 227–291.

A. Lainé: Précis d'analyse mathématique, Tome 1, p. 213.

H. Yoshida: 1986, Existence of exponentially unstable periodic solution and the non integrability of homogeneous Hamiltonian systems, *Physica D*, **21**

H. Yoshida: 1987, A criterium for the non existence of an additional integral in systems with an homogeneous potential, *Physica D*, **29**

# GENERALIZED GYLDEN–TYPE SYSTEMS IN

# UNIVERSAL DS–LIKE TR–VARIABLES

LUIS FLORÍA

*Grupo de Mecánica Celeste I, Departamento de Matemática Aplicada a la Ingeniería,
ETSII, Universidad de Valladolid. E – 47 011 Valladolid, Spain.*

Scheifele (1970) applied *Delaunay-Similar* (DS) elliptic Keplerian elements (with the *true anomaly* as the independent variable) to the $J_2$ Problem in Artificial Satellite Theory, making an element of the true anomaly. Deprit (1981) views Scheifele's TR-mapping as an extension of Hill's transformation from a 6-dimensional phase space to an enlarged, 8-dimensional one. To adapt this approach to elliptic-type two-body problems with a time-varying Keplerian parameter $\mu(t)$, Floría (1997, §3, §4) treated a *Gylden system* (Deprit 1983) and derived "Delaunay-Similar" variables via a TR-like transformation. Now we extend our treatment to *perturbed Gylden systems*, and modify the TR-map to deal with *any kind of two-body orbit*. We work out our generalization and the resulting variables within a *unified pattern* whatever the type of motion, in terms of *universal functions* (Stiefel & Scheifele 1971, §11; Battin 1987, §4.5, §4.6) and auxiliary angle-like parameters.

In *extended polar nodal variables* $(r, \theta, \nu, t; p_r, p_\theta, p_\nu, p_0)$, we formulate

$$\mathcal{H} \equiv \mathcal{H}(r, -, -, t; p_r, p_\theta, p_\nu, p_0; \varepsilon) = \mathcal{H}_0(r, t; p_r, p_\theta)$$
$$+ \sum_{j=0}^{2} V_j(t; p_\theta, p_\nu, p_0; \varepsilon) r^{-j} + p_0.$$

In this *perturbed Gylden system*, $\mathcal{H}_0$ is the Gylden Hamiltonian. The perturbation can be expanded in powers of a small parameter $\varepsilon$. Using the Stumpff $c_n$ and Battin $U_n$ *universal functions*, and *universal parameters* $s(r, t; \Phi, L, G, N)$ and $f(r, t; \Phi, L, G, N)$, a generator based on Deprit (1981, Formula [23]) defines a TR–like mapping, from extended polar nodal variables to universal, generalized DS variables $(q_\Phi, q_L, q_G, q_N; \Phi, L, G, N)$, with

$$Q \equiv Q(r, t; \Phi, L, G, N; \varepsilon) = -2[L + V_0(t; G, N, L; \varepsilon)]$$
$$+ \frac{2[\mu(t) - V_1(t; G, N, L; \varepsilon)]}{r} - \frac{[\gamma^2 + 2V_2(t; G, N, L; \varepsilon)]}{r^2},$$
$$\gamma \equiv \gamma(\Phi, L, G, N), \quad \mu^* = \mu - V_1, \quad q = [\mu^*(1-e)]/(2\Lambda), \quad \Lambda = L + V_0,$$
$$\Gamma^2 \equiv \gamma^2 + 2V_2 = \mu^* q(1+e), \quad r = [q(1+e)]/[1 + e\cos f],$$
$$r = q + \mu^* e s^2 c_2 \left(2\Lambda s^2\right) = q + \mu^* e U_2(s, 2\Lambda).$$

*J. Henrard and S. Ferraz-Mello (eds.), Impact of Modern Dynamics in Astronomy,* 461–462.
© 1999 *Kluwer Academic Publishers.*

This TR–like transformation converts Hamiltonian $\mathcal{H}$ into

$$\tilde{\mathcal{H}} = \left(G^2 - \gamma^2\right) / \left(2\,r^2\right) + \Delta V_0 + (\Delta V_1/r) + \left(\Delta V_2/r^2\right)$$
$$- (\partial V_0/\partial t)\,[q\,s + \mu^*\,e\,U_3] + (\partial \mu^*/\partial t)\,s - (\partial V_2/\partial t)\,f/\Gamma\,,$$
$$\Delta V_j \equiv \tilde{V}_j - V_j\,, \quad V_j \equiv V_j\,(t; G, N, L; \varepsilon)\,, \quad \tilde{V}_j \equiv V_j\,(t; G, N, \tilde{p}_0; \varepsilon)\,,$$

where $\tilde{p}_0$ refers to $p_0$ in the *new* canonical variables. A *time transformation*, $dt = \tilde{f}d\tau$, with $\tilde{f} = r^2/\mathcal{G}$, $\mathcal{G} \equiv \mathcal{G}\,(\Phi, L, G, N)$, simplifies $\tilde{\mathcal{H}}$. The *Hamiltonian* with this pseudo–time is $\mathcal{K} = \tilde{\mathcal{H}}\tilde{f}$. These Hamiltonians depend on the momenta $(\Phi, L, G, N)$ and on the *non–DS variables* $(r, t, s)$, and should be expressed in the new DS–type variables. To this end, *implicit function results, inversion theorems and Fourier analysis of the two–body problem* must be applied. As in the case of classical DS variables, certain specifications of $\gamma$ and $\mathcal{G}$, as functions of the new momenta, reduce the first term of $\tilde{\mathcal{H}}$ and $\mathcal{K}$. Assumptions concerning the $V_j$–terms produce further simplifications.

Research partially supported by Junta de Castilla y León (Grant VA61/98).

## References

Battin, R. H.: 1987, *"An Introduction to the Mathematics and Methods of Astrodynamics."* AIAA Education Series, New–York.

Deprit, A.: 1981, *Celest. Mech.* **23**, 299–305.

Deprit, A.: 1983, *Celest. Mech.* **31**, 1–22.

Floría, L.: 1997, *Celest. Mech. and Dyn. Astron.* **68**, 75–85.

Scheifele, G.: 1970, *C. R. Acad. Sci. Paris*, série **A**, **271**, 729–732.

Stiefel, E. L. & Scheifele, G.: 1971, *"Linear and Regular Celestial Mechanics."* Springer–Verlag, Berlin–Heidelberg–New York.

# ORBITAL STRUCTURE OF THE TWO FIXED CENTRES PROBLEM

A. CORDERO

*Dpt. Matematica Aplicada. Universidad Politécnica de Valencia. Valencia. Spain, e-mail:*
*acordero@mat.upv.es*

J. MARTÍNEZ ALFARO

*Dpt. Matemàtica Aplicada. Facultat de Matemàtiques. Universitat de València. València.*
*Spain. e-mail: Jose.Mtnez.Alfaro@uv.es*

and

P. VINDEL

*Dpt. Matemàtiques. Universitat Jaume I. Castelló. Spain. e-mail: vindel@nuvol.uji.es*

The set of orbits of the Two Fixed Centres problem has been known for a long time (Charlier, 1902, 1907; Pars, 1965), since it is an integrable Hamiltonian system.

We consider a plane that contains the fixed masses. Denote by $\varphi$ the angle defined by this plane and the one that contains also the third body. The momentum $p_\varphi$ is a first integral of the system and when $p_\varphi$ is different from zero, the manifold generated by the generalized coordinates and momenta are two copies of the three-dimensional sphere $S^3$. If $p_\varphi = 0$, that is to say when the planet crosses the line joining both suns, the motion is restricted to a planar one. All the equilibrium points appears in this case and therefore the phase spaces are more complex. We restrict our attention to this case which has two degrees of freedom.

It is again a Bott-integrable Hamiltonian system. The set of periodic orbits of this systems can be studied from a subset of them, the Non-Singular Morse-Smale type orbits (see Casasayas, 1992). It is proved in Campos (1997) that a small perturbation of a Bott-integrable Hamiltonian system transforms it into a Non-Singular Morse-Smale system. The NMS periodic orbits belong to both the NMS system and the Hamiltonian one. Moreover, The NMS p.o. can be continued to nearly Hamiltonian systems. For instance, in our case to the Restricted Three Body Problem and in the study of the motion of a material point moving inside the gravitational field generated by two stars. This approximation is also useful when the motion of an artificial satellite around a spheroidal body is considered.

If $p_\varphi \neq 0$ the link generated by the NMS p.o. is the Hopf link. The orbital structure of the NMS type periodic orbits in the different cases of negative and positive energy is obtained (see Cordero *et al.*, 1997b). We analyze the dynamical and physical consequences of this structure.

In the physical space it is shown (see Cordero *et al.*, 1997a) that different regions of motion are delimited by limit curves corresponding to fixed points of the new Hamiltonian functions, all the possible orbits being classified. So, these NMS orbits become collision or limit orbits in the physical space.

As the Three Dimensional Two Fixed Points problem is considered an integrable approximation of the Three-Body Restricted problem, we study the equilibrium

*J. Henrard and S. Ferraz-Mello (eds.), Impact of Modern Dynamics in Astronomy, 463–464.*

points of this system that come from the previous one, and it is found that in this case the Hamiltonian is not separable in two new Hamiltonian functions. At this point, we deduce from the NMS type orbits found in the Planar Two Fixed Centres Problem the behaviour of this quasi-integrable problem. The NMS orbits that we are using as an approximation to the Restricted Three Body problem are defined by certain equilibrium points on the separated phases spaces. Five NMS orbits appeared associated to other five equilibrium points: three of them are collision orbits and the other two are limit orbits.

Working with the equations of motion in hyperbolic coordinates and substituting the conditions of the NMS orbits corresponding to collision we obtain two equilibrium points in the restricted problem, i.e., this coordinate transformation has regularized the problem, removing the singularities.

## References

Campos, B., J. Martínez Alfaro and P. Vindel: 1996, Orbital Structure in the Two Fixed Centres Problem, Proceedings of the Third International Workshop on Positional Astronomy and Celestial Mechanics, pp. 313-323.

Campos, B: 1997, Orbitas periódicas Anudadas en oscilaciones no lineales, Tesis Doctoral, Universitat Jaume I

Charlier, C.V.L.: 1902 *Die Mechanik des Himmels, Volume 1*. W. de Gruyter and Co, Volume 1, 1902; Volume 2, 1907.

Charlier, C.V.L.: 1907 *Die Mechanik des Himmels, Volume 2*. W. de Gruyter and Co

Cordero, A., J. Martínez Alfaro and P. Vindel: 1997a, Study of the orbits in the Two Fixed Centres Problem, Proceedings of the Forth International Workshop on Positional Astronomy and Celestial Mechanics, Peñiscola.

Cordero, A., J. Martínez Alfaro and P. Vindel: 1997b, NMS Periodic Orbits of the Two Fixed Centres Problem, Proceedings of the Forth International Workshop on Positional Astronomy and Celestial Mechanics, Peñiscola

Cordero, A.: 1997, Estudio analitico y geométrico del problema de los Dos Centros Fijos, Tesis de Licenciatura, Universidad Jaime I de Castellón

Pars, L.A.: 1965, *A Treatise on Analytical Dynamics*. Ed. Heinemann

# CHAOS IN THREE-BODY DYNAMICS: INTERMITTENCY, STRANGE ATTRACTOR, KOLMOGOROV-SINAI ENTROPY

ARTHUR D. CHERNIN

*Tuorla Observatory, University of Turku, 21500 Piikkiö, FINLAND,*
*Sternberg Astronomical Institute, Moscow University, 119899 Moscow, RUSSIA.*

The temporal structure of chaos in three-body dynamics is analyzed; the emphasis is made on a similarity and difference between three-body chaos and basic patterns of chaotic behaviour known in nonlinear physics.

1. With the use of homology mapping (Agekian and Anosova 1967), we study a set of computer models of thee-body systems in a stationary spherically symmetric potential well (Valtonen *et al.* (1994); the well confines the bodies, and because of this the system can generate fairly long time series. Typical time series reveal sequences of seemingly periodic motion and short bursts of strong chaos that appear in an irregular manner (Heinämäki *et al.* 1998). The quasi-ordered states are associated with hierarchical homology, and the quasi-period of the low-amplitude oscillations is very near the period of the temporary close binary in the system. The high-amplitude irregular states are mostly due to active three-body interplay when each of the bodies interacts with the two others with almost equal intensity. In the evolutionary history of most systems, these two extreme kinds of states alternate in an apparently random way producing together a non-stationary pattern of unpredictable behaviour.

The time behaviour of this type is similar in its appearance – and actually in its physical nature too – to the phenomenon of intermittency observed in the ocean flows. A time series close in shape to what we observe in three-body dynamics may also be found in laboratory hydrodynamical experiments with Rayleigh-Bènard flow. The chaotic behaviour with this shape of time series is classified as type-III intermittency.

2. We use the correlation integral method (Lehto *et al.* 1993) and find that the dimension $D$ of the time series generated by three-body systems is between 2 and 2.1 (Heinämäki *et al* 1998). It means that the number of the major physical parameters of the systems is 3; one can expect that they are the eccentricities of the binary orbit and that of the third body and also the ratio of semi-major axes of these two orbits, in hierarchical states of the systems. Two first of these parameters are related to the compact 2-dimensional homology hypersurface of the whole phase space. The third parameter corresponds to the infinite configuration coordinate in the whole non-compact phase space of the system.

For the system motions, the area of the homology hypersurface contracts with time onto the set of hierarchical states of lower dimensionality $D_H = D - 1 \leq 1.1$. The phenomenon of contraction (which is similar to the contraction of the whole phase space in dissipative systems) is seen from the fact that a typical system spends about 2/3 of its life-time in hierarchical states with temperary close binaries; these

*J. Henrard and S. Ferraz-Mello (eds.), Impact of Modern Dynamics in Astronomy, 465–466.*
© 1999 *Kluwer Academic Publishers.*

states attract almost all the trajectories, no matter where they start. The trajectories approach these states in highly irregular, chaotic way, which is similar to what is observed in the cases of strange attractors of dissipative systems.

However, some of the trajectories may not be trapped in the set of hierarchical states forever; they can leave it, and this proceeds also in a chaotic way. Such a behaviuor with transient trapping is different from what is observed in standard attractors and indicates one of the special features of three-body chaotic attractor. Because almost any three-body system ends its chaotic evolution with the decay, the system comes to the attracting set ultimately.

Note that the dimension of the time series in three-body dynamics proves to be near to one for the Lorenz dissipative system, introduced first in meteorology. A low dimension (close to 2) of the time series is an important sign of intermuttent chaos in nonlinear physics where this is considered as an evidence for low-dimension strange attractors that may be behind the time series.

One of the possible representations of the three-body strange attractor is given in a return map (Heinämäki *et al.* 1998). The object is cone or a piramid-like structure which can almost be reduced to a two-dimensional surface, except for the sparce loops that appear to avoid self-crossing of the trajectories.

3. An effective way to study the onset of chaos in dynamical systems involves a notion of the phase drop; it can be reformulated for three-body dynamics with the use of homology mapping. To quantify the process of deformation and spreading of the 'homology' drop in a coarse-grained description, we calculate the average exponential growth rate of the area occupied by the drop: it proves to be $h = 0.7 - 1.3$ in most of the observed cases of our computer simulations (Heinämäki *et al* 1999). The growth rate $h$ has a close analogue in the Kolmogorov-Sinai entropy, defined in a similar manner for the whole phase space. With this values of $h$, one can conclude that the state of developed chaos occurs in about one crossing time, in three-body dynamics.

The 'fine' structure of the homology drop reflects the behaviour of individual trajectories. The average growth rate of divergence of the trajectories $\sigma$ in the homology mapping which is a close analogue of the Lyapunov first exponent defined in the phase space. We finds that $\sigma = 0.4 - 1.5$. On the order of magnitude, $h$ and $\sigma$ are close, and the both prove to be close to the inverse time of correlation decay, $\tau_c$, estimated for tree-body chaos by Ivanov *et al.* (1995). The relation $h \approx \sigma \approx 1/\tau_c$ is characteristic for the 'standard' patterns of chaotic behaviour in nonlinear physics.

## References

Agekian, T.A. and Anosova, J.P.: 1967, *Sov. Astron.*, **44**, 1261.
Heinämäki P., Lehto H., Valtonen M.J., Chernin A.D.: 1998, *MNRAS*, **298**, 790.
Heinämäki P., Lehto H., Valtonen M.J.: Chernin A.D, 1999, *MNRAS*, in press
Ivanov A.V., Filistov E.A., Chernin A.D: 1995, *Astron. Rep.*, **39**, 368.
Lehto H.J., Czerny B., McHardy M.: 1993, *MNRAS*, **216**, 125.
Valtonen M.J., Mikkola S., Heinämäki, Valtonen H.:, 1994, *ApJS*, **95**, 69

The manufacturer's authorised representative in the EU is Springer
Nature Customer Service Centre GmbH, Europaplatz 3, 69115 Heidelberg,
Germany. If you have any concerns regarding our products, please
contact ProductSafety@springernature.com

Printed and bound by CPI Group (UK) Ltd, Croydon, CR0 4YY
23/04/2026
02095585-0009